MECHANICAL ENGINEERING AND MECHATRONICS HANDBOOK

MECHANICAL ENGINEERING AND MECHATRONICS HANDBOOK

D. Kumar

MERCURY LEARNING AND INFORMATION
Dulles, Virginia
Boston, Massachusetts
New Delhi

Publisher: David Pallai
MERCURY LEARNING AND INFORMATION
22841 Quicksilver Drive
Dulles, VA 20166
info@merclearning.com
www.merclearning.com
1-800-232-0223

D. Kumar. *Mechanical Engineering and Mechatronics Handbook*.
ISBN: 978-1-6839-2856-0

Library of Congress Control Number: 2022932254

222324321   Printed on acid-free paper in the United States of America.

Our titles are available for adoption, license, or bulk purchase by institutions, corporations, etc. For additional information, please contact the Customer Service Dept. at 800-232-0223(toll free).

All of our titles are available in digital format at *academiccourseware.com* and other digital vendors. The sole obligation of MERCURY LEARNING AND INFORMATION to the purchaser is to replace the book, based on defective materials or faulty workmanship, but not based on the operation or functionality of the product.

CONTENTS

1

STRESSES AND STRAINS

The subject of mechanics of solids is a study of

i.   the behavior of materials under various types of loads and moments, and

ii. the action of forces and their effects on structural and machine elements such as angle irons, circular bars, and beams.

The knowledge thus acquired provides a rational approach to all design problems, that is, it helps an engineer to design all types of machines and structures and suggest protective measures for the safe working conditions of such elements. The analysis, however, assumes that

i.   the material of the body is homogeneous and isotropic, and

ii. there are no internal stresses present in the material before the application of loads.

1.1. STRESS

Consider an initially straight metallic bar of constant cross-sectional area loaded at the ends by a pair of oppositely directed forces or loads. These external forces are collinear, coincide with the longitudinal axis of the bar and act through the centroid of each cross-section. The bar deforms (undergoes a change in length) and this deformation depends upon the external load, cross-sectional area, and material of the body. While a tensile force will elongate the bar and a compressive force will shorten it.

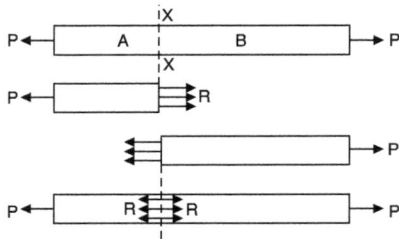

FIGURE 1.1

The cohesive forces between the molecules of the bar offer resistance to change in length, and this resistance is due to the strength of the bar material. The bar will remain in equilibrium if the resistance R equals the external load P. This implies that section X–X is offering the resistance R against a possible separation of the segments A and B. The resistance R against deformation is called **stress.**

Presuming resistance to be uniform across the section, the resistance per unit area of the section is called the *unit stress* or *intensity of stress*. Most commonly, the term *stress* is used to mean intensity of stress:

$$\text{Stress } \sigma = \frac{R}{A} = \frac{P}{A} \quad (\because R = P). \tag{1.1}$$

The units of stress correspond to those of load and area. Stress has the dimensions of FL^{-2} and is usually expressed in N/mm^2, N/m^2, or Pascal.

1 Pascal = 1 N/m^2.

Since Pascal is a very small unit, it is quite common practice to express stress in kPa, MPa, or GPa.

1 kPa = 10^3 Pa, 1 MPa = 10^6 Pa, and 1 GPa = 10^9 Pa.

Load and Stress: The external forces acting on any piece of material are said to constitute that which is known as a *load*. It represents the combined effect of external forces acting on a body. The resistance offered by the body against deformation caused by the load is called *stress*. The load is applied to the body whereas stress is induced in the material of the body.

Stress and Pressure have the same units (N/m^2); the difference between them is briefly stated below:

- Stress is encountered in solids. Its magnitude depends upon the direction of applied load with respect to the plane passing through the point under consideration. This magnitude is different on different planes passing through the point considered.

- Pressure is associated with fluids (liquids and gases) and it represents the force exerted per unit area due to the impact of fluid molecules on the walls of the container or on the body immersed in a fluid, and its value is the same at a point in a fluid.

1.2. TENSION, COMPRESSION, AND STRAIN

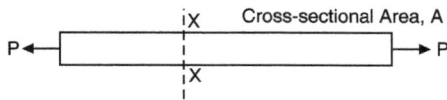

FIGURE 1.2 Bar in tension.

A structural member is said to be in tension when it is subjected to two equal and opposite pulls (Figure 1.2) and the member tends to elongate/increase in length. The stress so produced is called tensile stress and at any cross-section X–X is given as

$$\sigma_t = \frac{P}{A}.$$

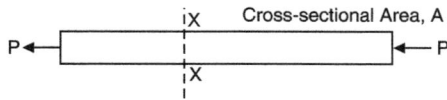

FIGURE 1.3 Bar in compression.

The hoisting ropes used in cranes and passenger elevators are the elements subjected to tensile loading. If the structural member is subjected to two equal and opposite pushes (Figure 1.3) and the member tends to shorten/decrease in length, the member is said to be in compression. The stress so produced is called compressive stress and at cross-section X–X, it is given as

$$\sigma_c = \frac{P}{A}.$$

When a body is acted upon by tensile or P compressive loading, its dimensions will increase or decrease along the line of action of load applied. This deformation (change in length) per unit original length is called *primary strain* or *longitudinal strain*:

$$\text{Strain} = \frac{\text{change in length}}{\text{original length}} \; ; \quad \varepsilon = \frac{\delta l}{l}. \tag{1.2}$$

The strain is usually expressed in units of millimeters per millimeter or meters per meter and consequently is dimensionless. It is taken positive if there is an extension of the bar, and negative if there is shortening.

1.3. ELASTIC LIMIT, HOOK'S LAW, AND MODULUS OF ELASTICITY

Experimental evidence shows that upon removal of external load, the force of resistance vanishes, and the body regains its original shape and size. However, such a situation exists only if the external loading is within a certain limit.

The loading limit under which the deformation entirely disappears on the removal of load is called the *elastic limit*. If the loading is large and the intensity of stress exceeds the elastic limit, the material gets into the plastic stage. Under that condition, the deformation does not entirely disappear and a residual deformation stays permanently even on the removal of load.

Hook's law states that when a material is loaded within the elastic limit, stress is directly proportional to strain. Mathematically,

$$\text{Stress} \propto \text{strain} \quad ; \quad \sigma \propto \varepsilon$$

$$\text{Stress} = \text{constant of proportionality} \times \text{strain,}$$

$$\text{or} \quad \sigma = E \ \varepsilon \quad ; \quad E = \frac{\sigma}{\varepsilon} = \frac{\text{stress}}{\text{strain}} \tag{1.3}$$

where the constant of proportionality E is called *Young's modulus* or *modulus of elasticity*. The Young's modulus has the same units as those of stress (N/m^2, Pa, kPa, and MPa) because the strain is a dimensionless quantity.

Young's modulus is a property of the material, and its value is independent of the magnitude of force, the geometrical configuration of the body, and the type of loading (tension or compression). For common engineering materials, Young's modulus of elasticity is

Material	Steel	Cast iron	Aluminum	Brass	Bronze
E, GPa	200–210	100–110	68–70	100–110	110–120

$$\text{Recalling that} \quad \varepsilon = \frac{\delta l}{l} \text{ and } \sigma = \frac{P}{A}, \text{ we get}$$

$$\frac{P}{A} = E\varepsilon = E\frac{\delta l}{l}. \tag{1.4}$$

$$\text{Change in length,} \quad \delta l = \frac{\sigma l}{E} = \frac{Pl}{AE}$$

EXAMPLE 1.1
A steel bar of 1.5 m long, 50 mm wide, and 20 mm thick is subjected to an axial tensile load of 120 kN. If the extension in the length of the bar is 0.9 mm, make calculations for the intensity of stress, strain, and modulus of elasticity of the bar material.

Solution: Area of cross-section $A = 50 \times 20 = 1000 \text{ mm}^2$,

$$\text{Intensity of stress}, \sigma = \frac{\text{load } P}{\text{area } A} = \frac{120 \times 10^3}{1000} = \mathbf{120 \ N/mm^2}$$

$$\text{Strain}, \varepsilon = \frac{\text{change in length}}{\text{original length}} = \frac{0.9}{1.5 \times 10^3} = \mathbf{0.0006}.$$

$$\text{Modulus of elasticity}, E = \frac{\text{stress } \sigma}{\text{strain } \varepsilon} = \frac{120}{0.0006}$$

$$= 2 \times 10^5 \ \text{N/mm}^2 = \mathbf{200 \ GN/m^2}$$

EXAMPLE 1.2

A hollow right circular cylinder is made of cast iron and has an outside diameter of 75 mm and an inside diameter of 60 mm. The cylinder measures 600 mm in length and is subjected to an axial compressive load of 50 kN. Neglecting any possibility of lateral buckling of the cylinder, determine the normal stress, and shortening in the length of the cylinder under this load. Take the modulus of elasticity of cast iron to be 100 GPa.

Solution: $E = 100 \ \text{GPa} = 100 \ \text{G N/m}^2 = 1 \times 105 \ \text{N/mm}^2$,

$$\text{Area of cross-section}, A = \frac{\pi}{4}(75^2 - 60^2) = 1589.62 \ \text{mm}^2$$

$$\text{Stress}, \sigma = \frac{\text{load}}{\text{area}} = \frac{50 \times 10^3}{1589.62} = 31.45 \ \text{N/mm}^2.$$

$$\text{Decrease in length}, \delta l = \text{strain} \times \text{original length}$$

$$= \frac{\sigma}{E} \times \frac{31.45 \times 600}{1 \times 10^5} = \mathbf{0.1887 \ mm}$$

EXAMPLE 1.3

The wire working on a railway signal is 6 mm in diameter and 250 m long. If the movement at the signal end is to be 15 cm, make calculations for the movement which must be given to the end of the wire at the signal box. Assume a pull of 1500 N on the wire and take the modulus of elasticity for the wire material as 2 × 10⁵ N/mm².

Solution: Pull on the wire will induce tensile stress given by

$$\sigma = \frac{P}{A} = \frac{1500}{\frac{\pi}{4}(6)^2} = 53.1 \text{ N/mm}^2$$

Strain in the wire, $\varepsilon = \dfrac{\sigma}{E} = \dfrac{53.1}{2 \times 10^5} = 26.55 \times 10^{-5}$

Increase in length of wire due to pull,

$$\delta l = \varepsilon \times l = 26.55 \times 10^{-5} \times (250 \times 10^3)$$

$$= 66.375 \text{ mm} \approx 6.64 \text{ cm}.$$

∴ Total movement which needs to be given at the signal box end

$$= 15 + 6.64 = \mathbf{21.64 \text{ cm}}$$

EXAMPLE 1.4

Make calculations for the minimum diameter of a steel wire which is required to raise a load of 6 kN. Presume that the wire can sustain a maximum stress of 120 MN/m². Also, calculate the extension in 3 m length of the wire. Take modulus of elasticity E = 200 GPa.

Solution: $E = 200 \text{ GPa} = 200 \text{ GN/m}^2 = 2 \times 10^5 \text{ N/mm}^2$;
$\sigma_{max} = 120 \text{ MN/m}^2 = 120 \text{ N/mm}^2$.

Tensile stress induced in the wire due to lifting of load,

$$\sigma = \frac{P}{A} = \frac{6 \times 10^3}{\frac{\pi}{4}d^2} = \frac{24 \times 10^3}{\pi d^2}.$$

But σ is limited to 120 N/mm². Therefore,

$$120 = \frac{24 \times 10^3}{\pi d^2} \quad ; \quad d^2 = \frac{24 \times 10^3}{\pi \times 120} = 63.694.$$

∴ Diameter of steel wire $d = \sqrt{63.694} = \mathbf{7.98 \text{ mm}}$.

Elongation of wire δl = strain × original length

$$= \frac{\sigma}{E} \times l = \frac{120}{2 \times 10^5} \times 3000 = \mathbf{1.8 \text{ mm}}.$$

EXAMPLE 1.5
A steel wire of 2 m long and 3 mm in diameter is extended by 0.75 mm due to the weight suspended from the wire. If the same weight is suspended from the brass wire, 2.5 m long and 2 mm in diameter, it is elongated by 4.65 mm. Determine the modulus of elasticity of brass if that of steel is 2×10^5 N/mm^2.

Solution: For the type of given loading, a change in the length of wire is given by;

$$\delta = \frac{Pl}{AE}.$$

For the steel wire,

$$0.75 = \frac{P \times 2000}{\frac{\pi}{4}(3)^2 \times (2 \times 10^5)} = 0.001415 \, P.$$

$$\therefore \text{Weight suspended, } P = \frac{0.75}{0.001415} = 530 \text{ N}$$

For the brass wire,

$$4.65 = \frac{530 \times 2500}{\frac{\pi}{4}(2)^2 \times E_b} = \frac{421974}{E_b}.$$

$\therefore$ Modulus of elasticity for the brass wire,

$$E_b = \frac{421974}{4.65} = \textbf{90747 N / mm}^2.$$

1.4. STRESS–STRAIN DIAGRAM

Stress–strain curve is a graphical plot of stress versus strain. These quantities are experimentally obtained by subjecting a metallic bar of a uniform cross-section to a gradually increasing tensile load until failure of the bar occurs. The test is conducted in a tensile testing machine on a test specimen having the appearance/configuration as shown in Figure 1.4. The specimen has collars provided at both ends for gripping it firmly in the fixtures of the machine.

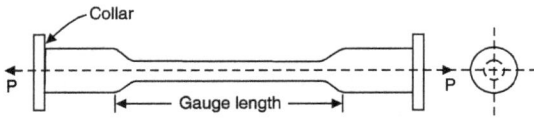

FIGURE 1.4 Test specimen.

The central portion of the test specimen is somewhat smaller than the end regions, and this central section constitutes the gauge length over which elongations are measured. An extensometer (dial gauge) is used to measure very small changes in length. After that vernier scale on the machine is used to measure extension. Load and extension are simultaneously recorded till the specimen breaks.

Stress is calculated by dividing the load by the original cross-sectional area of the test specimen. Strain is calculated by dividing the extension of a given length (gauge length) by the original unstrained length. That is

$$\text{Strain}, \varepsilon = \frac{\delta l}{l} \text{ and stress}, \sigma = \frac{P}{A},$$

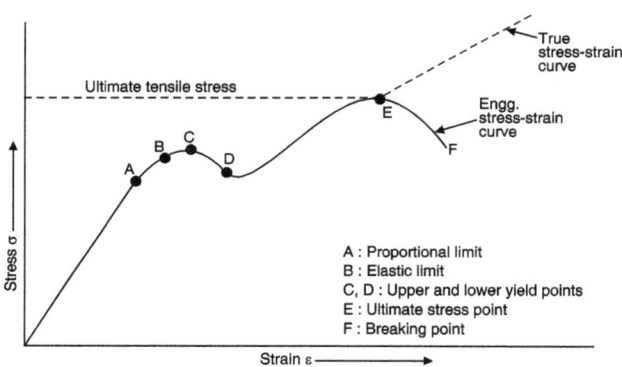

A : Proportional limit
B : Elastic limit
C, D : Upper and lower yield points
E : Ultimate stress point
F : Breaking point

FIGURE 1.5 Stress–strain curve.

where l is the gauge length and A is the original cross-sectional area.

Stress–strain curve is then plotted by having obtained numerous pairs of values of stress and strain; stress plots as ordinate and strain as abscissa on a graph taking suitable scale.

Figure 1.5 shows the typical behavior of stress–strain curve for mild specimen and its salient features are:

- **Proportional limit:** Stress is a linear function of strain and the material obeys Hook's law. This proportionality extends up to point A and this point is called *proportional limit* or *limit of proportionality*. 0-A is a straight-line portion of the curve and its slope represents the value of modulus of elasticity.
- **Elastic limit:** Beyond the **proportional** limit, stress and strain depart from straight-line relationship. The material, however, remains elastic up to state point B. The word elastic implies that the stress developed in the material is such that there is no residual or permanent deformation when the load is removed. Yet, the deformation is reversible or recoverable.

Stress at B is called the *elastic limit stress*; this represents the maximum unit stress to which a material can be subjected and is still able to return to its original form upon removal of load.

• **Yield point:** Beyond elastic limit, the material shows considerable strain even though there is no increase in load or stress. This strain is not fully recoverable, that is, there is no tendency of the atoms to return to their original positions. The behavior of the material is inelastic, and the onset of plastic deformation is called the yielding of the material. Yielding pertains to the region C-D and there is a drop in load at point D. Point C is called the *upper yield point*, and point D is the *lower yield point*. The difference between the upper and lower yield point is small and the quoted yield stress is usually the lower value.

• **Ultimate strength or tensile strength:** After yielding has taken place, the material becomes strain hardened (strength of the specimen increases) and an increase in load is required to take the material to its maximum stress at point E. Strain in this portion is about 100 times than that of the portion from 0 to D. Point E represents the maximum ordinate of the curve and the stress at this point is known either as *ultimate stress* or the *tensile stress* of the material.

• **Breaking strength:** In the portion EF, there is falling off the load (stress) from the maximum until fracture takes place at F. Point F is referred to as the fracture or breaking point and the corresponding stress is called the *breaking stress*.

The apparent fall in stress from E to F may be attributed to the fact that stress calculations are made on the basis of the original cross-sectional area. In fact elongation of the specimen is accompanied by a reduction in cross-sectional area and this reduction becomes significant near the ultimate stress. In case stress calculations are based on actual area, the curve would be seen to rise until a fracture occurs. For mild steel, the test piece breaks making a cup and cone type fracture; the two pieces can be joined together to find out the diameter (actual area) at the neck under the specimen breaks.

For many ductile materials other than mild steel, for example, aluminum and copper, no definite yield point is obtained. For such materials, the strain–strain curve plots are shown in Figure 1.6.

For brittle materials, like cast iron, no appreciable deformation is obtained and the failure occurs without yielding (Figure 1.7).

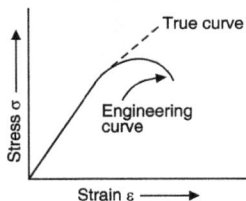

FIGURE 1.6

- **Proof stress:** Quite often it is desired to determine the stress at which a specified permanent extension takes place in a tensile test. The extension specified is usually 0.1, 0.2, or 0.5% of gauge length. Such a stress is known as proof stress.

  For determining 0.2% proof stress from stress–strain curve, a point G representing 0.2%, that is, 0.002 is marked on the strain axis. A line GH is then drawn parallel to the initial slope line OA. The stress at the point where this line cuts the curve is 0.2% proof stress.

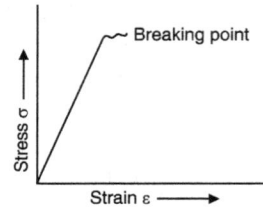

FIGURE 1.7

- **Working stress and safety factor:** Working stress is the allowable stress for design purposes. During the design of an element, it is to be kept in mind that actual stress developed in the element does not exceed the working stress. Frequently, such a stress is determined by dividing either the yield stress or the ultimate stress by a number termed the *safety factor.* The safety factor accounts for

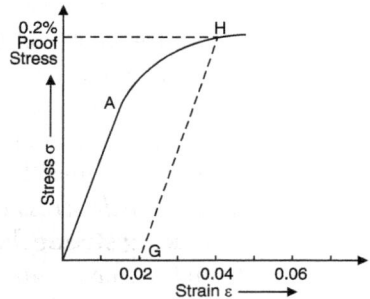

FIGURE 1.8

- internal flows in the material,
- stress concentration, and
- uncertainties about the magnitude and nature to which the machine element is subjected, etc.

  The value of safety factor depends upon the judgment and experience of the designer and is usually governed by:

- type of loading,
- reliability of material and manufacturing processes such as castings and forgings, and
- the extent of damage that will be caused if the machine element fails. The allowable or working stress used in design calculations is taken as

$$\text{Working stress} = \frac{\text{ultimate stress}}{\text{safety factor}}.$$

Material Classification: Materials are commonly classified as:

i. *Homogeneous and isotropic materials:* A homogeneous material implies that the elastic properties such as modulus of elasticity and Poisson's ratio of the material are the same everywhere in the material system. Isotropic means that these properties are not directional characteristics, that is, an isotropic material has the same elastic properties in all directions at any one point of the body.

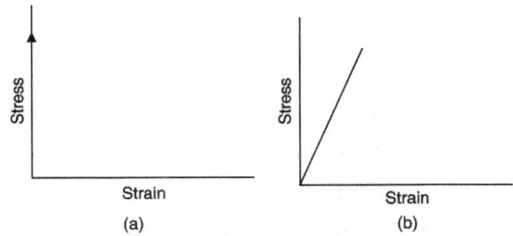

FIGURE 1.9 (a) Rigid and (b) linearly elastic materials.

ii. *Rigid and linearly elastic materials:* A rigid material is one that has no strain regardless of the applied stress. A linearly elastic material is one in which the strain is proportional to the stress.

iii. *Plastic and rigid-plastic materials:* For a plastic material, there is definite stress at which plastic deformation starts. A rigid-plastic

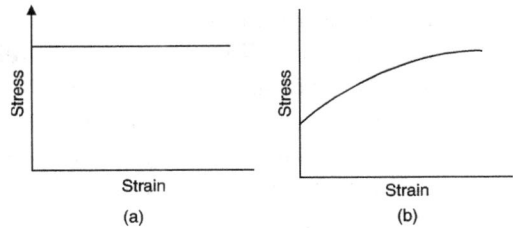

FIGURE 1.10 (a) Plastic and (b) rigid-plastic materials.

material is one in which elastic and time-dependent deformations are neglected. The deformation remains even after the release of stress (load).

iv. *Ductile and brittle materials:* A material that can undergo "large permanent" deformation in tension, that is, it can be drawn into wires is termed as ductile. A material that can be only slightly deformed without rupture is termed as brittle.

The ductility of a material is measured by the percentage elongation of the specimen or the percentage reduction in the cross-sectional area of the specimen when a failure occurs. If l is the original length and l' is the final length, then

$$\% \text{ increase in length} = \frac{l'-l}{l} \times 100.$$

The length l' is measured by putting together two portions of the fractured specimen.

Likewise, if A is the original area of cross-section and A' is the minimum cross-sectional area at fracture, then

$$\% \text{ age reduction in area} = \frac{A-A'}{A} \times 100.$$

A brittle material like cast iron or concrete has very little elongation and very little reduction in cross-sectional area. A ductile material like steel or aluminum has a large reduction in area and an increase in elongation. An arbitrary percentage elongation of 5% is frequently taken as the dividing line between these two classes of material.

EXAMPLE 1.6
The following data were recorded during tensile test made on a standard tensile test specimen:

Original diameter and gauge length = 15 mm and 60 mm; minimum diameter at fracture = 10 mm; distance between gauge points at fracture = 75 mm; load at yield point and at fracture = 40 kN and 45 kN; and maximum load that specimen could take = 70 kN.

Make calculations for (a) yield strength, ultimate tensile strength, and breaking strength, (b) percentage elongation and percentage reduction in the area after fracture, and (c) nominal and true stress and fracture.

Solution: Original area, $A_0 = \dfrac{\pi}{4}(15)^2 = 176.625 \text{ mm}^2$.

Final area, $Af = \dfrac{\pi}{4}(10)^2 = 78.5 \text{ mm}^2$ (at fracture).

(a) Yield strength $= \dfrac{40 \times 10^3}{176.625} = 226.47 \text{ N/mm}^2$

$\qquad\qquad = 226.47 \times 10^6 \text{ N/m}^2 = \textbf{226.47 MPa.}$

Ultimate tensile strength $= \dfrac{70 \times 10^3}{176.625} = 396.32 \text{ N/mm}^2$

$\qquad\qquad = 396.32 \times 10^6 \text{ N/m}^2 = \textbf{396.32 MPa.}$

Breaking strength $= \dfrac{45 \times 10^3}{176.625} = 254.78 \text{ N/mm}^2$

$\qquad\qquad = 254.78 \times 10^6 \text{ N/m}^2 = \textbf{254.78 MPa.}$

It may be noted that calculations for all the strengths are made on the basis of original area at the test section.

(b) Percentage elongation $= \dfrac{75 - 60}{60} \times 100 = \textbf{25\%.}$

$$\text{Percentage reduction in area } = \frac{176.625 - 78.5}{176.625} \times 100 = \mathbf{55.56\%}.$$

(c) Nominal stress $= \dfrac{\text{load at fracture}}{\text{original area}} = \dfrac{45 \times 10^3}{176.625} = \mathbf{254.78 \ N/mm^2}.$

The nominal strength and the breaking strength are synonymous

$$\text{True stress} = \frac{\text{load at fracture}}{\text{final area}} = \frac{45 \times 10^3}{78.5} = 573.25.$$

1.5. EXTENSION OF A TAPERED BAR

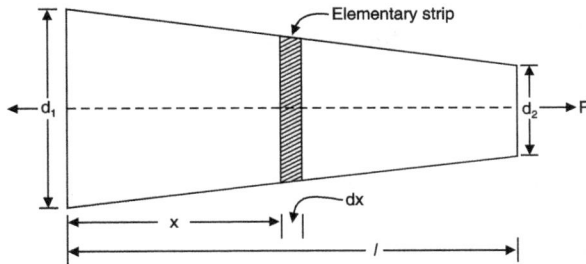

FIGURE 1.11.

Consider a circular bar that tapers uniformly from diameter d_1 at the bigger end to diameter d_2 at the small end and is subjected to axial tensile load P (Figure 1.11).

Let attention be focused on an elementary strip of length dx at distance x from the bigger end.

Diameter of the elementary strip,

$$d_x = d_1 - (d_1 - d_2)\frac{x}{l}$$
$$= d_1 - k\,x \text{ where } k = (d_1 - d_2)/l.$$

The cross-sectional area of the strip,

$$A_x = \frac{\pi}{4}d_x^2 = \frac{\pi}{4}(d_1 - k\,x)^2$$

Stress in the strip, $\sigma_x = \dfrac{P}{A_x} = \dfrac{P}{\dfrac{\pi}{4}(d_1 - k\,x)^2} = \dfrac{4P}{\pi(d_1 - kx)^2}.$ \hfill (1.5)

Strain in the strip, $\varepsilon_x = \dfrac{\sigma_x}{E} = \dfrac{4P}{\pi (d_1 - k\,x)^2 E}$. $\qquad\qquad$ (1.6)

Elongation of the strip, $\delta l_x = \varepsilon_x dx = \dfrac{4P\,dx}{\pi (d_1 - k\,x)^2 E}$. $\qquad\qquad$ (1.7)

The total elongation of this tapering bar can be worked out by integrating the above expression between the limits $x = 0$ to $x = l$:

$$\delta l = \int_0^l \frac{4P\,dx}{\pi (d_1 - kx)^2 E} = \frac{4P}{\pi E}\int_0^l \frac{dx}{(d_1 - k\,x)^2}$$

$$= \frac{4P}{\pi E}\left[\frac{(d_1 - k\,x)^{-1}}{(-1)\times(-k)}\right]_0^l = \frac{4P}{\pi Ek}\left[\frac{1}{d_1 - kx}\right]_0^l.$$

Putting the value of $k = \dfrac{(d_1 - d)}{l}$ in the above expression, we obtain:

$$\delta l = \frac{4Pl}{\pi E(d_1 - d_2)}\left[\frac{1}{d_1 - \dfrac{(d_1 - d_2)l}{l}} - \frac{1}{d_1}\right]$$

$$= \frac{4Pl}{\pi E(d_1 - d_2)}\left[\frac{1}{d_2} - \frac{1}{d_1}\right]. \qquad\qquad (1.8)$$

$$= \frac{4Pl}{\pi E(d_1 - d_2)}\times\frac{d_1 - d_2}{d_1 d_2} = \frac{4Pl}{\pi E d_1 d_2}$$

If the bar is of uniform diameter d throughout its length, then:

$$\delta l = \frac{4Pl}{\pi E d^2} = \frac{Pl}{\dfrac{\pi}{4}d^2 E} = \frac{PL}{AE}.$$

which is the same expression as derived earlier in Section 1.3.

EXAMPLE 1.7
A conical bar tapers uniformly from a diameter of 4–15 cm to a length of 40 cm. If an axial force of 80 kN is applied at each end, then it will determine the elongation of the bar. Take E = 200 GPa.

Solution: In a bar of uniformly tapering circular section, the elongation is prescribed by the relation:

$$\delta l = \frac{4Pl}{\pi E d_1 d_2}.$$

Given: $P = 80 \times 10^3$ N; $E = 200$ GPa $= 200$ GN/m$^2 = 2 \times 10^5$ N/mm^2; $d_1 = 40$ mm; $d_2 = 15$ mm; and $l = 400$ mm:

$$\therefore \quad \delta l = \frac{4 \times (80 \times 10^3) \times 400}{\pi (2 \times 10^5) \times (40 \times 15)} = \textbf{0.3397 mm.}$$

EXAMPLE 1.8

A tension test bar is found to taper from $(D + a)$ diameter to $(D - a)$ diameter. Show that the error involved in using mean diameter to calculate Young's modulus of elasticity is $\left(10 \dfrac{a}{D} \right)^2$ percent.

Solution: Let the bar elongate by δl when a tensile force P is applied to it. Then for a tapering bar:

$$\delta l = \frac{4Pl}{\pi E (D + a)(D - a)} = \frac{4Pl}{\pi E (D^2 - a^2)},$$

Young's modulus, $E = \dfrac{4Pl}{\pi (D^2 - a^2)\delta l}.$

This gives the actual value of Young's modulus.

Mean diameter $= \dfrac{1}{2}[(D + a) + (D - a)] = D.$

Let E' be the value of Young's modulus calculated with mean diameter,

$$E' = \frac{\text{stress}}{\text{strain}} = \frac{P \big/ \frac{\pi}{4} D^2}{\delta l / l} = \frac{4Pl}{\pi D^2 \delta l}$$

$$\text{Error} = E - E' = \frac{4Pl}{\pi (D^2 - a^2)\delta l} - \frac{4Pl}{\pi D^2 \delta l}$$

$$= \frac{4Pl}{\pi \delta l} \left[\frac{1}{D^2 - a^2} - \frac{1}{D^2} \right] = \frac{4Pl}{\pi \delta l} \frac{a^2}{(D^2 - a^2) D^2}$$

$$\therefore \text{Percentage error} = \left[\frac{4Pl}{\pi \delta l} \frac{a^2}{(D^2 - a^2)D^2} \div \frac{4Pl}{\pi (D^2 - a^2)\delta l} \right] \times 100$$

$$= \frac{a^2}{D^2} \times 100 = \left(\frac{\mathbf{10a}}{\mathbf{D}} \right)^2.$$

EXAMPLE 1.9

The cross-sectional area of a bar is given by $(100 + x^2/100)$ mm^2 where x in mm is the distance of the section from one end. Make calculations for an increase in length of the bar when a tensile load equivalent to 20 kN is applied on a length of 150 mm. The value of elastic modulus for the bar material is 200 GN/m^2.

Solution: Consider a small element of length dx at distance x from the small element. Due to tensile load applied at the ends, the elemental length dx elongates by a small amount, Δx, and

$$\Delta x = \frac{P\,dx}{AE} = \frac{P\,dx}{(100 + x^2/100)E}.$$

The total elongation of the bar is then worked out by integrating the above identity between the limits $x = 0$ and $x = l$

$$\delta l = \int_0^l \frac{P\,dx}{(100 + x^2/100)E}$$

$$= 100 \frac{P}{E} \int_0^{150} \frac{dx}{(100^2 + x^2)}$$

Recalling that $\int \dfrac{dx}{a^2 + x^2} = \dfrac{1}{a}\tan^{-1}\left(\dfrac{x}{a}\right)$, we get

$$\delta l = \frac{100\,P}{E} \times \frac{1}{100} \left[\tan^{-1}\left(\frac{x}{100}\right) \right]_0^{150}$$

$$= \frac{P}{E}(\tan^{-1} 1.5 - \tan^{-1} 0) = \frac{P}{E}(0.9823 - 0).$$

Substituting for $P = 20$ kN $= 20 \times 10^3$ N and $E = 200$ GN/m$^2 = 2 \times 10^5$ N/mm^2, we obtain

$$\delta l = \frac{20 \times 10^3}{2 \times 10^5} \times 0.9823 = \mathbf{0.09823\ mm}.$$

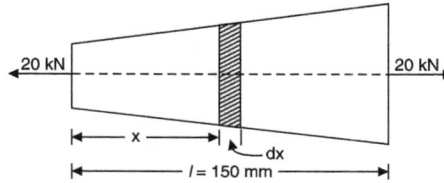

FIGURE 1.12.

1.6. EXTENSION OF BAR DUE TO SELF-WEIGHT

1.6.1. Bar of Uniform Section

Consider a bar of cross-sectional area A and length l hanging freely under its own weight (Figure 1.13). Let attention be focused on a small element of length dy at distance y from the lower end. If w is the specific weight (weight per unit volume) of the bar material, then total tension at section m–n equals the weight of the bar for the length y and is given by

$$P = w\,A\,y. \qquad (1.9)$$

As a result of this load, the elemental length dx elongates by a small amount Δx, and

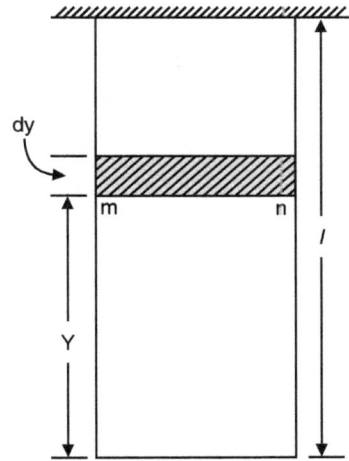

FIGURE 1.13.

$$\Delta x = \frac{P\,dy}{AE} = \frac{wAy}{AE}\;dy = \frac{w}{E}y\,dy. \qquad (1.10)$$

The total change in length of the bar due to self-weight is worked out by integrating the above expression between the limits $y = 0$ and $y = l$. Therefore,

$$\delta l = \int_0^l \frac{w}{E}y\,dy = \frac{w}{E}\left[\frac{y^2}{2}\right]_0^l = \frac{w}{E}\frac{l^2}{2}. \qquad (1.11)$$

If W is the total weight of the bar ($W = w\,A\,l$), then $w = W/Al$. In that case, total extension of the bar

$$\delta l = \left(\frac{W}{Al}\right)\frac{l^2}{2E} = \frac{Wl}{2\,AE}. \qquad (1.12)$$

Thus the total extension of the bar due to self-weight is equal to the extension that would be produced if one-half of the weight of the bar is applied at its end.

EXAMPLE 1.10

A rod of uniform cross-sectional area A is subjected to load P as shown in Figure 1.14. Set up the following expression for the displacement at the free end by taking the self-weight of the bar into account

$$l = \frac{w(a+b)^2}{2E} \frac{Pa}{AE},$$

where w is the specific weight and E is the modulus of elasticity of the rod material.

FIGURE 1.14.

Solution: Compression in the upper segment of length a due to load P

$$= \frac{Pa}{AE} \quad (-ve).$$

Extension in length $(a + b)$ of the rod due to self-weight

$$= \frac{Wl}{2\,AE}$$

$$= \frac{[wA(a+b)](a+b)}{2\,AE} = \frac{w(a+b)^2}{2\,E} \quad (+ve).$$

The net displacement at the free end of the rod is caused by the combined effect of extension due to self-weight and external load.

∴ Net displacement at a free end of the rod

$$= \frac{w(a+b)^2}{2E} - \frac{Pa}{AE}.$$

EXAMPLE 1.11

Determine the greatest length of mild steel wire of uniform cross-section that can be suspended vertically if the maximum stress is not to exceed 250 MN/m². Given gravitational acceleration $g = 9.81$ m/s² and density of steel $\rho = 7.85 \times 10^3$ kg/m³.

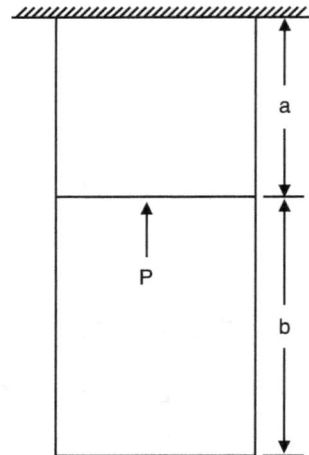

Solution: The maximum tension, that is, tensile force in the wire occurs at the suspension point and it equals the total weight of the suspended wire. If the cross-sectional area and the length of the wire are represented by A (m²) and l (m), respectively, then

Maximum tensile force, $P = w\,A\,l = \rho\,g\,A\,l = (7.85 \times 10^3 \times 9.81)\,Al$ (N).

Maximum stress, $\sigma_{max} = \dfrac{P}{A} = \dfrac{(7.85 \times 10^3 \times 9.81)Al}{A} = 77 \times 10^3 l$ (N/m²).

Since σ_{max} is not to exceed 250 MN/m², we have

$$250 \times 10^6 = 77 \times 10^3\, l.$$

∴ The maximum value of the length of wire,

$$l = \frac{250 \times 10^6}{77 \times 10^3} = 3.247 \times 10^3 M = \mathbf{3.247\ km}.$$

It is to be noted that this length is independent of the cross-sectional area.

EXAMPLE 1.12

An aerial copper wire ($E = 1 \times 105$ N/mm²) 40 m long has cross-sectional area of 80 mm² and weighs 0.6 N/m run. If the wire is suspended vertically, calculate

  a. the elongation of the wire due to self-weight,

  b. the total elongation when a weight of 200 N is attached to its lower end, and

  c. the maximum weight which this wire can support at its lower end if the limiting value of stress is 65 N/mm².

Solution:
  a. Weight of the wire $W = 0.6 \times 40 = 24$ N

The elongation due to self-weight is,

$$\delta = \frac{wl^2}{2E},$$

where w is the specific weight (weight per unit volume).
In terms of total weight $W = w\,Al$

$$\delta = \frac{Wl}{2AE} = \frac{24 \times (40 \times 10^3)}{2 \times 80 \times (1 \times 10^5)} = \mathbf{0.06\ mm}.$$

b. Extension due to weight P attached at the lower end,

$$\delta = \frac{Pl}{AE} \qquad \left[\because E = \frac{\sigma}{E} = \frac{P/A}{\delta/l} = \frac{Pl}{A\delta} \text{ or } \delta = \frac{Pl}{AE} \right]$$

$$= \frac{200 \times (40 \times 10^3)}{80 \times (1 \times 10^5)} = 1.0 \text{ mm}.$$

∴ Total elongation of the wire = 0.06 + 1.0 = **1.06 mm.**

c. Maximum limiting stress = 65 N/mm^2.

Stress due to self-weight equals that produced by a load of half its weight applied at the end. That is

Stress due to self-weight $= \dfrac{W/2}{A} = \dfrac{24/2}{80} = 0.15 \text{ N/mm}^2,$

Remaining stress = 65 − 0.15 = 64.85 N/mm^2.

∴ Maximum weight which the wire can support
    = 64.85 × 80 = **5188 N.**

1.7. PRINCIPLE OF SUPERPOSITION

Quite often a machine member is subjected to a number of forces acting on its outer edges (ends) as well as at some intermediate sections along its length. The forces are then split up and their effects are considered on individual sections. The resulting deformation is then given by the algebraic sum of the deformation of the individual sections. This is the *principle of superposition* which may be stated as

"The resultant elongation due to several loads acting on a body is the algebraic sum of the elongations caused by individual loads"

Or

"The total elongation in any stepped bar due to a load is the algebraic sum of elongations in individual parts of the bar"

Mathematically $\quad \delta l = \displaystyle\sum_{i=1}^{i=n} \delta l_i.$ $\qquad\qquad$ (1.13)

1.8. STRESSES IN BARS OF VARYING CROSS-SECTION

Consider a bar made up of different lengths and having different cross-sections as shown in Figure 1.15.

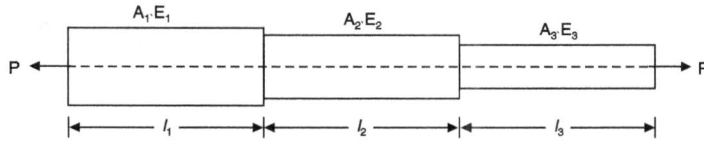

FIGURE 1.15 A varying cross-section bar.

For such a bar, the following conditions apply:

i. Each section is subjected to the same external pull or push.

ii. Total change in length is equal to the sum of changes of individual lengths

That is:
$$P_1 = P_2 = P_3 = P \text{ and}$$
$$\delta l = \delta l_1 + \delta l_2 + \delta l_3$$

$$= \frac{\sigma_1 l_1}{E_1} + \frac{\sigma_2 l_2}{E_2} + \frac{\sigma_3 l_3}{E_3}$$

$$= \frac{P_1 l_1}{A_1 E_1} + \frac{P_2 l_2}{A_2 E_2} + \frac{P_2}{A_3 E_3}. \tag{1.14}$$

If the bar segments are made of same material, then $E_1 = E_2 = E_3 = E$. In that case

$$\delta l = \frac{P}{E}\left[\frac{l_1}{A_1} + \frac{l_2}{A_2} + \frac{l_3}{A_3}\right]. \tag{1.15}$$

EXAMPLE 1.13
Two prismatic bars are rigidly fastened together and support a vertical load of 45 kN as shown in Figure 1.16. The upper bar is steel having a mass density of 7750 kg/m^3, length of 10 m, and cross-sectional area of 65 cm^2. The lower bar is brass having a mass density of 9000 kg/m^3, length of 6 m, and cross-sectional area of 50 cm^2. Determine the maximum stress in each material. For steel, E_s = 200 GN/m^2 and for brass E_b = 100 GN/m^2.

Solution: Refer to Figure 1.16. The maximum stress in the brass bar occurs at junction *BB*, and this stress is caused by the combined effect of 45 kN load together with the weight of brass bar:

Weight of brass bar, $W_b = \rho_b \, g \, V_b$
$$= 9000 \times 9.81 \times (6 \times 50 \times 10^{-4})$$
$$= 2648.7 \text{ N.}$$

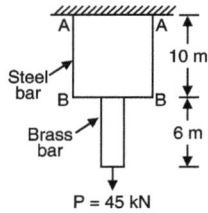

FIGURE 1.16.

$\therefore$ Stress at section BB, $\sigma_b = \dfrac{P + W_b}{A_b} = \dfrac{45000 + 4648.7}{50 \times 10^{-4}}$

$= 9,529,740 \text{ N/m}^2 = 9.53 \text{ MN/m}^2.$

The maximum stress in the steel bar occurs at section AA. Here the entire weight of steel and brass bars and 45 kN load gives rise to normal stress:

Weight of steel bar, $W_s = \rho_s\, g\, V_s = 7750 \times 9.81 \times (10 \times 65 \times 10^{-4})$
$= 4941.79 \text{ N}.$

$\therefore$ Stress at section AA,

$$\sigma_s = \frac{P + W_b + W_s}{A_s} = \frac{45000 + 2648.7 + 4941.79}{65 \times 10^{-4}}$$

$$= 8,090,845 = \mathbf{8.09\ MN/m^2}.$$

EXAMPLE 1.14

A straight bar of 450 mm long is 10 mm in diameter for the first 200 mm length and 20 mm in diameter for the remaining length. If the bar is subjected to an axial push of 10 kN, determine a decrease in the length of the bar.

Take modulus of elasticity of bar material, $E = 2 \times 10^5 \text{ N/mm}^2$.

Solution: Refer to Figure 1.17

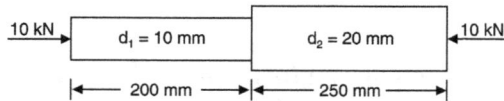

FIGURE 1.17.

$A_1 = \dfrac{\pi}{4}\,(10)^2 = 78.5 \text{ mm}^2;\ A_2 = \dfrac{\pi}{4}\,(20)^2 = 314 \text{ mm};$

$l_1 = 200 \text{ mm; and } l_2 = 250 \text{ mm}.$

Let δl_1 and δl_2 be the contractions in parts 1 and 2 of the bar, respectively. It may be recalled that

$$E = \frac{\sigma}{\varepsilon} = \frac{P/A}{\delta l/l} = \frac{Pl}{A\delta l} \quad \text{or} \quad \delta l = \frac{Pl}{AE}.$$

$$\text{Then, } \delta l_1 = \frac{P_1 l_1}{A_1 E_1} \quad \text{and} \quad \delta l_2 = \frac{P_2 l_2}{A_2 E_2}.$$

∴ Total decrease in length of the bar,

$$\delta l = \delta l_1 + \delta l_2 = \frac{P_1 l_1}{A_1 E_1} + \frac{P_2 l_2}{A_2 E_2}.$$

Both the segments are of the same material $(E_1 = E_2)$ and are subjected to the same push $(P_1 = P_2)$

$$\therefore \quad \delta l = \frac{P}{E}\left(\frac{l_1}{A_1} + \frac{l_2}{A_2}\right) = \frac{10 \times 10^3}{2 \times 10^5}\left(\frac{200}{78.5} + \frac{250}{314}\right)$$

$$= 0.05\,(2.548 + 0.796) = \textbf{0.1672 mm.}$$

EXAMPLE 1.15

A solid circular steel rod of 6 mm in diameter and 500 mm long is rigidly fastened to the end of a square brass bar 25 mm on a side and 400 mm long, the geometric axis of the bars lying along the same line. An axial tensile force of 5 kN is applied at each extreme end. Determine the elongation of assembly. For steel, E_s = 200 GPa and for brass E_b = 90 GPa.

Solution: Refer to Figure 1.18.

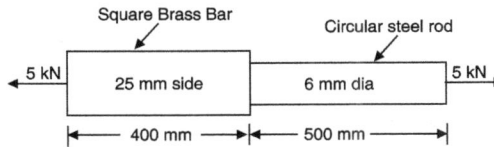

FIGURE 1.18.

$E_s = 200$ GPa $= 200$ GN/m$^2 = 2 \times 10^5$ N/mm^2,
$E_b = 90$ GPa $= 0.9 \times 10^5$ N/mm^2.

Let δl_b and δl_s be the elongations in brass and steel segments of the assembly. Then total increase in length of the assembly is

$$\delta l = \delta l_b + \delta l_s = \frac{P_b l_b}{A_b E_b} + \frac{P_s l_s}{A_s E_s}$$

$$= \frac{5 \times 10^3 \times 400}{(25 \times 25) \times (0.9 \times 10^5)} + \frac{5 \times 10^3 \times 500}{\frac{\pi}{4}(6)^2 \times (2 \times 10^5)}$$

$$= 0.035 + 0.442 = \textbf{0.477 mm.}$$

EXAMPLE 1.16
The bar shown in Figure 1.19 is subjected to an axial pull of 150 kN. Determine the diameter of the middle portion if stress there is limited to 125 N/mm². Proceed to determine the length of this middle portion if a total extension of the bar is specified as 0.15 mm. Take modulus of elasticity of bar material E = 2.5 × 105 N/mm².

Solution: Refer to Figure 1.19
Each segment of this composite bar is subjected to axial pull P = 150 kN.

(*i*) Stress in the middle portion, $\sigma_2 = \dfrac{\text{axial pull}}{\text{area}} = \dfrac{150 \times 10^3}{\dfrac{\pi}{4} d_2^{\,2}}$.

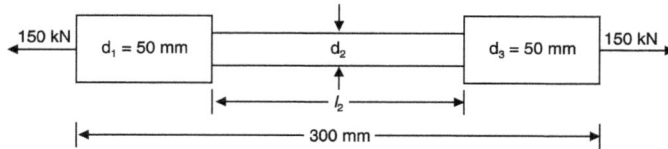

FIGURE 1.19.

Since stress is limited to 125 N/mm²,

$$125 = \frac{150 \times 10^3}{\dfrac{\pi}{4} d_2^{\,2}}; \quad d_2^{\,2} = \frac{150 \times 10^3}{\dfrac{\pi}{4} \times 125} = 1528.66.$$

∴ Diameter of the middle portion, $d_2 = \sqrt{1528.66} = $ **39.1 mm.**

(*ii*) Stress in the end portions, $\sigma_1 = \sigma_3 = \dfrac{150 \times 10^3}{\dfrac{\pi}{4}(50)^2} = 76.43$ N/m².

Total change in length of the bar,
= change in length of end portions + change in length of mid-portion

$$\delta l_1 = \frac{P_1 l_1}{A_1 E_1} + \frac{P_3 l_3}{A_3 E_3} + \frac{P_2 l_2}{A_2 E_2} = \frac{\sigma_1 l_1}{E_1} + \frac{\sigma_3 l_3}{E_3} + \frac{\sigma_2 l_2}{E_2}.$$

Here, $E_1 = E_2 = E_3 = 2 \times 10^5$ N/mm²; $\sigma_1 = \sigma_3 = 76.43$ N/m² and combined length $(l_1 + l_3) = (300 - l_2)$. Further, total extension of the bar is specified as 0.15 mm.

$$\therefore \quad 0.15 = \frac{76.43}{2 \times 10^5} \times (300 - l_2) + \frac{125 l_2}{2 \times 10^5}$$

or $\quad 0.15 \times 2 \times 105 = 22929 - 76.43 \, l_2 + 125 \, l_2$

or $30000 - 22929 = 48.57 \, l_2$

$\therefore$ length of the middle portion, $l_2 = \dfrac{30000 - 22929}{48.57} = \mathbf{145.58 \ mm}.$

EXAMPLE 1.17

A steel tie rod of 50 mm in diameter and 2.5 m long is subjected to a pull of 100 kN. Calculate the percentage change in extension produced under the same pull by boring the rod centrally for 1.2 m length; the bore being 25 mm in diameter. For steel modulus of elasticity is 2×10^5 N/mm^2.

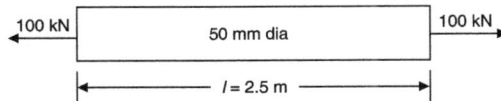

FIGURE 1.20.

Solution: When the rod is solid:

Area of cross-section, $A = \dfrac{\pi}{4}(50)^2 = 1962.5 \ \text{mm}^2$

Stress, $\quad \sigma = \dfrac{\text{load or pull}}{\text{area}} = \dfrac{100 \times 10^3}{1962.5} = 50.95 \ \text{N/mm}^2$

Elongation, $\delta l = \dfrac{\sigma}{E} l = \dfrac{50.95}{2 \times 10^5} \times 2500 = 0.637 \ \text{m}.$

When the rod is bored:

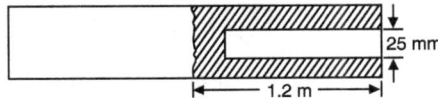

Area at the reduced section $= \dfrac{\pi}{4}(50^2 - 25^2) = 1471.9 \ \text{mm}^2$

Stress at the reduced section $= \dfrac{100 \times 10^3}{1471.9} = 67.94 \ \text{N/mm}^2$

Total elongation of the bored rod,

= elongation of solid part + elongation of bored part

$$= \frac{50.95}{2 \times 10^5} \times (2500 - 1200) + \frac{67.94}{2 \times 10^5} \times 1200$$

$$= 0.3312 + 0.4076 = 0.7388 \text{ mm.}$$

∴ Percentage increase in elongation

$$= \frac{0.7388 - 0.637}{0.637} \times 100 = \mathbf{15.98\%}.$$

EXAMPLE 1.18

A member *ABCD* of uniform diameter 200 mm has been subjected to point loads as shown in Figure 1.21. Determine the net change in the length of the bar. Take modulus of elasticity of the bar material as *E* = 200 GN/m².

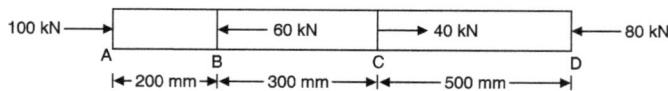

FIGURE 1.21.

Solution: Cross-sectional area of each segment

$$A = \frac{\pi}{4} \times (200 \times 10^{-3})^2 = 0.0314 \text{ m}^2.$$

The forces acting on each segment of the member are as indicated in the free-body diagram drawn below:

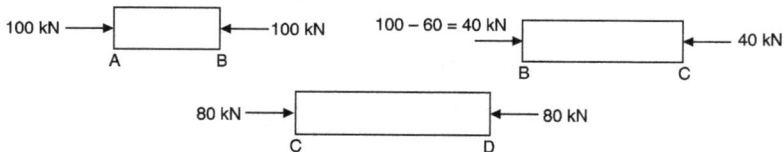

It is to be noted that each segment is subject to compressive forces and accordingly shorten in length. Recalling that extension/contraction in a bar is given by *Pl/AE*, we have

Compression of segment *AB*, $\delta l_1 = \dfrac{(100 \times 10^3) \times (200 \times 10^{-3})}{0.0314 \times (200 \times 10^9)}$ m.

Compression of segment BC, $\delta l_2 = \dfrac{(40 \times 10^3) \times (300 \times 10^{-3})}{0.0314 \times (200 \times 10^9)}$ m.

Compression of segment CD, $\delta l_3 = \dfrac{(80 \times 10^3) \times (400 \times 10^{-3})}{0.0314 \times (200 \times 10^9)}$ m.

Total contraction in length,

$$\delta l = \delta l_1 + \delta l_2 + \delta l_3$$

$$= \frac{1}{0.0314 \times (200 \times 10^9)} [20000 + 12000 + 32000]$$

$$= 1.019 \times 10^{-5} \text{ m} = \mathbf{0.01019 \text{ mm}}.$$

EXAMPLE 1.19
Refer to Figure 1.22 for the tapering bar of circular cross-section which is subjected to point loads as shown. Section AB has a uniform diameter of 60 mm and for section CD, the uniform diameter is 300 mm. Determine load P_3 for equilibrium and net change in length of the bar. Take modulus of elasticity of the bar material as $E = 200$ GPa.

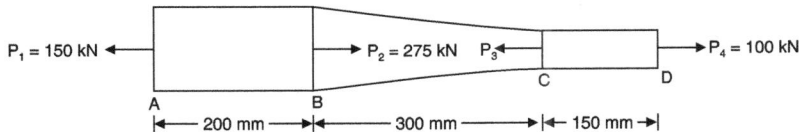

FIGURE 1.22.

Solution: Modulus of elasticity, $E = 200$ GPa $= 2 \times 10^5$ N/mm^2.
Considering equilibrium of forces along the axis of bar,

$$P_1 + P_3 = P_2 + P_4; \ 150 + P_3 = 275 + 100.$$

∴   Force $P_3 = 275 + 100 - 150 = 225$ kN.
The forces acting on each segment of the bar are as shown below:

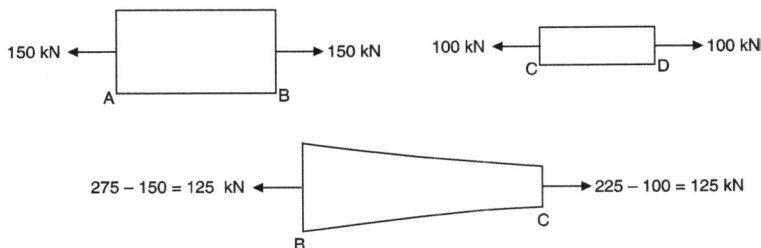

Recalling that extension/contraction for a uniform bar is given by Pl/AE and that for a tapering bar is $4Pl/\pi E\, d_1, d_2$, we have

Extension of segment AB, $\delta l_1 = \dfrac{(150 \times 10^3) \times 200}{\dfrac{\pi}{4}(60)^2 \times (2 \times 10^5)} = 0.053$ mm.

Contraction of segment BC, $\delta l_2 = \dfrac{4 \times \left(125 \times 10^3\right) \times 300}{\pi \times (2 \times 10^5) \times 60 \times 30} = 0.1327$ mm.

Extension of segment CD, $\delta l_3 = \dfrac{\left(100 \times 10^3\right) \times 150}{\dfrac{\pi}{4} \times (30)^2 \times (2 \times 10^5)} = 0.1063$ mm.

$\therefore$ Net change in length of member
$\delta l = \delta l_1 + \delta l_2 + \delta l_3 = 0.053 - 0.1327 + 0.1063 = \mathbf{0.0266}$ **mm.**

EXAMPLE 1.20
A member ABCD is subjected to point loads P_1, P_2, P_3, and P_4 as shown in Figure 1.23.

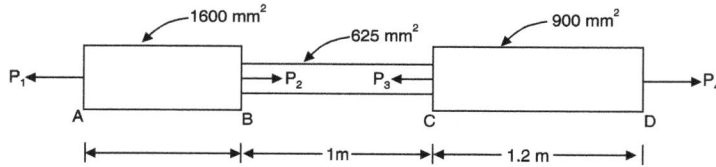

FIGURE 1.23.

Calculate the force P_3 necessary for equilibrium if $P_1 = 120$ kN, $P_2 = 220$ kN, and $P_4 = 160$ kN. Determine also the net change in length of the member. Take modulus of elasticity, $E = 200$ GN/m².

Solution: Modulus of elasticity, $E = 200$ GN/m² $= 2 \times 10^5$ N/mm².
Considering equilibrium of forces along the axis of the member:

$$P_1 + P_3 = P_2 + P_4;\ 120 + P_3 = 220 + 160.$$

$\therefore$ Force $P_3 = 220 + 160 - 120 = \mathbf{260}$ **kN.**
The forces acting on each segment of the member are shown in the free-body diagrams below:

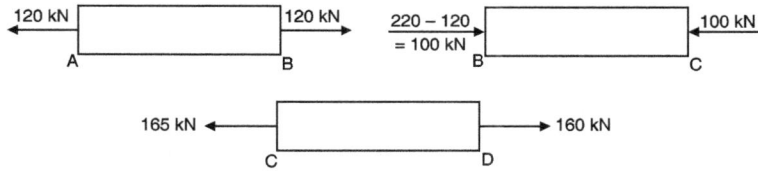

Recalling that extension/contraction in a bar is given by $\dfrac{Pl}{AE}$, we have

Extension of segment, $AB = \dfrac{(120 \times 10^3) \times (0.75 \times 10^3)}{1600 \times (2 \times 10^5)} = 0.28125$ mm.

Compression of segment, $BC = \dfrac{(100 \times 10^3) \times (1 \times 10^3)}{625 \times (2 \times 10^5)} = 0.8$ mm.

Extension of segment, $CD = \dfrac{(160 \times 10^3) \times (1.2 \times 10^3)}{900 \times (2 \times 10^5)} = 1.0667$ mm.

∴ Net change in length of the member:

$$\delta l = 0.28125 - 0.8 + 1.0667 = \textbf{0.54795 mm} \text{ (increase).}$$

EXAMPLE 1.21

A square bar of 25 mm side is held between two rigid plates and loaded by an axial pull equal to 300 kN as shown in Figure 1.24. Determine the reactions at end A and C and elongation of the portion AB. Take $E = 2 \times 10^5$ N/mm^2.

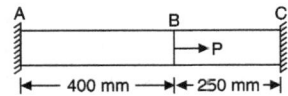

FIGURE 1.24.

Solution: Cross-section area of the bar, $A = 25 \times 25 = 625$ mm^2.

Since the bar is held between rigid supports at the ends, the following observations need to be made:

i. Portion *AB* will be subjected to tension and portion *BC* will be under compression.

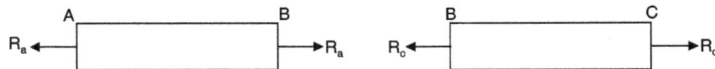

ii. Elongation in portion *AB* equals shortening in portion *BC*.

iii. Sum of reactions equals the applied axial pull, that is, $P = R_a + R_c$.

Invoking the relation: $\delta l = \dfrac{Pl}{A}$

$$\delta_{ab} = \frac{R_a \times 400}{625 \times 2 \times 10^5} \text{ and } \delta_{bc} = \frac{R_c \times 250}{625 \times 2 \times 10^5}.$$

Equating the changes in the length of the two portions

$$\frac{R_a \times 400}{625 \times 2 \times 10^5} = \frac{R_c \times 250}{625 \times 2 \times 10^5}; \quad R_c = 1.6\, R_a.$$

Substituting the value of $R_c = 1.6\,R_a$ in the identity $P = R_a + R_c$, we have

$$300 \times 10^3 = R_a + 1.6R_a; \quad R_a = \frac{300 \times 10^3}{2.6} = \mathbf{1.154 \quad 10^5\ N}$$

and $\quad R_c = 1.6\,R_a = 1.6 \times 1.154 \times 10^3 = \mathbf{1.846 \quad 10^5\ N}.$

EXAMPLE 1.22

A rod *ABCD* rigidly fixed at the ends *A* and *D* is subjected to two equal and opposite forces *P* = 25 kN at *B* and *C* as shown in Figure 1.25:
Make calculations for the axial stresses in each section of the rod.

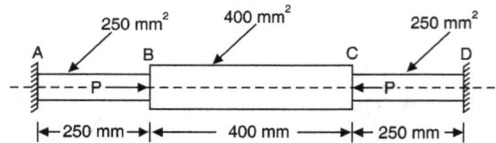

FIGURE 1.25.

Solution: The following observations need to be made:

i. Due to symmetrical geometry and load, reaction at each of the fixed ends will be the same both in magnitude and direction. That is $P_a = P_d = P_1$ (say).

ii. Segments *AB* and *CD* are in tension and the segment *BC* is in compression.

iii. End supports are rigid and therefore total change in length of the rod is zero. The forces acting on each segment will be as shown below:

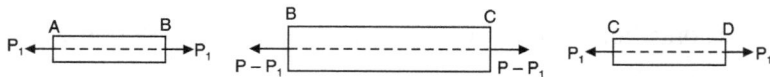

Invoking the relation, $\delta l = \dfrac{Pl}{AE}$

$$\delta_{ab} = \frac{P_1 \times 250}{250 \times E} = \frac{P_1}{E} \quad \text{(extension)},$$

$$\delta_{bc} = \frac{(P - P_1) \times 400}{400 \times E} = \frac{P - P_1}{E} \quad \text{(compression)},$$

$$\delta_{cd} = \frac{P_1 \times 250}{250 \times E} = \frac{P_1}{E} \quad \text{(extension)}.$$

∴ The net change in the length of the rod,

$$\frac{P_1}{E} - \frac{P \times P_1}{E} + \frac{P_1}{E} = 0 \quad \text{(Since ends are rigid)}$$

or $P_1 - P + P_1 + P_1 = 0$

or $P_1 = \dfrac{P}{3} = \dfrac{25}{3}$ kN

and $P - P_1 = 25 - \dfrac{25}{3} = \dfrac{50}{3}$ kN.

∴ Stress in segment AB and CD,

$$\sigma_{ab} = \sigma_{cd} = \frac{25 \times 10^3}{3 \times 250} = \textbf{33.33 N / mm}^2 \text{ (tensile)}.$$

Stress in segment BC,

$$\sigma_{bc} = \frac{50 \times 10^3}{3 \times 400} = \textbf{41.67 N / mm}^2 \text{ (compressive)}.$$

EXAMPLE 1.23

A uniform concrete slab weighing 1200 kN has been attached to two rods, whose lower ends are at the same level as shown in Figure 1.26. One rod is of aluminum and has a length of 1.5 m. The other rod is of steel and measures 2 m in length. Make calculations for the ratio of the areas of the rods if the slab is remain horizontal.

The modular ratio of steel to aluminum is 2.86.

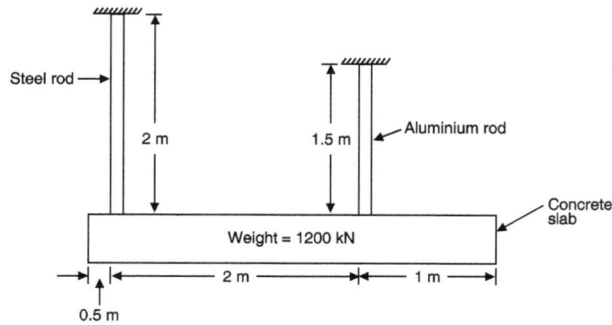

FIGURE 1.26.

Solution: Refer to the figure shown below for the free-body diagram of the slab:

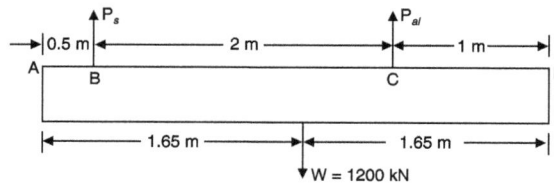

From equations of equilibrium,

$\Sigma F_y = 0$: $P_s + P_{al} = 1200$,

$\Sigma M = 0$:   Taking moments about point B, $1200 \, (1.65 - 0.5) - P_{al} \times 2 = 0$

or $P_{al} = \dfrac{1200 \times 1.15}{2} = 690$ kN

and $P_s = 1200 - P_{al} = 1200 - 690 = 510$ kN.

Since the slab is to remain horizontal, extensions of the two rods are equal. That is

$$\frac{P_s l_s}{A_s E_s} = \frac{P_{al} l_{al}}{A_{al} E_{al}},$$

∴. $$\frac{A_{al}}{A_s} = \frac{P_a}{P_s} \times \frac{E_s}{E_{al}} \times \frac{l_{al}}{l_s} = \frac{690}{510} \times 2.86 \times \frac{1.5}{2} = \mathbf{2.90}.$$

1.9. STRESSES IN COMPOSITE BARS

Quite often it becomes necessary to have a compound tie or strut (column) where two or more material elements are fastened together to prevent their uneven straining. The salient features of such a composite system are:

- The system extends (or contracts) as one unit when subjected to tensile (or compressive) load. This implies that the deformation (extension or contraction) of each element is the same.
- Strain, that is, deformation per unit length of each element is the same.
- Total external load on the system equals the sum of loads carried by the different materials comprising the composite system.

Consider a composite bar subjected to load P and fixed at the top as shown in Figure 1.27. Total load is shared by the two bars and as such

FIGURE 1.27.

$$P = P_1 + P_2 = \sigma_1 A_1 + \sigma_2 A_2. \qquad (1.16)$$

Further elongations in two bars are the same, that is, strains in the bars are equal. Thus

$$\varepsilon_1 = \varepsilon_2 \quad ; \quad \frac{\sigma_1}{E_1} = \frac{\sigma_2}{E_2},$$

$$\therefore \quad \frac{\sigma_1}{\sigma_2} = \frac{E_1}{E_2}. \qquad (1.17)$$

The ratio E_1/E_2 is called the *modular ratio*.

EXAMPLE 1.24

Two copper rods one steel rod lie in a vertical plane and together support a load of 50 kN as shown in Figure 1.28. Each rod is 25 mm in diameter, the length of the steel rod is 3 m, and the length of each copper rod is 2 m. If the modulus of elasticity of steel is twice that of copper, make calculations for the stress induced in each rod. It may be presumed that each rod deforms by the same amount.

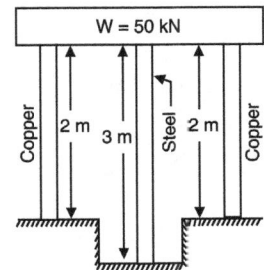

FIGURE 1.28.

Solution: Each rod deforms by the same amount and accordingly

$$\frac{\sigma_s}{E_s} l_s = \frac{\sigma_c}{E_c} l_c$$

$$\sigma_s = \frac{E_s}{E_c} \times \frac{l_c}{l_s} \times \sigma_c$$

$$= 2 \times \frac{2}{3} \sigma_c = 1.33 \sigma_c.$$

The division of load between the steel and copper rods is as follows:
total load = load carried by steel rod
  + rod carried by two copper rods

$$50 \times 10^3 = \sigma_s A_s + 2\sigma_c A_c$$

$$= 1.33\,\sigma_c \times \frac{\pi}{4}(25)^2 + 2\sigma_c \frac{\pi}{4}(25)^2 = 1633.78\,\sigma_c$$

$$\sigma_c = \frac{50 \times 10^3}{1633.78} = \mathbf{30.60\ N/mm^2}$$

$$\sigma_s = 1.33\,\sigma_c = 1.33 \times 30.60 = \mathbf{40.7\ N/mm^2}.$$

EXAMPLE 1.25

A beam weighing 500 N is held in a horizontal position by three wires. The outer wires are of brass of 1.2 mm diameter attached to each end of the beam. The central wire is of steel of 0.6 mm diameter attached to the middle of the beam. The beam is rigid and the wires are of the same length and unstressed before the beam is attached. Determine the stress induced in each of the wires. Take Young's modulus for brass as 80 GN/m^2 and for steel as 200 GN/m^2.

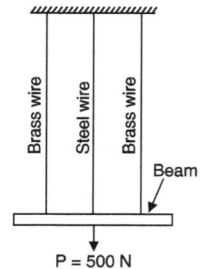

FIGURE 1.29.

Solution: Refer to Figure 1.29 for the arrangement of the wires and the beam.

If P_b denotes load taken by each brass wire and P_s denotes load taken by steel wire, then

i.
$$P = 2P_b + P_s$$
$$= 2\sigma_b A_b + \sigma_s A_s.$$

As the beam is horizontal, all the wires extend by the same amount. Further, since each wire is of the same length, the wires would experience the same amount of strain. Thus

$$\varepsilon_s = \varepsilon_b \quad ; \quad \frac{\sigma_s}{E_s} = \frac{\sigma_b}{E_s}$$

ii.
$$\sigma_s = \frac{E_s}{E_b} \times \sigma_b = \frac{200}{80}\sigma_b = 2.5\sigma_b.$$

From expression (*i*) and (*ii*),

$$500 = 2\sigma_b \times \frac{\pi}{4}(1.2)^2 + 2.5\sigma_b \times \frac{\pi}{4}(0.6)^2 = 2.9673\sigma_b.$$

That gives: $\sigma_b = \dfrac{500}{2.9673} = \mathbf{168.5\ N/mm^2}$

and $\qquad \sigma_s = 2.5\ \sigma_b = 2.5 \times 168.5 = \mathbf{421.25\ N/mm^2}.$

EXAMPLE 1.26

Two steel rods and one copper rod lie equidistant in a vertical plane and together support a load of 15 kN as shown in Figure 1.30. Each rod is 10 mm in diameter and these rods share the load equally. Determine the stress to which each rod would be subjected when the system is to support an additional load of 10 kN. All the rods undergo equal deformation and the modulus of elasticity of steel equals twice that of copper.

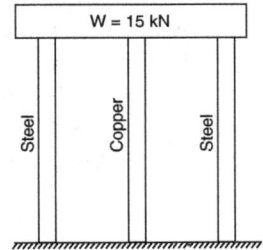

FIGURE 1.30.

Solution: Initial load carried by each rod

$$= \frac{15 \times 10^3}{3} = 5000\ \text{N}.$$

$$\text{Stress in each rod} = \frac{5000}{\dfrac{\pi}{4}(10)^2}\ \text{N/mm}^2 = 63.694\ \text{N/mm}^2.$$

Division of additional load is as follows,

i. additional load = load carried by copper rod + load carried by two steel rods

$$10 \times 10^3 = \sigma_c\ A_c + 2\sigma_s A_s.$$

The rods undergo equal deformation and are of the same length; their strains are equal and accordingly

$$\frac{\sigma_s}{E_s} = \frac{\sigma_c}{E_c}$$

$$\sigma_s = \frac{E_s}{E_c}\sigma_c = 2\sigma_c.$$

Expression (*i*) may then be written as

$$10 \times 10^3 = \sigma_c \times \frac{\pi}{4}(10)^2 + 2(2\sigma_c) \times \frac{\pi}{4}(10)^2 = 392.5\ \sigma_c$$

or $\sigma_c = \dfrac{10 \times 10^3}{392.5} = 25.48 \text{ N/mm}^2$ and $\sigma_s = 2 \times 25.48 = 50.96 \text{ N/mm}^2$.

∴ Total stress in copper = $63.694 + 25.48 = \mathbf{89.174 \text{ N/mm}^2}$.

Total stress in steel = $63.694 + 50.96 = \mathbf{114.694 \text{ N/mm}^2}$.

EXAMPLE 1.27

A concrete column, 250 mm × 250 mm in section, is reinforced by 8 longitudinal 15 mm diameter round steel bars. The column carries a compressive load of 300 kN. Make calculations for the loads carried by and compressive stresses produced in the steel bars and concrete. Take, $E_s = 200 \text{ GN/m}^2$ and $E_c = 15 \text{ GN/m}^2$.

Solution: Cross-sectional area of column = $250 \times 250 = 62{,}500 \text{ mm}^2$,

Area of steel bars, $A_s = 8 \times \dfrac{\pi}{4}(15)^2 = 1413 \text{ mm}^2$,

∴ Area of concrete = $62{,}500 - 1413 = 61{,}087 \text{ mm}^2$.

Each component (concrete and steel bars) shorten by the same amount under the compressive load, and therefore

strain in concrete = strain in steel,

$$\frac{\sigma_c}{E_c} = \frac{\sigma_s}{E_s},$$

where σ_c and σ_s are the stresses induced in concrete and steel, respectively.

or $\qquad \sigma_s = \dfrac{E_s}{E_c} \times \sigma_c = \dfrac{200 \times 10^9}{15 \times 10^9} \sigma_c = 13.33\,\sigma_c$.

Further,

the load carried by steel + load carried by concrete = total load on the column,

$$\sigma_s \times A_s + \sigma_s A_c = P,$$

$13.33\,\sigma_c \times 1413 + \sigma_c \times 61087 = 300 \times 10^3$.

Solution gives: $\qquad \sigma_c = 3.753 \text{ N/mm}^2$ and

$\qquad\qquad \sigma_s = 13.33 \times 3.753 = \mathbf{50.036 \text{ N/mm}^2}$.

The load carried by concrete,

$\qquad P_c = \sigma_c \times A_c = 3.753 \times 61087 = 229{,}259 \text{ N} \approx \mathbf{229.3 \text{ kN}}$.

Load carried by steel,

$\qquad P_s = \sigma_s \times A_s = 50.036 \times 1413 = 70{,}700 \text{ N} = \mathbf{70.7 \text{ kN}}$.

EXAMPLE 1.28

A mild steel tube of external diameter 80 mm and internal diameter 60 mm is enclosed centrally inside a brass tube of the same length and has an external diameter of 100 mm and an internal diameter of 80 mm. The ends of the tubes are brazed together and the composite arrangement is subjected to an axial pull of 25 kN. If the modulus of elasticity for steel and brass are 200 GPa and 100 GPa, respectively, and tube length is 30 cm. Make calculations for the load carried and shortening in the length of each tube.

Solution: Refer to Figure 1.31 for the composite arrangement.

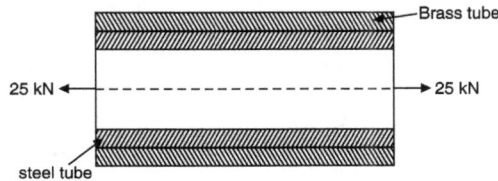

FIGURE 1.31.

$$E_s = 200 \text{ GPa} = 200 \text{ GN/m}^2 = 2 \times 10^5 \text{ N/mm}^2,$$
$$E_b = 100 \text{ GPa} = 1 \times 10^5 \text{ N/m}^2,$$

Area of steel tube, $A_s = \dfrac{\pi}{4}(80^2 - 60^2) = 2198 \text{ mm}^2,$

Area of brass tube, $A_b = \dfrac{\pi}{4}(100^2 - 80^2) = 2826 \text{ mm}^2.$

Since the tubes are brazed (rigidly fixed) together at the ends, each tube experiences the same amount of strain. Thus,

$$\frac{\sigma_s}{E_s} = \frac{\sigma_b}{E_b},$$

$$\therefore \sigma_s = \frac{E_s}{E_b} = \sigma_b = \frac{2 \times 10^5}{1 \times 10^5} \sigma_b - 2\sigma_b.$$

Further, the total axial load is shared by both the tubes. Therefore,

$$P = P_s + P_b = \sigma_s A_s + \sigma_b A_b$$

or $25 \times 10^3 = 2\sigma_b \times 2198 + \sigma_b \times 2826 = 7122\ \sigma_b$ $\quad (\therefore\ \sigma_s = 2\sigma_b)$

$$\therefore \qquad \sigma_b = \frac{25 \times 10^3}{7122} = 3.462 \text{ N/mm}^2$$

and $\quad \sigma_s = 2\sigma_b = 2 \times 3.462 = 6.924$ N/mm^2.

Load carried by steel tube, $P_s = \sigma_s A_s$

$$= 6.924 \times 2198 = \mathbf{15{,}219 \text{ N.}}$$

Load carried by brass tube, $P_b = \sigma_b A_b$

$$= 3.462 \times 2826 = \mathbf{9783.6 \text{ N.}}$$

Shortening in each tube is the same and can be worked out from the relation:

$$\delta l = \sigma \, l/E,$$

$$\therefore \quad \delta l = \frac{\sigma_b l}{E} = \frac{3.462 \times (30 \times 10)}{1 \times 10^5} = \mathbf{0.010386 \text{ mm.}}$$

1.10. SHEAR STRESS, SHEAR STRAIN, AND MODULUS OF RIGIDITY

Stress and strain produced by a force tangential to the surface of a body are known as *shear stress and shear strain,* respectively.

Consider a rectangular block *ABCD* fixed at the bottom plane and subjected to tangential force *P* at the upper plane. Then

$$\text{shear stress, } \tau = \frac{\text{tangential force}}{\text{area of face } DCFE}.$$

$$= \frac{P}{bl}$$

Reaction at the bottom plane will be equal and opposite to the applied force *P*.

The distinction between normal stress and shear stress in one of direction becomes apparent if we consider a bar cut by a plane perpendicular to its axis. Whereas the normal stress σ_n is perpendicular to this plane, the shear stress τ acts along the plane.

The common examples of a system involving shear stress are riveted and welded joints, towing devices, punching operations, etc.

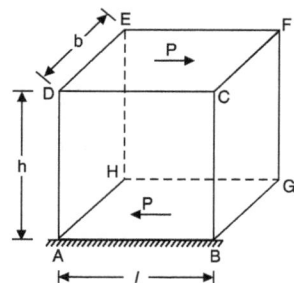

FIGURE 1.32.

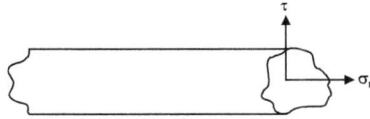

FIGURE 1.33.

Under the action of tangential force, the block $ABCD$ gets distorted and takes the shape $ABC'D'$ by deforming through an angle ϕ.

Since ϕ is very small,

$$\phi = \tan\phi = \frac{CC'}{BC}. \tag{1.18}$$

The angular deformation ϕ in radians represents the shear strain.

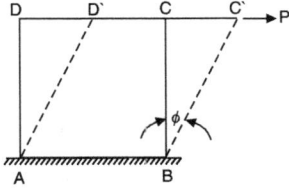

FIGURE 1.34.

With a gradual increase in tangential force, the shear stress increases gradually and so does the shear strain. That is

$$\tau \propto \phi$$

or $\quad \tau = C\phi \quad ; \quad C = \dfrac{\tau}{\phi}. \tag{1.19}$

The ratio of the shear stress to the shear strain is called *modulus of rigidity* or *modulus of elasticity in shear* and is usually denoted by C.

Since shear strain is dimensionless, the units of modulus of rigidly are the same as those of shear stress, *for example,* N/m^2 and GPa.

EXAMPLE 1.29
A single riveted lap joint as shown in Figure 1.35 is to transmit a force, $P = 5$ kN. Calculate the shear stress acting on the rivet at the junction where the rivet tends to shear. Take rivet diameter, $d = 12.5$ mm.

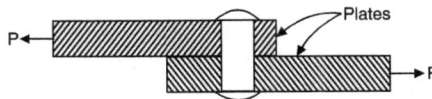

FIGURE 1.35.

Solution: Shear stress $= \dfrac{\text{tangential load}}{\text{area being sheared}} = \dfrac{5 \times 10^3}{\dfrac{\pi}{4}(12.5)^2} = \textbf{40.76 N / mm}^2.$

EXAMPLE 1.30
A steel punch can be worked to a compressive stress of 800 N/mm². Find the least diameter of hole which can be punched through a steel plate 10 mm thick if its ultimate shear strength is 350 N/mm².

Solution: Let d be the diameter of the hole in mm:

Area being sheared $= \pi d\, t = \pi d \times 10 = 10\ \pi d$ mm².

Force required to punch the hole,

$$= \text{ultimate shear strength} \times \text{area sheared}$$

$$= 350 \times 10\ \pi d = 3500\ \pi d\ (\text{N}).$$

The cross-sectional area of the hole $= \dfrac{\pi}{4} d^2$ mm²,

Compressive stress on the punch, $\sigma_c = \dfrac{3500}{\dfrac{\pi}{4} d^2} = \dfrac{14000}{d}$,

But σ_c is limited to 800 N/mm² and therefore $800 = \dfrac{14000}{d}$.

∴ Diameter of the hole, $d = \dfrac{14000}{800} = \mathbf{17.5\ mm}$.

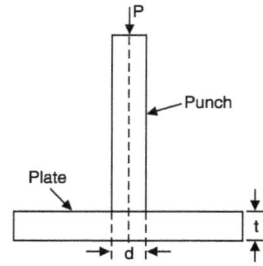

FIGURE 1.36.

1.11. COMPLIMENTARY SHEAR STRESS

Consider a rectangular block $ABCD$ subjected to shear stress of intensity τ at the forces AB and CD (Figure 1.37a).

If the block has unit thickness perpendicular to the plane of the paper, then shear forces acting on these faces will be $\tau \times AB$ or $\tau \times DC$. These forces constitute a couple of moment

$$= \text{force} \times \text{perpendicular distance}$$
$$= (\tau \times AB) \times BC \text{ or } (\tau \times DC) \times BC.$$

This couple tends to rotate the block in the clockwise direction. To keep the block in equilibrium, shear stress τ' must act on the other two faces of the block. The direction of these stresses has to be such that these provide

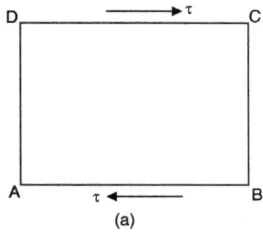

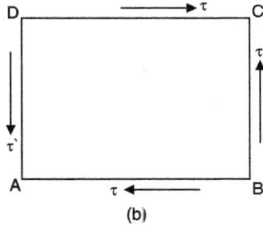

FIGURE 1.37.

an equal but opposite couple (Figure 1.37b). Thus, under equilibrium conditions

$$(\tau \times AB) \times BC = (\tau' \times BC) \times AB; \qquad \tau' = \tau.$$

The stress τ' is called *complimentary shear stress.* Thus, if a rectangular or square block is in equilibrium, shear stress in one plane is accompanied by equal complimentary shear stress in the other plane. The state of shear depicted in Figure 1.37(b) is called *Simple* or *pure shear* as no other type of stress is acting.

Let attention be focused on a cubical element $ABCD$ of each side l (Figure 1.38). The block is in simple shear, that is, the faces AB and CD are subjected to shear stress of intensity τ, whereas the faces DA and BC are subjected to complimentary shear stress τ' of equal magnitude.

Considering equilibrium of the free-body diagram of the triangular portion ABD, Shear forces on face $DA =$
$$\tau' \times (l \times l) = \tau\, l^2 \qquad (\because\ \tau' = \tau),$$
Shear force on face $AB= \tau \times (l \times l) = \tau\, l^2$.

The resultant of these forces is $\sqrt{2}\ \tau l$, and this is balanced by force σ_n acting normal to diagonal BD.

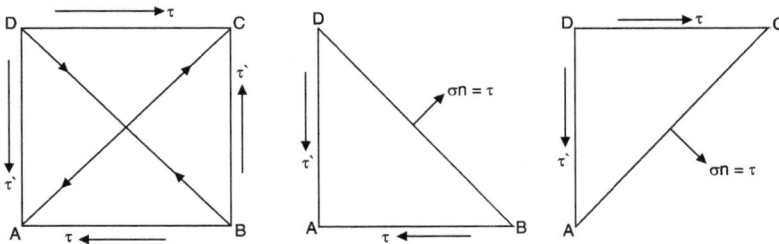

FIGURE 1.38.

The sectional area of the face, $BD = \sqrt{l^2 + l^2} = \sqrt{2}\, l.$

$\therefore$ Normal stress on diagonal BD,

$$\sigma_n = \frac{\sqrt{2}\,\tau\, l}{\sqrt{2}\, l} = \tau.$$

Likewise, equilibrium of free-body diagram of triangular portion DCA would give normal stress on diagonal AC

$$\sigma_n = \tau.$$

The above analysis shows that when an element is in a state of simple shear, tensile (along diagonal AC) and compressive (along diagonal BD) stresses are equal in magnitude to the shear stress that act on planes inclined at 45° compared to the planes of simple shear.

1.12. HYDROSTATIC STRESS, VOLUMETRIC STRAIN, AND BULK MODULUS

When a body is immersed in a fluid to a large depth, the body gets subjected to equal external pressure at all points on the body. This external pressure is compressive in nature and is called *hydrostatic stress*.

The hydrostatic stress causes a change in the volume of the body, and this change of volume per unit volume is called a volumetric strain, ε_v.

$$\varepsilon_v = \frac{\Delta V}{V}.$$

Consider a cuboid of sides x, y, and z subjected to hydrostatic stress and let its sides change by dx, dy, and dz, respectively. Then change in volume,

$$\Delta V = (x + dx)(y + dy)(z + dz) - xyz.$$

Neglecting product of small changes,

$$\Delta V = (x\,y\,z + x\,y\,dz + y\,z\,dx + x\,z\,dy) - x\,y\,z$$

$$= y\,z\,dx + x\,z\,dy + x\,y\,dz.$$

∴ Volumetric strain,

$$\varepsilon_v = \frac{\Delta V}{V} = \frac{y\,z\,dx + x\,z\,dy + x\,y\,dz}{x\,y\,z}$$

$$= \frac{dx}{x} + \frac{dy}{y} + \frac{dz}{z} \qquad (1.20)$$

$$= \varepsilon_x + \varepsilon_y + \varepsilon_z.$$

Apparently, volumetric strain equals the sum of the linear normal strains in x, y, and z directions.

Within elastic limits, the hydrostatic stress is directly proportional to volumetric strain. Mathematically, hydrostatic stress $\propto$ volumetric strain

= constant of proportionality × volumetric strain

or $$\sigma_v = K\varepsilon_v \quad ; \quad K = \frac{\sigma_v}{\epsilon_v}. \tag{1.21}$$

The constant of proportionality K is called the *bulk modulus* or *modulus of volume expansion* of the material. Physically, the bulk modulus K is a measure of the resistance of a material to change of volume without change of shape or form.

EXAMPLE 1.31
A solid steel cube of 1 m side is immersed in water to a depth of 600 m. Make calculations for the decrease in the length of its sides and in its volume. For steel, the bulk modulus is 135 GPa and for water, mass density is 1000 kg/m^3.

Solution: Hydrostatic pressure at 600 m depth is

$$\sigma_v = w\,h = \rho\,g\,h$$
$$= 1000 \times 9.81 \times 600 = 5.886 \times 10^6 \text{ N/m}^2.$$

Volumetric strain $\varepsilon_v = \dfrac{\sigma_v}{K} = \dfrac{5.886 \times 10^6}{135 \times 10^9} = 43.55 \times 10^{-6}.$

$\therefore$ Change in volume dV = volumetric strain × original volume

$$= (43.55 \times 10^{-6}) \times 1 = 43.55 \times 10^{-6} \text{ m}^3$$
$$= \textbf{43555 mm}^3 \text{ (decrease)}.$$

The volumetric strain ε_v is given approximately by the sum of the linear normal strains in x, y, and z directions. That is

$$\varepsilon_v = \varepsilon_x + \varepsilon_y + \varepsilon_z.$$

Here, $\varepsilon_x = \varepsilon_y = \varepsilon_z$

$$\therefore \quad \varepsilon_x = \frac{\varepsilon_v}{3} = \frac{43.55 \times 10^{-6}}{3} = 14.518 \times 10^{-6}.$$

$\therefore$ Change in length of the side of cube,

$$\delta x = \text{linear strain} \times \text{original length}$$
$$= (14.518 \times 10^{-6}) \times 1 = 14.518 \times 10^{-6} \text{ m}$$
$$= \textbf{0.014518 mm} \text{ (decrease)}.$$

1.13. POISSON'S RATIO

When a rectangular bar is subjected to tensile load along a longitudinal dimension, it not only elongates but simultaneously there also occurs shortening in its breadth and thickness. Likewise, when tensile force is applied to a circular bar of uniform diameter, there occurs a simultaneous increase in length and shortening of diameter.

The strain in the direction of applied load is called *longitudinal* or *primary strain* and the strain in the perpendicular (transverse) direction is called *lateral* or *secondary strain:* Thus for a rectangular bar of length l, breadth b, and thickness t, the longitudinal strain is $\delta l/l$ and the lateral strain is $\delta b/b$ and $\delta t/t$. For a circular bar of length l and diameter d, the longitudinal strain equals $\delta l/l$ and the lateral (diametral) strain is $\delta d/d$. It is to be noted that the longitudinal and the lateral strains are of opposite nature.

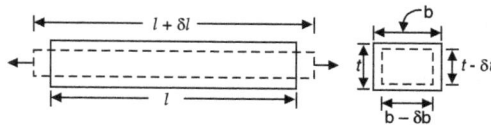

FIGURE 1.39.

When the deformation of the bar is within the elastic limit, the ratio of the lateral strain to the longitudinal strain is constant for a given material. This ratio is called *Poisson's ratio* and is denoted by μ. The value of μ lies between 0.25 and 0.33 for most of the engineering materials.

EXAMPLE 1.32
During the compression test, a metallic bar of 50 mm × 50 mm in cross-section was subjected to an axial compressive load of 500 kN. Measurements showed that there was a 0.04 mm increase in thickness and 0.5 mm contraction in length over a gauge length of 200 mm. Determine the values of Poisson's ratio and modulus of elasticity.

Solution: Longitudinal strain $= \dfrac{\delta l}{l} = \dfrac{0.5}{200} = 0.0025,$

Lateral strain $= \dfrac{\delta t}{t} = \dfrac{0.04}{50} = 0.0008,$

Poisson's ratio, $\mu = \dfrac{\text{lateral strain}}{\text{longitudinal strain}} = \dfrac{0.0008}{0.0025} = \mathbf{0.32,}$

Stress $= \dfrac{\text{load}}{\text{area}} = \dfrac{500 \times 10^3}{50 \times 50} = 0.2 \times 10^3 \,\text{N/mm}^2,$

Modulus of elasticity, $E = \dfrac{\text{stress}}{\text{longitudinal strain}} = \dfrac{0.2 \times 10^3}{0.0025} = \mathbf{0.8 \quad 10^5 \ N/mm^2}$.

EXAMPLE 1.33

A steel bar of 2 m long and 20 mm × 10 mm in cross-section is subjected to a tensile load of 20 kN along its longitudinal axis. Make calculations for changes in length, width, and thickness of the bar stating whether it is increasing or decreasing. Take modulus of elasticity as 2×10^5 N/mm^2 and Poisson's ratio 0.3.

Solution: Longitudinal strain

$$\frac{\delta l}{l} = \frac{\text{stress}}{\text{modulus of elasticity}}$$

$$= \frac{P/A}{E} = \frac{P}{AE} = \frac{20 \times 10^3}{(20 \times 10) \times (2 \times 10^5)} = 0.5 \times 10^{-3}.$$

Change in length δl = longitudinal strain × original length
$$= (0.5 \times 10^{-3}) \times (2 \times 10^3) = \mathbf{1.0 \ mm} \ \text{(increase)}.$$
Lateral strain = Poisson's ratio × longitudinal strain
$$= 0.3 \times (0.5 \times 10^{-3}) = 0.15 \times 10^{-3}.$$

The lateral strain equals $\delta b/b$ and $\delta t/t$.

∴ Change in breadth $\delta b = b \times$ lateral strain $= 20 \times (0.15 \times 10^{-3}) = \mathbf{3 \times 10^{-3} \ mm}$ (decrease).

Change in thickness $\delta t = t \times$ lateral strain $= 10 \times (0.15 \times 10^{-3}) = \mathbf{1.5 \times 10^{-3} \ mm}$ (decrease).

1.14. RELATION BETWEEN ELASTIC CONSTANTS *E*, *K*, AND *C*

1.14.1. Relation between *E*, *K*, and μ

Consider a cubical element subjected to volumetric stress σ which acts simultaneously along the mutually perpendicular *x*-, *y*-, and *z*-direction.

The resultant strains along the three directions can be worked out by taking the effect of individual stresses.

Strain in the *x*-direction,

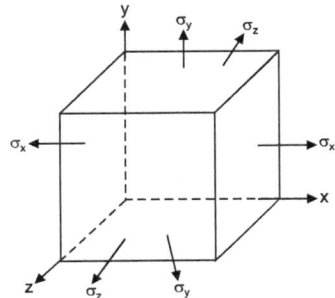

FIGURE 1.40.

εx = strain in x-direction due to σ_x − strain in x-direction due to σ_y
  − strain in x-direction due to σ_z

$$= \frac{\sigma_x}{E} - \mu\frac{\sigma_y}{E} - \mu\frac{\sigma_z}{E}. \tag{1.22}$$

But $\sigma_x = \sigma_y = \sigma_z = \sigma$,

$$\therefore\quad \varepsilon_x = \frac{\sigma}{E} - \mu\frac{\sigma}{E} - \mu\frac{\sigma}{E} = \frac{\sigma}{E}(1-2\mu).$$

Likewise, $\varepsilon_y = \dfrac{\sigma}{E}(1-2\mu)$ and $\varepsilon_z = \dfrac{\sigma}{E}(1-2\mu)$. (1.23)

Volumetric strain, $\varepsilon_v = \varepsilon_x + \varepsilon_y + \varepsilon_z = \dfrac{3\sigma}{E}(1-2\mu).$ (1.24)

Now, bulk modulus $K = \dfrac{\text{volumetric stress}}{\text{volumetric strain}} = \dfrac{\sigma}{\dfrac{3\sigma}{E}(1-2\mu)} = \dfrac{E}{3(1-2\mu)},$

$$\therefore\quad E = 3K(1-2\mu). \tag{1.25}$$

1.14.2. Relation between *E*, *C*, and μ

Consider a cubic element *ABCD* fixed at the bottom face and subjected to a shearing force at the top face. The block experiences the following effects due to this shearing load:

- shearing stress τ is induced at the faces *DC* and *AB*.
- Complimentary shearing stress of the same magnitude is set up on the faces *AD* and *BC*.
- The block distorts to a new configuration *ABC′D′*.
- The diagonal *AC* elongates (tension) and diagonal *BD* shortens (compression).

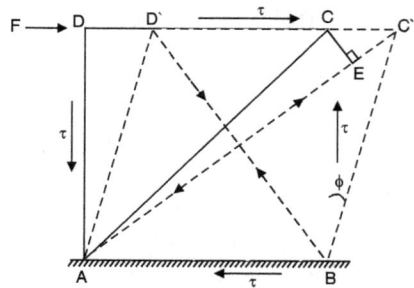

FIGURE 1.41.

Longitudinal strain in diagonal AC

$$= \frac{AC' - AC}{AC}$$

$$= \frac{AC' - AE}{AC} = \frac{EC'}{AC},$$ (1.26)

where CE is perpendicular from C onto AC'.

Since extension CC' is small, $\angle AC'B$ can be assumed to be equal to $\angle ACB$ which is 45°.

Therefore,

$$EC' = CC' \cos 45° = \frac{CC'}{\sqrt{2}}.$$

$$\text{Longitudinal strain} = \frac{CC'}{\sqrt{2}AC} = \frac{CC'}{\sqrt{2} \times \sqrt{2}BC} = \frac{CC'}{2BC}.$$

From triangle, BCC': $\dfrac{CC'}{BC} = \tan\phi.$

$$\therefore \quad \text{Longitudinal strain} = \frac{\tan\phi}{2} = \frac{\phi}{2},$$ (1.27)

where $\phi = \dfrac{CC'}{BC}$ represents the shear strain.

In terms of shear stress τ and modulus of rigidity C, shear strain = τ/C:

$$\therefore \quad \text{Longitudinal strain of diagonal, } AC = \frac{\tau}{2C}$$ (1.28)

The strain in diagonal AC is also given by
= strain due to tensile stress in AC
    − strain due to compressive stress in BD

$$= \frac{\tau}{E} - \left(-\mu\frac{\tau}{E}\right) = \frac{\tau}{E}(1 + \mu).$$ (1.29)

From expressions (1.27) and (1.28)

$$\frac{\tau}{2C} = \frac{\tau}{E}(1 + \mu)$$

$$\text{or } E = 2C\,(1 + \mu).$$ (1.30)

1.14.3. Relation between *E*, *C*, and *K*

With reference to the relations (1.25) and (1.30) derived above,

$$E = 2C(1 + \mu) = 3K(1 - 2\mu).$$

To eliminate μ from these two expressions for E, we have

$$\mu = \frac{E}{2C} - 1 \quad \text{and} \quad E = 3K\left[1 - 2\left(\frac{E}{2C} - 1\right)\right]$$

or
$$E = 3K\left[1 - \left(\frac{E}{C} - 2\right)\right] = 3K\left[3 - \frac{E}{C}\right] = 9K - \frac{3KE}{C}$$

or
$$E + \frac{3KE}{C} = 9C \quad ; \quad E\left(\frac{C + 3K}{C}\right) = 9K$$

or
$$E = \frac{9KC}{C + 3K}.$$

Accordingly,
$$E = 2C(1 + \mu) = \frac{9KC}{C + 3K}. \tag{1.31}$$

EXAMPLE 1.34

A tensile load of 56 kN was applied to a bar of 30 mm diameter with 300 mm gauge length. Measurements showed a 0.12 mm increase in length and the corresponding 0.0036 mm contraction in diameter. Make calculations for the Poisson's ratio and the values of three moduli (elastic constants).

Solution: Linear strain, $\varepsilon_l = \dfrac{\delta l}{l} = \dfrac{0.12}{300} = 0.0004,$

Diametral (lateral) strain, $\varepsilon_d = \dfrac{\delta d}{d} = \dfrac{0.0036}{30} = 0.00012,$

Poisson's ratio, $\mu = \dfrac{\text{lateral strain}}{\text{linear strain}} = \dfrac{0.00012}{0.0004} = 0.3.$

Normal stress, $= \dfrac{56 \times 10^3}{\dfrac{\pi}{4}(30)^2} = 79.26 \text{ N/mm}^2.$

Modulus of elasticity $E = \dfrac{\text{normal stress}}{\text{linear strain}} = \dfrac{79.26}{0.0004} = 1.9815 \quad 10^5 \text{ N/mm}^2.$

From the relations $E = 2C\,(1 + \mu) = 3K\,(1 - 2\mu)$, we have

$$C = \frac{E}{2(1 + \mu)} = \frac{1.9815 \times 10^5}{2(1 + 0.3)} = 0.762 \quad 10^5 \text{ N/mm}^2,$$

$$K = \frac{E}{3(1 - 2\mu)} = \frac{1.9815 \times 10^5}{3(1 - 2 \times 0.3)} = 1.65 \quad 10^5 \text{ N/mm}^2.$$

EXAMPLE 1.35

A bar of 2 cm × 4 cm in cross-section and 40 cm long is subjected to a radial tensile load of 70 kN. It is found that the length increases by 0.176 mm and the lateral dimension of 4 cm decreases by 0.0044 mm. Find (a) Young modulus, (b) Poisson's ratio, (c) change in volume of the bar, and (d) Bulk modulus.

Solution: Longitudinal strain

$$\epsilon_x = \frac{\delta l}{l} = \frac{0.176 \times 10^{-1}}{40} = 1.1 \times 10^{-4}.$$

Lateral strain, $\epsilon_y = \dfrac{0.0044 \times 10^{-1}}{4} = 1.1 \times 10^{-4}.$

Normal stress $= \dfrac{\text{load}}{\text{area}} = \dfrac{70 \times 10^3}{2 \times 4} = 8.75 \times 10^3 \text{ N/cm}^2 = 87.5 \text{ N/mm}^2.$

a. Modulus of elasticity, $E = \dfrac{\text{normal stress}}{\text{longitudinal strain}} = \dfrac{87.5}{4.4 \times 10^{-4}}$

$$= 19.886 \text{ N/mm}^2.$$

b. Poisson's ratio, $\mu = \dfrac{\text{lateral strain}}{\text{longitudinal strain}} = \dfrac{1.1 \times 10^{-4}}{4.4 \times 10^{-4}} = 0.25 \text{ N/mm}^2.$

c. Strain in the 2 cm side direction (z-direction)

$$\mu = \frac{\epsilon_z}{\epsilon_x}; \epsilon_z = \mu \times \epsilon_x = 0.25 \times 4.4 \times 10^{-4} = 1.1 \times 10^{-4}.$$

Volumetric strain, $\epsilon_v = \epsilon_x + \epsilon_y + \epsilon_z$
$$= 4.4 \times 10^{-4} + 1.1 \times 10^{-4} + 1.1 \times 10^{-4} = 6.6 \times 10^{-4}.$$
∴   Change in volume of the bar

$$\delta V = V \times \epsilon_v = (2 \times 4 \times 40) \times 6.6 \times 10^{-4} = \mathbf{0.2112\ cm^3}.$$

d.   From the relation, $E = 3K(1 - 2\mu)$, we have

Bulk modulus, $K = \dfrac{E}{3(1 - 2\mu)} = \dfrac{19.886 \times 10^4}{3(1 - 2 \times 0.25)} = \mathbf{13.257\quad 10^4\ N/mm^2}.$

EXAMPLE 1.36
A vertical circular copper bar of 20 mm diameter and 3 m long carries a tensile load of 200 kN. Calculate the elongation, decrease in diameter, and the volumetric strain.
   Take $E = 105$ N/mm^2 and $\mu = 0.25$.

Solution: Longitudinal strain $\dfrac{\delta l}{l} = \dfrac{\text{stress}}{\text{modulus of elasticity}}$

$$= \frac{P/A}{E} = \frac{P}{AE} = \frac{200 \times 10^3}{\dfrac{\pi}{4}(20)^2 \times 10^5} = 6.37 \times 10^{-3}.$$

Change in length = longitudinal strain × original length

$$= (6.37 \times 10^{-3}) \times (3 \times 10^3) = 19.11 \text{ mm (increase)}.$$

Lateral strain = Poisson's ratio × longitudinal strain

$$= 0.25 \times (6.37 \times 10^{-3}) = 1.5925 \times 10^{-3}.$$

The lateral strain equals $\dfrac{\delta d}{d}$.

∴   Change in diameter δd = lateral strain × original diameter
$$= (1.592 \times 10^{-3}) \times 20 = 0.03185 \text{ mm (decrease)}.$$

Now, volume of circular bar $V = \dfrac{\pi}{4}d^2 l.$

Upon differentiation: $\delta V = \dfrac{\pi}{4}(2d\delta\, dl + d^2\delta l).$

$$\text{Volumetric strain } \frac{\delta V}{V} = \frac{\frac{\pi}{4}(2d\delta\,d\,l + d^2\delta\,l)}{\frac{\pi}{4}d^2l} = 2\frac{\delta\,d}{d} + \frac{\delta\,l}{l}$$

$$= 2\,(-1.5925 \times 10^{-3}) + (6.37 \times 10^{-3}).$$

The −ve sign with $\dfrac{\delta\,d}{d}$ stems from the fact that whereas the length increases with tensile load, there is a decrease in diameter:

$$\frac{\delta V}{V} = -3.185 \times 10^{-3} + 6.37 \times 10^{-3} = 3.185 \times 10^{-3}.$$

EXAMPLE 1.37
A circular rod of 100 mm diameter and 500 mm length is subjected to a tensile force of 1000 kN. Determine the modulus of rigidity, bulk modulus, and change in volume if Poisson's ratio = 0.3 and Young's modulus = 2 × 10⁵ N/mm².

Solution: From the relations, $E = 2C\,(1 + \mu) = 3\,K\,(1 - 2\mu)$, we have

Modulus of rigidity, $C = \dfrac{E}{2(1+\mu)} = \dfrac{2 \times 10^5}{2(1+0.3)} = \mathbf{0.769}\;\; \mathbf{10^5\,N/mm^2}.$

Bulk modulus, $K = \dfrac{E}{3(1-2\mu)}\dfrac{2 \times 10^5}{3(1-2\times 0.3)} = \mathbf{1.667}\;\;\mathbf{10^5\,N/mm^2}.$

Normal stress, $\sigma = \dfrac{W}{A}\dfrac{1000 \times 10^3}{\frac{\pi}{4}(100)^2} = 127.388\ \text{N/mm}.$

Linear (longitudinal) strain, $\dfrac{\delta l}{l}\dfrac{\text{normal stress}}{\text{Young's modulus}} = \dfrac{127.388}{2 \times 10^5} = 0.000637.$

Diametral (lateral) strain, $\dfrac{\delta\,d}{d} = \mu\dfrac{\delta l}{l} = 0.3 \times 0.000637 = 0.0001911.$

Now, the volume of a circular rod $V = \dfrac{\pi}{4}d^2l.$

Upon differentiation, $\delta V = \dfrac{\pi}{4}[2d\delta d\, l + d^2\delta l]$.

Volumetric strain, $\dfrac{\delta V}{V} = \dfrac{\dfrac{\pi}{4}(2d\delta\,dl + d^2\delta l)}{\dfrac{\pi}{4}d^2 l} = \dfrac{2\delta d}{d} + \dfrac{\delta l}{l}$.

Substituting the values of $\dfrac{\delta l}{l}$ and $\dfrac{\delta l}{l}$ as calculated above, we have

$$\frac{\delta V}{V} = 2(-0.0001911) + 0.000637 = 0.0002548.$$

The −ve sign with $\delta d/d$ stems from the fact that whereas the length increases with tensile force, there is a decrease in diameter:

$\therefore$ Change in volume, $\delta V = 0.0002548 \times \left[\dfrac{\pi}{4}(100)^2 \times 500\right] = \mathbf{1000.09\ mm^3}$.

EXAMPLE 1.38

A rectangular block of 250 mm × 150 mm × 100 mm is subjected to axial loads as follows: 600 kN tensile in the direction of its length; 900 kN compressive on the 250 mm × 150 mm faces, and 750 kN tensile on 250 mm × 100 mm faces.
Assuming Young's modulus, $E = 2 \times 10^5$ N/mm^2 and Poisson's ratio, $\mu = 0.25$ determine:

a. strains in the direction of each force,

b. change in volume of the block due to the loads specified, and

c. modulus of rigidity and bulk modulus for the material of block.

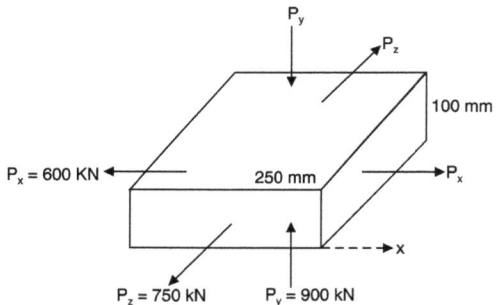

FIGURE 1.42.

Solution: Volume of block, $V = 250 \times 150 \times 100 = 375 \times 10^4$ mm^3
Stresses in the three directions are

$$\sigma_x = \frac{600 \times 10^3}{150 \times 100} = 40 \text{ N/mm}^2 \text{ (tensile)},$$

$$\sigma_y = \frac{900 \times 10^3}{250 \times 150} = 24 \text{ N/mm}^2 \text{ (compressive)},$$

$$\sigma_z = \frac{750 \times 10^3}{250 \times 100} = 30 \text{ N/mm}^2 \text{ (tensile)}.$$

The resultant strains along the three directions can be worked out by taking the effect of individual stresses.

Strain in the x-direction:

$$E_x = \text{strain due to } \sigma_x - \text{strain due to } \sigma_y - \text{strain due to } \sigma_z$$

$$= \frac{\sigma_x}{E} - \mu \frac{\sigma_y}{E} - \mu \frac{\sigma_z}{E}$$

$$= \frac{40}{2 \times 10^5} - \frac{0.25(-24)}{2 \times 10^5} - \frac{0.25 \times 30}{2 \times 10^5}$$

$$= \frac{40 + 6 - 7.5}{2 \times 10^{-5}} = 19.25 \times 10^{-5}.$$

Likewise,

$$\varepsilon_y = \frac{\sigma_y}{E} - \mu \frac{\sigma_x}{E} - \mu \frac{\sigma_z}{E}$$

$$= \frac{24}{2 \times 10^5} - \frac{0.25 \times 40}{2 \times 10^5} - \frac{0.25 \times 30}{2 \times 10^5}$$

$$= \frac{-24 - 10 - 7.5}{2 \times 10^{-5}} = 20.75 \times 10^{-5}$$

and

$$\varepsilon_z = \frac{\sigma_z}{E} - \mu \frac{\sigma_x}{E} - \mu \frac{\sigma_y}{E}$$

$$= \frac{30}{2 \times 10^5} - \frac{0.25 \times 40}{2 \times 10^5} - \frac{0.25(-24)}{2 \times 10^5}$$

$$= \frac{30 - 10 + 6}{2 \times 10^{-5}} = 13 \times 10^{-5}.$$

For a rectangular body subjected to three mutually perpendicular forces, volumetric strain is given by

$$\varepsilon_v = \varepsilon_x + \varepsilon_y + \varepsilon_z$$

$$= (19.25 - 20.75 + 13) \times 10^{-5} = 11.5 \times 10^{-5}.$$

$\therefore$ Change in volume, $\delta \, V = \varepsilon_v \times V = 11.5 \times 10^{-5} \times (375 \times 10^4) = \mathbf{431.25 \ cm^3}$.

From the relations $E = 2C\,(1+\mu) = 3K\,(1-2\mu)$, we have

Modulus of rigidity, $C = \dfrac{E}{2(1+\mu)} \ \dfrac{2 \times 10^5}{2(1+0.25)} = \mathbf{0.8} \ \ \mathbf{10^5 \ N/mm^2}$.

Bulk modulus, $K = \dfrac{E}{3(1-2\mu)} = \dfrac{2 \times 10^5}{3(1-2 \times 0.25)} = \mathbf{1.33} \ \ \mathbf{10^5 \ N/mm^2}$.

EXAMPLE 1.39

Calculate the value of Poisson's ratio for a material block if its volume is to remain constant under (a) uni-axial stress condition, (b) equal bi-axial stress condition, and (c) equal tri-axial stress condition. Comment on the results.

Solution:

a. For uni-axial stress (stress σx is applied along x-direction)

$$\varepsilon_x = \frac{\sigma_x}{E} \quad ; \quad \varepsilon_y = -\mu \frac{\sigma_x}{E} \quad \text{and} \quad \varepsilon_z = -\mu \frac{\sigma_x}{E}$$

$$\therefore \ \varepsilon_v = \frac{\delta V}{V} = \varepsilon_x + \varepsilon_y + \varepsilon_z = \frac{\sigma_x}{E} - \mu \frac{\sigma_x}{E} - \mu \frac{\sigma_x}{E} = \frac{\sigma_x}{E}(1-2\mu).$$

Since the change in volume is zero,

$$\frac{\sigma_x}{E}(1-2\mu) = 0 \quad ; \quad 1-2\mu = 0 \quad \therefore \mu = 0.5.$$

b. For equal bi-axial conditions, stress σ_x is applied x-direction and stress σ_y is applied along y-direction.

Further, $\sigma_x = \sigma_y = \sigma$ (say)

$$\varepsilon_x = \frac{\sigma_x}{E} - \mu \frac{\sigma_y}{E} = \frac{\sigma}{E} - \mu \frac{\sigma}{E}$$

$$\varepsilon_y = \frac{\sigma_y}{E} - \mu \frac{\sigma_x}{E} = \frac{\sigma}{E} - \mu \frac{\sigma}{E}$$

$$\varepsilon_z = -\mu \frac{\sigma_x}{E} - \mu \frac{\sigma_y}{E} = -2\mu \frac{\sigma}{E}$$

$$\therefore \qquad \varepsilon_V = \frac{\delta V}{V} = \varepsilon_x + \varepsilon_y + \varepsilon_z = \frac{2\sigma}{E}(1-2\mu).$$

Since the change in volume is zero,

$$\frac{2\sigma}{E}(1-2\mu) = 0;\ 1-2\mu = 0 \qquad \therefore \mu = 0.5.$$

c. For tri-axial stress conditions, stresses σ_x, σ_y, and σ_z are applied along x, y, and z directions, respectively. Further, $\sigma_x = \sigma_y = \sigma_z = \sigma$ (say)

$$\varepsilon_x = \frac{\sigma_x}{E} - \mu\frac{\sigma_y}{E} - \mu\frac{\sigma_z}{E} = \frac{\sigma}{E} - 2\mu\frac{\sigma}{E}$$

$$\varepsilon_y = \frac{\sigma_y}{E} - \mu\frac{\sigma_x}{E} - \mu\frac{\sigma_z}{E} = \frac{\sigma}{E} - 2\mu\frac{\sigma}{E}$$

$$\varepsilon_z = \frac{\sigma_z}{E} - \mu\frac{\sigma_x}{E} - \mu\frac{\sigma_y}{E} = \frac{\sigma}{E} - 2\mu\frac{\sigma}{E}$$

$$\therefore \qquad \varepsilon_V = \frac{\delta V}{V} = \varepsilon_x + \varepsilon_y + \varepsilon_z = \frac{3\sigma}{E}(1-2\mu).$$

Since the change in volume is zero,

$$\frac{3\sigma}{E}(1-2\mu) = 0 \quad;\quad 1-2\mu = 0 \quad \therefore \quad \mu = 0.5E.$$

Comment: Poisson's ratio μ is independent of the type of loading and this has to be so because μ is a property of the material.

1.15. TEMPERATURE STRESSES

When a member is subjected to temperature variations, its dimensions get changed. There occurs expansion, that is, an increase in dimensions when temperature increases. Conversely, with a fall in temperature, contraction occurs and dimensions of the member decrease. The change in length due to temperature variation is given by

$$\delta l = l\,\alpha\,\Delta t, \tag{1.32}$$

where l is the length of the member, α is the coefficient of thermal expansion, and Δt is the change in temperature. Then
    thermal strain

$$= \frac{l \propto \Delta t}{l} = \propto \Delta t. \tag{1.33}$$

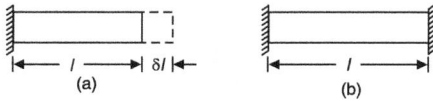

FIGURE 1.43.

If expansion is allowed to take place freely [Figure 1.43(a)], there occurs only thermal strain and no stresses are developed in the material of the member. However, if the ends of the member are fixed to rigid supports, that is, the member is not allowed to expand [Figure 1.43(b)], compressive stresses are developed in the material. Such stresses are known as *thermal* or *temperature stresses:*

thermal stress = modulus of elasticity × thermal strain

$$= E \, \alpha \, \Delta t. \tag{1.34}$$

If the end supports yield by an amount δ, then net elongation is $(l \, \alpha \Delta t - \delta)$. Accordingly

$$\text{thermal strain} = \frac{la \, \Delta t - \delta}{l}, \tag{1.35}$$

$$\text{normal thermal stress} = E\left(\frac{la \, \Delta t - \delta}{l}\right). \tag{1.36}$$

Note: The temperature strains and stresses due to temperature rise are compressive in nature, and due to fall in temperature are tensile.

1.15.1. Effect of temperature change in a composite bar

Consider the temperature rise of a composite bar consisting of two members; one of steel and the other of brass rigidly fastened to each other. If allowed to expand freely:

expansion of brass bar: $AB = l \, \alpha_b \, \Delta t$,
expansion of steel bar: $AC = l \, \alpha_s \, \Delta t$.

Since the coefficient of thermal expansion of brass is greater than that of steel, the expansion of brass will be more.

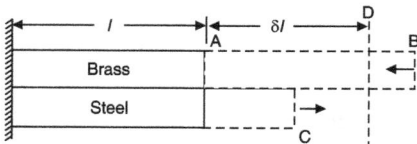

FIGURE 1.44.

But the bars are fastened together and accordingly both will expand to the same final position represented by DD with a net expansion of composite system AD equal to δl. To attain this position, the brass bar is pushed back and the steel bar is pulled. Obviously, compressive stress will be induced in the brass bar and tensile stress will be developed in the steel bar. Under equilibrium state compressive force in brass = tensile force in steel

$$\sigma_c A_c = \sigma_s A_s. \tag{1.37}$$

Corresponding to the brass rod:
Reduction in elongation, $DB = AB - AD = l\,\alpha_b\,\Delta t - \delta l$

$$\text{Strain, } \varepsilon_b = \frac{la_b\Delta t - \delta l}{l} = a_b\Delta t - \varepsilon,$$

where $\varepsilon = \delta l/l$ is the actual strain of the composite system.
Corresponding to steel rod:
extra elongation, $CD = AD - AC$
$$= \delta l - l\,\alpha_s\,\Delta t$$

$$\text{strain, } \varepsilon_s = \frac{\delta l - la_s\Delta t}{l} = \varepsilon - a_s\,\Delta t.$$

Adding ε_b and ε_s, we get

$$\varepsilon_b + \varepsilon_s = (\alpha_b - \alpha_s)\,\Delta t. \tag{1.38}$$

It may be pointed out that the nature of the stresses in the bars will get reversed if there is a reduction in the temperature of the composite system.

EXAMPLE 1.40

A steel rod of length 4 m and diameter 20 mm is being stayed between two plates at a temperature of 20°C. Make calculations for the force exerted by the rod after it has been heated to 60°C.

    (i)   when the plates do not yield;

    (ii) when the yielding at the two ends is 1 mm.

    Take as $= 12 \times 10^{-6}$ per degree centigrade and $E_s = 2 \times 10^5$ N/m².

Solution: Increase in length due to rise in temperature

$$\delta l = l\,\alpha\,\Delta t = 4000 \times (12 \times 10^{-6}) \times (60 - 20) = 1.92 \text{ mm}.$$

i.   Plates do not yield (contraction is fully prevented)

$$\text{Strain, } \varepsilon = \frac{\delta l}{l} = \frac{1.92}{4000} = 0.00048.$$

Thermal stress $\sigma = \varepsilon\,E_s = 0.00048 \times (2 \times 10^5) = 96 \text{ N/mm}^2$.
Force exerted by the rod $= \sigma\,A$

$$= 96 \times \frac{\pi}{4}(20)^2 = 30144 \text{ N}.$$

ii. Plates yield by 1 mm (expansion of 1 mm is permitted). Increase in length which is prevented $= 1.92 - 1 = 0.92$ mm.

The corresponding thermal strain and stress are

$$\varepsilon = \frac{0.92}{4000} = 0.00023,$$

$$\sigma = 0.00023 \times (2 \times 10^5) = 46 \text{ N/mm}^2.$$

Force exerted by the rod $= 46 \times \frac{\pi}{4} \left(\frac{20}{4} \right)^2 = \textbf{14444 N.}$

EXAMPLE 1.41
A 30 m length of the railway is to be so laid that there is no stress in the rails at 10°C. Determine the stress induced in the rails at 60°C if
   (i)   there is no allowance for expansion;
   (ii)   there is an expansion allowance of 8 mm per rail;
   (a)   what should be the expansion allowance if the stress in the rail is to be zero at 60°C temperature?;
   (b)   what is the maximum allowable temperature to have no stress in the rails if expansion allowance is 12 mm per rail?

Take $E = 2 \times 10^5$ N/mm^2 and $\alpha = 12 \times 10^{-6}$/°C.
Solution: Increase in length due to rise in temperature

$$\delta l = l\,\alpha\,\Delta t = (30 \times 10^3) \times (12 \times 10^{-6}) \times 50 = 18 \text{ mm.}$$

i.   When there is no allowance for expansion,

$$\text{Strain } \in = \frac{\delta l}{l} = \frac{18}{30 \times 10^3} = 0.6 \times 10^{-3}.$$

Thermal stress $\sigma = \varepsilon E$

$$= (0.6 \times 10^{-3}) \times (2 \times 10^5) = \textbf{120 N/mm}^2.$$

ii.   When there is an expansion allowance of 8 mm per rail increase in length which is prevented

$$= 18 - 8 = 10 \text{ mm.}$$

The corresponding thermal strain and stress are

$$\in = \frac{10}{30 \times 10^3} = 0.333 \times 10^{-3}.$$

$$\sigma = (0.333 \times 10^{-3}) \times (2 \times 10^5) = \textbf{66.6 N/mm}^2.$$

   (b)   Expansion allowance for no stress in the rails
              = increase in length due to rise in temperature
              = **18 mm.**

(c) Maximum temperature to have no stress in the rails with expansion allowance of 12 mm:

$$l \, \alpha \, \Delta t = 10$$
$$(30 \times 10^{-3}) \times (12 \times 10^{-6}) \times (t - 10) = 12$$

or $$0.36 \, (t - 10) = 12.$$

That gives: $$t = \frac{12}{0.30} + 10 = \textbf{43.33°C}.$$

EXAMPLE 1.42

A steel bar of 30 mm uniform diameter is heated to 75°C and then clamped at ends with the help of two fixtures 3.5 m apart and left to cool to room temperature of 20°C. At that instant, the distance between the fixtures was found to be 1 mm shorter than that at 75°C. Determine the stress in the bar when it has cooled down to room temperature and the reaction at the fixtures.

Take, $E_s = 210$ GN/m² and $\alpha_s = 1.1 \times 10^{-5}$°C.

Solution: Free contraction (decrease in length) due to temperature drop

$$\delta l = l \, \alpha \, \Delta t = 3500 \times 1.1 \times 10^{-5} \times (75 - 20) = 2.1175 \text{ mm.}$$

The decrease of 1 mm distance between the fixtures amounts to contraction (yielding) allowed:

∴ contraction in length which is prevented = 2.1175 − 1 = 1.1175 mm,

thermal strain, $\varepsilon = \dfrac{1.1175}{3500} = 0.0003193,$

thermal stress, $\sigma = \varepsilon \, E_s = 0.0003193 \times (2.1 \times 10^5) = 67.05$ N/mm².

Force exerted by the bar $= \sigma \, A = 67.05 \times \dfrac{\pi}{4}(30)^2 = 47371$ N.

Reaction at the supports equals the force exerted by the bar
∴ Reaction at supports = 47371 N = **47.371 kN.**

EXAMPLE 1.43

A composite bar consisting of three segments and the geometrical configuration shown below is rigidly attached to the end supports. Determine the stress induced in each segment if the temperature of the composite bar is raised by 50°C. Consider the supports to be rigid and take

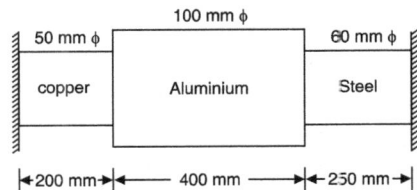

FIGURE 1.45.

For copper: E_c = 100 GPa   and a_c = 16 × 10^{-6}/°C.
For aluminum: E_{al} = 70 GPa   and a_{al} = 20 × 10^{-6}/°C.
For steel: E_s = 200 GPa   and a_s = 12 × 10^{-6}/°C.

Solution: Increase in length of composite bar due to temperature rise

$$(dl)_t = (\alpha_c l_c + \alpha_{al} l_{al} + \alpha_s l_s) \times \Delta t$$
$$= (16 \times 10^{-6} \times 0.2 + 20 \times 10^{-6} \times 0.4 + 12 \times 10^{-6} \times 0.25) \times 50$$
$$= (3.2 + 8 + 3) \times 10^{-6} \times 50 = 7.1 \times 10^{-4} \text{ m.}$$

Decrease in length due to reactive forces

$$(dl)_p = P\left[\left(\frac{1}{AE}\right)_c + \left(\frac{1}{AE}\right)_{al} + \left(\frac{1}{AE}\right)_s\right]$$

$$= P\left[\frac{0.2}{\frac{\pi}{4}(0.05)^2 \times (100 \times 10^9)} + \frac{0.4}{\frac{\pi}{4}(0.1)^2 \times (70 \times 10^9)} + \frac{0.25}{\frac{\pi}{4}(0.06)^2 \times (200 \times 10^9)}\right]$$

$$= \frac{P}{10^9}[1.019 + 0.728 + 0.442] = 2.189 \times 10^{-9} P.$$

From compatibility condition: $(dl)_t = (dl)_p$
∴ $7.1 \times 10^{-4} = 2.189 \times 10^{-9} P.$

That gives: $P = \dfrac{7.1 \times 10^{-4}}{2.189 \times 10^{-9}} = 324349 \text{ N/mm}^2.$

∴   Stress in copper segment $\sigma_c = \dfrac{324349}{\frac{\pi}{4}(50)^2} = \mathbf{165.27 \text{ N/mm}^2.}$

Stress in aluminum segment $\sigma_{al} = \dfrac{324349}{\frac{\pi}{4}(100)^2} = \mathbf{43.32 \text{ N/mm}^2.}$

Stress in steel segment $\sigma_s = \dfrac{324349}{\frac{\pi}{4}(60)^2} = \mathbf{114.77 \text{ N/mm}^2.}$

EXAMPLE 1.44

A thin steel ring (tire) is shrunk on a rigid wooden wheel 1.2 m in diameter. If the stress in the steel tire is limited to 120 MN/m², determine (a) the minimum internal diameter of the tire and (b) the least temperature to which the tire must be heated above that of a wheel. For the tire material, E = 200 GN/m² and α = 12 × 10^{-6}/°C.

Solution:

i. Let d be the diameter of the ring at normal temperature and D be the corresponding diameter when heated. When the tire cools back,

$$\text{decrease in circumference} = \pi D - \pi d.$$

This decrease is prevented as the tire is tightly fixed onto the wheel

$$\text{strain prevented} = \frac{\pi D - \pi d}{\pi d} = \frac{D - d}{d}.$$

Stress developed in the tire $= E \times$ strain prevented

$$\therefore \quad 120 \times 10^6 = 200 \times 10^9 \times \frac{1.2 - d}{d}$$

or $\quad 120 \times 10^6 \, d = 240 \times 10^9 - 200000 \times 10^6 \, d,$

$$d = \frac{240 \times 10^9}{(200000 + 120) \times 10^6} = \mathbf{1.199 \text{ m}}.$$

ii. Let t be the least temperature to which the tire must be heated above that of a wheel. Then final circumference of the ring after heating

= circumference at room temperature + increase in circumference due to temperature rise,

$$\pi D = \pi d + \pi d \, \alpha \, t$$

or $\quad \alpha \, t = \dfrac{\pi D - \pi d}{\pi d} = \dfrac{D - d}{d}$

$$\therefore \quad t = \frac{D - d}{\alpha \, d} = \frac{1.2 - 1.199}{12 \times 10^{-6} \times 1.199} = \mathbf{69.5^\circ C}.$$

REVIEW QUESTIONS

A. Conceptual and conventional questions

1. Differentiate between the concepts of (*i*) load and stress, (*ii*) stress and pressure, and (*iii*) stress and strength.

2. State and explain Hook's law. Distinguish between the limit of proportionality and elastic limit.

3. Sketch stress–strain diagram for ductile material and explain its salient features. How does stress–strain curve for brittle materials differ from this curve?

4. Explain the following terms in the context of stress–strain curve for mild steel:

    *   elastic limit and limit of proportionality,
    *   upper and lower yield points, and
    *   ultimate point and breaking point.

5. State the principle of superposition and point out its utility.

6. Define direct, shear, and volumetric stresses, and the corresponding strains.

7. Define and explain the terms: longitudinal strain, lateral strain, and Poisson's ratio.

8. Set up a relation for the volumetric strain of a rectangular or cylindrical bar subjected to axial force in terms of longitudinal strain and Poisson's ratio. $\{\varepsilon v = \varepsilon\,(1 - 2\,\mu)\}.$

9. Define volumetric strain. Derive an expression for it in terms of linear normal strains in x, y, and z directions.

10. What is meant by simple shear or plane shear system? Show that when an element is in a state of simple shear, tensile and compressive stresses equal in magnitude to the simple shear stresses act on planes inclined at 45° to the planes of simple shear.

11. Derive the following relations for the elastic constants for an isotropic material:

    i. $E = 2\,C\,(1 + \mu) = 3K\,(1 - 2\mu),$

    ii. $E = \dfrac{9CK}{3K + C},$ and

    iii. $\mu = \dfrac{3K - 2C}{6K + 2C},$

    where the symbols have their usual meanings.

12. What do you mean by thermal stresses and strains? Set up an expression for the thermal stresses developed in a composite bar.

13. A composite bare constituting of iron and copper in parallel is heated. What would be the nature (tensile or compressive) of thermal stresses induced in the two bars?

14. During a tensile test on a mild steel specimen, 40 mm diameter and 200 mm long, the following data were obtained:

Extension at 40 kN load = 0.0304 mm; yield load = 161 kN and length of specimen at fracture = 249 mm.

Determine (a) modulus of elasticity, (b) percentage elongation, and (c) yield point stress. [**Ans.** 2.05×10^5 N/mm^2, 24.5%, 128.18 N/mm^2]

15. A mild steel bar of 20 mm × 20 mm in cross-section is subjected to axial forces as shown below: If $E = 2 \times 10^5$ N/mm^2, then determine change in length of mid-portion BC. [**Ans.** 0.45 mm]

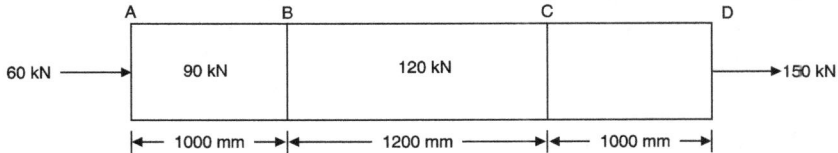

FIGURE 1.46.

16. A steel rod of 50 mm diameter and 2.5 m length is subjected to an axial pull of 100 kN. Determine the length for which a 20 mm diameter hole be drilled so that a 10% increase in total extension is achieved. For steel, the modulus of elasticity is 2×10^5 N/mm^2.

[**Ans.** 1.315 m]

17. A load of 300 kN is applied on a short concrete column 250 mm × 250 mm. The column is reinforced by steel bars of a total area of 5600 mm^2. If the modulus of elasticity of steel is 15 times that of concrete, find the stresses in concrete and steel.

If the stress in concrete should not exceed 4 N/mm^2, find the area of steel required so that column may support a load of 600 kN.

[**Ans.** 2.13 N/mm^2, 31.95 N/mm^2, 6250 mm^2]

18. A brass tube of 60 mm outside diameter completely encloses a steel bar of 40 mm diameter. The composite system measures 300 mm in length and carries an axial thrust which induces a thrust equal to 50 N/mm^2 in the brass tube. Make calculations for the (a) stress developed in steel bar, (b) magnitude of compressive force, and (c) change in length of the composite bar. Take $Es = 210$ GPa and $Eb = 105$ GPa.

[**Ans.** 100 N/mm^2, 204.1 kN; 0.1428 mm]

19. Three bars made of copper, zinc and aluminum are of equal length and have cross-sectional areas 500, 750, and 1000 mm², respectively. They are rigidly connected at their ends. If this composite system is subjected to a longitudinal pull of 2.5×10^5 N, make calculations for the proportion of load carried on each bar and the induced stresses. Take $E_c = 1.3 \times 10^5$ N/mm², $E_z = 1 \times 10^5$ N/mm², and $E_{al} = 0.8 \times 10^5$ N/mm². [**Ans.** $P_c = 0.7378 \times 10^5$ N, $P_z = 0.8522 \times 10^5$ N, $P_a = 0.9090 \times 10^5$ N, $\sigma_c = 147.72$ N/mm², $\sigma_z = 113.63$ N/mm², $\sigma_a = 90.9$ N/mm²]

20. It is required to punch a 20 mm diameter hole in a 15 mm thick steel plate. Presuming that the ultimate shear stress of mild steel is 350 N/mm², make calculations for the force required to punch the hole and the compressive stress on the punch. [**Ans.** 329.7 kN, 1050 N/mm²]

21. A steel bar of 2 m long, 20 mm wide, and 15 mm thick is subjected to an axial pull of 30 kN in the direction of its length. If Poisson's ratio is 0.25, make calculations for the change in volume. Take $E = 2 \times 10^5$ N/mm². [**Ans.** 150 mm³]

22. Determine a decrease in the volume of a solid sphere of 250 mm diameter when it is subjected to the uniform hydrostatic pressure of 80 N/mm². Take $E = 2.1 \times 10^5$ N/mm² and $\mu = 0.3$ [**Ans.** 3738 mm³]

23. A vertical circular copper bar 20 mm diameter and 3 m long carry a tensile load of 200 kN. Calculate the elongation, decrease in diameter, and the volumetric strain. Take $E = 1 \times 10^5$ N/mm² and $\mu = 0.25$. [**Ans.** 19.11 mm, 0.03185 mm, 3.185×10^{-3}]

24. A copper rod 15 mm diameter, 0.8 long is heated through 50°C. What is the extension when free to expand? Suppose the expansion is prevented by gripping it at both ends, find the stress, its nature, and the force applied by the grips when
    (i) the grips do not yield
    (ii) one grip yields back by 0.5 mm.
    Take $\alpha_c = 18.5 \times 10^{-6}$/°C, $E_c = 1.25 \times 10^5$ N/mm².

25. A 3 m long rod has a cross-sectional area that tapers uniformly from 50 mm diameter to 30 mm diameter. When an axial load of 50 kN is applied to it, the extension measured over its length is stated to be 0.3635 mm. Determine modulus of elasticity for the rod material. Further, make calculations for the bulk modulus and shear modulus if the Poisson's ratio is 0.25. [**Ans.** 200 GPa, 133.3 GPa, and 80 GPa]

B. Fill in the blanks with appropriate word(s)

1. The deformation per unit length in the direction of load is called.

2. The loading limit under which the deformation entirely disappears on removal of load is called _____

3. The slope of the linearly elastic portion of stress strain diagram is a measure of _____.

4. Young's modulus of elasticity for a perfectly rigid body is _____.

5. Stress and strain produced by a force tangential to the surface of a body are known as _____ and _____ respectively.

6. _____ is defined as the ratio of shear stress to shear strain.

7. The hydrostatic stress causes a change in of the body.

8. The value of Poisson's ratio lies between _____ and _____ for most of the engineering materials.

9. If both the modulus of elasticity and the shear modulus of a metal are doubled, the Poisson's ratio of the metal will _____.

10. The temperature strains and stresses due to temperature rise are in nature.

Answers:
1. Linear or longitudinal strain; **2.** Elastic limit; **3.** Modulus of elasticity; **4.** Shear stress, **5.** Shear strain; **6.** Modulus of rigidity; **7.** Volume; **8.** 0.25–0.5; **9.** Remain unaffected; **10.** Compressive.

C. Multiple choice questions

1. Stress represents the
   (a) external force acting on the body
   (b) pressure setup within the body material
   (c) force by which the material of the body opposes the deformation
   (d) resistance per unit is to deformation by internal forces.

2. Measurements have been made for Young's modulus of elasticity for mild steel specimen both in tension and compression. The ratio E_t/E_c will be approximately equal to

   (a) 0.5   (b) 0.75   (c) 1.0   (d) 1.25.

3. Hook's law holds good up to
   (a) a limit of proportionality   (b) elastic limit
   (c) yield point   (d) breaking point.

4. Young's modulus of elasticity for a perfectly rigid body is
   (a) zero   (b) unity
   (c) infinity   (d) some finite non-zero value.

5. The deformation of the bar under its own weight as compared to that when subjected to direct load equal to its own weight will be
   (a) the same   (b) one-fourth   (c) half   (d) double.

6. A tapering bar (diameters of the end sections being d_1 and d_2) and a bar of uniform cross-section d and the same length are subjected to the same axial pull. Both bars will have the same extension if d is equal to

   (a) $\dfrac{d_1+d_2}{2}$ (b) $\sqrt{d_1 d_2}$ (c) $\sqrt{\dfrac{d_1 d_2}{2}}$ (d) $\sqrt{\dfrac{d_1+d_2}{2}}$.

7. The value of Poisson's ratio depends upon
   (a) nature of load; tensile or compressive
   (b) magnitude of the load
   (c) material of test specimen
   (d) cross-section and dimension of the test piece.

8. The elastic constants E, G, and K are related by the expression

   (a) $E = \dfrac{9GK}{3K+G}$ (b) $\dfrac{GK}{2K+G}$

   (c) $E = \dfrac{GK}{2K+3G}$ (d) $\dfrac{3GK}{K+2G}$

9. For a given material, the modulus of rigidity is 100 GPa and the Poisson's ratio is 0.25. The value of modulus of elasticity in GPa is
   (a) 12.5   (b) 150   (c) 200   (d) 250.

10. In a composite system subjected to temperature rise and with ends constrained to remain together, the component having a lower value of the coefficient of linear expansion will experience a
    (a) tensile stress
    (b) compressive stress
    (c) zero value of stress
    (d) tensile or compressive depending upon the loading system.

11. The temperature stress is a function of
    1. coefficient of linear expansion
    2. temperature rise
    3. modulus of elasticity.
    Which of the statements given above are correct?
    (a) 1 and 2 only
    (b) 1 and 3 only
    (c) 2 and 3 only
    (d) 1, 2, and 3.

12. Match List 1 (elastic properties) with List II (nature of strain produced)

List I	**List II**
A Young's modulus	1. Shear strain
B Modulus of rigidity	2. Normal strain
C Bulk modulus	3. Transverse strain
D Poisson's ratio	4. Volumetric strain

Answers:
1. (d)   **2.** (c)   **3.** (a)   **4.** (c)   **5.** (c)   **6.** (b)
7. (c)   **8.** (a)   **9.** (d)   **10.** (a)   **11.** (d)   **12.** A–2; B–1; C–4; D–3.

SHEAR FORCE AND BENDING MOMENT

A beam is a structural member of which longitudinal dimension is large compared to the transverse dimension. The beam is supported along its length and is acted upon by a system of loads at right angles to its axis. Due to external loads and couples, shear force and bending moment develop at any section of the beams. For the design of beams, information about the shear force and bending moment are desired. Accordingly, it is appropriate to learn about the variation of shear force and bending moment along the length of the beam.

2.1. SHEAR FORCE AND BENDING MOMENT

Consider a simply supported beam acted upon by different point loads as shown in Figure 2.1. The loads are transferred to supports and for equilibrium conditions:

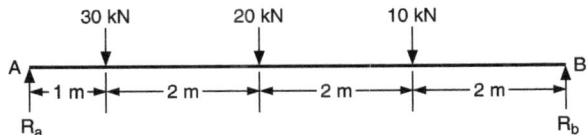

FIGURE 2.1.

$$R_a + R_b = W_1 + W_2 + W_3 = 30 + 20 + 20 = 60 \text{ kN.}$$

Also ΣM_a (moments about support A) = 0. That gives

$$R_b \times 7 = 10 \times 5 + 20 \times 3 + 30 \times 1 = 140$$

$$R_b = \frac{140}{7} = 20 \text{ kN and } R_a = 60 - 20 = 40 \text{ kN.}$$

Imagine the beam to be cut at an arbitrary section xx at distance $x = 4$ m from the end A and draw separately the free-body diagrams of the two portions (Figure 2.2).

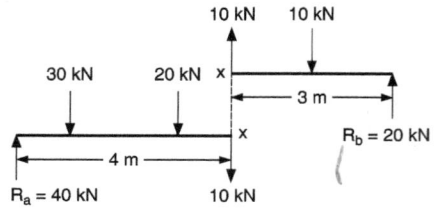

FIGURE 2.2.

- Considering equilibrium of forces on each portion of the beam, the net resultant vertical forces are

$$F_{left} = 30 + 20 - 40 = 10 \text{ kN} \qquad \text{(downward)}$$
$$F_{right} = 10 - 20 = -10 \text{ kN} \qquad \text{(upward)}$$

It is to be noted that forces on the left and right sides of section xx are equal in magnitude but opposite in direction.

Obviously, section xx is subjected to a force of 10 kN which is trying to shear the beam. The force of 10 kN is called as *shear force* at section xx.

"Shear force at a section in a beam is the force that is trying to shear off the section. It is obtained as algebraic sum of all the forces acting normal to the axis of beam; either to the left or to the right of the section."

- Considering equilibrium of moments on each portion of the beam:

$$M_{left} = 40 \times 4 - 30 \times 3 - 20 \times 1 = 50 \text{ kNm} \qquad \text{(clockwise)}$$
$$M_{right} = 20 \times 3 - 10 \times 1 = 50 \text{ kNm} \qquad \text{(anti-clockwise)}$$

It is to be noted that moments on the left and right sides of section xx are equal in magnitude but opposite in direction. Obviously, section xx is acted upon by a moment of 50 kNm. Since this moment is trying to bend the beam, it is called the *bending moment*.

"Bending moment at a section in a beam is the moment that tends to bend the beam and is obtained as algebraic sum of moment of all the forces about the section, acting either to the left or to the right of the section."

Sign Conversion: The following sign conventions are normally adopted for the shear force and bending moment.

FIGURE 2.3.

i. Shear force is taken positive if it tends to move the left portion upward with respect to the right portion.

ii. Bending moment is taken positive if it tends to sag (concave upward) the beam, and it is taken negative if it tends to hog (concave down) the beam.

The shear force and bending moment vary along the length of the beam, and this variation is represented graphically. The plots are known as shear force and bending moment diagrams. In these diagrams, the abscissa indicates the position of the section along the beam, and the ordinate represents the value of SF and BM, respectively. These plots help determine the maximum value of each of these quantities.

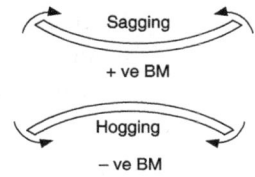

FIGURE 2.4.

2.2. TYPES OF BEAMS AND LOADS

The shear force and the bending moment that develop at any cross section of the beam depending on the types of beams and the types of loads acting on them.

Beam is generally classified as follows:

- Cantilever beam (Figure 2.5a): A beam having its one end fixed or built-in and the other end free to deflect. There is no deflection or rotation at the fixed end.
- Fixed beam (Figure 2.5b): A beam having both of its ends fixed or built-in.
- Simply supported beam (Figure 2.5c): A beam made to freely rest on supports which may be knife edges or rollers. The term "freely supported" implies that these supports exert only forces but no moments on the beam. The horizontal distance between the supports is called span.
- Overhanging beam (Figure 2.5d): A beam having one or both ends extended over the supports. The end portion or portions extend in the form of cantilever beyond the support/supports.
- Continuous beam (Figure 2.5e): A beam provided with more than two supports. Further such a beam may or may not have overhang.

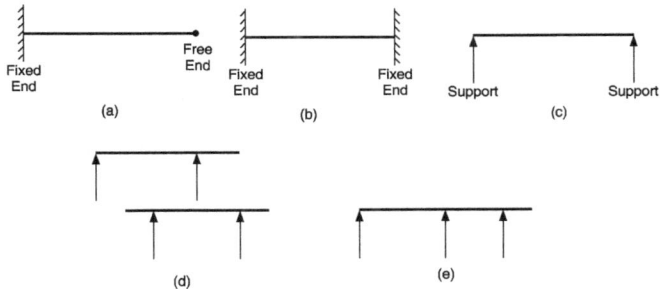

FIGURE 2.5.

The different types of loads acting on a beam are as follows:

- **Concentrated load:** The load acts at a point on the beam. This point load is applied through a knife edge.
- **Uniformly distributed load:** The load is evenly distributed over a part or the entire length of the beam. The total *udl* is assumed to act at the center of gravity of the load. The *udl* is expressed as N/m length of beam.

- **Uniformly varying load:** The load of which intensity varies linearly along the length of beam over which it is applied.
- A beam may be loaded by a couple of which magnitude is expressed as Nm.

A beam may carry any one of the above load systems or combination of two or more loads at a time.

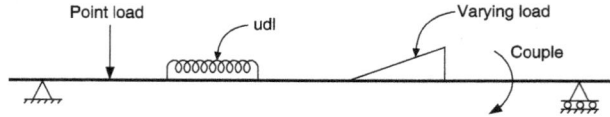

FIGURE 2.6.

2.3. RELATION BETWEEN LOAD INTENSITY, SF AND BM

Consider a beam subject to any type of transverse load of the general form shown in Figure 2.7. From the beam, an element of length dx at a distance x from left end is isolated and its free body diagram is drawn as shown in Figure 2.7. Since the element is of extremely small length, the loading over the beam can be considered to be uniform and equal to w kN/m. The element is subject to shear force F on its left-hand side and shear force $(F + dF)$ on its right-hand side. Further, the bending moment M acts on the left side of the element and it changes to $(M + dM)$ on the right side.

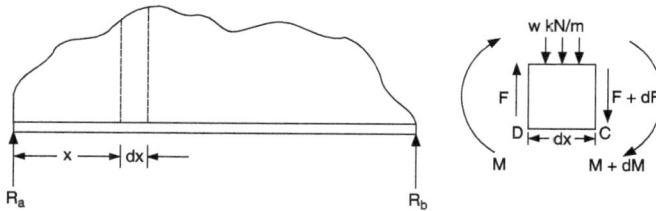

FIGURE 2.7.

Taking moments about point C on the right side,

$$\Sigma M_c = 0: \quad M - (M + dM) + F \times dx - (w \times dx) \times \frac{dx}{2} = 0.$$

The *udl* is considered to be acting at its *CG*

$$dM = Fdx - \frac{w(dx)^2}{2} = 0.$$

The last term consists of the product of two differentials and can be neglected.

$$\therefore \qquad dM = Fdx \text{ or } F = \frac{dM}{dx}.$$

Thus, the shear force is equal to the rate of change of bending moment with respect to x. Applying the condition $\Sigma F_y = 0$ for equilibrium, we obtain

$$F - w \, dx - (F + dF) = 0$$

or $$w = \frac{dF}{dx}.$$

That is the intensity of loading is equal to rate of change of shear force with respect to x.

2.4. SHEAR FORCE AND BENDING MOMENT DIAGRAMS FOR A FEW STANDARD CASES

The method of drawing shear force and bending moment diagrams for a few standard cases is explained below.

2.4.1. Cantilever Subjected to Concentrated Load at Free End

Consider a cantilever of length l carrying weight W at its free end.

At a section x-x at any distance x from the free end.

Shear force $\Sigma F = -W$

Apparently the shear force

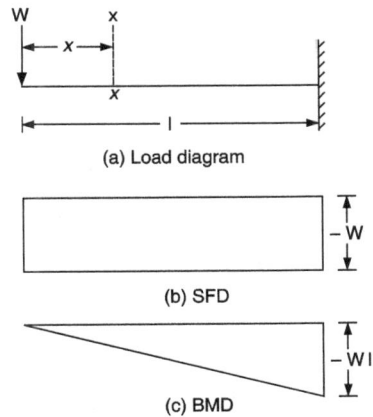

(a) Load diagram

i. is negative as the left segment of the beam moves downward.

ii. does not vary with x and has a constant value (equal to load W) throughout the beam length

(b) SFD

Bending moment $BM = -Wx$

Apparently the bending moment

i. is negative as it tends to hog the beam (concave downward)

(c) BMD

ii. has a linear variation with distance x from the free end.

FIGURE 2.8.

At $x = 0 \quad BM = 0$

At $x = l \quad BM = -W\,l.$

Thus, bending moment diagram is a straight line which gives maximum value at the fixed end.

The *SF* and *BM* diagrams are shown in Figure 2.8.

2.4.2. Cantilever Subjected to *udl* Over its Entire Span

Consider a cantilever of length l carrying uniformly distributed load of intensity w/unit length over the entire span.

At a section x-x at any distance x from the free end,

$$\text{Shear force } SF = -wx.$$

Apparently the shear force

i. is negative as the left segment of the beam moves downward

ii. has a linear variation with distance x from the free end

$$\text{At } x = 0 \quad SF = 0$$
$$\text{At } x = l \quad SF = -wl.$$

Thus, the SF diagram is a straight line with maximum value at the fixed end.

$$\text{Bending moment } BM = -wx \times \frac{x}{2} = -w$$

$\frac{x^2}{2}$ *udl* is considered to be acting at its center.

Apparently, the bending moment

i. is negative as its tends to hog the beam (concave downward)

ii. has a parabolic variation with distance x from the free end.

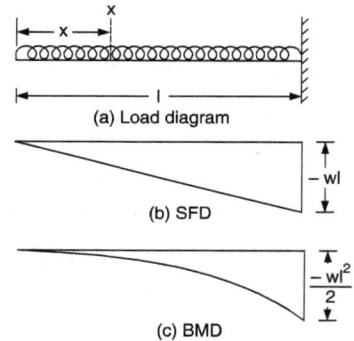

(a) Load diagram

(b) SFD

(c) BMD

FIGURE 2.9.

$$\text{At } x = 0 \quad BM = 0$$
$$\text{At } x = l \quad BM = \frac{wl^2}{2}.$$

The magnitude of BM has a faster variation with x and accordingly its variation conforms to concave parabola.

The SF and BM diagrams are shown in Figure 2.9.

2.4.3. Simply Supported Beam Subjected to Concentrated Load

Consider a simply supported beam AB with span l and subjected to a concentrated load W at a point C that lies at distance a from the left and support A.

$$\text{Now, } R_a + R_b = W$$

and taking moments about end support B,

$$R_a \times l = W_b; \qquad R_a = \frac{wb}{l}$$

$$R_b = W - R_a = W - \frac{wb}{l} = \frac{W(l-b)}{l} = \frac{Wa}{l}.$$

Let x be the distance of any section considered from the end support A. Then For portion AC $(0 < x < a)$,

$$SF = R_a = \frac{Wb}{l} \qquad \text{(constant)}$$

$$BM = R_a x = \frac{Wb}{l}x \qquad \text{(linear variation)}$$

At $x = 0$ (left support) : $\quad BM = 0$

At $x = a$ (under the load) : $BM = \dfrac{Wab}{l}.$

For portion CB $(a < x < l)$

$$SF = R_a - W = \frac{Wb}{l} - W = W\left(\frac{b}{l} - 1\right) = \frac{-Wa}{l} \qquad \text{(constant)}$$

$$BM = R_a x - W(x - a) = \frac{Wb}{l}x - W(x - a) \qquad \text{(linear variation)}$$

At $x = a$ (under the load):

$$BM = \frac{Wb}{l} \times a - W(a - a) = \frac{Wab}{l}$$

At $x = l$ (right support):

$$BM = \frac{W\,b}{l} \times l - W(l - a)$$

$$= Wb - Wb = 0.$$

The SF and BM diagrams for the entire beam are shown in Figure 2.10.

In case, the beam carries concentrated load at mid-span, then $a = b = \dfrac{l}{2}$

and accordingly SF and BM under the load are as follows:

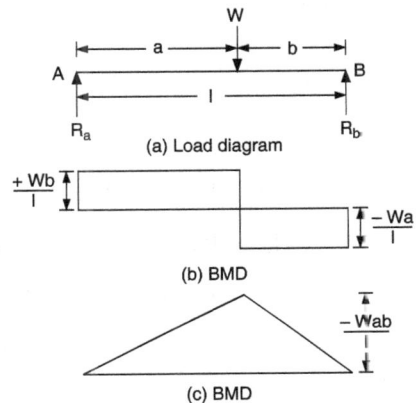

(a) Load diagram

(b) BMD

(c) BMD

FIGURE 2.10.

$$SF = \frac{Wb}{l} = \frac{W \times \frac{l}{2}}{l} = \frac{W}{2}$$

$$BM = \frac{W_{ab}}{l} = \frac{W \times \frac{l}{2} \times \frac{l}{2}}{l} = \frac{Wl}{4}.$$

2.4.4. Simply Supported Beam Subjected to *udl* Over its Entire Span

Consider a simply supported beam of span l carrying uniformly distributed load of intensity w/unit length over the entire span.

Because of symmetrical loading, each vertical reaction is equal to half the total load on the span. That is

$$R_a = R_b = \frac{wl}{2}.$$

At a section at distance x from the left and support A,

$$\text{Shear force } SF = Ra - wx = \frac{wl}{2} - wx \qquad \text{(linear variation)}$$

At $x = 0$; $SF = \dfrac{wl}{2}$

At $x = l$; $SF = \dfrac{wl}{2} - wl = \dfrac{-wl}{2}.$

Apparently, the shear force changes uniformly from $\dfrac{wl}{2}$ at end A to $\dfrac{-wl}{2}$ at end B.

Further, the shear force will be zero at $x = \dfrac{1}{2}$ (mid span).

$$\text{Bending moment } BM = Rax - wx \times \frac{x}{2} = \frac{wl}{2}x - \frac{wx^2}{2} \text{ (parabolic variation)}$$

The *udl* is considered to be acting at its *CG*.

At $x = 0$; $\qquad BM = 0$

At $x = \dfrac{1}{2}$; $\qquad BM = \dfrac{wl}{2} \times \dfrac{l}{2} - \dfrac{wl^2}{8} = \dfrac{wl^2}{8}$

At $x = l$; $\qquad BM = \dfrac{wl}{2} \times l - \dfrac{wl^2}{2} = 0$

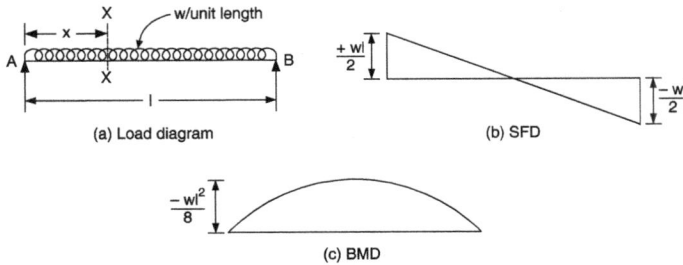

(a) Load diagram

(b) SFD

(c) BMD

FIGURE 2.11.

A reduction in the value of *BM* becomes faster as *x* increases, and accordingly the bending moment variation conforms to convex parabola. Further, it is to be noted that maximum bending moments occurs at the mid-span where shear force is zero.

EXAMPLE 2.1

Draw the shear force and bending moment diagrams for the beam loaded and supported as shown in Figure 2.12.

Solution: The line of action of the reaction will be at right angles to the roller base at end *A*. The reaction at a hinge can have two components acting in the horizontal and the vertical directions.

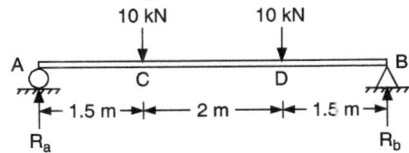

FIGURE 2.12.

Since there is no horizontal external force acting on the beam, the reaction at the hinged end *B* will be only in the vertical direction.

Due to symmetry of loading,

$$R_a = R_b = \frac{10+10}{2} = 10 \text{ kN}.$$

Shear force

SF at $A = 10$ kN
SF just on left of $C = 10$ kN
SF just on right of $C = 10 - 10 = 0$
SF just on left of $D = 0$
SF just on right of $D = 0 - 10 = -10$ kN
SF just on left of $B = -10$ kN
SF just on right of $B = -10 + 10 = 0$

Bending moment. Taking a section at distance x from end A and considering forces on left side.

Portion AC:

$$M = R_a \times x = 10x \qquad \text{(linear variation)}$$

At $x = 0$: $M_a = 0$

At $x = 1.5$ m : $M_c = 10 \times 1.5 = 15$ kNm

Portion CD:

$$M = R_a \times x - 10(x - 1.5)$$
$$= 10x - 10x + 15 = 15 \text{ kNm.}$$

The bending moment remains constant at 15 kNm within the portion CD.

Portion DB:

$$M = R_a \times x - 10(x - 1.5) - 10(x - 3.5)$$
$$= 10x - 10x + 15 - 10x + 35$$
$$= -10x + 50 \qquad \text{(linear variation)}$$

At $x = 3.5$ m:

$$M_a = -10 \times 3.5 + 50 = 15 \text{ kNm}$$

At $x = 5$ m:

$$M_b = -10 \times 5 + 50 = 0.$$

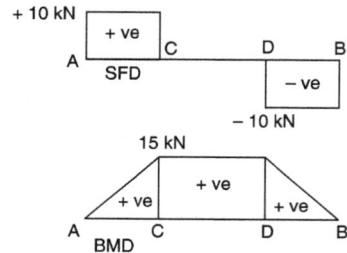

FIGURE 2.13.

The variation of shear force and bending moment for the entire length of the beam is depicted in Figure 2.13.

EXAMPLE 2.2

Construct the shear force and bending moment diagram for the cantilever beam loaded as shown in Figure 2.14.

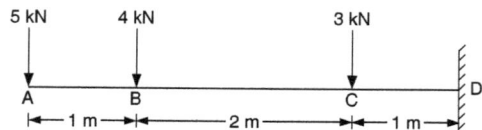

FIGURE 2.14.

Solution: For shear force calculations, consider any section at distance x from the free end A

At $x = 0$: $SF = -5$ kN.

The shear force is being taken −ve because it tends to move the left portion downward with respect to the right portion.

At $x = 1$ m

just left of B : $SF = -5$ kN just right of B : $SF = -5 - 4 = -9$ kN

At $x = 3$ m

just left of C : $SF = -9$ kN just right of C : $SF = -9 - 3 = -12$ kN.

Bending moment

Portion AB: Imagine a section between A and B, and at distance x from end A. Then

$$M_x = -5\,x\,(\text{linear variation})$$

At $x = 0$: $M_a = 0$

At $x = 1$ m : $M_b = -5 \times 1 = -5$ kNm.

Portion BC: Consider the section to be between B and C, and at distance x from end A. Then

$$M_x = -5\,x - 4(x-1) \quad (\text{linear variation})$$

At $x = 1$ m : $M_b = -5 \times 1 - (1-1) = -5$ kNm as calculated above

At $x = 3$ m : $M_c = -5 \times 3 - 4(3-1) = -23$ kNm.

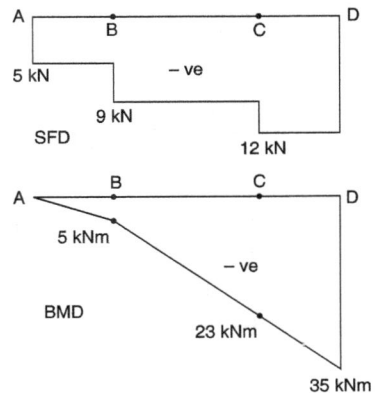

FIGURE 2.15.

Portion CD: Consider the section to be between C and D, and at distance x from end A.
Then

$$M_x = -5x - 4(x-1) - 3(x-3) \qquad (\text{linear variation})$$

At $x = 3$ m : $M_c = -5 \times 3 - 4(3-1) - 3(3-3) = -23$ kNm

At $x = 4$ m : $M_d = -5 \times 4 - 4(4-1) - 3(4-3) = -35$ kNm.

The shear force and the bending moment for the entire beam are shown in Figure 2.15.

EXAMPLE 2.3

Construct the shear force and bending moment diagrams for the cantilever beam loaded as shown in *Figure 2.16*.

Solution: For shear force calculations *for portion AB*, take section at distance x from end A.

$$SF = -10 - 10x \ (\text{linear variation})$$

At $x = 0$; $\quad SF = -10$ kN

At $x = 1$m (just to left of point B);

$$SF = -10 - 10 = -20 \text{ kN}$$

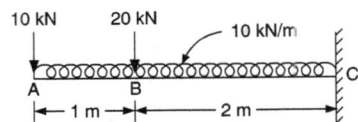

FIGURE 2.16.

For *portion BC*, again we consider a section at distance x from the end A,

$$SF = -10 - 20 - 10x \qquad (\text{linear variation})$$

At $x = 1$ m (just to left of point B);
$SF = -10 - 20 - 10 = -40$ kN

At $x = 3$ m (fixed end);
$SF = -10 - 20 - 10 \times 3 = -60$ kN.

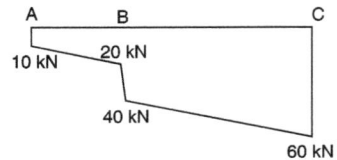

FIGURE 2.17.

The shear force diagram indicating the values of shear force at salient points is as shown in Figure 2.17.

(*b*) For bending moment for *portion AB*, take section at distance x from the free end A.

$$BM = -10x - 10x \times \frac{x}{2} \qquad \text{(parabolic variation)}$$

The *udl* is taken to be acting at its CG

At $x = 0$; $\qquad BM = 0$

At $x = 1$ m; $\qquad BM = -10 \times 1 - 10 \times 1 \times \dfrac{1}{2} = -15$ kNm.

For portion BC, again we consider a section at distance x from the end A

$$BM = -10x - 20(x - 1) - 10x \times \frac{x}{2}$$

At $x = 1$ m : $\qquad BM = -10 - 20(1 - 1) - 10 \times 1 \times \dfrac{1}{2} = -15$ kNm

At $x = 3$ m (fixed end):

$$BM = -10 \times 3 - 20(3 - 1) - 10 \times 3 \times \frac{3}{2}$$

$$= -30 - 40 - 45 = -115 \text{ kNm.}$$

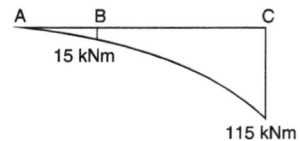

The bending moment diagram indicating the value of bending moment at salient points is as shown in Figure 2.18.

FIGURE 2.18.

EXAMPLE 2.4

Determine the reactions and construct the shear force and bending moment diagrams for the beam loaded as shown in Figure 2.19. Also find the point of contraflexure, if any.

Solution: A point of contraflexure is a point where bending moment is zero. From conditions of static equilibrium ($\Sigma V = 0$ and $\Sigma M = 0$), we have

$$R_a + R_b = 2 \times 2 + 10 + 2 = 16 \qquad (i)$$
$$-2 \times 2 \times 10 + R_a \times 9 - 10 \times 5 + R_b \times 1 = 0; \ 9R_a + R_b = 90 \qquad (ii)$$

The *udl* is considered to be concentrated at its *CG*.

From expression (*i*) and (*ii*), $R_a = 9.25$ kN and $R_b = 6.75$ kN

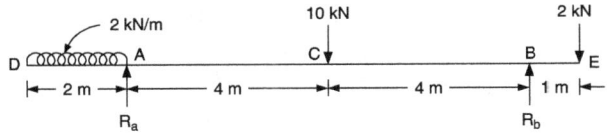

FIGURE 2.19.

Shear Force:

At $D = 0$

Just left of $A = -2 \times 2 = -4$ kN ; Just right of $A = -4 + 9.25 = 5.25$ kN
Just left of $C = 5.25$ kN ; Just right of $C = 5.25 - 10 = -4.75$ kN
Just left of $B = -4.75$ kN ; Just right of $B = -4.75 + 6.75 = 2$ kN
Just left of $E = 2$ kN ; Just right of $E = 2 - 2 = 0$ kN

Bending moment

$$M_D = 0.$$

At distance x from D (within portion DA)

$$M_x = -2x \times \frac{x}{2} = -x^2$$

$\therefore$ M (at $x = 1$m) $= 1$ kNm and M (at $x = 2$m) $= -4$ kNm
$M_A = -4$ kNm
$M_C = -2 \times 2 \times 5 + 9.25 \times 4 = -20 + 37 = 17$ kNm.

Apparently there is a point of contraflexure between A and C as bending moment changes sign between A and C.

Bending moment at x between A and C with x measured from D

$$M_x = -4(x - 1) + 9.25(x - 2) = 5.25x - 14.5$$
$\therefore$ $5.25x - 14.5 = 0$ for point of contraflexure

That gives $x = \dfrac{14.5}{5.25} = 2.76$ m

$$M_B = -2 \times 1 = -2 \text{ kNm.}$$

(considering the segment EB from right-hand side)

Since bending moment at C is +ve and at B is −ve, there is also a point of contraflexure between C and B.

Bending moment at distance x measured from end E toward left,

$$M_x = -2x + 6.75(x - 1)$$
$$= 4.75x - 6.75$$
$\therefore$ $4.75x - 6.75 = 0$

for point of contraflexure

That gives $x = \dfrac{6.75}{4.75} = 1.42$ m.

The shear force and bending moment diagrams for the entire beam are shown in Figure 2.20 along with position of points of contraflexure.

EXAMPLE 2.5

A cantilever beam has been loaded as shown in Figure 2.21. Draw the shear force and bending moment diagrams, and locate the position of the point of contraflexure.

Solution: Consider any section at distance x from the free end

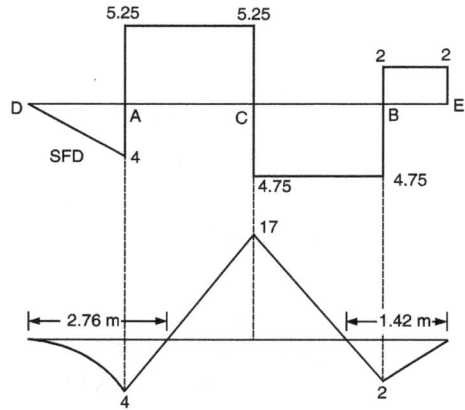

FIGURE 2.20.

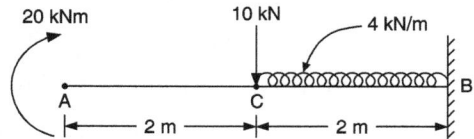

FIGURE 2.21.

Shear force:
Portion AC: The shear force is zero in this section.

   Portion CB: $SF = 10 + 4(x - 2)$ (linear variation)

   At $x = 2$ m : $SF = 10$ kN

   $x = 4$ m : $SF = 10 + 4(4 - 2) = 18$ kN.

Bending moment:
Portion AC: The bending moment has a constant value of 20 kNm

Portion CB: $M_x = 20 - 10(x - 2) - 4(x - 2) \times \dfrac{(x - 2)}{2}$ \hfill (i)

At $x = 2$ m : $M_c = 20 - 10(2 - 2) - 4 \times \dfrac{(2 - 2)^2}{2} = 20$ kNm

$x = 4$ m : $M_b = 20 - 10(4 - 2) - 4 \times \dfrac{(4 - 2)^2}{2}$

$= 20 - 20 - 4 \times 2 = -8$ kNm.

Since the bending moment changes in portion *CB*, the location of the point of contraflexure can be determined by setting expression (*i*) equal to zero. That is

$$20 - 10(x - 2) - 2(x - 2)^2 = 0.$$

Upon simplification, we get $x^2 + x - 16 = 0$

$$\therefore x = \frac{-1 \pm \sqrt{1 + 64}}{2}$$

$$= 3.53 \text{ m.}$$

Thus, the point of contraflexure is located at 3.53 m distance from the free end.

The variation of shear force and bending moment for the entire length of the beam is depicted in Figure 2.22.

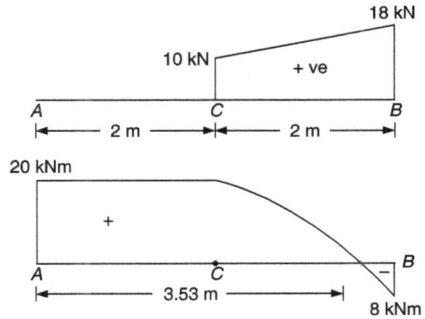

FIGURE 2.22.

EXAMPLE 2.6

Draw the shear force and bending moment diagram for the beam loaded and supported as shown in Figure 2.23.

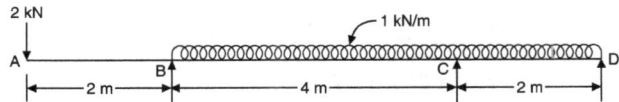

FIGURE 2.23.

Solution: From conditions of equilibrium ($\Sigma V = 0$ and $\Sigma M = 0$), we have

$$R_b + R_c = 2 + (4 + 2) \times 1 = 8. \tag{i}$$

Taking moments about end point A (clockwise moments positive),

$$-R_b \times 2 + 6 \times (2 + 3) - R_c \times 6 = 0.$$

The *udl* is considered to be concentrated at its *CG*

$$R_b + 3R_c = 15. \tag{ii}$$

Solving expression (*i*) and (*ii*),

$$R_b = 4.5 \text{ kN and } R_c = 3.5 \text{ kN.}$$

Shear force

At point A	:	$SF = -2$ kN
Just left of B	:	$SF = -2$ kN
Just right of B	:	$SF = -2 + 4.5 = 2.5$ kN
Just left of C	:	$SF = 2.5 - 4 \times 1 = -1.5$ kN
Just right of C	:	$SF = -1.5 + 3.5 = 2$ kN
At point D	:	$SF = 2 - 2 \times 1 = 0$

Bending moment: Considering any section at distance x from end A
Portion AB

$$M_x = -2 \times x = -2x \qquad \text{(linear variation with } x\text{)}$$

when $x = 0$: $M_a = 0$

$x = 2$ m : $M_b = -2 \times 2 = -4$ kNm.

Between B and C

$$M_x = -2x + 4.5 \times (x - 2) - \{(x - 2) \times 1\} \times \frac{x - 2}{2}$$

$$= -2x + 4.5(x - 2) - \frac{1}{2}(x - 2)^2$$

when $x = 2$ m : $M_b = -2 \times 2 = -4$ kNm

$x = 6$ m : $M_c = -2 \times 6 + 4.5(6 - 2) - \frac{1}{2}(6 - 2)^2$

$$= -12 - 18 - 8 = -2 \text{ kNm.}$$

Due to *udl*, the bending moment variation between B and C will be parabolic.

Between C and D

$$M_x = -2x + 4.5(x - 2) - \frac{1}{2}(x - 2)^2 + 3.5(x - 6)$$

when $x = 6$ m : $M_c = -2 \times 6 + 4.5(6 - 2) - \frac{1}{2}(6 - 2)^2 + 3.5(6 - 6)$

$$= -12 + 18 + 8 + 0 = -2 \text{ kNm}$$

$x = 8$ m : $M_d = -2 \times 8 + 4.5(8 - 2) - \frac{1}{2}(8 - 2)^2 + 3.5(8 - 6)$

$$= -16 + 27 - 18 + 7 = 0.$$

Due to *udl*, the bending moment variation between C and D will be parabolic. The shear force and bending moment for the entire beam are shown in Figure 2.24. The shear force changes sign at point E and its location with respect to point B is

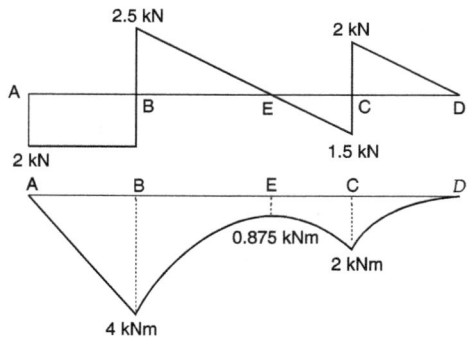

FIGURE 2.24.

$$\frac{2.5}{BE} = \frac{1.5}{EC}$$

$$\frac{2.5}{7} = \frac{1.5}{4 - y}$$

$$10 - 2.5\,y = 1.5\,y;\ y = 2.5 \text{ m.}$$

The point E is, thus, located at distance $x = 2 + 2.5 = 4.5$ m from point A
Bending moment at point E

$$M_e = -2 \times 4.5 + 4.5(4.5 - 2) - \frac{1}{2}(4.5 - 2)^2$$
$$= -9 + 11.25 - 3.125$$
$$= -0.875 \text{ kNm}.$$

EXAMPLE 2.7

Draw shear force and bending moment diagram for the cantilever beam loaded as shown in Figure 2.25.

Solution: Consider any section of distance x from end A.

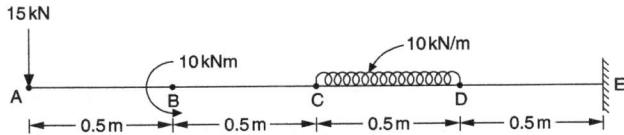

FIGURE 2.25.

Shear Force

   (*i*) **Portion AC**

At A : $SF = -15$ KN

This shear force is taken −ve because it tends to move the left point downward with respect to right portion.

Since there is no load on the beam in the portion AC, the shear force is constant at −15 kN from A to B.

   (*ii*) **Portion CD**

$$SF = -15 - 10 \times (x - 1)$$

At point C $(x = 1)$: $SF = -15 - 10\,(1 - 1) = -15$ kN
At point D $(x = 1.5)$: $SF = -15 - 10\,(1.5 - 1) = -20$ kN.

Since there is *udl* in the region CD, the shear force changes linearly for −15 kN to −20 kN in span CD of the beam.

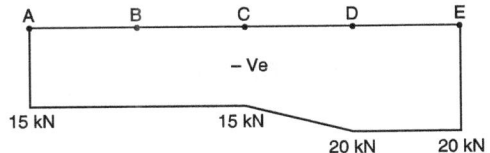

FIGURE 2.26.

   (*iii*) **Portion DE**

Since there is no load on the beam in the portion DE, the shear force is constant at −20 kN from D to E.

There shear force for the beam is then plotted as shown in Figure 2.26.

Bending Moment:

$$M_a = 0$$
$$M_b = -15 \times 0.5 - 10 = -17.5 \text{ kNm}$$
$$M_c = -15 \times 1 - 10 = -25 \text{ kNm}$$

At any cross section xx in portion CD and at distance x from end A,

$$M_x = -15x - 10 - 10(x - 1) \times \frac{x - 1}{2}$$

$$= -15x - 10 - 5(x - 1)^2$$

At C $(x = 1)$: $M_c = -15 \times 1 - 10 - 5(1 - 1)^2 = -25$ kNm
At D $(x = 1.5)$: $M^d = -15 \times 1.5 - 10 - 5(1.5 - 1)^2$

$$= -22.5 - 10 - 1.25 = -33.75 \text{ kN.}$$

Due to udl, the variation of bending moment in portion CD of the beam will be parabolic

$$M_e = -15 \times 2 - 10 - 10 \times 0.5 \times 0.75$$
$$= -30 - 10 - 3.75 = -43.75 \text{ kN.}$$

Since there is no load on the beam in portion DE, the bending moment will vary linearly from −33.75 kN at point D to −43.75 kN at point E.

The bending moment for the entire beam will then plot as shown in Figure 2.27.

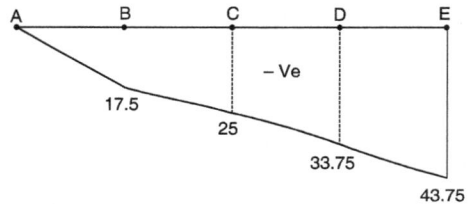

FIGURE 2.27.

EXAMPLE 2.8

A simply supported beam is loaded and supported as shown in the figure given below: Draw the shear force and bending moment diagrams and deter-mine the magnitude and location of maxi-mum bending moment.

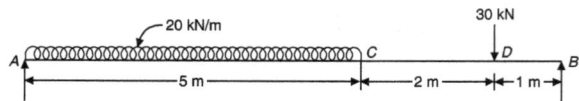

FIGURE 2.28.

Solution: Considering equilibrium of beam

$$\Sigma F_y = 0 : \quad R_a + R_b = 20 \times 5 + 30 = 130$$
$$\Sigma M = 0 : \text{Taking moments about point } A \text{ (clockwise +ve)}$$

$$-R_b \times 8 + 30 \times 7 + 20 \times 5 \times \frac{5}{2} = 0$$

or $$R_b = \frac{210 + 250}{8} = 57.5 \text{ kN}$$

and $$R_a = 130 - 57.5 = 72.5 \text{ kN.}$$

Consider any section at distance x from end A.

Shear Force:

Portion AC: $\qquad SF = R_a - wx = 72.5 - 20x \qquad$ (Linear variation)

At point A, $\quad x = 0 \quad$ and $\quad SF = 72.5$ kN

At point B, $\quad x = 5 \quad$ and $\quad SF = 72.5 - 20 \times 5 = -27.5$ kN.

Portion CD: Since there is no load between CD, the shear force remains constant at -27.5 kN just up to left of point D.

SF just on right side of $D = -27.5 - 30 = -57.5$ kN.

Portion DB: Since there is no load between DB, the shear force remains constant at -57.5 kN just up to left of point B.

SF just on right side of support $B = -57.5 + 57.5 = 0$.

The point of zero shear force as measured from end A and lying between AB can be worked out from the relation

$$72.5 - 20x = 0; x = \frac{72.5}{20} = 3.625 \text{ m.}$$

Bending Moment:

Portion AC: $\qquad BM = R_a x - \dfrac{wx^2}{2}$

$$= 72.5x - 10x^2 \qquad \text{(Parabolic variation)}$$

At point A, $\quad x = 0$ and $BM = 0$

At point B, $\quad x = 5$ and $BM = 72.5 \times 5 - 10 \times 5^2 = 112.5$ kNm

The bending moment is maximum at $x = 3.625$ m where shear force is zero.

Maximum bending moment $= 72.5 \times 3.625 - 10 \times 3.625^2 = 262.81 - 131.41 = 131.4$ kN m

Portion CD: As there is no loading in this section, the bending moment is a linear variation, and at point $D(x = 7)$, it will have the value.

$$= 72.5 \times 7 = 20 \times 5 \times \left(\frac{5}{2} + 2\right) = 57.5 \text{ kNm.}$$

Portion DB: As there is no loading in this section, the bending moment is a linear variation and it is dropped from 57.5 kNm at point D to zero at support point B.

The variations in shear force and bending moment for the entire beam are shown in Figure 2.29.

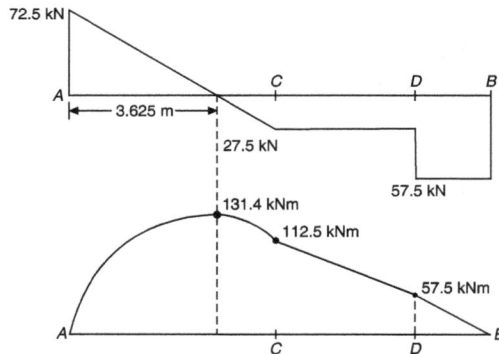

FIGURE 2.29.

EXAMPLE 2.9

A simply supported beam with 8 m span is loaded as shown in the figure given below:

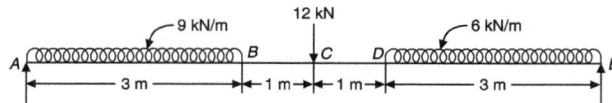

FIGURE 2.30.

Draw the shear force and bending moment diagrams. Also determine the magnitude and position of maximum bending moment on the beam.

Solution: Considering equilibrium of beam

$$\Sigma F_y = 0: \quad R_a + R_b = (9 \times 3) + 12 + (6 \times 3) = 57 \text{ kN}$$
$$\Sigma M = 0 : \text{Taking moments about end point } A \text{ (clockwise moments +ve)}$$

$$27 \times 1.5 + 12 \times 4 + 18 \times 6.5 - R_e \times 8 = 0.$$

The *udl* is considered to be concentrated at *CG*.

$$R_e = \frac{27 \times 1.5 + 12 \times 4 + 18 \times 6.5}{8}$$
$$= \frac{40.5 + 48 + 117}{8} = 25.69 \text{ kN}$$

and $\quad R_a = 57 - 25.69 = 31.31 \text{ kN.}$

Shear Force:
Portion AB: Consider any section at distance x from end support A

$$SF = 31.31 - 9x \qquad \text{(Linear variation)}$$

At point A, $x = 0$ and $SF = 31.31$ kN
At point B, $x = 3$ m and $SF = 31.31 - 9 \times 3 = 4.31$ kN.

Portion BCD: The shear force remains constant at 4.32 kN between B and just left of C.

Just right of CD; $SF = 4.31 - 12 = 7.69$ kN.

The shear force remains constant at 7.69 kN between and just left of D.

Portion DE: Consider any section between DE and at distance x from end support A.

$$SF = 31.31 - 9 \times 3 - 12 - (x - 5) \times 6$$
$$= 22.31 - 6x$$

At point D, $x = 5$ m
and $SF = 22.31 - 6 \times 5 = 7.69$ kN
At just left of point E, $x = 8$ m
and $SF = 22.31 - 6 \times 8 = 25.69$ kN
At point E, $SF = 25.69 - 25.69 = 0$.

Bending Moment:
Portion AB: Consider any section between AB at distance x from the end support A.

$$BM = 31.31x - 9\frac{x^2}{2} \qquad \text{(Parabolic variation)}$$

At point A, $x = 0$ and $BM = 0$
At point B, $x = 3$ m
and $BM = 31.31 \times 3 - \dfrac{9 \times 3^2}{2} = 53.43$ kN m.

Portion BCD:

At point C, $BM = R_a \times 4 - (9 \times 2) \times (1.5 + 1)$
                $= 31.31 \times 4 - 67.5$
                $= 57.74$ kN m

At point D, $BM = R_a \times 5 - (9 \times 3) \times (1.5 + 2)$
                $= 31.31 \times 5 - 94.5 - 12$
                $= 50.05$ kN m.

Portion DE: Consider any section within *DE* at distance *x* from the end support *A*.

$$BM = 31.31x - (9 \times 3) \times (x - 1.5) - 12 \times (x - 4) - 6x(x - 5) \times \frac{x - 5}{2}$$

$$= 31.31x - 27 \times (x - 1.5) - 12(x - 4) - 3(x - 5)^2$$

BM at *D* (at *x* = 5)
$$= 31.31 \times 5 - 27(5 - 1.5) - 12(5 - 4) - 3(5 - 5)^2$$
$$= 50.05 \text{ kN m}$$

BM at *E* (at *x* = 8)
$$= 31.31 \times 8 - 27(8 - 1.5) - 12(8 - 4) - 3(8 - 5)^2 = 0.$$

Since there is *udl* in the segment *DE*, the variation in bending moment is parabolic.

The variation in shear force and bending moment for the entire beam are as shown in Fig 2.31.

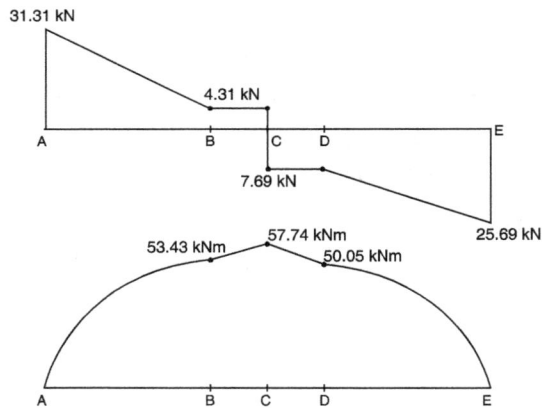

FIGURE 2.31.

EXAMPLE 2.10

Explain clearly the terms shearing force and bending moment for a beam.

Draw the shear force and bend-ing moment diagrams for the sim-ply supported overhanging beam shown in the figure given below: Also locate the position of the points of contraflexure, if any

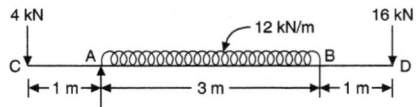

FIGURE 2.32.

Solution: Considering equilibrium of beam

$$\Sigma F_y = 0: \qquad R_a + R_b = 4 + 12 \times 3 + 16 = 56 \text{ kN}$$
$$\Sigma M = 0: \text{Taking moments about point } A \text{ (clockwise +ve)}$$

$$16 \times 4 - R_b \times 3 + 12 \times 3 \times \frac{3}{2} - 4 \times 1 = 0$$

or $\qquad R_b = \dfrac{64 + 54 - 4}{3} = 38 \text{ kN and } R_a = 56 - 38 = 18 \text{ kN.}$

Shear Force:

Portion CA: At point C,

$SF = -4$ kN and it remains constant upto just left of point A because there is no loading between this segment.

SF just on right side of $A = -4 + 18 = 14$ kN

Portion AB: Consider any section at distance x from point C.

$$SF = -4 + 18 - (x - 1) \times 12 = 26 - 12x$$

SF just on left side of B, *i.e.*, at $x = 4$ m

$$SF = 26 - 12 \times 4 = -22 \text{ kN}$$

SF just right side of $B = -22 + 38 = 16$ kN.

The shear force will be zero at $x = \dfrac{26}{12} = 2.1666$ m

Portion BD: As there is no loading in this section, the shear force remains constant at 16 kN just up to left of D, and just to right of D, the shear force will become

$$16 - 16 = 0.$$

Bending Moment:

Portion CA: BM at point $C = 0$

$$BM \text{ at point } A = -4 \times 1 = -4 \text{ kNm.}$$

The bending moment from 0 at A and -4 kNm at B will have linear variation as there is no loading in this segment.

Portion AB: Consider any section at distance x from the end point C.

$$BM = -4x + 18(x - 1) - 12(x - 1) \times \frac{x - 1}{2}$$

$$= -4x + 18x - 18 - 6(x - 1)^2$$
$$= 14x - 18 - 6(x - 1)^2 \qquad\qquad (i)$$

At point A, $x = 1$ m

and $BM = 14 \times 1 - 18 - 6(1-1)^2 = -4$ kNm.

The bending moment at $x = 2.166$ m (point E)
$$= 14 \times 2.166 - 18 - 6(2.166 - 1)$$

At point B, $x = 4$ m

and $BM = 14 \times 4 - 18 - 6(4-1)^2 = -16$ kNm.

The variation of bending moment from -4 kN m at point A to -16 kN m at point B will be parabolic as there is *udl* in this segment.

Portion BD: As there is no loading in this section, the bending moment will have a linear variation and it will rise from -16 kNm at point B to zero at end point D.

The variations is shear force and bending moment for the entire beam, which are shown in Figure 2.33.

Points of Contraflexure: The bending moment is zero at the points of contraflexure.

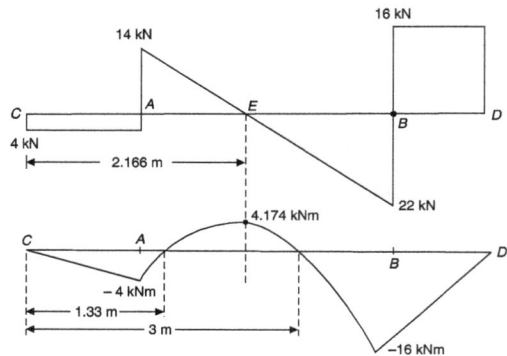

FIGURE 2.33.

Since the bending moment changes sign in portion AB, the location of the points of contraflexure can be determined by setting expression (i) equal to zero. That is

$$14x - 18 - 6(x-1)^2 = 0.$$

Upon simplification: $3x^2 - 13x + 12 = 0$

The solution of this quadratic equation gives

$$x = \frac{13 \pm \sqrt{(-13)^2 - 4 \times 3 \times 12}}{2 \times 3} = \frac{13 \pm 5}{5} = 3 \text{ m and } 1.33 \text{ m}.$$

Thus, the points of contraflexure are located at 1.33 m and 3 m from the end point C.

EXAMPLE 2.11

A horizontal beam 10 m long carries a uniformly distributed load of 8 kN/m together with concentrated loads of 40 kN at the left end and 60 kN at the right end. The beam is supported at two points 6 m, so chosen that reaction is the same at the each support. Determine the

position of props and show the variation of shear force and bending moment over the entire length of the beam.

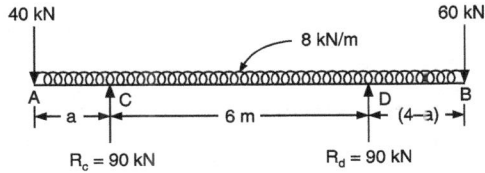

FIGURE 2.34.

Solution: Refer to Figure 2.34 for the beam loaded and supported as per the statement. Let the prop C be at distance a from end A.

Then the prop D is at distance $(4 - a)$ from end B.

Total load on the beam $= 40 + 60 + (10 \times 8) = 180$ kN. Since reaction is the same at each support,

$$R_c = R_d = \frac{180}{2} = 90 \text{ kN.}$$

Taking moments about end A,

$$60 \times 10 + (8 \times 10) \times \frac{10}{2} = 90 \times a + 90(6 + a)$$

or $\qquad 600 + 400 = 90a + 540 + 90a$

$\therefore \qquad a = \dfrac{(600 + 400) - 540}{180} = 2.55$ m.

Thus, the left support is 2.55 m from A, and the right support is $(4 - 2.55) = 1.45$ m from B.

Shear force:

SF at $A = -40$ kN

SF just on left side of $C = -40 - 8 \times 2.55 = -60.40$ kN

SF just on right side of $C = -60.40 + 90 = 29.60$ kN

SF just on left side of $D = 29.60 - 8 \times 6 = -18.40$ kN

SF just on right side of $D = -18.40 + 90 = 71.60$ kN

SF just on left side of $B = 71.60 - 8 \times 1.45 = 60$ kN

SF just on right side of $B = 60 - 60 = 0$

The point of zero shear stress as measured from end and lying between CD can be worked out from the equation:

$$-40 + 90 - 8x = 0; x = \frac{50}{8} = 6.25 \text{ m.}$$

Bending moment:

BM at $A = 0$

$$BM \text{ at } C = -40 \times 2.55 - (8 \times 2.55) \times \frac{2.55}{2} = -128 \text{ kNm}$$

$$BM \text{ at } D = -40 \times 8.55 - (8 \times 8.55) \times \frac{8.55}{2} + 90 \times 6$$

$$= -342 - 292.4 + 540 = -94.4 \text{ kNm}.$$

BM at a distance of 6.25 m from A,

$$= -40 \times 6.25 - (8 \times 6.25) \times \frac{6.25}{2} + 90 \times (6.25 - 2.55)$$

$$= -250 - 156.25 + 333 = -73.25 \text{ kNm}.$$

The variation of shear force and bending moment length of the beam is depicted in Figure 2.35.

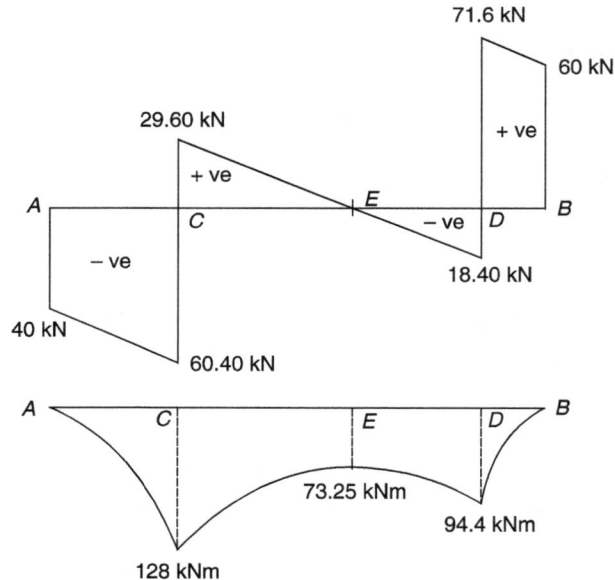

FIGURE 2.35.

EXAMPLE 2.12

A horizontal beam *AB* of span 10 m carries a uniformly distributed load of intensity 160 N/m and a point load of 400 N at the left end *A*. The beam is supported at a point *C* which is 1m from *A* and at *D* which is on the right half of the beam. If the point of contraflexure is at the mid-point of the beam, the distance of support at *D* from the end

B of the beam is determined. Proceed to draw the shear force and bending moment diagrams for the arrangement.

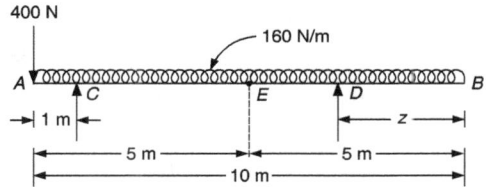

FIGURE 2.36.

Solution:

The bending moment is zero at the point of contraflexure. Therefore,

$$M_e = 0 = -400 \times 5 - 160 \times 5 \times \frac{5}{2} + R_c \times 4 \qquad \text{(left half of beam)}$$

The *udl* is taken to be acting at its *CG*.

or $4 R_c = 2000 + 2000$; $R_c = 1000$ N.

Applying the condition $\Sigma F_y = 0$ for equilibrium of beam, we have

$$R_c + R_d = 400 + 160 \times 10 = 2000$$
$$\therefore \quad R_d = 2000 - R_c = 2000 - 1000 = 1000 \text{ N.}$$

Again taking moments about the point of contraflexure *E*,

$$M_e = 0 = -R_d \times (5 - z) + 160 \times 5 \times \frac{5}{2} \qquad \text{(right half of beam)}$$

$1000 \times (5 - z) = 2000$; $z = $ **3 m.**

Thus, the support *D* is at a distance of 3 m from end *B*.

Shear Force

Portion AC:

At *A*: $SF = -400$ N

Just left of *C*	:	$SF = -400 - 160 \times 1 = -560$ N
Just right of *C*	:	$SF = -560 + 1000 = +440$ N
Just left of *D*	:	$SF = 440 - 160 \times 6 = -520$ N
Just right of *D*	:	$SF = -520 + 1000 = 480$ N
At point *B*	:	$SF = 480 - 160 \times 3 = 0.$

*** The shear force changes sign between the sections *CD*. The location of the point of zero shear stress can be obtained from the relations:

$$-400 - 160x + 1000 = 0; x = 3.75 \text{ m.}$$

Bending moment: Considering any section at distance x from end *A*,

Portion AC:

$$M_x = -400x - 160x \times \frac{x}{2} = -400 \times - 80x^2$$

when $\qquad x = 0 : M_a = 0$

$\qquad\qquad x = 1\text{m} : M_c = -400 - 80 = -480$ Nm.

Between C and D

$$M_x = -400x - 160x \times \frac{x}{2} + R_b(x - 1)$$

$$= -400x - 80x^2 + 1000(x - 1).$$

when $\quad x = 1$ m $\quad : M_b = -400 \times 1 - 80 \times 1^2 + 1000(1 - 1) = -480$ Nm

$\qquad x = 3.75$ m $: M = -400 \times 3.75 - 80 \times 3.75^2 + 1000(3.75 - 1) = 125$ Nm

$\qquad x = 5$ m $\quad : M_e = -400 \times 5 - 80 \times 5^2 + 1000(5 - 1) = 0$

$\qquad x = 7$ m $\quad : M_d = -400 \times 7 - 80 \times 7^2 + 1000(7 - 1) = -720$ Nm.

Between D and B:

$$M_x = -400x - 80x^2 + 1000 \times (x - 1) + 1000(x - 7)$$

At $x = 7$ m : $\quad M_d = -400 \times 7 - 80 \times 7^2 + 1000 \times (7 - 1) + 1000(7 - 7)$

$\qquad\qquad\qquad = -720$ Nm

$\qquad x = 10$ m : $\quad M_b = -400 \times 10 - 80 \times 10^2 + 1000(10 - 1) + 1000(10 - 7)$

$\qquad\qquad\qquad = -4000 - 8000 + 9000 + 3000 = 0.$

The shear force and bending moment for the entire beam are shown in Figure 2.37.

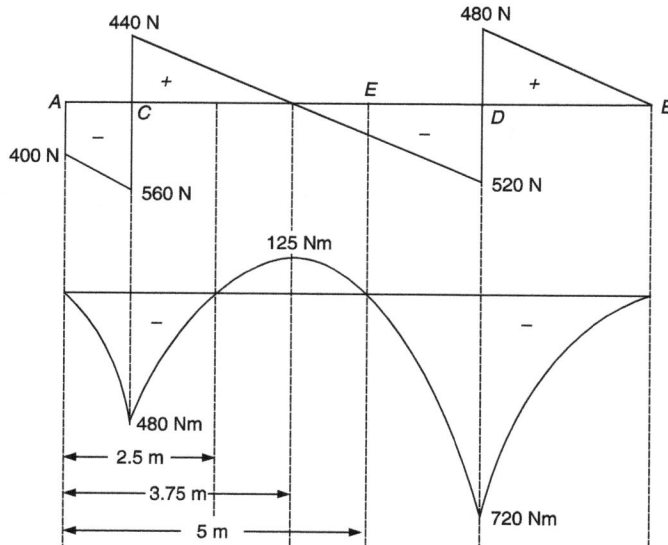

FIGURE 2.37.

EXAMPLE 2.13

A girder 10 m long rests on two supports with equal overhangs on either side and carries a uniformly distributed load of 20 kN/m over the entire length. Calculate the overhangs if the maximum bending moment, positive or negative, is to be as small as possible. Proceed to draw the shear force and bending moment diagrams for the arrangement.

Solution: Refer to Figure 2.38 for the space diagram of the loaded girder. The overhand on each side has been indicated as *a*.

Due to symmetrical arrangement, the total load on the beam will be shared equally between the two supports.

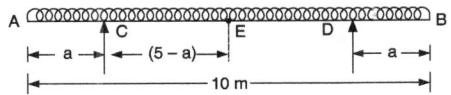

FIGURE 2.38.

$$\therefore \qquad R_c = R_d = \frac{20 \times 10}{2} = 100 \text{ kN}.$$

The maximum positive moment would occur at the mid span (point E) and the maximum negative would occur at the supports. Since these moments are stated to be equal in magnitude, we have

$$(20 \times a) \times \frac{a}{2} = 100(5 - a) - (20 \times 5) \times \frac{5}{2}.$$

Simplification gives $a^2 + 10a - 25 = 0$

$$\therefore \qquad a = \frac{-10 + \sqrt{10^2 - 4 \times 1 \times (-25)}}{2} = 2.07 \text{ m}$$

Shear force:
SF at $A = 0$
SF just on left of $C = -2.07 \times 20 = -41.40$ kN
SF just on right of $C = -41.40 + 100 = +58.60$ kN
SF at mid-span (point E) $= 58.60 - 20(5 - 2.07) = 0$

Bending moment: Taking a section at distance x from end A and considering forces on left-hand side.

Portion AC:

$$M = -(20 \times x) \times \frac{x}{2} = -10x^2 \qquad \text{(parabolic variation)}$$

At $x = 0$: $M_a = 0$
At $x = 2.07$ m : $M_c = -10 \times (2.07)^2 = -42.84$ kNm.

Portion CD:

$$M = -(20 \times x) \times \frac{x}{2} + R_c(x - a)$$

$$= -10x^2 + 100(x - 2.07) \qquad \text{(parabolic variation)}$$

At $x = 2.07$m $: M_c = -10 \times (2.07)^2 + 100(2.07 - 2.07) = -42.84$ kNm

At $x = 5$m $\quad : M^e = -10 \times 5^2 + 100(5 - 2.07) = 43$ kNm.

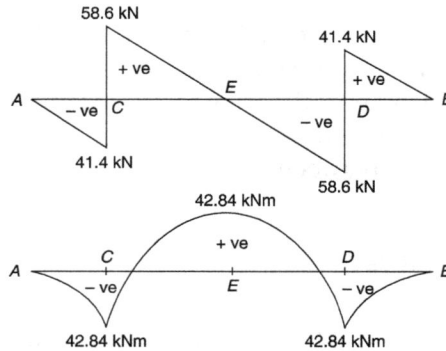

FIGURE 2.39.

The slight variation in the magnitude of bending moment at the support (point *B*) and at the center (point *E*) is due to rounding off.

For locating the position of the point of contraflexure, we have

$$-10x^2 + 100(x - 2.07) = 0$$

or $\qquad x^2 - 10x + 20.7 = 0$

$\therefore \qquad x = \dfrac{10 \pm \sqrt{10^2 - 4 \times 20.7}}{2} = 2.97$ m and 7.07 m.

The shear force and the bending moment diagrams for the entire span of the girder are shown in Figure 2.39.

Note: The *SF* and *BM* for the right half have been drawn making use of symmetry.

EXAMPLE 2.14

The shear force diagram for a simply supported beam of 10m span is shown in Figure 2.40. Draw the corresponding load diagram. Proceed to calculate the maximum bending moment and its location.

Solution:

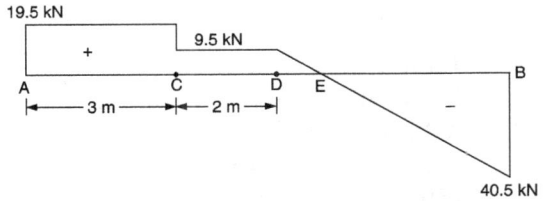

FIGURE 2.40.

i. Shear force of 19.5 kN at end A is equivalent to 19.5 kN reaction at end A.

ii. No variation in SF for the span $AC = 3$ m suggests that there is no load on the beam in this section.

iii. At C, there is an abrupt change in SF. Hence at B, there must be a concentrated vertical load of magnitude $(19.5 – 9.5) = 10$ N.

iv. No variation in SF for the span $CD = 2$ m suggests that there is no load on the beam in this region.

v. Since SF changes linearly from 9.5 kN to −40.5 kN in span DB of beam, there should be a uniformly distributed load in this region.

Intensity of load = slope of shear force diagram in span DB

$$= \frac{9.5-(-40.5)}{5} = 10 \text{ kN/m}$$

vi. Shear force of 40.5 kN at end B is equivalent to 40.5 kN reaction at end B.

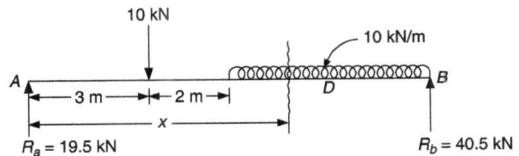

FIGURE 2.41.

Obviously then the load diagram for the given beam is plotted as shown in Figure 2.41.

Maximum bending moment occurs at a point where shear force is zero. That is

$$19.5 - 10 - 10(x - 5) = 0$$

∴ $$x = 5.95 \text{ m from end } A.$$

Bending moment at the chosen section,

$$M = 19.5x - 10(x - 3) - 10 \times (x - 5) \times \frac{x-5}{2}$$

$$M_{max} = 19.5 \times 5.95 - 10\,(5.95 - 3) - 10(5.95 - 5) \times \frac{(5.95-5)}{2}$$

$$= 116.02 - 29.50 - 4.51 = \textbf{82.01} \text{ kNm.}$$

EXAMPLE 2.15

Figure 2.42 shows the shear force diagram of a beam ABCD. Find the position of the supports and the magnitude of reactions.

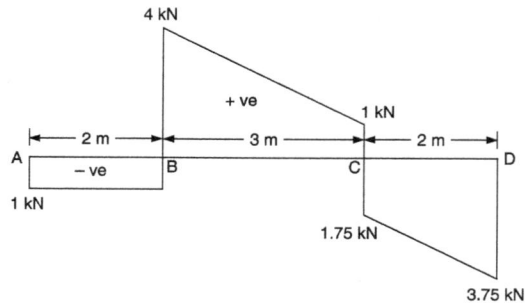

FIGURE 2.42.

Solution:

i. Corresponding to shear force of 1 kN at A, there must be point load of 1 kN at A.

ii. SF is constant from A to B, and obviously, there is no load on the beam in this region.

iii. SF changes sign at B, and accordingly, there will be a support at B with

$$\text{Reaction } R_b = 1 + 5 = 5 \text{ kN}$$

iv. Shear force consists of inclined line for the portion BC and that indicates the presence of *udl* in this portion.

$$\text{Intensity of load} = \text{slope of shear force diagram in span } BC$$

$$= \frac{4-1}{3} = 1 \text{ kN/m}$$

v. A point C, the SF changes from +1 kN to −1.75 kN, and accordingly, there is a point load at C.

$$\text{Magnitude of point load at } C = 1 + 1.75 = 2.75 \text{ kN}$$

vi. Shear force changes linearly form 1.75 kN to 3.75 kN in span CD of beam; there should be a uniformly distributed load in this region.

$$\text{Intensity of load} = \text{slope of shear force diagram in span } CD$$

$$= \frac{3.75-1.75}{2} = 1 \text{ kN/m}$$

vii. Shear force of −3.75 kN at end D is equivalent to 3.75 kN reaction at end D.

The load diagram for the given beam is plotted as shown in Figure 2.43

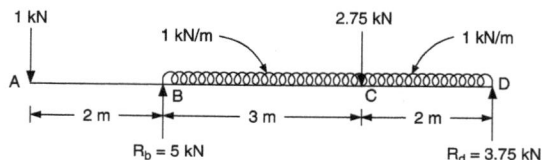

FIGURE 2.43.

EXAMPLE 2.16

The bending moment diagram of a simply supported beam is given below in Figure 2.44. Calculate the support reactions of the beam.

Solution:

i. Linear variation of bending moment in the sections AC, CD, and DB suggests that there is no load on the beam in these sections.

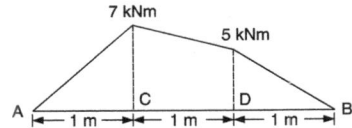

FIGURE 2.44

ii. Change in the slope of the bending moment at points C and D is indicative of the fact that there must be concentrated vertical loads at these points.

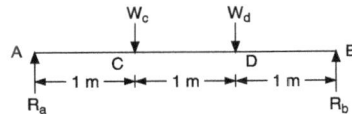

FIGURE 2.45.

With reference to Figure 2.45,
   R_a and R_b are the reactions at A and B
   W_c and W_d are the concentrated point loads at C and D.
   Moment at $C : R_a \times 1; R_a = 7$ kN
   Moment at $D : R_a \times 2 - W_c \times 1 = 5$

   or $7 \times 2 - W_c \times 1 = 5; W_c = 9$ kN.

Moment at $B : R_a \times 3 - W_c \times 2 - W_d \times 1 = 0$

   or $7 \times 3 - 9 \times 2 - W_d \times 1 = 0; W_d = 3.$

Considering equilibrium of forces in the vertical directions,

$$R_a + R_b = W_c + W_d$$
$$\therefore \quad R_b = W_c + W_d - R_a = 9 + 3 - 7 = 5 \text{ kN.}$$

Thus, the reactions at the end supports are:

$$R_a = 7 \text{ kN and } R_b = 5 \text{ kN.}$$

REVIEW QUESTIONS

A: Conceptual and conventional questions:

1. Define a beam. What is a cantilever, a fixed beam, and an overhang beam?

2. List the various types of loads to which a beam can be subjected.

3. Define shear force and bending moment.

4. What sign conventions are normally adopted while plotting the shear force and bending moment diagrams?

5. What are sagging and hogging moments?

6. Describe how you would proceed to draw shear force and bending moment in case of a

   i. simply supported beam loaded with a uniformly distributed load over its entire span

   ii. concentrated load being applied at the mid span of a simply supported beam.

7. What is the point of contraflexure?

8. Draw the shear force and bending moment diagrams for the simply supported overhanging beam loaded as shown in Figure 2.46. Also determine the location of the points of contraflexure, if any.

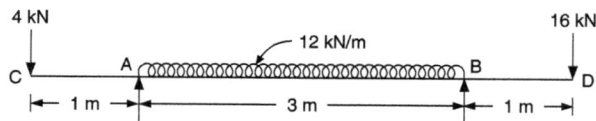

FIGURE 2.46.

[**Ans.** $R_a = 18$ kN, $R_b = 38$ kN, $x = 1.33$ m and 3.0 m from C]

9. Determine the reactions and construct the shear force and bending moment diagrams for the simply supported beam loaded as shown in Figure 2.47. Also determine the position and magnitude of maximum bending moment.

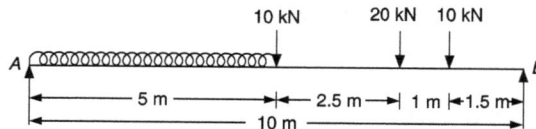

FIGURE 2.47.

[**Ans.** $M_{max} = 120.25$ kNm at 4.9 m from A]

10. For a symmetrically loaded overhang beam shown in Figure 2.48, make calculations for the value of load W such that the bending moment becomes zero at the mid-span of the beam.

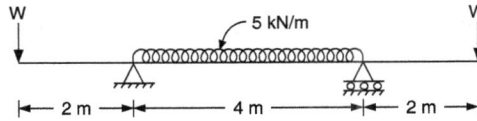

FIGURE 2.48.

[**Ans.** $W = 5$ kN]

11. Draw the shear force and bending moment diagrams for the beam loaded and supported as shown in Figure 2.49.

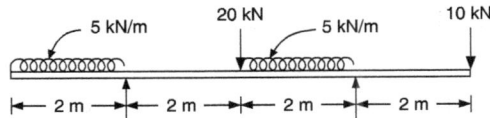

FIGURE 2.49.

12. Draw the shear force and bending moment diagram for the beam loaded as shown in Figure 2.50.

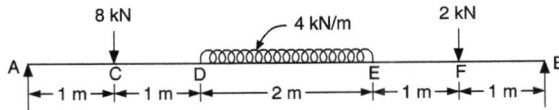

FIGURE 2.50.

13. Draw the shear force and bending moment diagrams for a cantilever beam loaded as shown in Figure 2.51 given below:

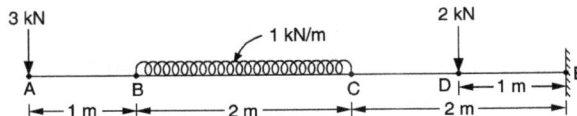

FIGURE 2.51.

Locate the position for maximum bending moment and determine its value.

B. Fill in the blanks with appropriate word/words

1. The shear force at certain section of the beam is stated to be zero. The bending moment at that section will be _____.

2. The bending moment for a certain portion of the beam is constant. For that section, shear force would be _____.

3. The shear force and bending moment diagrams for a cantilever beam carrying a concentrated load at free end will respectively be _____ and _____.

4. For a cantilever beam carrying several concentrated loads, the shear force is maximum at the _____.

5. A beam is said to be _____ if it has more than two supports.

6. The point of contraflexure in a loaded beam refers to the section where _____.

7. A simply supported beam has equal overhanging lengths and carries equal concentrated loads at ends. Bending moment over length between the supports is _____.

Answers:
1. either minimum or maximum; **2.** zero; **3.** rectangle and triangle; **4.** support; **5.** continuous; **6.** the bending moment changes sign; **7.** a non-zero constant.

C. Multiple choice questions

1. A simply supported beam of span l and carrying a load W concentrated at the mid span will have a maximum bending moment of

   (a) $\dfrac{Wl}{8}$ **(b)** $\dfrac{wl}{4}$ **(c)** $\dfrac{wl}{2}$ **(d)** Wl

2. For a beam of length l simply supported at its ends and carrying uniformly distributed load w per unit length, maximum bending moment occurs at the center of beam and is given by

   (a) $\dfrac{wl^2}{2}$ **(b)** $\dfrac{wl^2}{4}$ **(c)** $\dfrac{wl^2}{8}$ **(d)** $\dfrac{wl^2}{16}$

3. A simply supported beam carries two equal concentrated loads W at a distance $l/3$ from either support. The value of maximum bending moment anywhere in the section will be

(a) $\dfrac{Wl}{2}$ (b) $\dfrac{Wl}{3}$ (c) $\dfrac{Wl}{4}$ (d) $\dfrac{Wl}{6}$

4. If a simply supported beam carries a uniformly distributed load on the entire span, the shear force

(a) has a maximum value at the mid span.
(b) has a linear variation along the span.
(c) is constant at all the sections.
(d) is not zero at any section of the beam.

5. The maximum bending moment in a simply supported beam of length l loaded by a concentrated load W at the mid-point is given by

(a) Wl (b) $\dfrac{1}{2}Wl$ (c) $\dfrac{1}{4}Wl$ (d) $\dfrac{1}{8}Wl$

6. For the cantilever beam loaded as shown in figure given below:

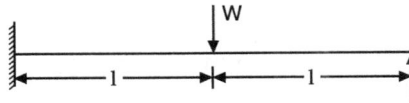

FIGURE 2.52.

The maximum bending moment will occur at
(a) mid-span of beam *i.e.*, under the load
(b) the free end supported with a prop
(c) between the fixed end and the mid-span
(d) the fixed end
(e) between mid-span and the propped free end.

7. The figure given below shows the shear force diagram for the beam $ABCD$
   The bending moment in the portion BC of the beam

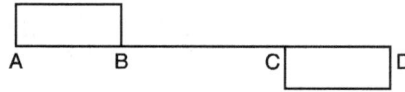

FIGURE 2.53.

(a) is non-zero constant.
(b) is zero.
(c) varies linearly from B to C.
(d) varies parabolically from B to C.

8. A simply supported beam is loaded as shown in the figure given below

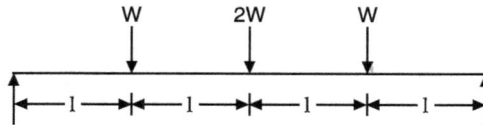

FIGURE 2.54.

The maximum shear force in the beam will be
(a) zero   (b) W (c) $2W$ (d) $4W$

Answers:

1. (b)	**2.** (c)	**3.** (b)	**4.** (b)
5. (a)	**6.** (a)	**7.** (a)	**8.** (c)

CHAPTER 3

INTERNAL COMBUSTION ENGINES

3.1. HEAT ENGINES

A machine or device which derives heat from the combustion of fuel and converts part of this energy into mechanical work is called a heat engine. Heat engines are broadly classified into internal combustion engines and external combustion engines.

An internal combustion engine is a reciprocating heat engine in which fuel mixed with the correct amount of air is burned inside a cylinder. The gaseous products of combustion form the working substance which makes the piston move and produces mechanical work at the engine crankshaft. In contrast, the combustion of fuel in steam engines is external. Combustion takes place on the fire grate of the boiler, and the heat energy of fuel, thus, released is used to convert water into steam. The steam is then led to the steam engine or steam turbine where work is produced. Obviously, the working substance (steam) is generated in a boiler that is outside the power-producing device.

Compared to steam engines, the IC engines are noted for

— high overall efficiency: the efficiency of IC engines ranges from 30 to 35% whereas the efficiency of steam engines lies between 15 and 20%,

— compact and small size,

— low weight to power ratio,

— easy and quick starting: in steam engines, firing of the boiler and generation of steam takes sufficient time, and

— less maintenance and operating costs.

The important applications of IC engines are as follows:

- road vehicles, locomotives, ships, and aircraft. As such IC engines enable passengers and cargos to cross lands, oceans, and skies,

- portable stand by units for power generation in case of scarcity of electric power, and

- extensively used in farm tractors, lawn movers, concrete mixing devices, and motorboats.

3.2. CLASSIFICATION OF IC ENGINES

The IC engines are classified on the basis of the following systems and their variations:

- *Number of strokes required for the completion of one cycle*

  i. Two-stroke engines in which the engine cycle is completed in two strokes of the piston, that is, in one revolution of crankshaft.

  ii. Four-stroke engines in which the engine cycle is completed in four strokes of the piston, that is, in two revolutions of crankshaft.

- *Thermodynamic cycle:* The thermodynamic cycles commonly used are as follows:

  i. Constant volume combustion (Otto) cycle: Most of the petrol and gas engines work on this cycle.

  ii. Constant pressure combustion (diesel) cycle: Low-speed diesel engines work on this cycle.

  iii. Mixed or limited pressure (dual) cycle: The high-speed diesel engines work on this cycle.

- *Ignition system:* The following two methods are used for the ignition of fuel.

  i. Spark ignition: Petrol engines use a spark for the ignition of compressed charge (mixture of air and petrol), and the spark may be produced by magneto or battery.

  ii. Compression ignition: diesel engines have a high compression ratio. The resulting high temperature is utilized to burn the fuel.

- *Kind of fuel used:*

  i. Light oil engines using kerosene or petrol. Petrol engines fall under this category.

  ii. Heavy oil or diesel oil engines: The oil used may be crude oil or mineral oil.

  iii. Gas engines: The gas used may be coal gas, producer gas, blast furnace gas, or coke oven gas.

  iv. By-fuel engines: The gas is used as the main fuel and liquid fuel is used for starting purposes.

- *Number and arrangement of cylinders*

  i. In-line engines: all the cylinders are arranged with their axes parallel, and they transmit power to a single crankshaft.

  ii. V-engines: the engines contain two banks of cylinders connected to the same crank and crankshaft. The crankshaft length for V-type engines is half of that for in-line engines.

  iii. Radial engines: the cylinders are arranged radially and are connected to a single crankshaft. Radial engines occupy less floor area and have little balancing problem.

- *Fuel supply system*

  i. Carburetor engines: mixture of petrol and air is prepared in the carburetor and is supplied to the engine during suction stroke.

  ii. Solid injection or airless injection: a fuel pump is used to inject the fuel into diesel engines.

  iii. Air injection: fuel is supplied, under pressure, to the engine cylinder of diesel engines by using compressed air.

- *Cooling system*

  i. Water-cooled engines in which the heat from the cylinder walls is transferred to cooling water which is kept circulating in the water jackets provided in the cylinder block. The water picks up heat and is taken to the radiator where the heat is transferred to the surrounding air. The water is repeatedly returned to the engine after being cooled in the radiator. Medium- and large-sized engines and automobile engines use the water cooling system.

ii. Air cooling in which the heat from the cylinder walls is directly transferred to the surrounding air. Air cooling is generally employed for small-capacity engines like scooters and motorcycle engines.

- *Lubrication system:* Lubrication system refers to the act of reducing friction by introducing a substance (called lubricant) between the mating parts of the engine.

  i. Splash lubrication system suitable for small-capacity engines with moderate speed and bearing loads.

  ii. Pressure lubrication system used for heavy-duty engines.

- *Governing system* (speed control under variable load)

  i. Quality control engines in which composition of mixture (air–fuel ratio) is changed by admitting more or less fuel in accordance with variation in load on the engine. This method is used in diesel engines.

  ii. Quantity control engines in which the air–fuel mixture has a constant composition. However, the quantity of the mixture supplied is changed in accordance with load on the engine. This is used in petrol and gas engines.

- *Valve location:*

  i. overhead-valve engine

  ii. side-valve engine

- *Speed:* Engines having speeds above 900 rpm are called high-speed engines, and less than 400 rpm are called slow-speed engines.

- *Field of application:*

  i. Stationary engines used for small- and medium-capacity electric power plants, concrete mixers, and pumping units

  ii. Mobile engines installed in motor vehicles, airplanes, and ships.

3.3. ENGINE PARTS AND THEIR FUNCTIONS

- **Cylinder and cylinder head:** The cylinder is the main body of the engine wherein direct combustion of fuel takes place. The cylinder is a stationary component, and the piston reciprocates inside it. The cylinder head closes one end of the cylinder, and it is usually cast as one piece and is bolted to the top of the cylinder. It contains the valve seats and ports, and supports the valves and valve-actuated mechanism. Cylinders are usually made of ordinary cast iron. However, for heavy-duty engines, alloy steels are used. In the interest of weight saving, particularly in airplanes, use is made of aluminum and magnesium alloys.

- **Piston and piston rings:** A piston is a metal cup with its crown facing the combustion space. The function of the piston, together with the rings, is to confine the gases in the combustion space and, thus, transmit the full force of expansion to the connecting rod and crankshaft. The piston also acts as a bearing for the small end of the connecting rod. Pistons are usually made of gray iron or of aluminum alloys for high-speed engines. Aluminum has the advantage of low density (the density of aluminum is about two-fifths that of cast iron).

The leakage of gases between the walls of the piston and cylinder is prevented by means of three to six cast iron rings which may be square or rectangular in cross-section. These rings are inserted into the grooves provided on the piston. There are usually two sets of piston rings:

i. upper piston rings (called the compression rings) provide gas-tight seal and prevent the leakage of high-pressure gas.

ii. lower piston rings (called oil rings) provide an effective seal and prevent the leakage of oil into the cylinder head.

- **Connecting rod:** The connecting rod transmits the force given by the piston to the crank, causing it to turn and, thus, convert the reciprocating motion of the piston into rotary motion of the crankshaft. The rotary motion is required to make the wheels turn, a cutting blade spin, or a pulley rotate.

The connecting rod connects the piston at one end and the crank at the

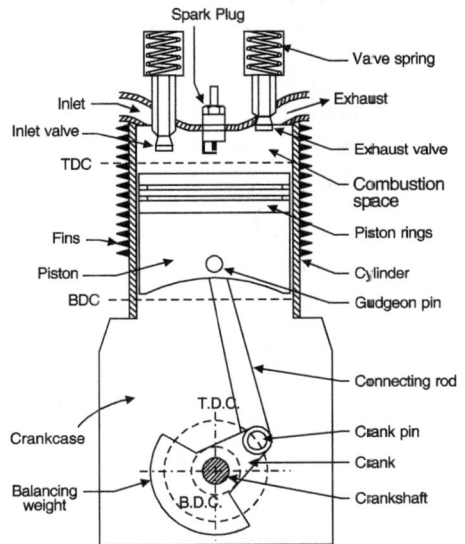

FIGURE 3.1 Nomenclature of engine components.

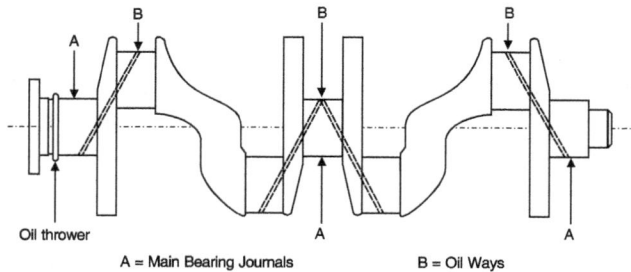

Oil thrower

A = Main Bearing Journals   B = Oil Ways

FIGURE 3.2 Crankshaft.

other end. The piston end is called the small end, and the crank end is called the big end. The connection at the small end is made by a pin called the gudgeon pin, wrist pin, or piston pin. At the big end, the connecting rod embraces the crank arm by a pin named crank pin.

The connecting rod is usually a small forging of I-section which provides the maximum stiffness with minimum weight. It is normally tapered along its length so as to provide a smaller cross-sectional area toward the small end. It has also a passage for the transfer of lubricating oil from the big-end bearing to the gudgeon pin.

- **Crank and crankshaft:** The reciprocating motion of the piston is converted into rotary motion by the connecting rod and crank mechanism. All the auxiliary mechanisms of the engine having mechanical transmission are geared in one way or the other to the crankshaft and obtain their motive power from it. The shape of the crankshaft, that is, the mutual arrangement of the cranks depends on the number and arrangement of cylinders and the turning order of the engine. Figure 3.2 shows a typical crankshaft layout for a four-cylinder engine.

  Both the crank and crankshaft are steel forgings machined to a smooth finish. The crankshaft mounts in bearings and can rotate freely. For smooth running, the crankshafts are perfectly balanced, both statically and dynamically.

- **Crankcase:** The engine cylinder, piston, and crankshaft are housed in the crankcase which also serves as an oil sump for the storage of the lubricating oil. The oil level is checked with the help of an oil stick or dip stick.

  The crankcase is generally made of cast iron.

- **Camshaft and valve mechanism:** The camshaft operates the intake and exhaust valves through the cams, cam followers, pushrods, and rocker arms. On a four-stroke engine, the inlet and exhaust valves operate once per cycle, that is, in two revolutions of the crankshaft. Consequently, the camshaft is driven by the crankshaft at exactly half its rotational speed.

The crankshaft material is commonly a steel forging with journals and cam faces case hardened.

The valves are usually mushroom-shaped (known as poppet valves) with conical seating surfaces. The poppet valve is practically universal on all modern car and commercial vehicle engines. The face of the valve and its seat on the cylinder is very accurately ground at an angle of 30° or 45°. Steel containing a small percentage of nickel and chromium is the usual valve material. Such an alloy has good heat-resisting qualities and is considered good enough to withstand high temperatures, mechanical forces, corrosive, and erosive effects of the high-velocity cylinder gases.

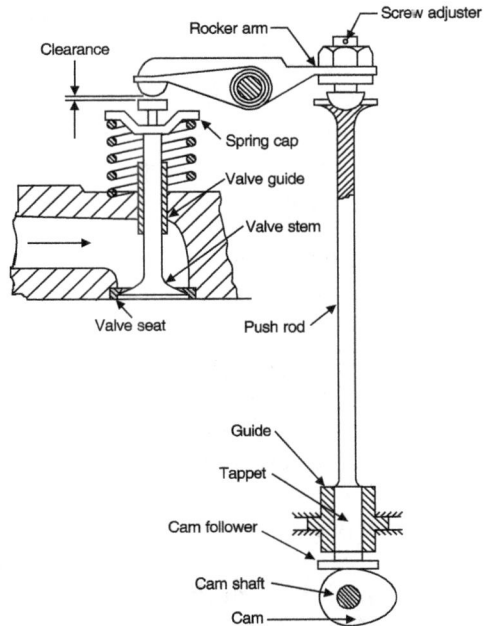

FIGURE 3.3 Overhead-valve mechanism.

The valves may be provided at the top or on the side of the engine cylinder. Figure 3.3 shows a typical overhead-valve assembly with the principle parts named. The cam lifts the pushrod through the cam follower, and the push rod actuates the rocker arm lever at one end. The other end of the rocker arm then gets depressed and that opens the valve. The valve returns to its seating by the spring after the cam has rotated. The valve stem moves in a valve guide which acts as a bearing. Some clearance is provided between the rocker arm and valve stem to take care of valve expansion during the running of engine, and it can be adjusted by the screw adjuster. The exhaust valve usually has a greater clearance as it runs hotter.

- *Flywheel:* The flywheel is a heavy and thoroughly balanced disk fitted onto the end of the crankshaft. It stores excess energy during the power stroke and returns this stored energy for use during the auxiliary strokes. Thus, it serves to reduce cyclic variations of speed and ensures uniform rotation of the crankshaft. The heavier the flywheel, the more uniform and stable will be the operation of the engine.

Flywheels are generally manufactured from cast iron or cast steel.

• **Governor:** A governor is used to adjust the power output from an engine in conformity with the external load and accordingly make the engine operate at a constant speed. The task is accomplished by regulating the quantity of charge in petrol engines and the amount of fuel in diesel engines.

All types of engines generally have the components described above. The components used either only for diesel engines or for petrol engines are given as follows:

Fuel pump and injector: In diesel engines, fuel pumps are used to deliver the correct quantity of fuel at the precise instant required for a wide range of loads and speeds. The nozzle atomizes the fuel and distributes it into the combustion chamber of the engine. Atomization of fuel means breaking the fuel stream into mist-like spray. Atomization ensures that each particle of fuel is surrounded by air needed for combustion and that assists in rapid and successful burning of fuel. Atomization is made possible by the high velocity of fuel through the nozzle which is due to the high pressure created by the pump.

Carburetor and spark plug: The carburetor delivers to the gas and petrol engines a combustible mixture of air and fuel in a condition that can be easily and efficiently burned in the engine cylinder. The process of mixture formation is called carburation.

Toward the end of the compression stroke, the combustible mixture is ignited by a spark plug which has to spark several thousand times a minute under a wide range of temperatures. Each cylinder is provided with its own spark plug screwed into the lid.

3.4. FOUR-STROKE PETROL ENGINE

A cycle is a sequence of operations constantly repeated, and "four-stroke" refers to the number of strokes of the piston required to complete one cycle. Refer to Figure 3.4 for the arrangement of different parts of a four-stroke cycle system. The piston reciprocating inside a cylinder is connected to the crankshaft through the connecting rod and the crank. The inlet (suction) and outlet (exhaust) valves are housed in the cylinder head. The cylinder head is also provided with an electric spark plug.

All events of the cycle namely suction, compression, combustion, expansion, and exhaust are completed in two revolutions of the crankshaft.

The salient features of the four strokes in a petrol engine are given as follows:

1. *Intake or suction stroke:* (Figure 3.5a) Initially, the piston is at top-dead center (TDC) position, the inlet valve is open, and the outlet valve is closed. The piston moves downward toward bottom-dead center (BDC) position, and the pressure inside the cylinder is reduced to a value below the atmospheric pressure. The vacuum, thus, created causes the charge to rush in and fill the space vacated by the piston. The charge consists of a mixture of air and petrol prepared by the carburetor.

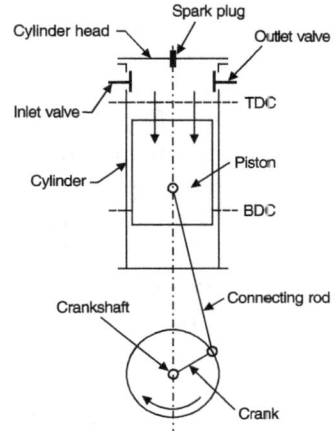

FIGURE 3.4 Four-stroke cycle system.

The suction continues till the piston reaches its BDC position. The piston has now made one stroke, and the crankshaft has turned through 180°C, that is, has made half the revolution.

2. *Compression stroke:* (Figure 3.5b) Both the valves (inlet and outlet) are closed and the movement of the piston is from BDC to TDC position. The charge inside the cylinder is compressed to the clearance volume; the volume decreases and there is a continuous rise both in temperature and pressure of the charge. A majority of the petrol engines use compression ratios between 5 to 1 and 8 to 1. Toward the end of compression, the approximate values of pressure and temperature are 6–12 bar and 250–300°C, respectively.

3. *Working, expansion, or power stroke:* (Figure 3.5c) When the piston reaches TDC position, the charge is ignited by causing an electric spark between the electrodes of a spark plug which is located in the cylinder head. During combustion, the chemical energy of the fuel is released and there is a rise both in pressure and temperature of the gases at almost constant volume. The temperature of the gases increases to about 1800-2000 °C, and the pressure reaches 30–40 bar.

With both valves closed, the gases at increased pressure and temperature expand, push the piston down the cylinder, and work is done by the system. The reciprocating motion of the piston is subsequently converted into the rotary motion of the crankshaft by connecting rod and crank. It is the rotary motion that is required to make wheels run, a cutting blade spin, or a pulley rotate.

During expansion, there is an increase in volume of the gases, and the pressure drops to as low as 3 bar.

4. *Exhaust Stroke:* (Figure 3.5d) The inlet valve remains closed but the exhaust valve opens when the piston reaches BDC position toward the completion of the power stroke. The pressure falls slightly above atmospheric pressure at constant volume. The piston moves upward from BDC to TDC, and this upward movement of the piston pushes the spent-up gases into the atmosphere through the exhaust valve and the exhaust manifold. Much of the noise associated with automobile engines is due to high exhaust velocity.

The exhaust stroke completes the cycle; the engine cylinder is ready to suck the fresh charge inside the cylinder once again, and the cycle is repeated.

Since the beginning of the suction stroke, the piston has made four strokes inside the cylinder: two up and two down. During the same period, the crank has turned two revolutions. Thus, for a four-stroke cycle, there is only one power stroke for every two revolutions of the crankshaft.

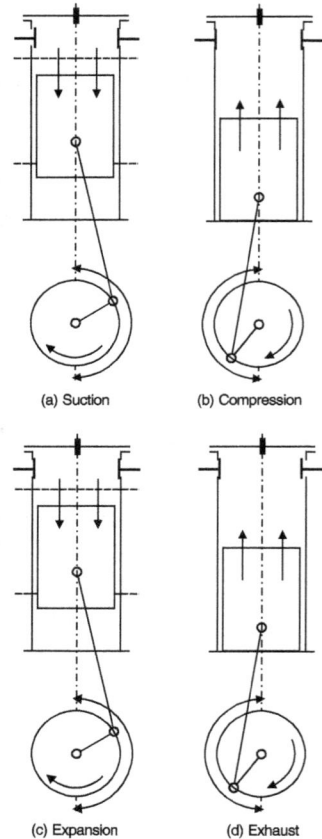

(a) Suction   (b) Compression

(c) Expansion   (d) Exhaust

FIGURE 3.5 Operation of a four-stroke cycle petrol engine.

Theoretical and Actual *p–V* Diagrams: The following assumptions have been made while carrying out the above operations:

i. Suction and exhaust are at atmospheric pressure.

ii. Opening and closing of the valves (both inlet and outlet) are instantaneous and at dead centers.

iii. Compression and expansion processes are isentropic, that is, reversible adiabatic.

iv. The combustion of fuel takes place instantaneously at constant volume.

v. There is a sudden drop in gas pressure to the atmospheric pressure at the end of the expansion stroke.

Such a theoretical operation of the cycle can be represented by the *p–V* plot as shown in Figure 3.6a.

However, an actual cycle deviates considerably form the hypothetical one because of the following reasons:

i. For efficient suction, the pressure inside the cylinder at suction is less than the atmospheric pressure. This pressure difference, called intake depression, is needed to overcome the resistance to the flow of charge through the restricted area of the inlet passages.

ii. For proper exhaust of burned gases, there has to be a net positive pressure from inside of the cylinder to outside. Accordingly, the exhaust line does not coincide with atmospheric pressure but is slightly above it.

iii. Compression and expansion are not isentropic. Friction is always present, and there is a considerable interchange of heat between the charge and the cylinder walls.

iv. There is always a time lag between the ignition of charge and its actual combustion. Consequently, the combustion does not take place at constant volume and the pressure rise is not along a straight line.

v. Opening and closing of the valves take some time and are never instantaneous. This is reflected in rounding off the corners of the p–V plot.

vi. Ignition of charge and opening of the valves are never at dead centers. These events occur at some degree on either side of dead centers to get better charging and scavenging (pushing out of burned gases) performance.

vii. At high temperature, there is all likelihood of dissociation of products of combustion such as CO_2 and H_2O (steam). This splitting is an endothermic process and requires heat. Further, the value of exponent γ decreases with a rise in temperature. Due to these aspects, the pressure and tem-

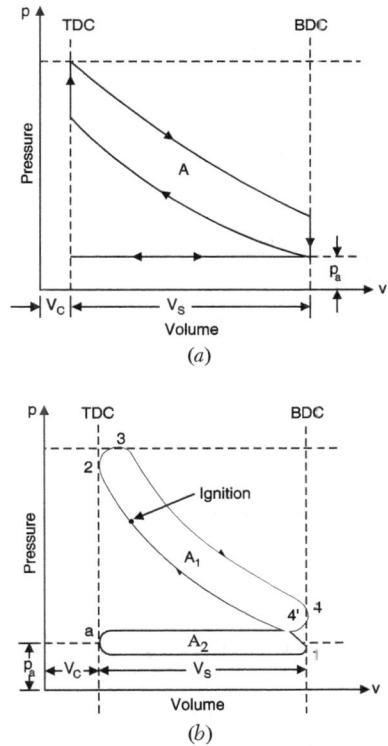

FIGURE 3.6 Theoretical and actual p–V diagram for four-stroke petrol engine.

perature attained during the actual cycle are lower than the theoretical values.

When these modifications are taken into account, the p–V plot takes the form as shown in Figure 3.6(b). The suction line a–1 lies below the atmospheric pressure line, and the exhaust 1–a lies slightly above the atmospheric line. The area enclosed by the exhaust and suction lines is called a negative loop or pumping loop.

This represents the work required for the admission of fresh charge and for the removal of burned gases. Network is obtained by subtracting the pump loss from the gross output Network per cycle $= (A_1 - A_2)$

The area $(A_1 - A_2)$ is always less than area A of the theoretical p–V diagram.

3.5. FOUR-STROKE DIESEL ENGINE

The engine completes its working cycle in four strokes of the piston, uses diesel oil as fuel, and therefore, is known as a four-stroke diesel engine. Figure 3.7 shows the different parts of the engine and illustrates its principle of operation. The piston reciprocates inside a cylinder and is connected to the crankshaft through the connecting rod and the crank. The inlet (suction) and outlet (exhaust) valves are housed in the cylinder head. The cylinder head is also provided with a nozzle for injecting the fuel.

The sequence of individual strokes is as follows:

1. *Intake or suction stroke:* Initially, the piston is at top-dead center (TDC) position, the inlet valve is open, and the outlet valve is closed. The piston moves downward toward BDC, and the pressure inside the cylinder is reduced to a value below the atmospheric pres-

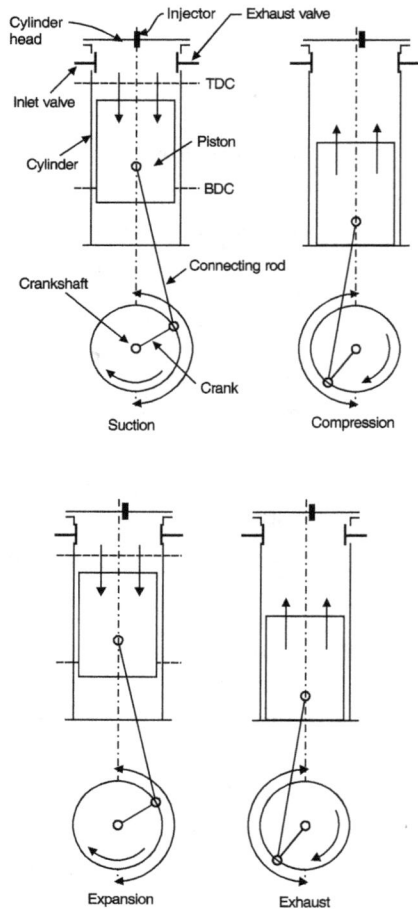

FIGURE 3.7 Operation of a four-stroke cycle diesel engine.

sure. The vacuum, thus, created causes the air from the atmosphere to rush in and fill the space vacated by the piston. The suction stroke is completed when the piston reaches the BDC position. The piston has then made one stroke, and the crankshaft has turned through 180°, that is, has made half the revolution.

2. *Compression stroke:* Both the inlet and outlet valves are closed and the movement of the piston is from BDC to TDC position (Figure 3.7 b). The air inside the cylinder is compressed to clearance volume; the volume decreases and there is a continuous rise both in temperature and pressure of the air. A majority of the diesel engines use compression ratios between 15:1 and 20:1. Toward the end of compression, the approximate values of pressure and temperature are 60 bar and 600 °C, respectively.

3. *Working expansion or power stroke:* When the piston reaches the TDC position, a fine spray of diesel is injected into the combustion space containing the high-temperature compressed air. The fuel vapors are raised to self-ignition temperature, and combustion occurs at approximately constant pressure.

The atomization of fuel and its supply to the combustion space can also be accomplished by compressed air supplied from compressed air bottles. The air entering the combustion space is so regulated that the pressure theoretically remains constant during the burning period.

With both valves closed, the combustion products at increased pressure and temperature push the piston down the cylinder with a large force. Expansion of the gases takes place, and work is done by the system. The reciprocating motion of the piston is subsequently converted into the rotary motion of the crankshaft by connecting rod and crank.

The expansion stroke gets completed as the piston reaches its BDC position. During expansion, there is an increase in the volume of the gases and the pressure drops.

4. *Exhaust stroke:* The inlet valve remains closed while the exhaust valve opens. The piston moves upward from BDC to TDC position, and this upward movement of the piston pushes the spent-up gases into the atmosphere through the exhaust valve and the exhaust manifold.

The exhaust stroke completes the cycle, then the engine cylinder is ready to suck fresh air inside the cylinder once again, and the cycle is repeated. Much of the noise associated with automobile engines is due to the high velocity of exhaust gases.

Since the beginning of the suction stroke, the piston has made four strokes inside the cylinder: two up and two down. During the same period, the crank has turned two revolutions. Thus, for a four-stroke cycle, there is only one power stroke for every revolution of the crankshaft.

The following assumptions have been made while carrying out the above operations.

i. Suction and exhaust are at atmospheric pressure.

ii. Opening and closing of the valves (both inlet and outlet) are instantaneous and at dead centers.

iii. Compression and expansion processes are isentropic, that is, reversible adiabatic.

iv. The combustion of fuel takes place at constant pressure during a small part of the expansion stroke.

v. There is a sudden drop in gas pressure to atmospheric pressure at the end of the expansion stroke.

Such a theoretical operation of the cycle can be represented by the p–V plot as shown in Figure 3.8(a).

However, the actual cycle differs considerably from the hypothetical one due to the reasons mentioned in Section 3.4. When those modifications are taken into account, then the plot takes the form as shown in Figure 3.8(b). The area enclosed by the exhaust and suction lines is called a negative loop or pumping loop, and the actual pressure inside the cylinder is slightly less than the atmospheric pressure during suction stroke and slightly higher than the atmospheric pressure during exhaust stroke (the corners are rounded off in the actual p–V plot) represents the work required for the admission of fresh charge and for the removal of burned gases. Network is obtained by subtracting the pump loss from the gross output

Network per cycle = $(A_1 - A_2)$

The area $(A_1 - A_2)$ is always less than area A of the theoretical p–V diagram.

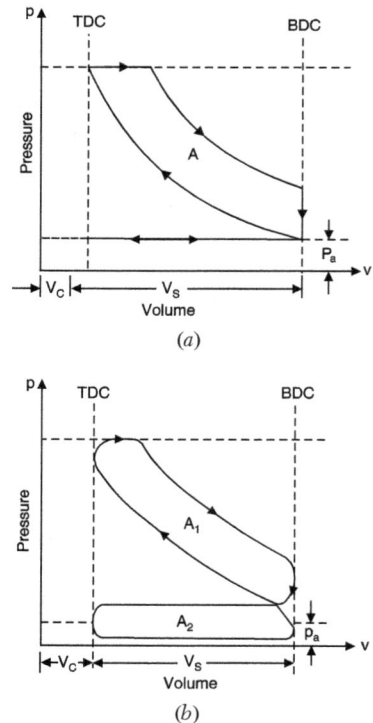

FIGURE 3.8 Theoretical and actual p–V diagram for four-stroke diesel engine.

3.6. COMPARISON BETWEEN PETROL AND DIESEL ENGINES

Below are given the differences between the construction and operation of a diesel engine and a petrol engine of similar capacity and number of cylinders.

i. *Basic cycles:* The petrol engine works on the Otto cycle whereas a diesel engine works on the diesel cycle.

ii. *Fuel used:* Petrol engine is a light oil engine and uses gasoline or petrol as fuel. Diesel engine is a heavy oil engine and uses diesel as fuel. The diesel has a high self-ignition temperature compared to petrol

iii. *Induction of fuel:* Mixture of petrol and air in required strength is prepared in the carburetor and inducted into the engine cylinder during the suction stroke of a petrol engine.

During the suction stroke of a diesel engine, only air from the atmosphere is sucked into the engine cylinder. A fuel pump is used to inject fuel directly into the combustion space where it meets the air which has been compressed.

iv. *Compression ratio:* A majority of petrol engines use a compression ratio between 5 to 1 and 8 to 1. The upper limit is fixed by anti-knock rating of fuels.

The compression ratio in diesel engines lies between 15:1 and 20:1; the upper limit is fixed because of an increase in the weight of the engine with an increase in compression ratio.

v. *Thermal efficiency:* For the same compression ratio, the efficiency of a diesel engine is lower than that of a petrol engine. However, this aspect is of not much practical significance since the petrol engines work with a compression ratio not exceeding 8 whereas diesel engines can safely have a compression ratio as high as 20. A high compression ratio for a diesel engine is a must not only for high efficiency but also to prevent the diesel knock – the phenomenon of uncontrolled and rapid combustion.

vi. *Ignition:* Petrol engines use a spark plug to ignite the charge (mixture of air and petrol) after it has been compressed. The combustion of fuel in diesel engines is due to the high temperature of compressed air.

vii. *Weight:* The cylinder walls of diesel engines have to be made thicker to sustain the high pressures attained due to higher compression ratios. The weight of a diesel engine amounts to between 30 and 50% more than that of a petrol engine giving the same power output.

viii. *Speed:* The petrol engines are high-speed engines due to light weight, and the diesel engines run at comparatively low speeds due to heavy weight.

ix. *Vibration and noise:* The vibration and noise level are higher with a diesel engine because of higher maximum pressure.

x. *Load control:* The petrol engines are quantity control engines. The air–fuel mixture prepared by the mixture being inducted is controlled by the throttle valve in accordance with load on the engine.

The diesel engines are quality control engines. The composition of the mixture (air–fuel ratio) is changed by admitting more or less fuel with variation in engine load. The fuel pump regulates the supply of fuel injected into the combustion space.

xi. *Power:* A petrol engine can be designated to give a better mean effective pressure mainly due to its improved combustion efficiency.

xii. *Cost:* The initial cost of a diesel engine is higher, mainly due to the cost of the fuel pump. Since diesel is cheaper than petrol, the running cost of a diesel engine works out to be low. Further, the experience of most operators is that a diesel engine requires less maintenance than a petrol engine. However, a larger battery is necessary for a diesel engine.

xiii. *Fire risk:* This is minimized in diesel engines owing to the higher ignition point of the fuel used.

xiv. *Applications:* Petrol engines are used in cars, scooters, and motor cycles.

Diesel engines are used in heavy-duty vehicles like trucks, buses, and locomotive engines.

3.7. TWO-STROKE SYSTEM

The working cycle is completed in two strokes of the piston or in one revolution of the crankshaft as against two crankshaft revolutions in a four-stroke cycle engine. The preparatory strokes (suction and exhaust) are combined with the working strokes (compression and expansion). The following two methods have been used to accomplish the desired objective.

i. Providing a separate pump outside the engine cylinder to compressible charge (air–fuel mixture from carburetor or air alone from the atmosphere) before forcing it into the cylinder. The pump is an integral part of the engine and gets its motive power from the engine itself. The arrangement is referred to as *two-channel system* and is used for large capacity multicylinder engines.

ii. Crankcase compression system where the crankcase works as an air pump as the piston moves up and down. The charge (air–fuel mixture or air alone) is compressed by the pumping action of the underside of the piston before being supplied to the engine cylinder. The

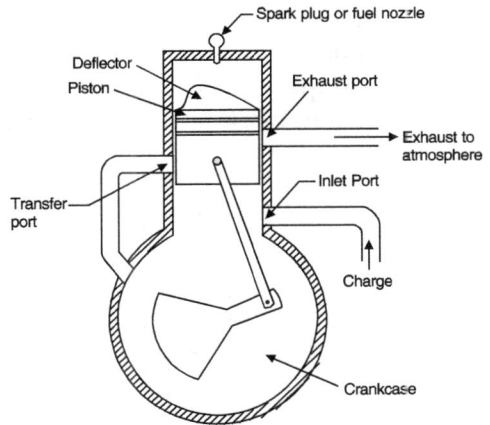

FIGURE 3.9 Two-stroke cycle system.

arrangement is referred to as *three-channel system* and is commonly used for single-cylinder small power engines such as scooters and motorcycle engines.

Construction: Figure 3.9 shows the arrangement of a typical three-port engine employing crankcase compression. The piston which is closely fitted in the cylinder is connected to the crankshaft through connecting rod and crank. The top of the piston is usually crown-shaped and that assists in sweeping the spent-up gases toward the exhaust port with the help of fresh charge. The engine employs ports against valves as provided in a four-stroke system. These ports are cut in the cylinder walls and are three in number: the transfer port, inlet, or induction port, and the exhaust port. The inlet and exhaust ports are located on one side, and the transfer port is provided on the other side. The cylinder top is provided with an electric spark plug-in a petrol engine or a nozzle for injecting the fuel in a diesel engine.

Working: The charge is led to the crankcase through the inlet port. The charge consists of a mixture of air and petrol prepared by the carburetor in the case of the petrol engine. The diesel engine admits only fresh air through the atmosphere. The transfer port takes the compressed charge from the crankcase to the engine cylinder. The spent-up gases are discharged to the atmosphere through the exhaust port. The closing and opening of the ports are controlled by up and down motion of the piston inside the cylinder. The piston crown helps prevent the loss of incoming fresh charge (charge being carried with the spent-up gases) and uses its momentum for exhausting only the burned gases. *Scavenging* is the term applied to the process of forcing the burned gases through the exhaust port by deflecting fresh charge across the cylinder.

Sequence of events: Figure 3.10 explains the working principle and sequence of events in a two-stroke cycle system.

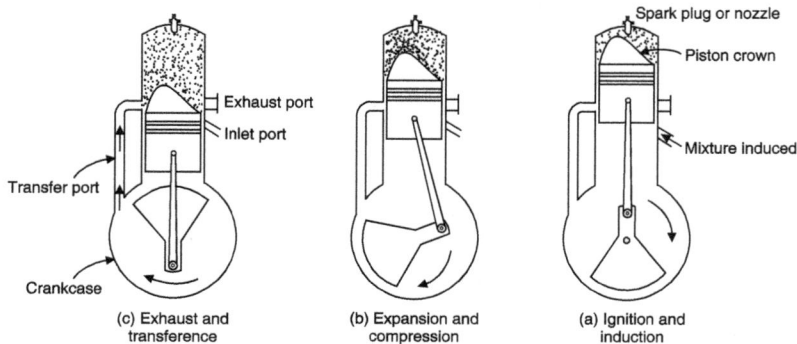

FIGURE 3.10 Working of two-stroke cycle engine.

Ignition and induction: In Figure 3.10(a), the piston occupies the almost TDC position toward the end of the compression stroke. The compressed charge is being ignited by providing a spark, or fuel is being injected into the hot compressed air. The combustion of fuel occurs, and thermal energy is released. There occurs a rise both in the pressure and temperature of combustion products.

At the same time, a partial vacuum (pressure lower than atmosphere) exists in the crankcase, and the fresh charge is being inducted into the crankcase through the inlet port which is uncovered by the piston.

Expansion and compression: (Figure 3.10b). The high-pressure gases push the piston down, expansion takes place and power is developed. With the downward movement of the piston, the charge in the crankcase gets compressed by the underside of the piston to a pressure of about 1.4 bar absolute.

After completion of about 80% of the expansion stroke, the piston uncovers the exhaust port. Some of the combustion products which are still above atmospheric pressure escape to the atmosphere. On its further downward motion, the piston uncovers the transfer port and allows the slightly compressed charge from the crankcase to be admitted into the cylinder via the transfer port.

Exhaust and transference: (Figure 3.10c). The piston lies at its BDC position. The expanded gases are escaping through the exhaust port, and simultaneously, the slightly compressed charge from the crankcase is being forced into the engine cylinder through the transfer port. The charge strikes the deflector on the piston crown, rises to the top of the cylinder, and pushes out most of the burned gases. During this scavenging action, a part of the fresh charge is likely to leave with the exhaust gases. The cylinder is completely filled with the fresh charge, although it is somewhat diluted due to its mixing with the burned gases.

When the piston moves upward from its BDC position, it first covers the transfer port and stops the flow of fresh charge into the cylinder. A little later, the exhaust port too gets covered and actual compression of the charge begins and continues till the piston reaches TDC position. The cycle of the engine is, thus, completed within two strokes of the piston (one up and one down) and one revolution of the crankshaft.

3.8. COMPARISON BETWEEN TWO-STROKE AND FOUR-STROKE ENGINES

Merits of two-stroke engines

i. For the same power output, a two-stroke engine is simple in design, easy to manufacture, and operate.

ii. A two-stroke cycle engine gives one working stroke for each revolution of the crankshaft. The four-stroke cycle engine gives one working stroke for every two revolutions of the crankshaft. As such, a two-stroke engine develops theoretically twice the power developed by four-stroke engine for the same engine speed and cylinder volume.

iii. The number of working strokes is twice that in a four-stroke engine. Consequently, the turning moment is uniform and, hence, the need only for a lighter flywheel.

iv. Less friction loss due to the absence of suction and exhaust strokes, consequently high mechanical efficiency. The absence of cams, camshaft, and rockers also contributes toward high mechanical efficiency.

v. Simpler construction and mechanism because of no valve and valve mechanism. The ports are easy to design, and they are covered and uncovered by the movement of piston itself.

vi. A two-stroke engine occupies less space, needs lighter foundations, and requires few spare parts.

vii. The reversing of a two-stroke engine can be achieved by a simple reversing gear mechanism.

viii. The initial cost of a two-stroke engine is less due to light weight and the absence of a valve mechanism.

ix. The two-stroke engines are much easier to start.

Demerits of two-stroke engines:

i. Scavenging (driving out of burned gases) is not complete due to the short time available for exhaust. This results in the dilution of fresh charge.

ii. Exhaust and inlet parts are uncovered (open) simultaneously during a certain period. Some fresh charge is likely to escape without giving any work output.

iii. Thermal efficiency of a two-stroke engine is likely to be lower due to some charge escaping without burning and poor scavenging. Consequently, fresh charge is diluted which results not only decrease in performance but also slow running, low combustion pressure, and poor efficiency.

iv. For the same stroke and clearance volume, the effective compression ratio is lower in a two-stroke engine. This too lowers the engine efficiency.

v. More wear and tear of moving parts due to double the number of power strokes.

vi. The piston gets overheated due to firing in each revolution of crankshaft. Higher temperatures make the cooling and lubrication requirements quite severe.

vii. Greater consumption of lubricating oil due to high operating temperatures.

viii. Noisy exhaust due to sudden release of burned gases.

3.9. APPLICATIONS OF TWO-STROKE ENGINES

Two-stroke engines are generally used where low cost, compactness, and light weight are the major considerations such as in scooters, motorcycles, mopeds, and other light vehicles. The two-stroke opposed piston diesel engines are quite suitable for marine installations (ship propulsion) where the engine room is small.

3.10. PERFORMANCE ANALYSIS OF IC ENGINE

Engine performance is an indication of the degree of success with which the engine does the job assigned to it, that is, the conversion of chemical energy contained in the fuel into useful mechanical work. The engine performance is expressed in terms of certain parameters and a study needs to be made on how those parameters are affected by the operating conditions, design concepts, and modifications. The desired engine performance depends on the power requirements of the apparatus or vehicle to which the engine has to be connected. A test run on the engine may be made

- to find the power output under certain conditions,

- to determine the fuel economy, and

- to obtain detailed information about the various losses and for preparing heat balance.

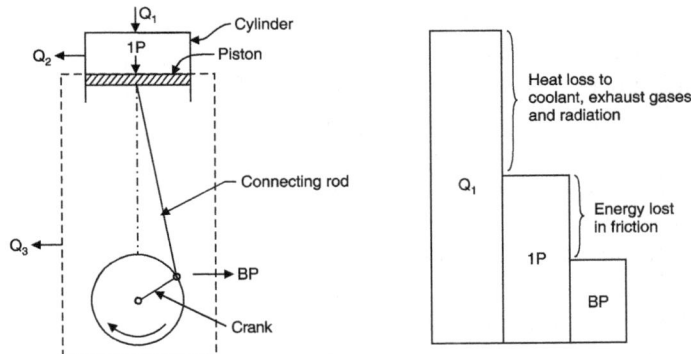

FIGURE 3.11 Energy flow in an IC engine.

Refer to Figure 3.11 for the energy flow through an IC engine.

When the mixture of air and fuel supplied is burned inside the engine cylinder, the chemical energy gets converted into heat energy. The total heat, thus, generated cannot be converted into work, and a substantial part is

a. lost to cooling water flowing through the cylinder jackets,

b. carried by exhaust gases, and

c. lost to surroundings by way of radiation.

The remaining part of the energy exerts a force on the piston which moves and mechanical work is done. The power developed inside the engine cylinder due to the combustion of fuel is called *indicated power* (*IP*).

A part of the indicated power is lost due to friction between the piston and cylinder as well as in the bearings. The net output available at the output shaft is called *brake power* (*BP*).

The difference between indicated power and brake power is called *friction power* (*FP*).

3.11. MEASUREMENT OF INDICATED POWER AND BRAKE POWER

The *indicated power* is usually measured by taking the indicator diagram with a device called an indicator. An indicator diagram is the graphical representation of the pressure–volume changes during the working of the cycle.

The indicator consists of a piston-cylinder assembly that communicates with the cylinder head of the engine being tested. The piston slides inside the cylinder, and the piston rod is connected to the straight line linkage through spring of proper stiffness. The spring controls the piston movement which is caused by any variation in pressure inside the engine cylinder. The higher the spring stiffness, the lesser the magnification is, and the lower the spring constant, the greater the magnification is. The movement of the indicator piston is transferred to a drum by the linkage at the end of which is attached with a stylus (pencil). The stylus moves in the vertical direction in proportion to the piston movement and records the gas pressure. The drum is made to rotate about its axis by a pulley and a cord connected to the engine piston through a reducing mechanism. The motion of the cord is proportional to the piston stroke.

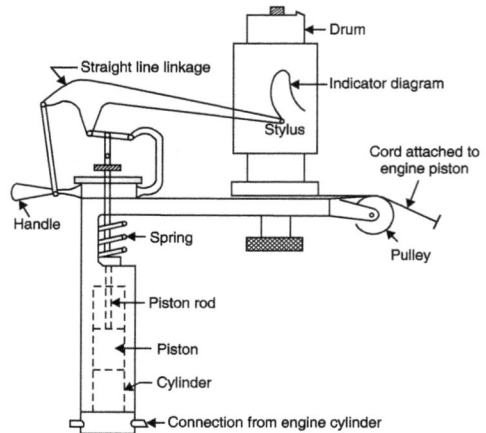

FIGURE 3.12 Engine indicator.

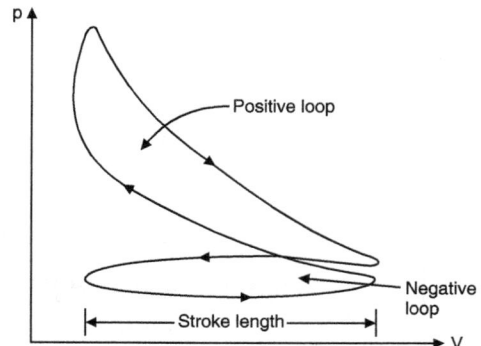

FIGURE 3.13.

Obviously, a diagram would be obtained on a strip of paper wrapped around the rotating drum. The vertical movement of the stylus and the horizontal movement of the cord combine to produce a closed figure known as an indicator diagram. The area of the diagram is a measure of the work developed, and the length of the diagram is proportional to the swept volume.

The indicator diagram has a positive loop (area between the compression and expansion lines) and a negative loop (area between the suction and exhaust lines). The positive loop gives the gross work done by the piston during the cycle, and the negative loop represents the pump loss due to admission of fresh charge and removal of exhaust gases.

Net work done per cycle = area of positive loop – area of negative loop

The area of the indicator card can be measured either with a planimeter or by the ordinate method. The indicated mean effective pressure (p_{mi}) is then given by

$$p_{mi} = \frac{\text{area of indicator diagram}}{\text{base length of indicator diagram}} \times \text{spring constant}$$

The mean effective pressure represents the constant pressure if it is acted over the full length of the stroke would produce the same amount of work as is actually developed by the engine during a cycle.

If the area is in mm^2, base length is in mm, and the spring constant is in N/mm^2 per mm of vertical movement of the indicator stylus, then mean effective pressure works out in N/mm^2.

Further, if p_{mi} = indicated mean effective pressure, N/m^2

A = area of the piston, m^2

L = length of stroke, m

and N = speed of the engine, rpm

Then, force on the piston = $p_{mi} \times A$ (newton)

work done per stroke = $p_{mi} LA$ (Nm)

work done per minute $p_{mi} LAN$ (Nm/min)

In a two-stroke cycle, the cycle is completed in two strokes of the piston or in one revolution of the crankshaft. In a four-stroke cycle, the engine cycle gets completed in four strokes of the piston or two revolutions of the crankshaft. Then

work done/minute = $p_{mi} LAN$ for two-stroke engine

$$= p_{mi} LA \frac{N}{2} \text{ for four-stroke engine}$$

Indicated power, $IP = \dfrac{p_{mi} LA(N \text{ or } N/2)}{60 \times 10^3} \times n \, kW$

where n is the number of cylinders in the engine.

The *brake power* (the useful power available at the engine crankshaft) is determined with the help of a dynamometer coupled to the engine crankshaft. A rope brake dynamometer consists of one or more ropes wrapped around the flywheel of the engine of which brake power is to be measured. The ropes are spaced evenly across the width of the rim by means of U-shaped wooden blocks located at different points of the rim of the flywheel. The upward ends of the ropes are connected together and attached to a spring balance, and the downward ends are kept in place by a dead weight. The rotation of the flywheel produces frictional force and the rope tightens. Consequently, a force is induced in the spring balance. The generation of heat is enormous and that necessitates a cooling arrangement for the brake.

The power of the engine is absorbed as frictional heat between the rope and pulley surface.

Let W = dead weight at the end of the rope, N

S = spring balance reading, N

N = engine speed, rpm

D = diameter of the brake wheel, m

d = diameter of the rope, m

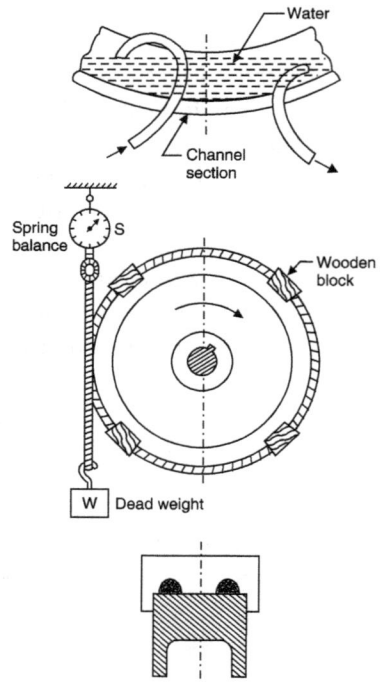

FIGURE 3.14 Rope brake dynamometer.

Then: effective radius of brake drum = $\dfrac{D+d}{2}$

$$\text{Torque, } T = (W - S) \times \frac{D+d}{2}$$

Work done per revolution = torque × angle turned per revolution = $2\,\pi T$

Work done per minute = $2\pi T \times N$

Brake power $(BP) = 2\pi T \times \dfrac{N}{60}\ N\ \text{Nm/s or W}$

$$= \frac{2\pi NT}{60 \times 10^3}\ \text{kW}$$

3.12. PERFORMANCE PARAMETERS

The engine performance is expressed in terms of certain parameters which have been enumerated and defined below:

Mechanical efficiency: The ratio of brake power to indicated power is called *mechanical efficiency.*

$$\text{Mechanical efficiency } \eta_m = \frac{BP}{IP}$$

The mechanical efficiency is a measure of the mechanical perfection of the engine or its ability to transmit the power developed in the cylinder to the drive shaft. The mechanical efficiency essentially depends on the design of the engine, its condition and load, and its value varies from 75 to 90 % for different types of engines.

Thermal efficiency: Thermal efficiency of an engine is the indicator of the conversion of heat supplied into work energy. It is based either on indicated power or on brake power. The indicated thermal efficiency of an IC engine is the ratio of the heat converted into indicated work to the heat energy in fuel supplied to the engine.

Indicated thermal efficiency

$$\eta_{ith} = \frac{\text{indicated power in watt}}{m_f \times CV}$$

where

mf = mass of fuel supplied

CV = lower calorific value of fuel in kJ/kg

Brake thermal efficiency is the ratio of the heat converted into useful shaft work to the heat supplied to the engine in fuel.

Brake thermal efficiency,

$$\eta_{bth} = \frac{\text{brake power in watt}}{m_f \times CV}$$

From the expression for η_{ith} and η_{bth} as given above, we can write

$$\frac{n_{bth}}{\eta_{ith}} = \frac{BP}{IP} = \eta_m$$

$$\eta_{bth} = \eta_m \times \eta_{ith}$$

The value of brake thermal efficiency usually varies from 25 to 30 % for spark-ignition engines and 30 to 40 % for compression ignition engines.

Relative efficiency: The relative efficiency or efficiency ratio is defined as the ratio of indicated thermal efficiency to the thermal efficiency of a theoretical reversible cycle operating with the same compression ratio. The relative efficiency for various cycles usually ranges from 60 to 80 %.

Volumetric efficiency: The engine output depends upon the maximum amount of charge taken in during the suction stroke. Volumetric efficiency is essentially an indication of the breathing capacity of the engine, and it is defined as the ratio of actual volume of the charge (reduced to NTP) drawn in during the suction stroke to the swept volume of the engine.

The average value of the volumetric efficiency of an engine ranges from 70 to 80%.

Specific output: The specific output of an engine is defined as the brake power per unit of piston displacement.

$$\text{specific output} = \frac{BP}{A \times L} = \text{constant} \times p_{mb} \times \text{rotational speed}$$

Thus, for a given displacement of the piston, the brake power can either be increased by increasing the speed of the engine or by increasing the brake mean effective pressure.

Specific fuel consumption: The fuel consumption and the evaluation of fuel cost are an important factor considered to have an indication of the relative economy of the engine under test when compared with test results on other engines. The criterion of economical power production is the specific fuel consumption (sfc) which is defined as *the mass of fuel required to be supplied to an engine to develop unit* kW *power per hour.* That is

$$sfc = \frac{m_f}{BP} \text{ kg/kWh}$$

The value of *sfc* is approximately 1.47 kg/kWh for diesel engines and 1.85 kg/kWh for petrol engines.

Air–fuel ratio: It is defined as the ratio of the mass of air to the mass of fuel in the air–fuel mixture. This ratio remains practically constant for petrol engines over a wide range of operation. However, for diesel engines, the air–fuel ratio varies with load as the mass of air remains constant but the mass of fuel changes with load on the engine.

Specific weight: It is an indication of the engine bulk and is defined as the weight of engine in kg per unit of brake power developed.

3.13. MORSE TEST

The Morse test is applicable to multicylinder engines and is used for determining their indicated power without any elaborate arrangement in the context of indicators employed to plot the pressure–volume diagram. The test consists in rendering inoperative, in turn, each cylinder of the engine by shorting the spark plug if petrol engine is under test or by disconnecting an injector in case of diesel engine. The speed falls because of the loss of power with one cylinder cut out but is restored by reducing the load.

The procedure for conducting the Morse test on a multicylinder petrol engine is outlined as follows:

i.   Start the engine with the throttle fully open and with all cylinders working, take the dynamometer readings, and note the engine speed.

ii. Keeping the throttle opening fixed, render the first cylinder inoperative by shorting its spark plug. This will cause the engine speed to decrease. Restore the speed back to original value by adjusting the dynamometer load and note down the new dynamometer reading.

iii. Make the first cylinder operative and cut out the second, third, and fourth cylinder in turn.

iv. Repeat the procedure taking care that speed remains constant and the throttle opening is undisturbed while the test is being conducted.

When all of the four cylinders are working,

$$(BP)_4 = (IP)_4 - (FP)_4 \qquad\qquad (i)$$

When one cylinder is cut out, its contribution toward power development is lost and the speed of the engine falls. The brake load is reduced so that the speed of the engine is restored to its original value. This is to main frictional power (FP) constant which is supposed to be independent of load and directly proportional to speed. Now only three cylinders will deliver power, but still, all the four cylinders will consume friction power. That is

$$(BP)_3 = (IP)_3 - (FP)_4 \qquad\qquad (ii)$$

From expressions (i) and (ii),

$$(BP)_4 - (BP)_3 = (IP)_4 - (IP)_3$$

Thus, the difference between the measured brake power with four and three plugs will be the *IP* of the cylinder rendered inoperative. Similarly, the

IP of each cylinder can be found separately. The addition of *IP* of different cylinders would then give the *IP* of the engine as a whole.

The following errors are associated with the determination of indicated power by the Morse test:

i. The cutting off a cylinder affects the distribution of charge to the cylinders, and accordingly, the breathing capacity (volumetric efficiency) of the engine becomes different.

ii. The cutting off a cylinder causes pulsations in the exhaust manifold and that affects the performance of the engine.

iii. The pumping and mechanical friction losses are assumed to be same whether a particular cylinder is working or not, and this is not essentially true as became evident from the test runs made by Recardo.

3.14. HEAT BALANCE SHEET

The engine is supplied with a certain quantity of energy in fuel form. Only a fraction of this energy is converted into useful mechanical work which is available at the engine crankshaft and the remainder is lost. It is the purpose of energy balance to trace the energy distribution, that is, to determine how the input energy is distributed, where the energy leaves and how much. If during the test, it is revealed that at certain loading conditions, a certain form of energy loss is excessive, then a careful examination of that form is needed.

A heat balance test of an engine indicates the following items: each expressed in heat units.

Credit:

Heat released by fuel $Q_f = m_f \times CV$

where m_f is the fuel consumption and *CV* is the lower calorific value of fuel

Debit:

1. Heat equivalent of useful work or power available at the crankshaft (Qb)

2. Heat loss to cooling water

$$Q_w = m_w \times c_w \times \Delta T$$

where m_w is mass of cooling water, c_w = specific heat of water, and ΔT is the rise in the temperature of cooling water.

3. Heat carried by exhaust gases

$$Q_g = m_g \times c_g \times (t_g - t_a)$$

where m_g is the mass of exhaust gases, c_g is the mean specific heat of exhaust gases, t_g is the temperature of exhaust gases, and t_a is the temperature of ambient air.

The mass of exhaust gases is equal to the sum of air consumption of the engine and the fuel burned at the same time.

4. Unaccounted and radiation loss

This loss cannot be measured directly and is calculated by subtracting the items Q_b, Q_w, and Q_g from the item Q_f. That is

$$Q_{rad} = Q_f - (Q_b + Q_w + Q_g).$$

It may be pointed out that the heat equivalent of friction power is dissipated in cooling loss, exhaust loss, and radiation loss. Friction energy is a hidden part of these measurements, and accordingly, the separate inclusion of friction loss is omitted from the energy balance of internal combustion engines.

Heat in fuel is approximately distributed as follows:

$\frac{1}{3}$ in brake power produced, $\frac{1}{3}$ in cooling water, and $\frac{1}{3}$ in exhaust gases.

EXAMPLE 3.1

A 4-cylinder, two-stroke cycle petrol engine develops 30 kW at 2500 rev/min. The mean effective pressure on each piston side is 8 bar and mechanical efficiency is 80%. Calculate (*a*) the diameter and stroke of each cylinder if the stroke to bore ratio is 1.5 (*b*) the fuel consumption if the brake thermal efficiency is 28% and the fuel used has a calorific value of 43,900 kJ/kg.

Solution: Indicated power $IP = \dfrac{\text{brake power}}{\text{mechanical efficiency}} = \dfrac{30}{0.8} = 37.5 \text{ kW}$

Indicated power per cylinder $= \dfrac{37.5}{4} = 9.375 \text{ kW}$

The indicated power per cylinder is also given by

$$IP = 100 \frac{p_m LAN}{60}$$

where p_m is the mep in bar, L and A are the stroke length in m and piston area in m^2, and n is the number of working strokes per minute.

For a two-stroke engine, $n = N$ (the engine speed in rev/min)

$$\therefore 9.375 = \frac{100 \times 8 \times 15 D \times \dfrac{\pi}{4} D^2 \times 2500}{60} \qquad (\because L = 1.5)$$

Solution gives $D^3 = 2.388 \times 10^{-4}$ or $D = 0.0638$ m
$\therefore D = \mathbf{63.8}$ mm and $L = 63.8 \times 1.5 = \mathbf{95.7}$ mm

(b) Brake thermal efficiency $\eta_{bth} = \dfrac{BP}{m_f \times CV}$

That is, $0.28 = \dfrac{30}{m_f \times 43900}$; $m_f = 2.44 \times 10^{-3}$ kg/s $= 8.786$ kg/h

The brake specific fuel consumption is then

$$b\,s\,f\,c = \frac{m_f \text{ in kg/hr}}{BP \text{ in kW}} = \frac{8.786}{30} = \mathbf{0.293} \text{ kg/kW/h.}$$

EXAMPLE 3.2

The following particulars refer to a 4-cylinder, 4-stroke petrol engine;
bore = 65 mm; and stroke length = 95 mm
torque developed = 64 Nm when engine turns 3000 rev/min
clearance volume = 63 cm^3 for each cylinder
relative efficiency = 0.5 and calorific value of petrol = 42,000 kJ/kg
Make calculations for brake mean effective pressure and the fuel con-
sumption in kg/h.

Solution: Swept volume per cylinder, $V_s = \dfrac{\pi}{4} d^2 l$

Compression ratio, $r = \dfrac{V_s + V_c}{V_c} = \dfrac{315.08 + 63}{63} = 6$

Air standard efficiency, $\eta = 1 - \dfrac{1}{r^{\gamma-1}} = 1 - \dfrac{1}{(6)^{1.4-1}} = 0.512$

Then, Brake thermal efficiency = relative efficiency × air standard efficiency

$$\eta_{bth} = 0.5 \times 0.512 = 0.2562$$
$$\text{Brake power, } BP = 2\,\pi NT = 2\pi \times 3000 \times 64 = 1205760 \text{ Nm/min}$$
$$= 20096 \text{ Nm/s} \simeq 21.1 \text{ kW}$$

Brake power per cylinder $= \dfrac{21.1}{4} = 5.025$ kW

The brake power per cylinder is also given as follows:

$$BP/\text{cylinder} = \frac{100 p_m L\, A\, n}{60}$$

where p_m is mep in bar, L and A are the stroke length (m) piston area (m^2), respectively, n is the number of working strokes per minute.

For a 4-stroke engine, $n = \dfrac{N}{2}$ (N is the engine speed in rev/min)

$$\therefore \qquad 5.025 = \frac{100 p_m \times 0.095 \times \dfrac{\pi}{4}(0.065)^2 \times \dfrac{3000}{2}}{60}$$

Solution gives Brake mean effective pressure $p_m = 6.38$ bar

(b) Brake thermal efficiency, $\eta_{bth} = \dfrac{BP}{m_f \times CV}$

That is, $0.2562 = \dfrac{21.1}{m_f \times 42000}$; $m_f = 1.96 \times 10^{-3}$ kg/s = **7.06 kg/h**

EXAMPLE 3.3

The following data were recorded during a test run made on a single-cylinder, four-stroke engine having a compression ratio of 6.

Bore and stroke = 10 cm and 12.5 cm, respectively
Dead load and spring balance reading = 60 N and 20 N
Effective radius of brake drum = 40 cm
Fuel consumption = 1.2 kg/h
Calorific value of fuel = 42,500 kJ/kg
If the engine turns 2000 rev/min, and the indicated mean effective pressure is 0.25 MPa, determine its
(a) indicated power and brake power
(b) mechanical, overall, and relative efficiencies.

Solution: (a) For a four-stroke engine, the indicated power is given by

$$IP = p_{mi} LA \left(\frac{N}{2} \right) = \frac{\left(0.25 \times 10^6 \right) \times 0.125 \times \dfrac{\pi}{4}(0.1)^2 \times \dfrac{2000}{2}}{60 \times 10^3} = 4.088 \text{ kW}$$

$$\text{Brake power } BP = 2\pi NT = \frac{[2\pi \times 2000] \times [(60-20) \times 0.4]}{60 \times 10^3} = \textbf{3.349 kW}$$

(b) Mechanical efficiency

$$\eta_m = \frac{BP}{IP} = \frac{3.349}{4.088} = \textbf{0.879 or 81.9 \%}$$

$$\text{Overall efficiency } \eta_0 = \frac{\text{heat equivalent of brake power}}{\text{heat supplied in fuel}}$$

$$= \frac{BP}{m_f \times CV} = \frac{3.349 \times 10^3}{\left(\dfrac{1.2}{3600}\right) \times 42500} = \textbf{0.2364 or 23.64 \%}$$

Air standard efficiency,

$$\eta_{air} = 1 - \frac{1}{r^{\gamma-1}} = 1 - \frac{1}{(6)^{1.4-1}} = 0.5116$$

∴ Relative efficiency,

$$\eta_r = \frac{\eta_0}{\eta_{air}} = \frac{0.2364}{0.5116} = \textbf{0.462 or 46.2 \%}$$

EXAMPLE 3.4

A four-stroke petrol engine having six cylinders is to operate with compression ratio 6 and deliver 300 kW brake power when running at 2400 rev/min. Determine
(a) bore and stroke of the engine,
(b) fuel consumption in kg/hr, and
(c) specific fuel consumption in kg/kWh.
The following data may be assumed:
stroke = 1.25 times the bore; mechanical efficiency = 0.8
indicated mean effective pressure = 10 bar, relative efficiency = 0.5, and
calorific value of fuel used = 44 MJ/kg

Solution: (a) Indicated power = $\dfrac{\text{brake power}}{\text{mechanical efficiency}} = \dfrac{300}{0.8} = 375$ kW

Indicated power per cylinder = $\dfrac{375}{6} = 62.5$ kW

In terms of indicated mean effective pressure p_{mi}' the indicated power per cylinder is given by

$$IP = p_{mi} LA \frac{N}{2} \text{ for a four-stroke engine}$$

$$\therefore \qquad 62.5 = \frac{\left(10 \times 10^5\right) \times 1.25D \times \frac{\pi}{4}D^2 \times \frac{2400}{2}}{60 \times 10^3} = 19625\,D^3$$

That gives, $\qquad D^3 = \dfrac{62.5}{19625} = 0.003185$

or $\qquad D = (0.003185)^{\frac{1}{3}} = 0.1474 \text{ m} = 14.74 \text{ cm}$

and $\qquad L = 1.25 \times 14.74 = 18.425 \text{ cm}$

Thus, cylinder bore = **14.74** cm and stroke = **18.425** cm

(b) Air standard efficiency $\eta_{air} = 1 - \dfrac{1}{r^{\gamma-1}} = 1 - \dfrac{1}{6^{1.4-1}} = 0.512$

On the basis of indicated thermal efficiency, $\eta_r = \dfrac{\eta_{ith}}{\eta_{air}}$

$\therefore \qquad \eta_{ith} = 0.5 \times 0.512 = 0.256$

Using the relation, $\qquad \eta_{ith} = \dfrac{IP}{m_f \times CV}$, we have

$$10.256 = \frac{375 \times 10^3}{\left(\dfrac{m_f}{3600}\right)\left(44 \times 10^6\right)}$$

$\therefore \qquad$ Fuel consumed/hr, $m_f = \dfrac{375 \times 10^3 \times 3600}{0.256 \times 44 \times 10^6} = \textbf{119.85}$ kg/hr

(c) Specific fuel consumption

$$sfc = \frac{\text{fuel consumed/hr}}{\text{brake power}} = \frac{119.85}{300} \approx \textbf{0.4 kg/kWh}$$

EXAMPLE 3.5

A 4-stroke six-cylinder engine has a bore of 80 mm and stroke of 100 mm. While running at a mean speed of 12.5 m/s, its fuel consumption is 20 kg/hr and develops a torque of 150 Nm. Assuming a clearance volume of 75 cm^3 per cylinder, determine,

(a) brake power and brake mean effective pressure
(b) brake thermal efficiency if the calorific value of fuel used is 42.5 MJ/kg
(c) relative efficiency on the basis of brake thermal efficiency.

Solution: Mean piston speed $S = 2\,LN$

$$\therefore \qquad 12.5 = 2 \times 0.1 \times \frac{N}{60} \; ; N = 3750 \text{ rev/min}$$

(a) Brake power, $BP = 2\pi NT = \dfrac{2\pi \times 3750 \times 150}{60 \times 10^3} = \mathbf{58.875}$ **kW**

In terms of brake mean effective pressure (pmb), the brake power for an engine working a four-stroke cycle is

$$BP = p_{mb}\, LA\left(\frac{N}{2}\right) \times \text{ number of cylinders}$$

$$58.875 = \frac{p_{mb} \times 0.1 \times \dfrac{\pi}{4}(0.08)^2 \times \dfrac{3750}{2}}{60 \times 10^3} \times 6$$

Solution gives $\qquad p_{mb} = 625000 \text{ N/m}^2 = \mathbf{6.25}$ **bar**

(b) Brake thermal efficiency,

$$\eta_{bth} = \frac{\text{heat equivalent of brake power}}{\text{heat supplied in fuel}} = \frac{\text{brake power}}{m_f \times CV}$$

$$= \frac{58.875 \times 10^3}{\left(\dfrac{20}{3600}\right) \times \left(42.5 \times 10^6\right)} = \mathbf{0.249} \text{ or } \mathbf{24.9}\ \%$$

(c) Swept volume $V_S = \dfrac{\pi}{4}d^2 l = \dfrac{\pi}{4}(8)^2 \times 10 = 502.4 \text{ cm}^2$

Clearance volume, $V_C = 75 \text{ cm}^3$

Compression ratio, $r = \dfrac{V_s + V_C}{V_C} = \dfrac{502.4 + 75}{75} \approx 7.7$

Air standard efficiency,

$$\eta_{air} = 1 - \frac{1}{r^{\gamma-1}} = 1 - \frac{1}{(7.7)^{1.4-1}} = 0.558$$

$\therefore$ Relative efficiency $\eta_r = \dfrac{\eta_{bth}}{\eta_{air}} = \dfrac{0.249}{0.559} = 0.446$ or **44.6 %**.

EXAMPLE 3.6

Following data were obtained during the trial of a two-cylinder, two-stroke engine:

Bore and stroke = 100 mm and 150 mm

Area of positive and negative loops of indicator diagram = 6 cm^2 and 0.25 cm^2

Length of the indicator = 6 cm

Spring constant = 3.8 bar per cm

Net brake load and effective drum radius = 235 N and 0.42 m

Fuel consumption = 4.5 kg/h

If the engine turns 1600 rev/min and the fuel used has a calorific value of 43.5 MJ/kg, determine

(a) indicated power and brake power

(b) mechanical and thermal efficiencies.

Solution: Net area of indicator diagram = 6 − 0.25 = 5.75 cm^2

a. Indicated mean effective pressure,

$$p_{mi} = \frac{\text{area of indicator diagram}}{\text{length of indicator diagram}} \times \text{spring constant}$$

$$= \frac{5.75}{6} \times 3.8 = 3.642 \text{ bar}$$

For a two-stroke engine, the indicated power is given by

$$IP = (p_{mi} LAN) \times \text{number of cylinders}$$

$$= \frac{\left(3.642 \times 10^5\right) \times 0.15 \times \dfrac{\pi}{4} \times 0.1)^2 \times 1600}{60 \times 10^3} \times 2 = \textbf{22.87 kW}$$

Brake power, $BP = 2\pi NT = \dfrac{2\pi \times 1600 \times (235 \times 0.42)}{60 \times 10^3} = \textbf{16.43 kW}$

(b) Mechanical efficiency,

$$\eta_m = \frac{BP}{IP} = \frac{16.43}{22.87} = \textbf{0.722 or 72.2\%}$$

Indicated thermal efficiency,

$$\eta_{ith} = \frac{IP}{m_f \times CV} = \frac{22.87 \times 10^3}{\left(\dfrac{4.5}{3600}\right) \times \left(43.5 \times 10^6\right)} = \textbf{0.42 or 42\%}$$

Brake thermal efficiency,

$$\eta_{bth} = \frac{BP}{m_f \times CV} = \frac{16.52}{\left(\dfrac{4.5}{3600}\right) \times \left(4.35 \times 10^6\right)} = 0.303 \text{ or } 30.3\%$$

η_{bth} could also be worked out using the relation:

$$\eta_{bth} = \eta_m \times \eta_{ith} = 0.722 \times 0.42 = 0.303 \text{ or } 30.3\%$$

EXAMPLE 3.7

A four-cylinder four-stroke petrol engine with bore 60 mm and stroke 90 mm was tested at full throttle at constant speed. The fuel supply was fixed at 0.082 kg/min, and the spark plugs of the four cylinders were successively short-circuited without change of speed; the brake torque is being correspondingly adjusted. The brake power was measured with a dynamometer having a torque arm of 35 cm, and the measurements of brake loads at a rated speed of 3000 rev/min were as follows:

> **With all cylinders working - - 160 N**
> **With cylinder No. 1 cutout - - 113.5 N**
> **With cylinder No. 2 cutout - - 109 N**
> **With cylinder No. 3 cutout - - 106.5 N**
> **With cylinder No. 4 cutout - - 120 N**

Taking calorific value of fuel used as 44 MJ/kg, determine the engine torque; brake mean effective pressure and bake thermal efficiency; specific fuel consumption; mechanical efficiency; and indicated mean effective pressure.

(*b*) If the engine was tested in an atmosphere of 1 bar and 15° C and the air–fuel ratio was 15:1, determine the volumetric efficiency of the engine.

Solution: Torque, $T = W \times R = 160 \times 0.35 = \mathbf{56}$ Nm

$$\text{Brake power, } BP = 2\pi NT = \frac{2\pi \times 3000 \times 56}{60 \times 10^3} = 17.584 \text{ kW}$$

$$BP \text{ per cylinder} = \frac{17.584}{4} = 4.396 \text{ kW}$$

In terms of brake mean effective pressure p_{mb}, the brake power per cylinder can be expressed as follows:

$$BP = p_{mb} \, LA \, \frac{N}{2}$$

$$\therefore \quad 4.396 = \frac{p_{mb} \times 0.09 \times \dfrac{\pi}{4}(0.06)^2 \times \dfrac{3000}{2}}{60 \times 10^3} = 6.36 \times 10^{-6} \, p_{mb}$$

That gives $p_{mb} = \dfrac{4.396}{6.36 \times 10^{-6}} = 6.91 \times 10^5 \text{ N/m}^2 = \textbf{6.91}$ bar

Brake thermal efficiency,

$$\eta_{bth} = \frac{\text{brake power}}{m_f \times CV} = \frac{17.584 \times 10^3}{\left(\dfrac{0.082}{60}\right) \times \left(44 \times 10^6\right)} = \textbf{0.292 or 29.2 \%}$$

Specific fuel consumption,

$$sfc = \frac{\text{fuel consumption per hr}}{\text{brake power}} = \frac{0.082 \times 60}{17.584} = \textbf{0.279} \text{ kg/kWh}$$

Indicated load $= (160 - 113.5) + (160 - 109) + (160 - 106.5) + (160 - 120)$
$$= 46.5 + 51 + 53.5 + 40 = 191 \text{ N}$$

Mechanical efficiency,

$$\eta_m = \frac{\text{brake load}}{\text{indicate load}} = \frac{160}{191} = \textbf{0.838 or 83.8 \%}$$

Indicated mean effective pressure,

$$p_{mi} = \frac{p_{mb}}{\eta_m} = \frac{6.91}{0.838} = \textbf{8.246} \text{ bar}$$

(b) Mass flow rate of air $= 0.082 \times 15 = 1.23$ kg/m
Then from characteristic equation $pV = mRT$,

Volume of air drawn, $V = \dfrac{1.23 \times 287 \times (273 + 15)}{1 \times 10^5} = 1.017 \text{ m}^3\text{/min}$

Swept volume of the engine,

$$V_S = \left[\frac{\pi}{4} \times (0.06)^2 \times 0.09 \times \frac{3000}{2}\right] \times 4 = 1.526 \text{ m}^3\text{/min}$$

∴   Volumetric efficiency,

$$\eta_v = \frac{1.017}{1.526} = \textbf{0.666 or 66.6\%}$$

EXAMPLE 3.8

In a test on a single-cylinder oil engine with bore 30 cm, stroke 45 cm, and working on the four-stroke cycle, the following data were obtained:

Net brake load and effective drum diameter = 1650 N and 150 cm
Engine speed and duration of test = 200 rpm and 1 h
Fuel consumption = 7.5 kg with calorific value 42 MJ/kg
Inlet and outlet temperatures of cooling water = 15°C and 60°C
Total mass of jacket cooling water = 450 kg
Total air consumption = 350 kg
Inlet air and exhaust gas temperatures = 20°C and 300°C
Mean specific heat of exhaust gases = 1.025 kJ/kg K
If the indicator diagram gives a mean effective pressure equal to 6 bar, calculate the indicated and brake power, the mechanical efficiency, the thermal efficiencies and prepare a heat balance chart for the engine trial.

Solution: Heat supplied in fuel

$$Q = \frac{7.6}{3600} \times (42 \times 10^3) = 88.67 \text{ kJ/s}$$

(a) Indicated power, $IP = p \, LA \, \dfrac{N}{2} \times$ number of cylinders

$$= \frac{6 \times 10^5 \times 0.45 \times \dfrac{\pi}{4}(0.30)^2 \times \dfrac{200}{2}}{60 \times 10^3} \times 1 = 31.79 \text{ kW}$$

Brake power, $BP = 2\pi NT = \dfrac{2\pi \times 200 \times (1650 \times 0.75)}{60 \times 10^3} = 25.9$

(b) Mechanical efficiency,

$$\eta_m = \frac{BP}{IP} = \frac{25.9}{31.79} = 0.815$$

Indicated thermal efficiency,

$$\eta_{ith} = \frac{31.79 \times 10^3}{\left(\dfrac{7.5}{3600}\right) \times \left(42 \times 10^5\right)} = \mathbf{0.36}$$

Brake thermal efficiency

$$\eta_{bth} = \frac{25.9 \times 10^3}{\left(\dfrac{7.5}{3600}\right) \times \left(42 \times 10^6\right)} = \mathbf{0.296}$$

(c) 1. Heat equivalent of brake power,

$Q_b = 27.63$ kJ/s

2. Heat lost to cooling water,

$$Q_w = m\, c_p\, dT = \frac{450}{3600} \times 4.186 \times (60 - 15) = 23.55 \text{ kJ/s}$$

3. Heat lost in exhaust gases

$$Q_g = m\, c_p\, dT = \frac{350 + 7.5}{3600} \times 1.025 \times (300 - 20) = 28.5 \text{kJ/s}$$

4. Heat lost by radiation and unaccounted for (by difference)

$Q_{rad} = 88.67 - (25.9 + 23.55 + 28.50)$

$= 10.72$ kJ/s

The heat balance account on second basis may then be tabulated as indicated below:

Credit	kJ/s	percent	Debit	kJ/s	percent
Heat supplied in fuel	88.67	100	Brake power	25.90	29.21
			Cooling water	23.55	26.56
			Exhaust gases	28.50	32.14
			Radiation and unaccounted for	10.72	12.09
	88.67	100		88.67	100

3.15. ELECTRIC AND HYBRID VEHICLES

An electric vehicle is a vehicle that uses one or more electric motors for propulsion. Electric motors are the replacement for internal combustion engines that generated power by burning a mix of fuel and air. The electric motor is powered by a large battery pack placed in the car. Depending on its size, the battery can last for a range of 100–450 km, and then be recharged.

There are three main types of electric vehicles classified by the degree that electricity is used as their energy source.

1. Battery-electric vehicle (BEV): These are fully electric vehicles with rechargeable batteries and do not have a petrol/diesel engine, fuel tank, and exhaust pipe. The batteries are charged from an external source.

2. Plug-in hybrid electric vehicle: Here recharging of the battery is done through both regenerative braking, and plugging to an external source of electric power.

3. Hybrid electric vehicle (HEV): The term hybrid refers to technology that uses two or more distinct power sources for the propulsion of a vehicle. The HEVs are powered both by the gasoline engine and electric motors. The electric energy needed for recharging the battery is generated by the car's own braking system. At start, the HEVs use the electric motor, and as the load or speed rises, the gasoline engine comes into operation. The two motors are controlled by an internal computer which ensures the best economy for the driving conditions.

The BEVs have the following advantages over vehicles run with gasoline engines as the power source:

- electric motors react quickly, they are very responsive and have very good torque,

- electricity is cheaper than gasoline and so low running cost,

- few parts; maintenance is less frequent and less expensive,

- quiet in operation and faster acceleration, and

- quicker repair and replacement of electric motor.

It is predicted that by the year 2040, electric vehicles will become the main stream, and cities will become less noise polluted and will also have much cleaner air.

REVIEW QUESTIONS

A. Conceptual and conventional questions

1. Differentiate between internal and external combustion engines.

2. List the various advantages of IC engines over external combustion (steam) engines.

3. Mention the various applications of IC engines.

4. How IC engines are classified?

5. State the purpose of the following parts of an IC engine: spark plug, piston rings, crank and crankshaft, camshaft and valve mechanism, fly wheel.

6. What is the function of compression and oil control rings provided on the piston of an IC engine?

7. Mention the material used for the following components of an IC engine: cylinder block, piston, piston rings, connecting rod, crank and crankshaft, valves, and flywheel.

8. Sketch the overhead-valve mechanism of an IC engine and label its different parts. How does it operate and what function does it perform?

9. Describe, with neat sketches, the sequence of events in the working of a four-stroke petrol/diesel engine.

10. Give a neat sketch of the theoretical and actual p–V diagrams for a four-stroke petrol/diesel engine. Describe briefly the factors which account for deviations between these two plots.

11. Give as many differences as you can between the construction and operation of a compression ignition engine and a petrol engine.

12. Compare a diesel engine with a petrol engine with special reference to maximum pressure, efficiency, power to weight ratio, cost, and load control. Give explanations.

13. On a two-stroke engine, is there a power stroke in each revolution of the crankshaft?

14. How does a two-stroke engine differ from a four-stroke engine?

15. Explain, with neat sketches, the sequence of events in the working of a two-stroke petrol/diesel engine.

16. Mention the relative merits and demerits of two-stroke engines when compared with four-stroke engines.

17. Define indicated power, brake power, and various efficiencies as applied to spark ignition and compression ignition engines.

18. Define specific fuel consumption and state its importance.

19. Sketch and explain the rope brake arrangement for determining the brake power of an engine.

20. Describe the Morse test for determining the indicated power of a multicylinder engine, state the assumptions made.

21. A four-cylinder four-stroke petrol engine develops 15 kW while running at 1000 rev/min. The mean effective pressure as estimated from the indicator card is 5.5 bar. Determine the bore and stroke of the engine, if the length of stroke is 1.5 times the bore.   [**Ans.** 88.5 mm, 132.75 mm]

22. The following data pertain to a test run on a four-stroke, four-cylinder spark-ignition engine.

Bore and stroke	:	6 cm and 9.5 cm
Clearance volume	:	53.5 cm^3
Heating value of fuel	:	44.1 MJ/kg
Relative efficiency based on brake thermal efficiency	:	50 %

When tested on load and running at 2800 rev/min, the engine develops 60 Nm torque. Taking mechanical efficiency is equal to 80 %, determine the specific fuel consumption and brake mean effective pressure.

[**Ans.** 0.4 kg/kWh, 7.02 bar]

23. The following data refer to a test on a single-cylinder oil engine working on four-stroke cycle:

Diameter of brake wheel and rope diameter	:	60 cm and 3 cm
Dead load and spring balance reading	:	250 N and 50 N
Bore and piston stroke	:	10 cm and 15 cm
Speed of engine	:	400 rev/min
Length and area of indicator diagram	:	6 cm and 4 cm^2
Spring stiffness	:	12 N/cm^2 per cm
Fuel consumption	:	0.32 kg/kWh
Calorific value of fuel	:	43950 kJ/kg

Determine brake power, indicated power, mechanical efficiency, and indicated thermal efficiency of the engine. [**Ans.** 2.64 kW, 3.14 kW, 84%, 30.4%]

24. A two-stroke diesel engine has a bore 100 mm and a stroke of 150 mm. Running at a mean speed of 5m/s, it develops a torque of 60 Nm. The engine has a mechanical efficiency of 80 percent, and the indicated thermal efficiency is 40%.

Assuming a calorific value of fuel used as 45000 kJ/kg, determine the indicated power, the indicated mean effective pressure, and the fuel consumption per kWh of brake power. **[Ans.**7.85 kW, 4 bar, 0.25 kg/kWh]

25. A four-stroke petrol engine with four cylinders, coupled to a hydraulic dynamometer was tested at full throttle at constant speed. For each cylinder, the bore and stroke length were 80 mm and 100 mm, respectively. Fuel was supplied at the rate of 5.5 kg/hr and the plugs of the four cylinders were successively short-circuited without the change of speed. The power measurements were as follows.

With all cylinders working	---	14.8 kW
With cylinder 1 cut-off	---	10.2 kW
With cylinder 2 cut-off	---	10.4 kW
With cylinder 3 cut-off	---	10.5 kW
With cylinder 4 cut-off	---	10.3 kW

The clearance volume for each cylinder was 100 cm^3, and the fuel used had a calorific value of 42000 kJ/kg. You are required to determine the mechanical efficiency, the indicated thermal efficiency, the air standard efficiency, and the relative efficiency. **[Ans.** 83.1%, 27.7%; 51.2%; 54%]

26. During the trial of a single-cylinder, four-stroke oil engine, the following results were obtained:

Cylinder bore and piston stroke	:	20 cm and 40 cm
Torque and speed	:	407 Nm and 250 rpm
Indicated mean effective pressure	:	6 bar
Oil consumption and its heating value	:	4 kg/hr and 43 MJ/kg
Cooling water flow rate	:	4.5 kg/min
Air used per kg of fuel	:	30 kg
Rise in temperature of cooling water	:	45°C
Temperature of exhaust gases	:	420°C
Room temperature	:	20°C

Mean specific heat of water and hot gases: 4.18 and 1 kJ/kg K

Determine the indicated power, brake power, and draw up the heat balance sheet for the test in kJ/hr **[Ans.** 15.71 kW, 10.65 kW]

B. Fill in the blanks with appropriate word/words

1. A four-stroke engine theoretically operates on _____ cycle.

2. The compression ratio of an IC engine is the ratio of _____ to _____.

3. The extreme positions of the piston in the cylinder of an IC engine are called _____.

4. The compression ratio varies from _____ in SI engines and from _____ in CI engines.

5. A two-stroke engine employs _____ cut in the cylinder walls instead of _____.

6. In a four-stroke cycle engine, the cycle is completed in _____ crank revolutions.

7. The top of the piston in two-stroke engine is _____.

8. A _____ provides the connection between small end of connecting rod and piston.

9. The specific fuel consumption is defined as the mass of fuel required to be supplied to an engine to develop _____

10. An engine indicator is used to measure the _____ of an engine.

Answers:
1. Otto; **2.** total cylinder volume, clearance volume; **3.** dead centers; **4.** 6-8, 16-20; **5.** ports, valves; **6.** two; **7.** crown-shaped; **8.** gudgeon; **9.** unit kW power per hour; **10.** indicated power.

C. Multiple choice questions

1. Which is a false statement in the context of the parts of an IC engine?

    (*a*) the piston is connected to the small end of connecting rod by a gudgeon pin

    (*b*) crank pin connects the big end of connecting rod to crank arm

    (*c*) the reciprocating motion of the piston is converted to the rotary motion by the connecting rod and crank mechanism

    (*d*) the crankshaft provides the power and direction to the suction and exhaust valves

2. What is the main shaft of an engine that controls the movement of piston?

(**a**) axle   (**b**) drive shaft   (**c**) crankshaft   (**d**) camshaft

3. What is mounted on the crankshaft to transform rotary motion into back-and-forth motion for actuating the valves?

(**a**) crank   (**b**) crank and slider   (**c**) camshaft   (**d**) flywheel

4. In four-stroke engines, the camshaft is connected to the crankshaft by gears or chain and rotates at

(**a**) half the crankshaft speed
(**b**) three-fourth the crankshaft speed
(**c**) equal the crankshaft speed
(**d**) double the crankshaft speed

5. The slotted grooves on oil rings for pistons help

(**a**) seal the cylinder
(**b**) minimize friction
(**c**) prevent the piston from wear
(**d**) provide an escape for the oil that the slots edge out from the cylinder walls

6. The number of working strokes per minute in case of four-stroke IC engine will be equal to

(**a**) $N/2$ (**b**) N (**c**) $2N$ (**d**) $4N$

where N represents the revolutions turned by the crankshaft in one minute

7. For the same compression ratio and heat supplied, the air standard efficiency of an Otto cycle compared to that of a diesel cycle is

(**a**) less   (**b**) more   (**c**) equal   (**d**) unpredictable

8. Which is not the common component between a petrol and a diesel engine?

(**a**) camshaft (**b**) exhaust silencer   (**c**) spray nozzle (**d**) dynamo

9. Which is not a part of petrol engine?

(**a**) valve mechanism   (**b**) fuel injector
(**c**) induction coil   (**d**) air filter

10. Which is the correct statement in the context of two-stroke engines?

    (*a*) compression ratio is always lower than that of a four-stroke cycle engine
    (*b*) there is only one valve for inlet and exhaust
    (*c*) charge enters the engine cylinder through ports only
    (*d*) a diesel engine cannot operate on two-stroke cycle

11. Compared to four-stroke cycle engine, a two-stroke cycle engine

    (*a*) can be easily started
    (*b*) has lesser shocks and vibrations
    (*c*) has lower fuel consumption
    (*d*) is smaller in size for the same output

12. Which aspect is not true for a two-stroke cycle engine when compared with a four-stroke cycle engine?

    (*a*) cycle is completed in one stroke of the piston
    (*b*) uniform turning moment and, hence, a lighter flywheel
    (*c*) theoretically develops twice the power
    (*d*) noisy exhaust
    (*e*) more consumption of lubricating oil

13. A two-stroke engine has a speed of 750 rpm. A four-stroke engine having an identical cylinder size runs at 1500 rpm. The theoretical output of the two-stroke engine will

    (*a*) be twice that of the four-stroke engine
    (*b*) be half that of the four-stroke engine
    (*c*) be the same as that of the four-stroke engine
    (*d*) depends upon whether it is SI or CI engine.

14. Scavenging air means

    (*a*) air used under compression
    (*b*) air used for forcing the burned gases out of the cylinder during the exhaust period
    (*c*) forced air for cooling the engine cylinder
    (*d*) burned air-containing combustion products

15. The top of the piston in two-stroke engine is

    (*a*) flat   (*b*) slanted   (*c*) crown-shaped   (*d*) convex shaped

16. Morse test in multicylinder engines is done to determine

    (*a*) air flow to the engine   (*b*) volumetric efficiency
    (*c*) indicated power   (*d*) mechanical efficiency

17. The thermal efficiency of a well-designed and properly designed diesel engine usually lies in the range

    (*a*) 10-15%   (*b*) 20-30%   (*c*) 35-40%   (*d*) 45-60%

18. What percentage of heat supplied in the form of fuel in a four-stroke engine is carried away by exhaust gases?

    (*a*) 5–8%   (*b*) 10–15%   (*c*) 20–35%   (*d*) 40–50%

Answers:
1. (*d*)   2. (*c*)   3. (*c*)   4. (*a*)   5. (*d*)   6. (*a*)   7. (*b*)   8. (*c*)
9. (*b*)   10. (*b*) 11. (*a*) 12. (*a*) 13. (*c*) 14. (*b*) 15. (*c*) 16. (*c*)
17. (*c*)   18. (*c*)

CHAPTER 4

REFRIGERATION AND AIR CONDITIONING

4.1. REFRIGERATION: WHAT IS IT?

Science of providing and maintaining temperatures below that of surroundings

Refrigeration and air conditioning are a fascinating branch of science that deals with the chilling or freezing of a substance by removing some of its heat. This artificial withdrawal of heat produces within the substance or within a space a temperature below the general temperature of its surroundings. Refrigeration essentially means continued abstraction of heat from a substance (perishable foods, drinks, and medicines) at the low-temperature level and then transfers this heat to another system at a high potential of temperature. To accomplish this, mechanical work must be performed to satisfy the second law of thermodynamics.

Air conditioning refers to the simultaneous control of temperature, humidity, cleanliness, and air motion within a confined region or space.

A brief review is given in this chapter about the basic principles of certain refrigeration systems and the properties of primary and secondary refrigerants used in them. Mention also has been made of the eco-friendly refrigerants which have become a necessity to prevent depletion of the ozone layer.

4.2. HEAT ENGINE, REFRIGERATOR, AND HEAT PUMP

A heat engine is a thermodynamic device used for the continuous production of work from heat when operating in a cyclic process. Both heat and work interactions take place across the boundary of this cyclically operating device. Essentially a heat engine takes heat from the combustion of fuel and converts part of this energy into mechanical work.

A heat engine is characterized by the following features:

- reception of heat Q_1 from a high-temperature source at T_1,
- partial conversion of heat received to mechanical work W,
- rejection of remaining heat Q_2 to a low-temperature sink at temperature T_2,
- cyclic/continuous operation, and
- working substance flowing through the engine.

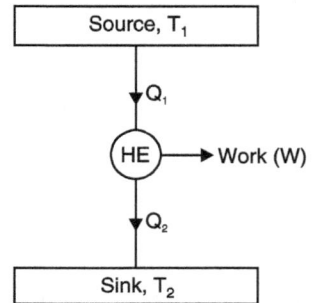

FIGURE 4.1 Energy interactions in a heat engine.

The performance of any machine is expressed as the ratio of "what we want" to "what we have to pay for." In the context of an engine, work is obtained at the expense of heat input. Accordingly, the performance of a heat engine is given by network output to the entire amount of heat supplied to the working medium, and this ratio is called **thermal efficiency**, η_{th}. (Thermal efficiency is a measure of the degree of useful utilization of heat received in a heat engine.)

$$\eta_{th} = \frac{\text{net work output}}{\text{total heat supplied}}$$

Application of the principle of energy conservation (First law) to the heat engine, which undergoes a cycle gives $W = Q_1 - Q_2$

$$\therefore \quad \eta_{th} = \frac{Q_1 - Q_2}{Q_1} = 1 - \frac{Q_2}{Q_1} \tag{4.1}$$

Obviously, the thermal efficiency of a heat engine operating between two thermal reservoirs is always less than unity. To increase the thermal efficiency, it is necessary to reduce Q_2 (heat rejected) with Q_1 (heat supplied) remaining constant. Thermal efficiency could be equal to unity if $Q_1 \to \infty$ and $Q_2 = 0$ which, however, cannot be realized in practice.

Refrigerators and heat pumps are **reversed** heat engines. The adjective reversed means operating backward. The direction of heat and work interactions are opposite to that of a heat engine, that is, work input and heat output. These machines (refrigerators and heat pumps) are used to remove heat from a body at a low-temperature level and then transfer this heat to another body at a high potential of temperature. When the main purpose of the machine is to remove heat from the cooled space, it is called a **refrigerator**. A refrigerator operates between the temperature of surroundings and a temperature below that of the surroundings. Refrigerators are essentially used to preserve food items and drugs at low temperatures.

The term **heat pump** is applied to a machine whose objective is to heat a medium that may already be warmer than its surroundings. A heat pump, thus, operates between the temperature of the surroundings and a temperature above that of the surroundings. Heat pumps are generally used to keep the rooms warm in winter.

The transfer of heat against a reverse temperature gradient in a refrigerator and heat pump is accomplished by supplying energy to the machine. A schematic representation of a heat pump and a refrigerator is shown in Figure 4.2.

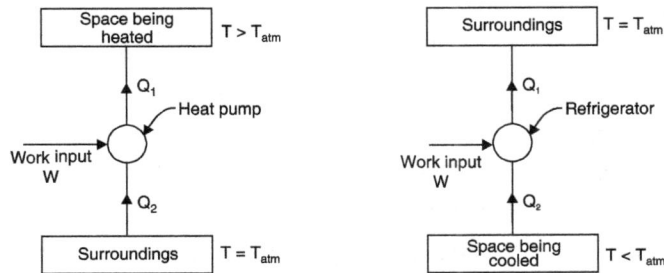

FIGURE 4.2 Functional difference between a heat pump and a refrigerator.

In the context of refrigerators and heat pumps, the performance is expressed in terms of **coefficient of performance (COP)** which represents the ratio of the desired effect to work input

$$COP = \frac{\text{desired effect}}{\text{work input}}$$

In a **refrigerator**, the desired effect is the amount of heat Q_2 extracted from the space being cooled, that is, the space at low temperature:

$$(COP)_{ref} = \frac{\text{heat extracted at low temperature}}{\text{work input}} = \frac{Q_2}{W}$$

From the principle of energy conservation,

$$W = Q_1 - Q_2$$

$$\therefore \quad (COP)_{ref} = \frac{Q_2}{Q_1 - Q_2}$$

For most of the refrigerating machines, the values of *COP* lie between 3 and 4, and the *COP* is greatest when temperature differences are least.

In a ***heat pump***, the desired effect is the amount of heat Q_1 supplied to the space being heated:

$$(COP)_{heat\ pump} = \frac{\text{heat rejected at high temperature}}{\text{work input}} = \frac{Q_1}{Q_1 - Q_2}$$

$$= 1 + \frac{Q_2}{Q_1 - Q_2} = 1 + (COP)_{ref}$$

Thus, the *COP* of a machine operating as a heat pump is higher than the *COP* of the same machine when operating as a refrigerator by unity.

Note: In the context of a refrigerator system:

1. The amount of heat extracted from the body at low temperature that is, the space being cooled is called *refrigerating effect*.

2. The *COP* of a refrigerator based on the theoretical values of refrigerating effect and work input is termed as *theoretical COP*. The theoretical refrigerating effect and work input are calculated by applying the laws of thermodynamics to the refrigeration cycle.

3. The *COP* of a refrigerator based on actual values of refrigerating effect and work input is termed as *actual COP*. The actual refrigerating effect and work input are obtained during the test run on a refrigerating plant.

4. The ratio of actual *COP* to theoretical *COP* is known as *relative COP.*

$$\text{Relative } COP = \frac{\text{actual } COP}{\text{theroretical } COP}$$

5. ***Refrigeration efficiency*** is defined as the ratio of *COP* of a cycle to the *COP* of a Carnot cycle operating in the same temperature range.

6. A single machine can fulfill both the functions of cooling and heating simultaneously. For example, cool a food storage space as a refrigerator and heat a water system as a heat pump.

4.3. RATING OR CAPACITY OF A REFRIGERATING UNIT

The refrigerating machines are usually rated in terms of their cooling capacity; the standard unit is *ton of refrigeration*.

One ton of refrigeration is defined as the refrigerating effect that freezes one ton (2000 pound mass) of liquid water during a period of 24 hours. The water is to be liquid at 0°C before and ice at 0°C after the process:

$$2000 \text{ pound mass} = \frac{2000}{2.205} = 907 \text{ kg.}$$

Enthalpy of fusion of water at 0°C = 333.43 kJ/kg

$$\therefore \quad 1 \text{ ton of refrigeration} = \frac{907 \times 333.43}{24 \times 60} = 210 \text{ kJ/min}$$
$$= 3.5 \text{ kJ/s} = 3.5 \text{ kW.}$$

A ton of refrigeration is, thus, not a unit of mass but a measure of the rate of cooling. The refrigerating capacity decides the mass flow rate of a given working substance (refrigerant) working under specified conditions:

$$\text{Mass flow rate of refrigerant} = \frac{\text{refrigeration capacity}}{\text{refrigerating effect per unit mass}}.$$

Quite often, power needed to produce a refrigeration effect equivalent of 1 ton of refrigeration is used as a measure to calculate the cost of operation or motor size of the refrigerating unit. Then,

$$\text{kW per ton of refrigeration} = \frac{3.5}{COP}.$$

Further, the ratio of heat removal rating (kJ/hr) of a refrigeration system to energy input (kWh) of the machine is called energy efficiency ratio (EER).

EXAMPLE 4.1

i. **The capacity of a refrigeration system is specified to be 12 tons. What is then the cooling rate of the machine?**

ii. **250 liters of drinking water is required per hour at 10°C. Would the use of 1.5 ton refrigerating system be justified if the available water is at 30°C?**

iii. **A refrigerating machine takes 1.25 kW and produces 25 kg/hr of ice at 0°C from water available at 30°C. Determine refrigerating effect, tonnage, and coefficient of performance of machine. Take**

Specific heat of water = 4.18 kJ/kg K.
Enthalpy of solidification of water from and at 0°C = 335 kJ/kg

Solution: (*i*) 1 ton of refrigeration $\equiv$ 3.5 kJ/s
$\therefore$ Cooling rate of machine = 12 × 3.5 = 42 kJ/s.

(*ii*) Refrigeration effect required for cooling the water

$$= mcp \, \Delta T = 250 \times 4.18 \times (30-10) = 20900 \text{ kJ/h.}$$

1 ton of refrigeration $\equiv$ 3.5 kJ/s = 12600 kJ/h.

$\therefore$ Tonnage required $= \dfrac{20900}{12600} = 1.658.$

As such, the use of 1.5 ton machine will not serve the purpose.

(*iii*) Refrigeration effect

$\equiv$ removal of heat from water at 30°C to convert into ice at 0°C

$$= m[cpw \, \Delta T + L] = 25[4.18(30-0) + 335] = 11510 \text{ kJ/h.}$$

1 ton of refrigeration $\equiv$ 3.5 kJ/s = 12600 kJ/h.

$\therefore$ Tonnage required $= \dfrac{11510}{12600} = 0.913$

$$COP = \frac{\text{refrigerating effect}}{\text{work input}}$$

$$= \frac{11510 / 3600}{1.25} = 2.558.$$

4.4. METHODS OF REFRIGERATION

Refrigeration is the technique of producing the cooling effect by abstraction/ withdrawal of heat so that temperature below that of surroundings is produced in a substance or within a space. The desired cooling effect can be produced by

1. Evaporation

When a liquid evaporates, it absorbs heat from the surroundings equivalent to its latent heat of vaporization, and that results in lowering the temperature of the surroundings. For example, we feel the cooling effect when there is the evaporation of a drop of spirit placed on the palm of hand. Likewise, the evaporation of moisture from the skin of a human body helps keep it cool. There is a common practice to cool the water for drinking purposes by keeping it in porous earthen pots. The water evaporates through the pores and that produces a cooling effect. The army people keep small water containers made of metal and covered with water-soaked namada; the walls of the metallic container get cooled and that cools the water kept inside it. In the refrigeration literature, there is mention of an experiment where a pump was used to create a partial vacuum over a container of ethyl ether. The liquid is then boiled by absorbing heat from the surrounding air. The cooling effect so created even produced a small amount of ice.

The principle of evaporative refrigeration is employed in desert (room) coolers. The dry atmospheric air is made to pass through water-soaked packings. When this water evaporates, it takes heat from the air causing it to cool.

With reference to Figure 4.3, a volatile liquid (liquid nitrogen, liquid carbon dioxide) contained in a flask evaporates and gets converted into gas. For evaporation, it absorbs heat from the chamber and a cooling effect is produced. The chamber is insulated to restrict the infiltration of heat from outside. The liquid N_2 and CO_2 are non-toxic and as such the liquid gas refrigeration finds application for keeping the perishable food articles cool when being transported.

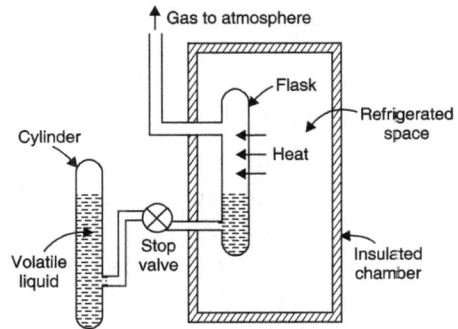

FIGURE 4.3 Liquid gas refrigeration.

2. Dissolution of salts in water

When certain salts are dissolved in water, they absorb heat and lower the temperature of water and create a sort of refrigeration bath for cooling substance. Sodium chloride lowers the water temperature up to $-20°C$ while calcium chloride up to $-50°C$. The salt can be recovered by evaporation of water from the solution.

The method of producing the cooling effect by the dissolution of salt in water could not become feasible for commercial purposes because (*i*)   the refrigeration effect produced is quite small and

(*ii*)   the process of regaining salt is cumbersome.

There has been a practice in France to produce cold drinks and liqueurs (a strong alcoholic drink with a sweet taste) by spinning long-necked bottles in water with dissolved saltpeter.

3. Ice refrigeration (change of phase)

The use of ice to refrigerate and, thus, preserve food goes back to prehistoric times and the ancient cultures of Chinese, Greeks, Romans, and Persians. Ice and snow were stored in caves or dugouts lined with straw or other insulating materials. This practice worked well down through the centuries, with ice houses remaining in use. Greater work was done on developing better insulation products for long-distance shipment of ice and the ice harvesting became a big business.

The natural ice or artificially produced ice is brought into contact with the substance to be cooled. The ice melts and the heat required for the melting of ice is supplied by the substance being cooled. The cooling effect produced by ice is

$$Q = \dot{m} \times h_{sf}$$

where $\dot{m}$ and h_{sf} are the rate of fusion of ice and enthalpy of fusion, respectively. At normal atmospheric pressure of 1.01325 bar (1 atm), h_{sf} is equal to 335 kJ/kg.

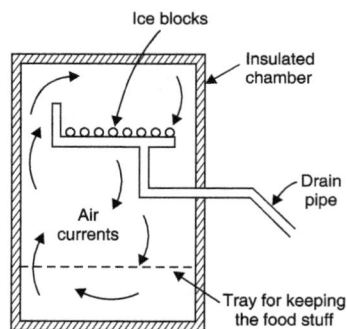

FIGURE 4.4 Ice refrigeration.

The ice refrigerator consists of a cabinet, which is completely insulated. The ice is kept in a container at the top, and a number of shelves are provided in the space below the ice container for storing the foodstuff. When air comes in contact with ice, it becomes cool, dense, and flows down over the shelves. It absorbs heat from the foodstuff which gets cooled. On absorption of heat, the air becomes warm. The warm air expands and returns back to the ice container from bottom, sides, and back of the cabinet. When this warm air flows past the ice, it gives its heat to the ice and gets cooled. On cooling, the air becomes dense and once again flows down over the food shelves. Temperatures in the range of 5 to 10° can be obtained with ice refrigeration. In case, temperatures below this range are required, and salt is mixed with ice and that results in reducing the temperature level to 0°C.

Ice refrigeration prevents dehydration and preserves the fresh appearance of eatable products like fruit and vegetables. However, ice has to be fed to the refrigerator and on and that is neither convenient nor economical. Further, water coming out of the refrigerator poses a problem in its disposal.

Although the ice-harvesting industry had grown immensely by the turn of the 20th century, pollution and sewage had begun to creep into natural ice, and eventually, breweries began to complain of tainted ice. This raised demand for more modern and consumer-ready refrigeration and ice-making machines.

4. Dry ice refrigeration (sublimation)

Solid carbon dioxide (called dry ice) has a peculiar characteristic that it changes from solid-state to vapor state without passing through the liquid state. During the change of phase, it absorbs heat equivalent to its latent heat of vaporization and produces a cooling effect. This process occurs when the solid is maintained below the triple point. Then,

$$Q = m\, h_{sv},$$

where h_{sv} is the enthalpy of evaporation.

At one atmospheric pressure, solid CO_2 produces 573 kJ/kg of refrigeration maintaining a temperature of $-78.5°C$. Dry ice is used to preserve foodstuff during transportation. The slabs of ice are usually packed on either side or on top of food packages in cartons. When dry ice evaporates, it absorbs heat from the foodstuff and preserves it in the frozen state.

The refrigeration methods (1) to (4) as mentioned above are called the **natural methods.** These methods are non-cyclic, and the temperatures attainable are limited. Further, there is continuous consumption of the refrigerating substance and that necessitates replenishment. However, these methods are sometimes convenient forms of cooling where small refrigeration is required such as in the laboratory and workshop.

5. Chemical methods

Here the heat required for the completion of the chemical reaction is taken from the substance being cooled.

The chemical method for producing a cooling effect cannot be followed on a commercial scale.

6. Air or gas refrigeration

Expansion of gas lowers its pressure and temperature. This cooling effect results without change in the phase of gas.

Consider air initially compressed is entropically from atmospheric conditions ($p_1 = 1$ atm and $T_1 = 15°C$) to 5 atm. The temperature after compression then would be

$$T_2 = T_1 \left(\frac{p_2}{p_1}\right)^{\frac{\gamma-1}{\gamma}} = 288\left(\frac{5}{1}\right)^{\frac{1.4-1}{1.4}} = 456.3 \text{ K.}$$

The compressed air may be next cooled to an initial (presume) temperature of 15°C in a heat exchanger without any loss of pressure. Then at state point 3, $T_3 = 288$ K and $p_3 = 5$ atm. Subsequently, the cooled high pressure can be expanded in a suitable device to the original pressure of 1 atm (state point 4). Then temperature after expansion will be

$$T_4 = T_3 \left(\frac{p_4}{p_3}\right)^{\frac{\gamma-1}{\gamma}} = 288\left(\frac{1}{5}\right)^{\frac{1.4-1}{1.4}} = 181.8 \text{ K} = -91.2°C.$$

The different air refrigeration systems use this thermodynamic principle for producing low temperatures.

7. Throttling process

Throttling is the expansion of fluid from high pressure to low pressure. This process occurs when fluid passes through an obstruction (partially opened valve or a small orifice) placed in the fluid flow passage.

Figure 4.5 shows the schematics of the porous plug experiment performed by Joule and Thomson in 1852. A stream of incompressible

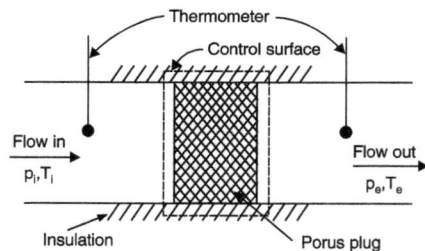

FIGURE 4.5 Schematic of porous plug apparatus.

fluid (gas) is made to pass steadily through a porous plug placed in an insulated and horizontal pipe. The upstream conditions of pressure p_i and temperature T_i are held constant, and the corresponding values at exit are measured. The friction of the narrow passage causes the pressure to drop, and accordingly, the exit pressure p_e is less than the intake pressure p_i.

A throttling process is characterized by the following features:

— no shaft work is involved,
— no heat interaction as the pipe is thermally insulated,

— no change in potential energy $(z_1 = z_2)$ as the pipe is placed horizontally, and
— negligible changes in kinetic energy.

With these stipulations, the steady flow energy equation

$$h_1 + \frac{V_1^2}{2} + gz_1 + q = h_2 + \frac{V_2^2}{2} + gz_2 + w_s$$

transforms to

$h_1 = h_2$, that is, enthalpy of fluid remains constant during throttling.

Thus, the throttling expansion process is an ***isenthalpic process***. If the fluid undergoing throttling behaves as an ideal gas for which $h = c_p T$, we get

$$c_p T_1 = c_p T_2;\ T_1 = T_2$$

Again for a perfect gas, internal energy is a function of temperature alone and equality of temperature implies that $u_1 = u_2$. Apparently, a throttling process takes place at constant enthalpy, constant temperature, and constant internal energy.

Throttling is an irreversible process and involves the degradation of energy and its dissipation in turbulence.

Joule–Thomson Coefficient, Inversion Point, and Inversion Curve

For real gases, enthalpy is a function of both temperature and pressure. As such even though enthalpy remains constant during throttling, the temperature needs not to remain the same. Experimental test runs can be conducted by keeping upstream conditions constant but with different downstream pressures. This is achieved by having porous plugs of different sizes. The exit temperature of the fluid at different exit pressures is measured. Since the upstream pressure and temperature conditions are kept constant, the enthalpy of the fluid for all measured conditions of exit pressure and temperature would be constant. The results are plotted as a constant enthalpy (isenthalpy) curve on $T - p$ diagram. Several enthalpy curves can be obtained by repeating the experiments with several inlet conditions.

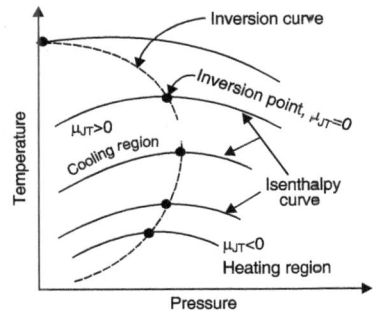

FIGURE 4.6 Isenthalpy and inversion.

The slope of an isenthalpic curve is called the Joule–Thomson coefficient, μ_{JT}. That is

$$\mu_{JT} = \left(\frac{\partial T}{\partial p} \right)_{h\ =\ \text{constant}}$$

This coefficient may be +ve, −ve, or zero. The point on the isenthalpic curve where $\mu_{JT} = 0$ is called the *inversion point*. Thus, the inversion point denotes the maximum value of temperature on T–p plot. The locus of all inversion points is called the *inversion curve*. The Joule–Thomson coefficient is positive on the left side of inversion curve, is zero at the inversion point, and is negative on the right side of inversion curve.

Throttling is always accompanied by pressure drop, that is, Δp is −ve. That leads to drop in temperature when μ_{JT} is +ve. Accordingly when a real gas is throttled at the condition such that its state lies to the left of inversion curve, the gas will get cooled. Thus, the region of $\mu_{JT} > 0$ represents the region of cooling. Likewise, when μ_{JT} is −ve, the temperature change is +ve, and therefore, the throttling of real gas would produce heating effect.

A knowledge of the inversion temperatures and inversion curves of real gases is of considerable importance in the design of refrigeration and liquefaction equipment. The use of positive values of Joule–Thomson coefficient is made in the liquefaction of gases such as air, nitrogen, and oxygen.

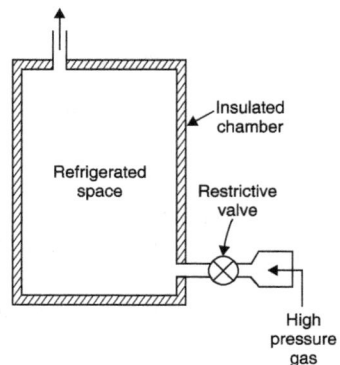

FIGURE 4.7 Refrigeration by throttling of a gas.

Refer to Figure 4.7 which shows a simplified arrangement of refrigeration by throttling of a gas. The throttling process occurs when the gas at high pressure with its temperature below its critical temperature is made to pass through a valve with a restricted opening. Upon throttling, the temperature of gas reduces and a cooling effect is produced.

8. Mechanical refrigeration

The natural and chemical methods have been successfully replaced by mechanical or heat-energy refrigeration techniques. In these methods, the heat is abstracted from the substance, or space (which is to be cooled) is pumped to a system (which is at high-temperature level) by taking energy from an external source as input to the refrigerating machine. The refrigeration system consists of a cycle of processes with the same quantity of working fluid (refrigerant) in continuous circulation.

The first commercial hand-operated refrigeration system was developed in the UK by Perkins. The system consisted of a hand-operated compressor, a water-cooled condenser, a throttle valve, and an evaporator. The working substance (ether) was compressed in the hand-operated compressor and then condensed in the water-cooled condensing unit. Thereafter, the liquid ether was throttled to low pressure and taken to the evaporator where heat was absorbed and a cooling effect was produced. The refrigerant ether was used again and again in the cyclic process with negligible wastage. Subsequent developments took place in the United States where the steam engine was used as a prime mover to drive the compressor. The scope got further widened with the development of electric motors and consequent high speeds of the compressor. Ether too was replaced by new working substances.

Mechanical refrigeration systems are broadly classified into

- air or gas refrigeration,
- vapor compression system,
- vapor absorption system, and
- steam jet refrigeration.

The energy needed for the mechanical systems is essential in the form of mechanical, electrical, or thermal. Due to world energy crisis, concerted efforts are being made by various agencies to develop refrigeration systems that utilize waste heat, solar energy, wind energy, and bio-energy for their functioning. A lot of research is being done to devise ways and means which make refrigeration systems more energy efficient.

9. Non-conventional refrigeration systems

- ***Thermoelectric*** cooling uses the Peltier effect to create a heat flux between the junctions of two different types of materials.

  When a direct current is made to pass through electrical junctions between unlike metals, heating or cooling effect is produced depending on the direction of flow of current. One junction gets heated and the other gets cooled. The cold junction is located inside the space desired to be cooled and the warmed junction lies outside in the surroundings. The hot and cold junctions get reversed when a change occurs in the direction of the flow of current. Apparently, the same setup can be used for heating and cooling of an insulated space.

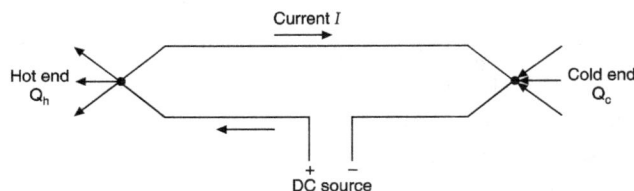

FIGURE 4.8 Thermoelectric cooling: Peltier effect.

With respect to Figure 4.8, the energy balance of the system is

$$Q_h - Q_c = EI$$

where E is the emf applied, I is the current, Q_h and Q_c are the actual amount of heat flows at the hot and cold junctions, respectively.

This effect is commonly used in camping and portable coolers, and for cooling electronic components and small instruments.

- **Magnetic refrigeration** (adiabatic demagnetization) is a cooling technology based on magnet caloric effect, an intrinsic property of magnetic solids. A strong magnetic field is applied to the refrigerant, forcing its various magnetic dipoles to align and putting these degrees of freedom of the refrigerant into a state of lowered entropy. A heat sink then absorbs the heat released by the refrigerant due to its loss of entropy. Thermal contact with the heat sink is then broken so that the system is insulated and the magnetic field is switched off. This increases the heat capacity of the refrigerant, thus, decreasing its temperature below the temperature of heat sink.

  The refrigerant is often a paramagnetic salt such as cerium magnesium nitrate, and the active magnetic dipoles are those of electron shells of the paramagnetic atoms.

- **Thermoacoustic** refrigeration uses sound waves in a pressurized gas to drive heat exchange.

- **Vortex tube** that operates with compressed air and is used for spot cooling. A high-pressure gas is allowed to expand through a nozzle fitted tangentially to a pipe. This causes a simultaneous discharge of cool air at the core and hot air at the periphery.

The mechanical refrigeration and the non-conventional refrigeration systems have been dealt with in detail at appropriate places.

4.5. APPLICATIONS OF REFRIGERATION

Refrigeration has played an important role in the growth and attainment of the present-day standard of living. Its various applications can be essentially grouped into the following categories:

    1. Domestic refrigeration deals with providing a low-temperature place for preserving eatables, drinks, and medicines. The use of refrigerators in our kitchens for the storage of fruits and vegetables has allowed us to add fresh salads to our diets year-round, and to store fish and meats for long periods.

2. Commercial refrigeration is related to refrigeration fixtures of the type used by restaurants and hotels, retail stores, and institutions for storing, displaying, processing, and dispensing perishable commodities of different types.

3. Industrial refrigeration is concerned with systems that are much larger in size and cooling capacity than those for commercial applications. Typical industrial applications are

- ice and large food plants,
- chilling and storage of foodstuffs including beverages, meat, poultry products, dairy products, vegetables, fruits and fruit juices, etc.,
- processing of food products and farm crops,
- processing of textiles, printing work, and photographic materials, and
- oil refineries and synthetic rubber manufacturing. In oil refineries, chemical manufacturing, and petrochemical plants, refrigeration is used to maintain processes at their required low temperatures.

Refrigeration is also used to liquefy gases like oxygen, nitrogen, propane, and methane. Further, in compressed air purification, it is used to condense water vapor from compressed air to reduce its moisture air content.

4. Transport refrigeration applies to refrigerated railway cars and trucks for local delivery and long-distance transport of temperature-sensitive foodstuffs and other materials.

5. Air conditioning refers to the simultaneous control of temperature, humidity, cleanliness, and air motion. The conditioning of a space done to provide comfortable and healthy conditions for the occupants is called comfort air conditioning. Residences, offices, theaters, and hospitals are the spaces air-conditioned for this purpose.

There are some manufacturing processes that need to be done under controlled environmental conditions. Even some of the sophisticated and precision instruments need controlled conditions for their effective working and upkeep. The conditioning done for this purpose is called industrial air conditioning, and this is concerned with the production of environment suitable for

- computer centers,
- pharmaceutical units,
- printing works and photographic products,
- precision devices and production shop laboratories, etc.

The applications cited above clearly indicate that refrigeration and air conditioning which was considered a luxury in the society a few decades ago have become the necessity of the present society and a tool for higher productivity.

4.6. DOMESTIC REFRIGERATOR

A domestic refrigerator serves to preserve food products (fruits and vegetables, meat and fish, milk and ice cream, cold drinks, etc.). The unit absorbs heat from these products and dissipates that heat to the surroundings by taking the power of a compressor.

The domestic refrigerator works on vapor compression refrigeration cycle, and consists of the following parts: compressor, condenser, capillary tube, and evaporator.

These components are schematically arranged as shown in Figure 4.9 and mounted on the refrigerator which has enamel painted metallic body with interior plastic lining of polystyrene.

The system works on the closed cyclic operation and transfers heat through a medium called refrigerant which is usually Freon-12. The refrigerant changes its phase when it passes through condenser and evaporator.

The sequence of operation of the refrigeration cycle is as follows:

1. *Reversible adiabatic compression (1-2)*: The refrigerant vapor at low pressure and temperature and preferably in the dry state is drawn from the evaporator during the suction stroke of the compressor. The compressor constricts the vapor raising its pressure and temperature.

The compression is of reciprocating type and is hermetically sealed which means that the compressor and electric motor are a single unit enclosed in a container.

2. *Constant pressure condensation (2-3)*: The vapor refrigerant at high pressure and temperature (state 2) coming from the compressor is pushed into the condenser coils which are painted black and are located on the back of the refrigerator. The hot refrigerant passes through these coils, which meets the cooler air of the kitchen and becomes a liquid.

3. *Throttling (3-4)*: The high-pressure refrigerant in the liquid state is throttled in the capillary tube. The capillary tube is a simple copper tube of low diameter and long length. This increases friction and that causes pressure drop leading to conversion of high-pressure liquid into low-pressure liquid.

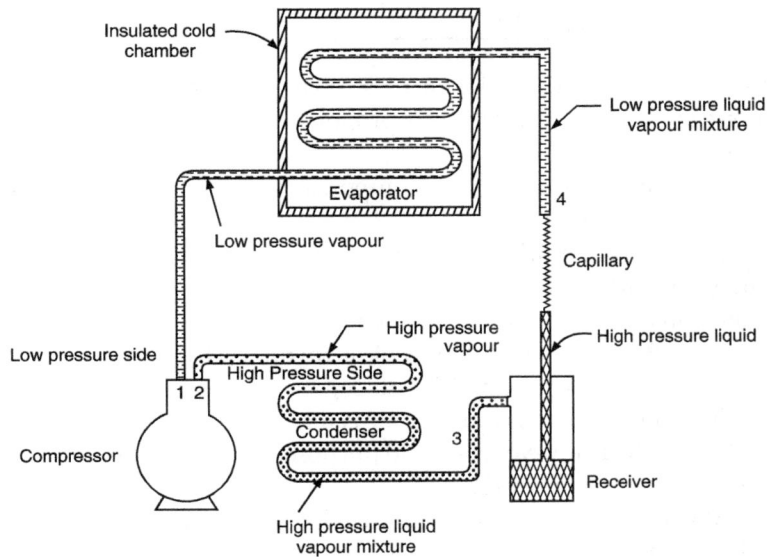

FIGURE 4.9 Vapor compression cycle for a domestic refrigerator.

4. *Constant pressure evaporation (4-1)*: The wet vapor after throttling passes through evaporator coils placed inside the refrigerator. The refrigerant absorbs heat from the food staff which gets cooled. The refrigerant itself vaporizes to gaseous state at constant pressure and flows to the compressor.

The cycle is completed and the process starts all over again.

The condenser and evaporator are the simple heat exchangers where the refrigerant changes the phase by rejecting heat (to the condenser) and accepting heat (from the evaporator). The unit has a thermostat that controls the cooling process by monitoring the temperature and then switching the compressor on or off. When the sensor senses that it is cold enough inside the refrigerator, it turns off the compressor. If too much heat is sensed, it switches the compressor and the cooling process begins again.

The domestic refrigerators are available in a wide range of sizes and designs and are specified by cooling capacity (refrigerating effect in tons), cooling in liters, overall dimensions (height, width, and depth), refrigerant used, and voltage range and power source (AC 230 V, 50 hertz). The refrigerant Freon-12 is now being replaced by HFC-134a which does not deplete the ozone layer.

4.7. PSYCHROMETRY

Psychrometry is the science of studying the thermodynamic properties of moist air and the use of these properties to analyze the conditions and processes involving moist air.

For many purposes, the composition of real air can be assumed to be a mixture of two components:

Standard dry air: Dry air is a mixture of number of gases such as oxygen, nitrogen, carbon dioxide, hydrogen, argon, etc. Oxygen and nitrogen are the main constituents with the following composition:

21% oxygen and 79% of nitrogen ..... by volume

23% oxygen and 77% nitrogen ..... by mass

The molecular mass of dry air is taken as 20.966 and gas constant is equal to 287 J/kgK. For psychrometric purpose, dry air is assumed to be a pure substance and not a mixture.

Water vapor: Air has affinity for water, and consequently, the atmospheric air always contains some water; water vapor content varies from 0 to 3% by mass.

The moist air is essentially a mixture of dry air and water vapor. The amount of water vapor present depends on the absolute pressure and temperature of the mixture.

For the **moist air**, which is a mixture of dry air and water vapor,

$$p_t = p_a + p_v,$$

where p_t = total pressure of moist air

p_a = partial pressure of dry air

p_v = partial pressure of water vapor.

The *saturated air mixture* is the mixture of dry air and water vapor in which the partial pressure of the water vapor is equal to its saturation pressure corresponding to the temperature of the mixture.

The *unsaturated air mixture* is the mixture of dry air and water vapor in which the partial pressure of the water vapor is less than its saturation pressure corresponding to the temperature of the mixture.

The *supersaturated air mixture* is a mixture of dry air and water vapor in which the partial pressure of the water vapor is greater than its saturation pressure corresponding to the temperature of the mixture.

4.8. PARTIAL PRESSURE AND DALTON'S LAW

Consider a homogeneous non-reacting mixture of ideal gases a, b, c, ..., etc., at temperature T, pressure p, and occupying a volume V (Figure 4.10).

Further, let it be presumed that each constituent of the mixture exists separately at temperature T and volume V, and pressures p_a, p_b, p_c, ... exerted by individual gases are measured separately. Each of these pressures would be less than the total pressure p of the mixture.

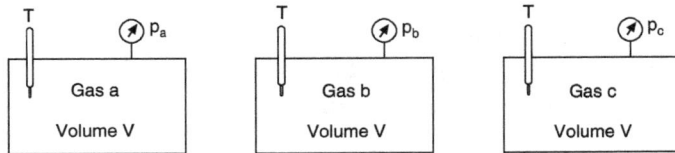

FIGURE 4.10.

When the equation of state for an ideal gas is applied to the gas mixture as well as the constituent gases, we have

For the mixture,

$$p_V = mRT$$

$$= nMRT = nR_{mol}T \tag{4.2}$$

where R_{mol} is the universal gas constant (R_{mol} = 8314 J/kg mole K) and M is the molecular mass.

For the constituent gases,

$$p_a V = n_a R_{mol}T$$

$$p_b V = n_b R_{mol}T$$

$$p_c V = n_c R_{mol}T, \text{ etc.}$$

Upon adding the components,

$$(p_a + p_b + p_c + ...) V = (n_a + n_b + n_c + ...) R_{mol}T$$

$$= nR_{mol}T \tag{4.3}$$

From expressions 7.4 and 7.5,

$$p = p_a + p_b + p_c + ...$$

$$p = \Sigma \, p_i \tag{4.4}$$

where $p_i = \dfrac{n_i R_{mol} T}{V}$ represents the pressure; the component i would exert if it alone occupied the volume V at temperature T. This is called the **partial pressure** of the ith component of the gas mixture. Thus,

Partial pressure is defined as the pressure that each individual component of a gas mixture would exert if it alone occupied the volume of the mixture at the same temperature.

Further, Equation 4.4 stipulates that the total pressure of a mixture of ideal gases is equal to the sum of the partial pressures of the individual gas components of the mixture.

This is known as **Dalton's law of partial pressures**.

The following relations are implicit in the Dalton's law:

$$t = t_a = t_b = t_c$$

$$V = V_a = V_b = V_c$$

and
$$m = m_a + m_b + m_c \qquad (4.5)$$

where t, V, and m, respectively, represent the temperature, volume, and mass. In terms of specific volume v,

$$m\, v = m_a\, v_a = m_b\, v_b = m_c\, v_c \qquad (4.6)$$

Combining expressions 7.7 and 7.8, we may write

$$\frac{m}{mv} = \frac{m_a}{m_a v_a} + \frac{m_b}{m_b v_b} + \frac{m_c}{m_c v_c}$$

or
$$\frac{1}{v} = \frac{1}{v_a} + \frac{1}{v_b} + \frac{1}{v_c} \qquad (4.7)$$

The reciprocal of specific volume is density, and so, we can write

$$\rho = \rho_a + \rho_b + \rho_c \qquad (4.8)$$

which means that density of the mixture is equal to the sum of the densities of the components.

4.9. SPECIFIC HUMIDITY, RELATIVE HUMIDITY, AND DEGREE OF SATURATION

Humidity refers to the dampness, that is, the water content of air. **Absolute humidity** represents the amount of water vapor actually present in the air, expressed as gram per cubic meter of air.

The **specific humidity** or **humidity ratio** or **moisture content** is the ratio of mass of water vapor to the mass of dry air in a given volume of the mixture. Consider a mixture consisting of m_a kg of dry air and m_v kg of water vapor contained in a vessel of volume V at total pressure p_t and temperature T. Then

Specific humidity, $\omega = \dfrac{m_v}{m_a}$.

Since both masses occupy volume V,

$$\omega = \frac{m_v / V}{m_a / V} = \frac{\rho_v}{\rho_a} = \frac{v_a}{v_v},$$

where ρ is the density and v is the specific volume

If both the vapor and dry air are considered as perfect gases, then from the characteristic gas equation,

$$m_a = \frac{p_a V}{R_a T} \quad \text{and} \quad m_v = \frac{p_v V}{R_v T}.$$

That gives $\omega = \dfrac{m_v}{m_a}$

$$= \frac{p_v V}{R_v T} \times \frac{R_a T}{p_a V} = \frac{R_a}{R_v} \times \frac{p_v}{p_a}$$

Taking $R_a = 287$ J/kgK and $R_v = 461$ J/kgK

$$\omega = \frac{287}{461} \frac{p_v}{p_a} = 0.622 \frac{p_v}{p_a} = 0.622 \frac{p_v}{p_t - p_v} \qquad (4.9)$$

The above relation shows that if the total pressure remains constant, the specific humidity is a function of partial pressure of water vapor only. This relationship has been obtained by assuming that behavior of water vapor is identical to that of an ideal gas. Such an assumption is quite valid at low pressure and normal humidity conditions.

The **relative humidity** is the ratio (expressed as a percentage) of the amount of water vapor actually present in a given volume of air to the maximum amount that the air could hold under the same pressure and temperature conditions.

Relative humidity

$$\phi = \frac{\text{mass of water vapor in a given volume}}{\text{mass of water vapor in the same volume}}$$

if saturated at the same temperature

With the assumption that vapors behave as perfect gases, we have

$$m_v = \frac{p_v V_v}{R_v T_v} \text{ and } m_{vs} = \frac{p_{vs} V_{vs}}{R_{vs} T_{vs}}$$

where vs subscript is used for saturated vapor

$$\therefore \quad \phi = \frac{m_v}{m_{vs}} = \frac{p_v V_v}{R_v T_v} \times \frac{R_{vs} T_{vs}}{p_{vs} V_{vs}}$$

Also $V_v = V_{vs}$; $R_v = R_{vs}$; and $T_v = T_{vs}$.
That gives

$$\phi = \frac{p_v}{p_{vs}} \tag{4.10}$$

where p_{vs} is the saturation pressure at the temperature of the mixture. This saturation pressure is obtained from the steam tables corresponding to the given temperature.

Apparently, the relative humidity can also be defined as the ratio of the partial pressure of water vapor in a given volume of mixture to the partial pressure of water vapor when the same volume of the mixture is saturated at the same temperature. This implies that ϕ is equal to unity for saturated air; 100 percent relative humidity means that air contains the maximum moisture it can hold.

Essentially, the term relative humidity compares the humidity of the given air with the humidity of saturated air at the same pressure and temperature. For saturated air, the relative humidity is 100 percent.

The **degree of saturation** μ represents the ratio of mass of water vapor associated with the unit mass of dry air to the mass of water vapor associated with the unit mass of dry air saturated at the same temperature:

$$\mu = \frac{\text{mass of water vapor with unit mass of dry air}}{\text{mass of water vapor with unit mass of dry saturated air}}.$$

This relation implies that μ represents the ratio of specific humidity of moist air to specific humidity of saturated air at the same temperature. That is

$$\mu = \frac{\omega}{\omega_s} = \frac{0.622 \dfrac{p_v}{p_t - p_v}}{0.622 \dfrac{p_{vs}}{p_t - p_{vs}}}$$

$$= \frac{p_v}{p_{vs}} \frac{p_t - p_{vs}}{p_t - p_{vs}} = \frac{p_v}{p_{vs}} \left[\frac{1 - \dfrac{p_{vs}}{p_t}}{1 - \dfrac{p_v}{p_t}} \right]. \tag{4.11}$$

The following observations can be made from the above relation:

i. If the air is dry, then $p_v = 0$, and therefore, $\mu = 0$.
    If the relative humidity $\omega = p_v / p_{vs} = 1$, then $p_v = p_{vs}$ and accordingly $\mu = 1$.
    Thus, the degree of saturation varies between 0 and 1.

ii. The degree of saturation is a measure of the capacity of the air to absorb moisture.
    When $\mu = 1$, then $\omega = \omega_s$
    which implies that air is holding the maximum amount of water vapor.

iii. The expression for the degree of saturation can be recast as follows:

$$\mu = \frac{p_v}{p_{vs}} \left[\frac{1 - \dfrac{p_{vs}}{p_t}}{1 - \dfrac{p_v}{p_{vs}} \times \dfrac{p_{vs}}{p_t}} \right] = \phi \left[\frac{1 - \dfrac{p_{vs}}{p_t}}{1 - \phi \dfrac{p_{vs}}{p_t}} \right]$$

or $\qquad \phi - \phi \dfrac{p_{vs}}{p_t} = \mu - \mu \phi \dfrac{p_{vs}}{p_t}$

or $\quad \phi \left[1 - \dfrac{p_{vs}}{p_t} + \mu \dfrac{p_{vs}}{p_t} \right] = \mu$

or $\quad \phi \left[1 - (1 - \mu) \dfrac{p_{vs}}{p_t} \right] = \mu$

$\therefore \qquad\qquad \phi = \dfrac{\mu}{1 - (1 - \mu) \dfrac{p_{vs}}{p_t}}$. $\qquad\qquad$ (4.12)

The difference between relative humidity and degree of saturation is usually less than 2%.

4.10. DRY-BULB TEMPERATURE AND WET-BULB TEMPERATURE

The **dry-bulb temperature** (*dbt*) is the normal temperature of an air–vapor mixture as indicated or recorded by any temperature measuring device placed in the mixture. This temperature is not affected by the moisture content in the mixture.

The **wet-bulb temperature** (*wbt*) of an air–vapor mixture is the temperature measured by a thermometer of which bulb is covered by a wick soaked in water. When the air passes over the wet wick, the moisture contained in the wick tends to evaporate. That produces a cooling effect at the bulb and an equilibrium temperature lower than that of the air stream is recorded.

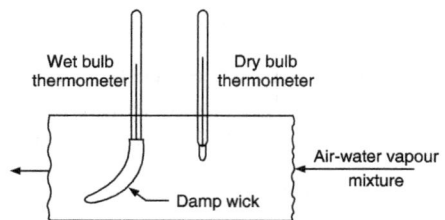

FIGURE 4.11 Dry- and wet-bulb temperature.

It is worthwhile to note that

i.   The wet-bulb temperature is lower than the dry-bulb temperature and the difference is known as the **wet-bulb depression**.

ii. The wet-bulb depression is greatest when the air is initially completely dry, that is, capable of absorbing a maximum amount of moisture.

iii. When the air is initially saturated, there will be no evaporation of water, and hence, the two thermometers will record equal temperatures. This implies that the depression is zero with 100% relative humidity.

iv. The dry- and wet-bulb temperatures are simultaneously measured by instruments called **psychrometers**.

A sling psychrometer consists of two identical mercury-in-glass thermometers mounted on a suitable frame and arranged with a swivel-mounted handle as shown in Figure 4.12. The temperature-sensing bulb of one of the thermometers is covered with a knitted

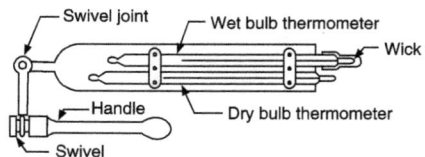

FIGURE 4.12 Sling psychrometer.

or woven cotton wick which is wetted with pure clean water. For better and accurate measurements of the wet-bulb temperature, a fast movement of air past the moistened wick is necessary. This is to ensure that the surrounding air does not cling to the moistened wick and that the air at the wet-bulb thermometer is always in immediate contact with the wet wick. The necessary air motion, 5 m/s to 10 m/s, is provided by rotating the psychrometer frame with the swivel-mounted handle. The readings are taken after swinging the psychrometer in a smooth circular path for 15 to 20 seconds. With a too-short duration, the temperature will not be depressed to its proper value. If the swinging period is too large, the wick will dry and the bulb temperature will not remain at its minimum value.

v. When dry- and wet-bulb temperatures are known, the other psychrometric properties can be determined by calculations.

vi. Many investigations have suggested different expressions to determine the partial pressure of water vapor in air from the wet- and dry-bulb temperature readings. Carrier's equation, as given below, is most widely used.

$$p_v = (p_{vs})_{wb} - \frac{(p_t - p_{vs})(dbt - wbt)}{1544 - 1.44wbt} \tag{4.13}$$

where $\quad p_v$ = partial pressure of water vapor

p_{vs} = partial pressure of water vapor when air is fully saturated

p_t = total (barometric pressure of moist air)

dbt and wbt = dry-bulb and wet-bulb temperature, respectively, in °C.

4.11. DEW-POINT AND ADIABATIC SATURATION TEMPERATURE

Dew point refers to the temperature at which the mixture becomes saturated, that is, the moisture (water vapor) present in the mixture begins to condense consequent to continuous cooling at constant pressure.

With reference to Figure 4.13, let point 1 represent the initial state of air–water vapor mixture. The vapor is superheated, the pressure equals the partial pressure of the superheated vapor and temperature corresponds to the temperature of the mixture. When the mixture is cooled at constant pressure along paths 1–3, the cooling continues until the vapor attains the saturated state at point 3. With further cooling, the vapor condenses, that is, the moisture is released. The state point 3 represents the dew-point temperature (dpt) of the air–water vapor mixture. The dew-point temperature, thus, corresponds to the saturation temperature of steam at the partial pressure

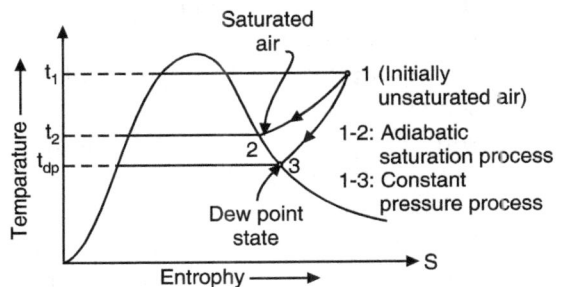

FIGURE 4.13 Concept of dew-point temperature.

of water vapor in the mixture. *For saturated air, the dry-bulb temperature, the wet-bulb temperature, and the dew-point temperature are same.*

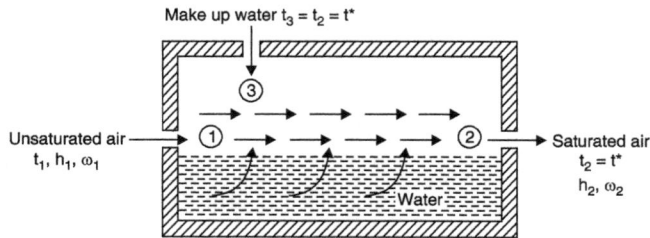

FIGURE 4.14 Concept of adiabatic saturation process.

The adiabatic saturation temperature (or the thermodynamic wet-bulb temperature) refers to the temperatures at which the air can be brought adiabatically to saturation state by the evaporation of water into a flowing stream.

Consider a stream of unsaturated air–vapor mixture flowing over a surface of water contained in a chamber that is sufficiently long and is perfectly insulated. Due to intimate contact between the unsaturated air and liquid water, some of the water evaporates and is carried by air resulting in an increase in its humidity. The heat required for evaporation comes both from the air–vapor mixture and the liquid water in the chamber. The process continues until the energy transferred from the air to the water is equal to the energy required to vaporize the water. By the time, air reaches the exit section, it becomes saturated and an equilibrium is established. The temperature of the saturated air at the exit section is known as ***adiabatic saturation temperature or the thermodynamic wet-bulb temperature***. The steady-state thermal equilibrium conditions are maintained by adding make up water steadily at the rate of evaporation. This make up water has to be at the adiabatic saturation temperature, that is, the temperature of the mixture at the exit section.

With reference to Figure 4.14, the adiabatic saturation process is represented by paths 1–2. During the adiabatic saturation process, the partial pressure of vapor increases, although the total pressure of the air–vapor mixture remains constant. The unsaturated air at dry-bulb temperature t_1 is cooled adiabatically to dry-bulb temperature t_2 which is equal to adiabatic saturation temperature t^*. For all practical purposes, the adiabatic saturation temperature is taken equal to wet-bulb temperature.

EXAMPLE 4.2

A metal beaker contains water initially at room temperature and the water is cooled by gradually adding ice water to it. When the water temperature reaches 13°C, the moisture from room air begins to condense on the beaker. Make calculations for the specific humidity and parts by mass of water vapor in the room air. Take
room air temperature = 22°C and barometric pressure = 1.01325 bar

Solution: From steam tables, the partial pressure of water vapor at *dpt* of 12.5°C

$$p_v = 1500 \text{ N/m}^2$$

partial pressure of dry air

$$p_a = 101325 - 1500 = 99825 \text{ N/m}^2$$

(*a*)   Specific humidity or humidity ratio,

$$\omega = \frac{m_v}{m_a} = 0.622 \frac{p_v}{p_a} = 0.622 \times \frac{1500}{99825} = \mathbf{0.009346} \text{ kg/kg of dry air}$$

(*b*)   Parts of mass of water vapor,

$$\frac{m_v}{m} = \frac{\omega}{1 + \omega} = \frac{0.009346}{1 + 0.00934} = \mathbf{0.00926} \text{ kg/kg of mixture}$$

EXAMPLE 4.3

The air supplied to an air-conditioned room is noted to be at temperature 20°C and specific humidity 0.0085. Corresponding to these conditions, determine the partial pressure of vapor, relative humidity, and dew-point temperature.

Take barometric or total pressure = 1.0132 bar

Solution: Specific humidity

$$\omega = 0.622 \frac{p_v}{p_a} = 0.622 \frac{p_v}{p_t - p_v}$$

Τηυσ, $0.0085 = 0.622 \dfrac{p_v}{1.0132 - p_v}$

∴ Partial pressure of vapor

$$p_v = \frac{1.0132 \times 0.0085}{0.622 + 0.0085} = \mathbf{0.01366} \text{ bar}$$

(*b*)   The relative humidity is defined as $\phi = p_v/p_{vs}$. From steam tables, the saturation vapor pressure at 20°C = 0.0234 bar.

Then, $\phi = \dfrac{0.01366}{0.0234} = 0.5837$ or **58.38%**

(*c*)   The dew-point temperature is the saturation temperature of water at a pressure of 0.01366 bar

DPT (from steam tables by interpolation)

$$= 11 + (12 - 11) \times \frac{(0.01366 - 0.01312)}{0.01401 - 0.01312} = 11 + 0.607 = \textbf{11.607°C}$$

EXAMPLE 4.4

10 gm of water vapor was removed from a given sample of one kg of atmospheric air at 40°C and 60% relative humidity. The temperature of air after the removal of moisture reduced to 30°C. Determine the humidity ratio, partial pressure of vapor, relative humidity, and dew-point temperature of this air (air after removal of moisture).

Take total atmospheric pressure as 101.325 kPa.

Solution: The relative humidity is defined as $\phi = p_v/p_{vs}$. From steam tables, the saturation vapor pressure at 40°C = 7.384 kPa

That gives

$$p_v = \phi\, p_{vs} = 0.6 \times 7.384 = 4.43 \text{ kPa}$$

Specific humidity or humidity ratio

$$\omega = 0.622\ \frac{p_v}{p_t - p_v}$$

$$= 0.622 \times \frac{4.43}{101.325 - 4.43}$$

$$= 0.0284 \text{ kg/kg of dry air}$$

$$= 28.4 \text{ gm/kg of dry air}$$

(*a*)   Since 10 gm of water vapor per kg has been removed,

Specific humidity = 28.4 − 10 = 18.4 gm/kg

$$= \textbf{0.0184} \text{ kg/kg of dry air}$$

(*b*)   At this state, the air is at 30°C with specific humidity 0.0184 kg/kg of dry air. Then

$$0.0184 = 0.622 \; \frac{p_v}{101.325 - p_v}$$

That gives partial pressure of water vapor,

$$p_v = \frac{101.325 \times 0.0184}{0.0184 + 0.622} = 2.92 \text{ kPa}$$

(c)   At 30°C,

$$p_{vs} = 4.246 \text{ kPa} \qquad \qquad \text{(from steam tables)}$$

∴ Relative humidity

$$\phi = \frac{p_v}{p_{vs}} = \frac{2.92}{4.246} = 0.688 \text{ or } 68.6\%$$

(d)   The dew-point temperature is the saturation temperature at the pressure of 2.92 kPa.

Then from steam tables

$$\text{DPT} = 24°\text{C}$$

4.12. PSYCHROMETRIC CHART

The subject which deals with the behavior of moist air is known as psychrometry, and the properties of moist air are called *psychrometric properties*.

Humidity calculations can be made by using the equations relating the dry- and wet-bulb temperatures to the humidity. The method, however, tends to be tedious, cumbersome, and time consuming. The key to humidity calculations is then provided by the *Psychrometric* or *Hygrometric chart* which graphically describes the relationship between the properties of moist air, that is, the dry-bulb, the wet-bulb, and dew-point temperatures of the mixture and its humidity. Figure 4.15 shows how these parameters are laid out on a typical psychrometric chart.

The psychrometric chart has the number of details and its salient aspects are

1. The dry-bulb temperature is taken as abscissa and specific humidity (i.e., moisture content) as ordinate.

The dry-bulb temperature lines are vertical and uniformly spaced. The specific humidity lines are horizontal and also uniformly spaced. The saturation curve is drawn by plotting the various saturation points at corresponding dry-bulb temperatures. The saturation curve represents 100 percent relative humidity at various dry-bulb temperatures. It also indicates the wet-bulb and dew-point temperatures.

2. The dew-point temperature lines are horizontal and non-uniformly spaced. At any point on the saturation curve, the dry- and dew-point temperatures are equal.

The wet-bulb temperature lines run diagonally to the right and their values are read at the left where these lines meet the 100 percent relative humidity line. These lines are inclined and straight but not uniformly spaced.

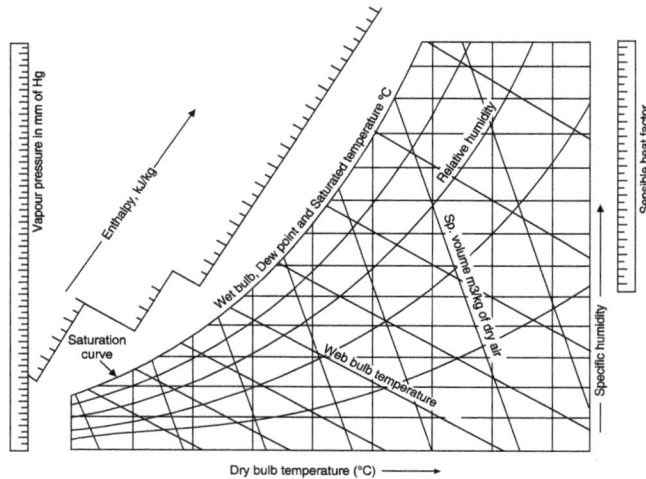

FIGURE 4.15 Psychrometric chart.

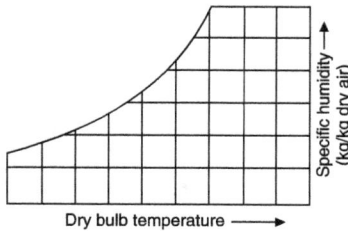

FIGURE 4.16 Dry-bulb and specific humidity lines.

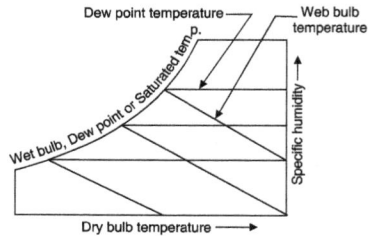

FIGURE 4.17 Dew-point and wet-bulb temperature lines.

3. The relative humidity lines curve upwards to the right with the percent values indicated on the lines themselves. The relative humidity curve depicts quantity of moisture actually present in the air as a percentage of the total amount possible at various dry-bulb temperatures and masses of vapor.

The specific volume (volume of air–vapor mixture per kg of dry air) lines are indicated by obliquely inclined straight lines. These lines are uniformly spaced and are drawn up to the saturation curve.

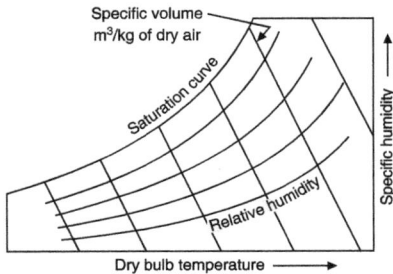

FIGURE 4.18 Relative humidity and specific volume lines.

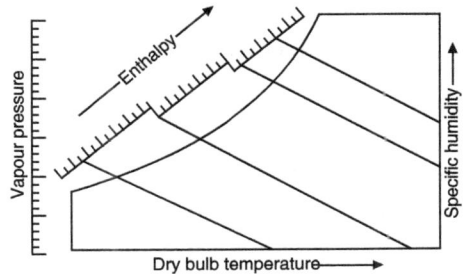

FIGURE 4.19 Enthalpy and vapor pressure lines.

4. The vapor pressure and enthalpy (total heat) lines are also scaled on the chart. The total heat at saturation temperature is represented by a diagonal system of co-ordinates. These inclined straight lines are uniformly spaced and are parallel to the wet-bulb temperature lines. The scale on the diagonal lines is separated from the body of the chart and is indicated above the saturation line.

Pressure of water vapor is shown in the scale on left and is the absolute pressure of steam in mm of mercury.

EXAMPLE 4.5

Atmospheric air at 1 bar pressure has 15°C wet-bulb temperature and 25°C dry-bulb temperature. With the help of a psychrometric chart, determine the salient psychrometric properties of the air.

Solution: Refer to Figure 4.20.

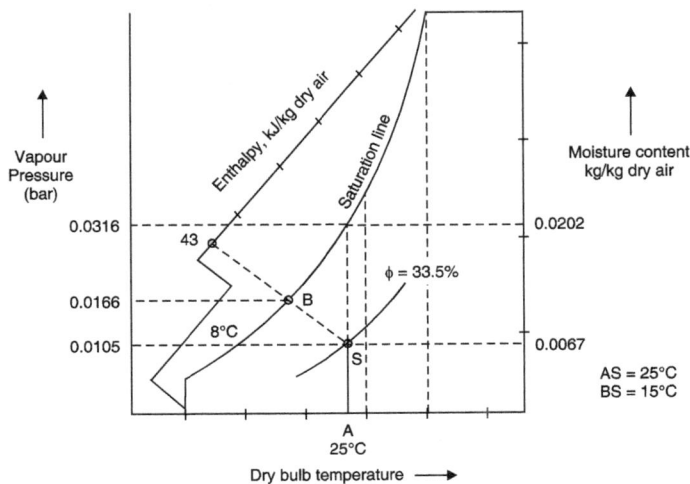

FIGURE 4.20.

Locate the state point S on the psychrometric chart. This is the point where a constant dry-bulb temperature line corresponding to 25°C intersects the wet-bulb temperature line corresponding to 15°C.

At this point, the following properties can be read:

i.   Relative humidity = 33.5%

ii. A horizontal line to the right from state point S would represent a constant humidity ratio line; its value being

$$\omega = 0.0067 \text{ kg of water/kg of dry air}$$

iii. A horizontal line to the left drawn from the state point S would intersect the saturation line; the temperature at the intersecting point gives the value of dew-point temperature equal to $dpt = 8°C$.

iv. The constant wet-bulb temperature line is produced to meet the enthalpy line; the enthalpy value approximates to 43 kJ/kg of dry air.

v.   A horizontal line drawn from the state point S meets the vapor pressure scale at 0.0105 bar. This represents the partial pressure of the vapor.

vi. Saturation pressure corresponding to $dbt = 25°C$ can be obtained by drawing a horizontal line to the left from the point where $dbt = 25°C$ line cuts the 100% relative humidity line. The saturation pressure value is 0.0316 bar.

vii. Saturation pressure corresponding to the wet-bulb temperature is obtained by drawing a horizontal line to the left from the point where $wbt = 15°C$ line cuts the 100% relative humidity line. Here, the pressure is scaled as 0.0166 bar.

viii. Humidity ratio corresponding 25°C dry-bulb temperature can be obtained by drawing a horizontal line to the right from the point where $dbt = 25°C$ line cuts the 100% relative humidity line. The corresponding value is scaled as 0.0202 kg of water/kg of dry air.

The degree of saturation is then calculated as follows:

$$\mu = \frac{\omega}{\omega_s} = \frac{0.0067}{0.0202} = 33\%.$$

EXAMPLE 4.6

The atmospheric air at 35°C dbt and 25°C wbt is cooled to 20°C. Using psychrometric chart, find (*i*) humidity ratio, (*ii*) relative humidity, and (*iii*) dew-point temperature. Comment whether or not there will be any condensation on the duct that supplies this cooled air to the conditioned space.

Solution: Locate the condition of air at 35°C *dbt* and 25°C *wbt* (Point 1) on the psychrometric chart and read there from

> Humidity ratio ω = 0.016 kg/kg of dry air
> Relative humidity φ = 0.45 or 45%
> Dew-point temperature *dpt* = 18.4°C

The condensation of moisture begins to take place when the dew-point temperature has been reached. Since the air is cooled only to 20°C which is higher than the dew-point temperature, there will be no condensation of moisture on the duct surface.

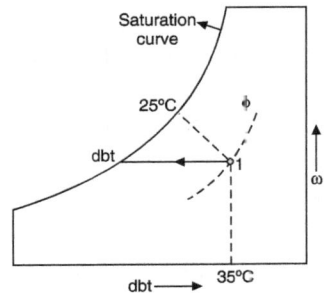

FIGURE 4.21.

4.13. PSYCHROMETRIC PROCESSES

The most common processes in air conditioning which involve the air–water vapor mixture are mentioned below and indicated on the psychrometric chart in Figure 4.22.

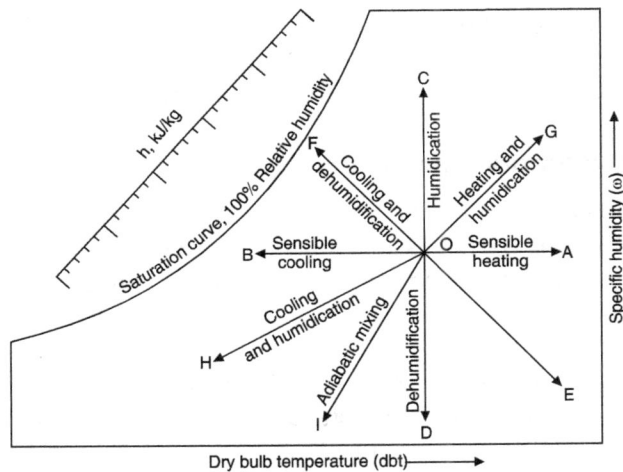

FIGURE 4.22 Psychrometric processes.

The initial state is represented by point 0. Heating and cooling processes are represented by horizontal lines while humidification and dehumidification plot as vertical lines. Humidification represents the process wherein the moisture is added to air, but its dry-bulb temperature is maintained constant. In dehumidification process, the moisture is removed from air without changing its dry-bulb temperature. The combination of these processes are set between them.

4.13.1. Sensible heating

The mixture is heated without any change in its moisture content. The process results when the mixture is made to pass over a surface of which temperature is above the dry-bulb temperature of the mixture. The heating surface may be the electric resistance-heating coils or steam passed through the coils or hot water passed through the coils.

With reference to Figure 4.22, the process of sensible heating is represented by horizontal line O-A that extends from left to right.

4.13.2. Sensible cooling

The mixture is cooled without any change in its moisture content. The process results when the mixture is made to pass over a surface of which temperature is below the dry-bulb temperature of the mixture. The cooling surface may be cooled water or gas flowing through coils or the refrigerant at low temperature in the coils of the evaporator of a vapor refrigeration system.

With reference to Figure 4.22, the process of sensible cooling is represented by horizontal line O–B that extends from right to left.

4.13.3. Humidification and dehumidification

Humidification represents the process wherein the moisture is added but its dry-bulb temperature is maintained constant. In dehumidification process, the moisture is removed from air without changing its dry-bulb temperature. These processes are obviously represented as vertical lines on the psychrometric chart.

With reference to Figure 4.22, it may be noted that in humidification process O–C, there is increase both in the specific humidity and relative humidity. However, in dehumidification process O–D, both the specific humidity and relative humidity decrease.

In practice, pure humidification and dehumidification processes are not possible. These are always accompanied by heating or cooling.

4.13.4. Heating and humidification

The process is achieved when the moist air is made to pass through spray water of which temperature is maintained at a temperature higher than the dry-bulb temperature of the air. The unsaturated air tends to become saturated and the heat of vaporization is absorbed from the spray water.

With reference to Figure 4.22, the process of heating and humidification is represented by line O-G, and it is to be noted that during this process:

i.   There is increase in specific humidity, dry- and wet-bulb temperatures, dew-point temperature, and enthalpy.

ii. The relative humidity may either increase or decrease.

The process of heating and humidification has practical application in winter air conditioning.

4.13.5. Cooling and dehumidification

The process takes place when the moist air is passed through a cooling coil of which effective surface temperature is lower than the dew-point temperature of the mixture.

With reference to Figure 4.22, the process of the cooling and dehumidification is represented by line O-F, and it is to be noted that during this process

i.   the dry-bulb temperature decreases,

ii. the air is cooled and condensation of moisture takes place, that is, it is dehumidification,

iii. there is decrease in specific humidity, and

iv. the relative humidity at outlet is generally higher than that at inlet.

The cooling and dehumidification process has practical application in summer air conditioning.

4.13.6. Adiabatic mixing

The process takes place when two streams of moist air having different specific humidities and enthalpies are allowed to mix without the addition or rejection of either heat or moisture, that is, adiabatically and at constant total moisture content.

The state of the resultant mixture lies on the straight line that joins the state of two streams on psychrometric chart. The location of final state on the straight line depends on the masses involved, and on the enthalpy and specific heat of each stream.

4.14. AIR CONDITIONING

Air conditioning is an artificial process that involves cooling as well as heating coupled with ventilation, filtration, and air circulation. It is essentially the process of treating air to control simultaneously its temperature, humidity, cleanliness, and distribution to meet the comfort requirements of the occupants of the conditioned space. The functioning of an air-conditioning system can be conceived as depicted in Figure 4.23.

Apart from the creation of an acceptable thermal environment (controlled temperature), control of humidity is of great importance both in humid and arid climates. Further, the air inside the conditioned space gets fouled due to absorption of pollutants from different sources and for human comfort, the indoor air has be purified.

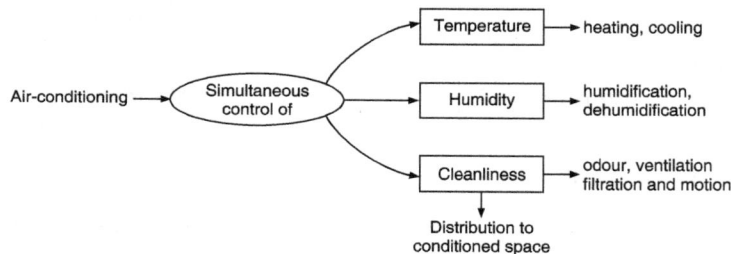

FIGURE 4.23 Functioning of air conditioning.

4.15. APPLICATIONS OF AIR CONDITIONING

Air conditioning which was once considered as luxury, has now become a necessity in our day-to-day life. The air conditioning has applications in diverse fields such as

i.   Residential and office buildings,

ii. Hospitals, cinema halls, and departmental stores,

iii. Libraries, museums, computer centers, and research laboratories,

iv. Transport vehicles:

    a. cars, buses, and rail coaches,

    b. aircrafts, space shuttles, and rockets,

    c. submarines.

iv. Printing, textile, and photographic products,

v. Food and process industries,

vi. Production shop laboratories, manufacture of materials, and precision devices.

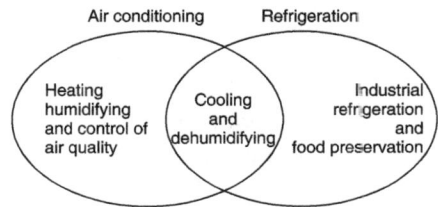

Air conditioning   Refrigeration

Heating humidifying and control of air quality / Cooling and dehumidifying / Industrial refrigeration and food preservation

FIGURE 4.24 Relationship of refrigeration and air conditioning.

Air conditioning essentially performs three services in the manufacture of precision metal parts. These services are

a. maintenance of uniform temperatures so that the metals neither expand nor contract,

b. control of humidity so that the rusting of metals is prevented, and

c. filtration of air so as to minimize dust. Cleanliness of air-conditioned space is absolutely essential where electronic components are being manufactured.

The fields of refrigeration and air conditioning are very closely inter-related as indicated in Figure 4.24.

4.16. COMFORT AIR CONDITIONING AND ITS TYPES

Comfort air conditioning deals with the creation of an optimum environmental condition conducive to human health, comfort, and efficiency. Air-conditioning systems in homes, offices, stores, restaurants, theaters, schools, and hospitals are of this type.

The comfort air-conditioning systems are generally classified into the following three categories:

- **Summer air conditioning:** These systems when properly designed and installed maintain the temperature and humidity of indoor air to a level at which persons feel comfortable. Essentially it involves reducing the air temperature and humidity (in humid tropics) by the process of cooling and dehumidification.

- **Winter air conditioning:** These systems are meant for the control of environmental conditions of indoor air so as to provide comfort in winter. Essentially it involves an increase in sensible heat and water content of air by the process of heating and humidification. The heating is done by furnaces or boilers fired with solid, liquid, or gaseous fuels.

- **Year-round air conditioning:** This system manifests in the control of temperature and humidity in an enclosed space for all times of the year; this is despite a change in the atmospheric conditions. Essentially the system comprises the heating and cooling equipment with associated components and automatic controls.

4.17. HUMAN COMFORT

Thermal comfort is a condition of mind which expresses satisfaction with thermal environment. It is the state where the person is entirely unaware of his surroundings, no consideration whether the space is too hot or too cold. Dissatisfaction with the thermal environment may be caused by the body as a whole (being too hot or cold) or by the unwanted heating or cooling of a particular part of the body (local discomfort).

Human comfort refers to the control of temperature and humidity of air and its circulation so that the resulting environment becomes human friendly; the state of environment where persons feel comfortable. Comfort is, however, a subjective quality; it is dependent on the preferences of an individual and varies with the age, sex, state of health, and clothing of a person.

4.18. WINDOW AIR CONDITIONER

An air-conditioning system is an assembly of different components and parts used to produce a comfortable cooling/heating conditions of air within a closed space.

The closed chamber may be a living room, a conference/seminar hall, or an auditorium/theater. Further, the requirement may be industrial air conditioning for a highly precision machine or for a research laboratory or for human comfort.

General human comfort conditions to be maintained fall in the range of

- temperature: 22°C to 25°C,
- relative humidity: 40% to 60%, and
- air velocity: 5 m/min to 8 m/min.

Besides these parameters, the standards of air purity in terms of freedom from dirt, dust, foul smell, and odor, and noise is also to be ensured for the conditioned space.

The basic elements of an air-conditioning system are

i. refrigerating plant,

ii. means for humidification or dehumidification of air,

iii. control system for automatic regulation of cooling or warming,

iv. fans for moving the air to and from the room,

v. filters for cleaning the air by removing dust and dirt particles, and

vi. supply and return ducts.

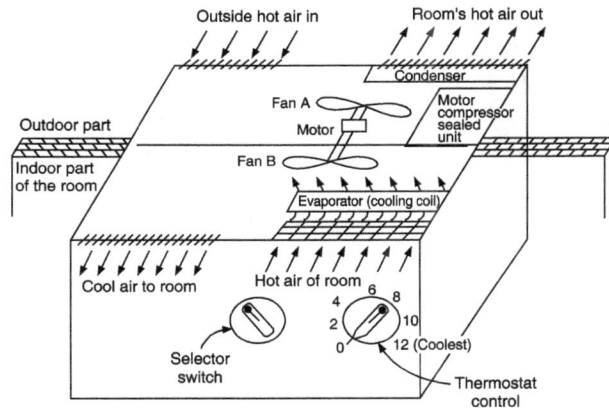

FIGURE 4.25 Window air conditioner.

Figure 4.25 shows the constructional details of a window air conditioner used for human comfort during summer. This is a self-contained machine because it houses all the components including evaporator and condenser in a common enclosure. The unit is mounted either in a window or on the wall of the room to be air-conditioned. Such units are available in cooling capacity from ½ to 3½ tons of refrigeration.

The main components of the machine comprise the following sub-assemblies:

(1) System assembly consisting of compressor, condenser, capillary, and evaporator units of the refrigerator system.

The hermetically sealed compressor is a compact unit containing both the compressor and motor mounted on a common shaft and encased in two-halves a dome-shaped casing which are joined together by circumferential welded joint. The lubrication of the compressor parts is done by the lubricating oil contained in the lower part of the dome.

The window air conditioners of small size generally have the condenser of the refrigerator air cooled, but in large sizes, the condenser may be water cooled, in which case pipe connections are needed.

The air-cooled finned type condenser is made up of copper tubes in the form of a coil and provided with aluminum fins.

The evaporator is in the form of coils made of copper and provided with aluminum fins. The capillary tubing is located between the condenser and the evaporator unit.

(2) Cabinet and grill assembly equipped with filtering unit. The filtering unit consists of oil filter or water filter and carbon filter. Oil or water filter cleans the dust particles while the carbon filter removes smell of different gases.

(3) Switch board panel assembly consists of selector switch and the thermostat control. The selector switch helps run the fan/compressor at low, medium, and high speeds, and the thermostat fixes the desired temperature.

(4) Outdoor and indoor fans which may be driven by the same motor or may be driven by separate motors.

The refrigerant unit employs Freon-22 or R-134 as the refrigerant.

The compressor motor unit, condenser, and outdoor fan are kept outdoor, that is, outside the room while the remaining components are placed indoor, that is, inside the room.

Specifications

A window air conditioner is normally specified by the following parameters:

- Capacity: 1, 1.5, 2 ton, etc.
- Overall dimensions: length × width × height
- Power supply: AC, 220–240 volts, 50 hertz
- Control: Site or remote

Working

When the power switch is put in the ON position, the motor-compressor unit starts running. The refrigerant vapors at low temperature and low pressure coming from the evaporator enter the compressor through suction line. The vapors are compressed and there is an increase both in temperature and pressure. These vapors are led to condenser through discharge line. The vapors condense rejecting their heat to the atmospheric air. The condensed vapors next enter the capillary tube, are throttled to low pressure on account of friction, and their temperature gets reduced to minimum operating temperature of the refrigerant cycle. The low-pressure liquid refrigerant now enters the evaporator, absorbs the latent heat of vaporization from the room air, and that results in the cooling of this air. This cooled air is directed through a ducted passage in the front cover grill and that provides comfortable cooling conditions in the air.

The refrigerant vapors leave the evaporator and enter the compressor during its suction stroke and that completes the working cycle.

REVIEW QUESTIONS

A. Conceptual and conventional questions

1. Define refrigeration and air conditioning.

2. State the difference between a refrigerator and a heat pump. How do these machines satisfy the second law of thermodynamics?

3. What is meant by COP? What value of COP is desirable, large or small and why?

4. Set up a relation for the COP of a heat pump and that of a refrigerator. Proceed to show that

$$(COP)_{\text{heat pump}} = 1 + (COP)_{\text{refrigerator}}$$

5. Define the following terms:
   (a) refrigerating effect   (b) relative COP
   (c) ton of refrigeration

6. Mention the various applications of refrigeration.

7. Describe, with a neat schematic arrangement, the working of a domestic refrigerator.

8. What are moist air and saturated air?

9. Define and explain the following terms in relation to psychrometry
   (a) dry-bulb, wet-bulb, and dew-point temperatures.
   (b) relative humidity and specific humidity

10. Establish the following expression for air–vapor mixture

$$\text{specific humidity } \omega = 0.622 \ \frac{p_v}{p_v - p_v}$$

where p_v is the partial pressure of water vapor and p_b is the barometric pressure.

11. Define and explain the concept of dew-point and adiabatic saturation temperature.

12. What is a sling psychrometer? Draw its neat sketch and explain its use.

13. What is a psychrometric chart? What information does it provide?

14. Name any five psychrometric processes and represent them on the psychrometric chart.

15. Define air conditioning and mention some of its applications.

16. What is meant by comfort air conditioning? Give brief description of its various types.

B. Fill in the blanks with appropriate word/words

1. A refrigeration system removes heat from a system at _____ temperature and transfers the same to a system at _____ temperature.

2. One ton of refrigeration is equivalent to _____ kW

3. The bank of tubes at the back of a domestic refrigerator of vapor compression type is the _____ tubes.

4. In a vapor compression refrigeration system, the capillary tube is located between _____ and _____

5. For psychrometric purpose, _____ is assumed to be a pure substance and not a mixture.

6. The _____ air is essentially a mixture of dry air and water vapor.

7. _____ humidity represents the amount of water vapor actually present in the air.

8. The wet-bulb temperature would be zero when the relative humidity is _____ percent.

9. The _____ is a measure of the capacity of air to absorb moisture.

10. If total pressure remains constant, the _____ humidity is a function of partial pressure of water vapor only.

11. The difference between the dry-bulb temperature and wet-bulb temperature is known as _____

12. The dry- and wet-bulb temperatures are measured by instruments called _____

13. On a psychrometric chart, the dry-bulb temperature is taken as _____ and specific humidity as _____ .

14. The _____ curve represents 100 percent relative humidity at various dry-bulb temperatures.

Answers:
1. low, high; **2.** 3.5 kW; **3.** condenser; **4.** condenser, evaporator; **5.** dry air; **6.** moist air; **7.** absolute; **8.** 100%; **9.** degree of saturation; **10.** specific; **11.** Wet-bulb depression; **12.** psychrometers; **13.** abscissa, ordinate; **15.** saturation.

C. Multiple choice questions

1. The capacity of a refrigerating machine is expressed as
   (a) lowest temperature attainable
   (b) rate of absorption of heat from the space being cooled
   (c) inside volume of the cabinet
   (d) gross weight of the machine in tons

2. One ton of refrigeration implies that the machine has a refrigerating effect equal to
   (a) 1 kJ/s   (b) 2.5 kJ/s   (c) 3.5 kJ/s   (d) 5 kJ/s

3. Which part of the vapor compression refrigeration cycle produces the refrigerating effect?
   (a) condenser   (b) throttle valve   (c) evaporator   (d) compressor

4. The domestic refrigerator has a refrigerating load of the order of
   (a) less than 0.25 ton   (b) between 0.5 and 1 ton
   (c) more than 1 ton   (d) more than 5 ton

5. Absolute humidity of air–vapor mixture at a particular temperature is defined as
   (a) mass of water vapor per kg of dry air in the mixture
   (b) mass of water vapor per m^3 of mixture
   (c) mass of water vapor per kg of saturated air
   (d) ratio of actual partial pressure of water vapor to the saturation pressure of water vapor at the same temperature

6. The humidity ratio or specific humidity of a given air–vapor mixture is
   (a) mass of water vapor per kg of dry air in the mixture
   (b) mass of water vapor per m^3 of mixture
   (c) mass of water vapor per m^3 of dry air in the mixture
   (d) mass of water vapor per kg of mixture

7. If m_a = mass of dry air and m_w = mass of water vapor in the air–water vapor mixture, then humidity ratio is given by

   (a) $\dfrac{m_w}{m_a}$ **(b)** $\dfrac{m_a}{m_w}$ **(c)** $\dfrac{m_w}{m_a + m_w}$ **(d)** $\dfrac{m_a}{m_a + m_w}$

8. The humidity ratio ω of air is expressed by the relation

   (a) $0.622 \dfrac{p_w}{p_a + p_w}$ **(b)** $0.622 \dfrac{p_a}{p_a + p_w}$ **(c)** $0.622 \dfrac{p_w}{p_t - p_w}$ **(d)** $0.622 \dfrac{p_a}{p - p_w}$

   where p_w is partial pressure of water vapor, p_a is partial pressure of dry air and p_t is the barometric pressure, that is, total pressure of moist air

9. Ratio of mass of water vapor in a given volume of mixture to the mass of water vapor in the same volume of saturated mixture at the same pressure and temperature is known as
   (a) specific humidity   **(b)** humidity ratio
   (c) relative humidity   **(d)** degree of saturation

10. The degree of saturation is defined as
    (a) mass of water vapor per unit volume of air–water vapor mixture
    (b) mass of water vapor per kg of dry air in the mixture
    (c) the ratio of mass of water vapor in a given volume of moist air to the mass of water vapor in the same volume of saturated air at the same temperature and pressure
    (d) the ratio of mass of water vapor associated with unit mass of dry air to the mass of water vapor associated with unit mass of dry air saturated at the same pressure and temperature.

11. Ratio of the specific humidity of a given volume of mixture to the specific humidity of the saturated mixture at the same temperature is called
    (a) relative humidity   **(b)** humidity ratio
    (c) percentage humidity   **(d)** degree of saturation

12. Which of the followings are synonyms?
    (a) absolute humidity and specific humidity
    (b) specific humidity and humidity ratio
    (c) humidity ratio and relative humidity
    (d) relative humidity and degree of saturation

13. The temperature of air–water mixture recorded by a thermometer when its bulb is covered with a wick saturated with water and placed in the air–water stream is called
    (a) wet-bulb temperature   **(b)** dew-point temperature
    (c) saturation temperature   **(d)** critical temperature

14. Wet-bulb depression represents the difference between
(**a**) dry-bulb temperature and wet-bulb temperature
(**b**) dry-bulb temperature and dew-point temperature
(**c**) dew-point temperature and saturation temperature
(**d**) adiabatic saturation temperature and dew-point temperature

15. The wet-bulb depression is zero when relative humidity is equal to
(**a**) zero   (**b**) 0.5   (**c**) 0.75   (**d**) 1.0

16. A 100 percent relative humidity of air implies that
(**a**) wet-bulb temperature is equal to the dew-point temperature
(**b**) dew-point temperature is equal to the saturation temperature
(**c**) saturation temperature is equal to the dry-bulb temperature
(**d**) dry-bulb, wet-bulb, dew-point, and saturation temperatures are equal

17. A sling psychrometer can measure
(**a**) absolute humidity   (**b**) specific humidity
(**c**) wet-bulb temperature   (**d**) dry- as well as wet-bulb temperature

18. Consider the following properties of air–water vapor mixture
1. dry-bulb temperature   **2.** Wet-bulb temperature
3. specific humidity   **4.** relative humidity

The psychrometric chart shows the relationships between
(**a**) 1 and 2 only   (**b**) 2 and 3 only
(**c**) 1, 2, and 4   (**d**) 1, 2, 3, and 4

19. In a psychrometric chart, the vertical lines parallel to the ordinate indicate
(**a**) dry-bulb temperature   (**b**) wet-bulb temperature
(**c**) specific humidity   (**d**) enthalpy of saturation

20. The uniformly spaced horizontal lines running parallel to the abscissa in a psychrometric chart indicate
(**a**) absolute humidity
(**b**) specific humidity
(**c**) dew-point temperature
(**d**) volume

21. On a psychrometric chart, the relative humidity lines are represented by
(**a**) uniformly spaced horizontal lines
(**b**) uniformly speed inclined lines
(**c**) non-uniformly spaced inclined lines
(**d**) curved lines

22. On a psychrometric chart, the constant wet-bulb temperature lines coincide with
    (a) constant relative humidity lines
    (b) constant enthalpy lines
    (c) constant dew-point temperature lines
    (d) constant volume lines

23. Consider the following factors:
    1. air motion
    2. dry-bulb temperature
    3. relative humidity
    4. wet-bulb temperature
    5. elevation of the plate and its latitude

    Which of these affect human comfort?
    (a) 1, 2, and 3   (b) 1, 2, 3, and 4
    (c) 2, 3, and 5   (d) 1, 2, 3, and 5

Answers:

1. (b)	**2.** (c)	**3.** (c)	**4.** (a)	**5.** (b)	**6.** (a)	**7.** (a)
8. (c)	**9.** (c)	**10.** (d)	**11.** (d)	**12.** (b)	**13.** (a)	**14.** (a)
15. (d)	**16.** (d)	**17.** (d)	**18.** (d)	**19.** (a)	**20.** (b)	**21.** (d)
22. (d)	**23.** (a)					

CHAPTER 5

FLUID MECHANICS

Fluid Mechanics is basically a study of the

i. physical behavior of fluids and fluid systems, and of the laws governing this behavior,

ii. action of forces on fluids and of the resulting flow pattern.

Fluid mechanics may be divided into three divisions:

1. *Hydrostatics* that studies the mechanics of fluids at absolute and relative rest; the fluid elements are free from shearing stresses.

2. *Kinematics* that deals with translation, rotation, and deformation of fluid elements without considering the force and energy causing such a motion.

3. *Dynamics* that prescribes the relations between velocities and accelerations and the forces which are exerted by or upon the moving fluids.

After reading this chapter, the reader would be able to

i. make distinction between the three states of matter (solid, liquid, and gas).

ii. understand the concept of a fluid, real and ideal fluid.

iii. make a brief review about the development of the exciting and fascinating subject of fluid mechanics.

iv. appreciate the unlimited practical applications of fluid mechanics.

5.1. SOLIDS, LIQUIDS, AND GASES

Matter exists in two principal forms: solid and fluid. Fluid is further subdivided into liquid and gas. Distinguishing features among these are as follows:

i. The solids, liquids, and gases exhibit different characteristics on account of their different molecular structure. Spacing and the latitude of the motion of molecules are large in a gas, small in a liquid, and extremely small in a solid. Accordingly, the inter-molecular bonds are very weak in a gas, weak in a liquid, and very strong in a solid. It is due to these aspects that solid is very compact and rigid in form, liquid accommodates itself to the shape of its container, and gas fills up the whole of the vessel containing it.

ii. For a given mass, the liquids have a definite volume irrespective of the size of the container. The variation of volume with temperature and pressure is insignificant. Liquid occupies the vessel fully or partially depending on its mass, and it forms a free surface with the atmosphere. The gas, however, expands to fill any vessel in which it is contained and does not form any free surface. Accordingly, it may be stated: *The solid has volume and shape; a liquid has volume but no shape; a gas has neither*.

iii. For all practical purposes, the liquids like solids can be regarded as incompressible. This means that pressure and temperature changes have practically no effect on their volume. The gases are, however, readily compressible fluids. They expand infinitely in the absence of pressure and contract easily under pressure. Nevertheless, density variation is small. For example in the flow of air in a ventilating system, the gas flow can also be treated as incompressible without involving any appreciable error.

When a gas can be readily condensed to a liquid, we call it a *vapor* such as steam and ammonia.

iv. The deformation due to normal and tangential forces for solids is such that within elastic limits, the deformation disappears and the solid body is restored to its original shape when the stress causing the deformation is removed. A fluid at rest can, however, sustain only normal stresses and deforms continuously when subjected to shear stress; no matter how small that shear stress may be. Even though the fluid comes to rest when the shear stress is removed, there is no tendency to restore the fluid body to its original shape or position.

Thus, fluid can offer no permanent resistance to shear force and possesses a characteristic ability to flow or change its shape. Flow means that the constituent fluid particles continuously change their positions relative to one another. This concept of fluid flow under the application of shear stress is illustrated in Figure 5.1. A fluid element occupying the initial position 011 continues to move or deform to new position 022, 033, etc., when shear stress τ is applied to it.

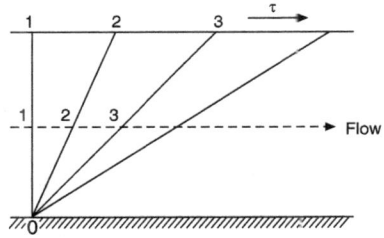

FIGURE 5.1 Fluid flow.

The tendency of continuous deformation of a fluid is called *fluidity*, and the act of continuous deformation is called *flow*.

The above discussion can be summed up as follows:

Liquid	Gas
i. A given mass of liquid has a definite volume independent of the size or shape of the container; however, it changes its shape easily and acquires the shape of its container.	i. A given mass of gas has no fixed volume, and it expands continuously to completely fill any container in which it is placed.
ii. A free surface is formed if the volume of the container is greater than that of the liquid.	iii. No free surface is formed.
iv. Liquids can be regarded as incompressible for all practical purposes.	v. Gases are readily compressible.
vi. Pressure and temperature changes have practically no effect on the volume of a liquid.	vii. A gas expands infinitely in the absence of pressure and contracts easily under pressure.
viii. Water, kerosene, petrol, etc. are liquids.	ix. Air, ammonia, carbon dioxide, etc. are gases.

5.2. IDEAL AND REAL FLUIDS

A fluid is said to be ideal if it is assumed to be both incompressible and inviscid (non-viscous). Further, an ideal fluid has no surface tension. For an inviscid fluid, viscosity is zero, and no frictional forces are set up even during fluid motion:

$$\mu = 0$$

$$\rho = \text{constant}; K = \frac{-dp}{dv/v} = \infty$$

$$\sigma = 0$$

Ideal fluids are imaginary and do exist in nature. However, most common fluids such as air and water have a very low value of viscosity and can be treated as ideal fluids for all practical purposes without introducing any appreciable error. Since water is incompressible, it is moved near to an ideal fluid than air.

Real or practical fluids have viscosity (μ), compressibility (K), and surface tension (σ). Whenever motion takes place, the tangential or shear forces always come into play due to viscosity and some frictional work is done.

5.3. SIGNIFICANCE OF FLUID MECHANICS

The subject of fluid mechanics encompasses a great many fascinating areas like

- design of a wide range of hydraulic structures (dams, canals, weirs, etc.) and machinery (pumps, turbines, and fluid couplings)

- design of a complex network of pumping and pipelines for transporting liquids; flow of water through pipes and its distribution to domestic service lines

- fluidic control devices both pneumatic and hydraulic

- design and analysis of gas turbines, rocket engines, conventional, and supersonic aircraft

- power generation from conventional methods such as hydroelectric, steam, and gas turbines, to newer ones involving magneto fluid dynamic

- methods and devices for the measurement of various parameters, for example, the pressure and velocity of a fluid at rest or in motion.

- study of men's environment in the subjects like meteorology, oceanography, and geology

- human circulatory system, that is, flow of blood in veins and the vital role, it plays in a variety of engineering applications.

The numerous natural phenomena such as the rain cycle, weather patterns, the rise of groundwater to the top of trees, winds, ocean waves, and currents in large water bodies are also governed by the principles of fluid mechanics.

Figure 5.2 briefly explains the significance of fluid mechanics and the vital role that it plays in a variety of engineering applications.

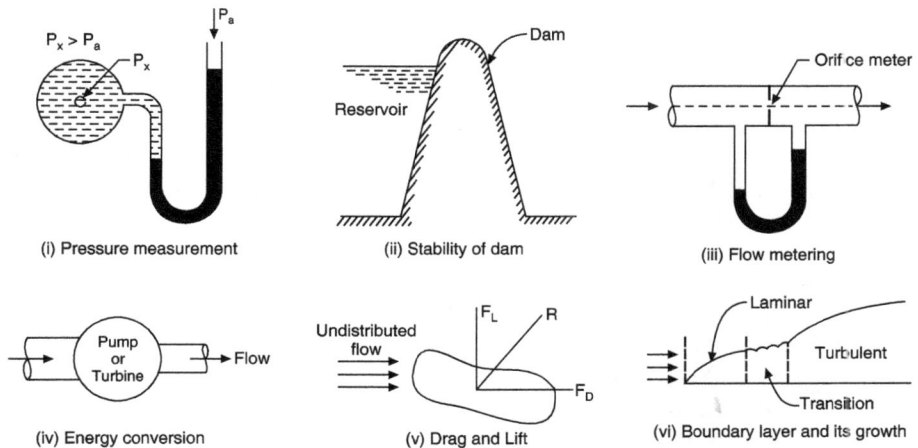

(i) Pressure measurement (ii) Stability of dam (iii) Flow metering

(iv) Energy conversion (v) Drag and Lift (vi) Boundary layer and its growth

FIGURE 5.2 Fluid mechanics applications.

The significance of fluid mechanics can be well judged by citing just one example of automobile drive where suspension is provided by pneumatic tires, road shocks are reduced by hydraulic shock absorbers, gasoline is pumped through tubes and later atomized, air resistance creates a drag on the vehicle as whole and the confidence that hydraulic brakes would operate when the vehicle is made to stop.

Undoubtedly, a study of the science of fluid mechanics is a must for an engineer so that he can understand the basic principles of fluid behavior and apply the same to flow situations encountered in engineering and physical problems.

5.4. FLUID PROPERTIES

Every fluid has certain characteristics by means of which its physical condition may be described. Such characteristics are called the **properties** of the fluid. Before an analyst of fluid flow problems can venture to formulate the physical principles governing the flow situation, he has to be thoroughly familiar with the physical properties of fluids. Toward that end, this section seeks to provide basic insight into the fluid properties and their behavior.

5.4.1. Specific weight, mass density, and specific gravity

a. *Specific weight* (w) of a fluid is its weight per unit volume.

$$w = \frac{W}{V} \tag{5.1}$$

where W is the weight of the fluid having volume V. The weight of a body is the force with which the body is attracted to the center of the earth. It is the product of its mass and the local gravitational acceleration, that is, $W = mg$. The value of g at sea level is 9.807 m/s^2 approximately. Since weight is expressed in Newton, the unit of measurement of specific weight is N/m^3. In terms of fundamental units, the dimensional formula of specific weight is $\left[\dfrac{F}{L^3}\right]$ or $\left[\dfrac{M}{L^2T^2}\right]$.

For pure water under standard atmospheric pressure of 760 mm of mercury at mean sea level and a temperature of 4°C, the specific weight is 9810 N/m^3. For sea water, the specific weight equals 10 000 – 10 105 N/m^3. The increased value of the specific weight of water is due to the presence of dissolved salts and suspended matter. The specific weight of petroleum and petroleum products varies from 6 350 to 8 350 N/m^3 and that of mercury at 0°C is 13 420 N/m^3. Air has a specific weight of 11.9 N/m^3 at 15°C temperature and at standard atmospheric pressure. The specific weight of fluid changes from one place to another depending upon changes in gravitational acceleration.

b. *Density* $(\rho$-pronounced rho) is a measure of the amount of fluid contained in a given volume and is defined as the mass per unit volume:

$$\rho = \frac{m}{V} \tag{5.2}$$

where m is the mass of fluid having volume V. Fluid mass is a measure of the ability of a fluid particle to resist acceleration and is approximately independent of its location on the earth's surface. The units of density correspond to those of mass and volume. The dimensional formula of density in fundamental units is $\left[\dfrac{M}{L^3}\right]$ or $\left[\dfrac{FT^2}{L^4}\right]$ and the corresponding units are kg/m^3 or N s^2/m^4.

The density of a fluid diminishes with a rise of temperature except for water which has a maximum value at 4°C. The mass density of water at 15.5°C is 1000 kg/m³, and for air at 20°C and at atmospheric pressure, the mass density is 1.24 kg/m³.

Relations 5.1 and 5.2 are valid only when the fluid medium fills the given volume completely without any blank space, that is, the fluid is a continuum. For a non-homogeneous fluid, these relations give average specific weight and density. To determine the absolute values of w and ρ at any point, the volume is regarded as tending to zero and the limit of the corresponding ratio is calculated:

$$w = \underset{V \to 0}{Lt}\; \frac{m}{V} = \frac{dW}{dV}$$

$$\rho = \underset{V \to 0}{Lt}\; \frac{m}{V} = \frac{dM}{dV}$$

The weight W and the mass m of a fluid are related to each other by the expression $W = mg$. Dividing this expression throughout by volume V of the fluid, we obtain:

$$\frac{W}{V} = \frac{m}{V}\, g \text{ or } w = \rho\, g \tag{5.3}$$

Equation 5.3 reveals that the specific weight w changes with location depending upon gravitational pull.

c. **Specific gravity** (s) refers to the ratio of specific weight (or mass density) of a fluid to the specific weight (or mass density) of a standard fluid. For liquids, the standard fluid is water at 4°C, and for gases, the standard fluid is taken either air at 0°C or hydrogen at the same temperature. Specific gravity is dimensionless and has no units.

A statement that the specific gravity of mercury is 13.6 implies that its weight (or mass) is 13.6 times that of the same volume of water. In other words, mercury is 13.6 times heavier than water.

d. **Specific volume** (v) represents the volume per unit mass of fluid; specific volume is the inverse of the mass density:

$$v = \frac{V}{m}\; ; v = \frac{1}{\rho} \tag{5.4}$$

The concept of specific volume is found to be practically more useful in the study of flow of compressible fluids, that is, gases.

EXAMPLE 5.1

2 liter of petrol weighs 14 N. Calculate the specific weight, mass density, specific volume, and specific gravity of petrol with respect to water.

Solution: 2 liter = 2×10^{-3} m^3

Specific weight is a measure of the weight per unit volume:

$$\therefore \qquad \text{Specific weight } w = \frac{14}{2 \times 10^{-3}} = \mathbf{7000 \ N/m^3}$$

Mass density is related to specific volume by the relation:

$$w = \rho \ g$$

$$\text{Mass density } \rho = \frac{w}{g} = \frac{7000}{9.81} = \mathbf{713.56 \ kg/m^3}$$

Specific volume v is the inverse of mass density

$$v = \frac{1}{\rho} = \frac{1}{713.56} = \mathbf{1.4 \times 10^{-3} \ m^3/kg}$$

$$\text{Specific gravity } s = \frac{\text{density of oil}}{\text{density of water}} = \frac{713.56}{1000} = \mathbf{0.7136}$$

EXAMPLE 5.2

If specific gravity of a liquid is 0.80, make calculations for its mass density, specific volume, and specific weight (weight density).

Solution: Specific gravity $= \dfrac{\text{mass density of liquid}}{\text{mass density of water}}$

$$\therefore \quad \text{Mass density of liquid } \rho = 0.80 \times 1000 = \mathbf{800 \ kg/m^3}$$

$$\text{Specific volume } v = \frac{1}{\rho} = \frac{1}{800} = \mathbf{1.25 \times 10^{-3} \ m^3/kg}$$

$$\text{Specific weight (weight density) } w = \rho \ g = 800 \times 9.81 = \mathbf{7848 \ N/m^3}$$

5.4.2. Viscosity

Viscosity is a property of the fluid by which it offers resistance to shear or angular deformation.

Experimental evidence indicates that when any fluid flows over a solid surface, the velocity is not uniform at any cross section; it is zero (no slip) at the solid surface and progressively approaches the free stream velocity in the fluid layers far away from the solid surface. This aspect of the velocity profile (a curve connecting the tips of velocity vectors) indicates the existence of some resistance to flow due to friction between a fluid layer and the solid surface, and between adjacent layers of fluid itself. Again the velocity gradient (the spatial rate of change of velocity du/dy) is large at the solid surface and gradually diminishes to zero with distance from the wall. Evidently, the resistance between the fluid and surface is greater when compared to that between the fluid layers themselves.

The resistance to flow because of internal friction is called ***viscous resistance***, and the property which enables the fluid to offer resistance to relative motion between adjacent layers is called the ***viscosity*** of fluid. Viscosity is thus a measure of resistance to relative translational motion of adjacent layers of fluid. This property is manifested by all the real fluids, and it distinguishes them from ideal or non-viscous fluids. Molasses, tar, and glycerine are examples of highly viscous liquids; the inter-molecular force of attraction between their molecules is very large, and consequently, they cannot be easily poured or stirred. Fluids like water, air, and petrol have a very small viscosity; they flow much more easily and rapidly and are called thin fluids.

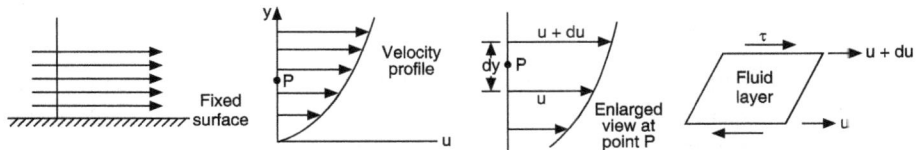

FIGURE 5.3 Velocity profile and viscosity concept.

Newton's Law of Viscosity

Consider two adjacent layers at an infinitesimal distance dy apart and moving with velocity u and $(u + du)$, respectively. The upper layer moving with velocity $(u + du)$ drags the lower layer along with it by exerting a force F. However, the lower layer tries to retard or restrict the motion of the upper layer by exerting a force equal and opposite to F. These two equal and opposite forces induce a shear or viscous resistance τ (pronounced tau) given by F/A where A is the contact area between the two layers. Experimental measurements have shown that the shear stress is proportional to the spatial rate of velocity normal to the flow:

$$\tau \propto \frac{du}{dy}; \ \tau = \mu \frac{du}{dy} \tag{5.5}$$

The term $\dfrac{du}{dy}$ is more usually called the velocity gradient at right angles to the direction of velocity itself. The proportionality constant μ (pronounced mew) in Equation 5.5 is a function of the fluid involved and is called the **coefficient of viscosity, absolute viscosity,** or **dynamic viscosity.** Equation 5.5 is first suggested by Newton and is referred to as **Newton's viscosity equation** or **Newton's law of viscosity.**

The following observations help appreciate the interaction between viscosity and velocity distribution:

- Maximum shear stresses occur where the velocity gradient is the largest, and the shear stresses disappear where the velocity gradient is zero.

- Velocity gradient at the solid boundary has a finite value. The velocity profile cannot be asymptotic to the boundary because that would imply an infinite velocity gradient and, in turn, infinite shear stress.

- Velocity gradient becomes less steep (du/dy becomes small) with distance from the boundary. Consequently, the maximum value of shear stress occurs at the boundary, and it progressively decreases with distance from the boundary.

Deformation of fluid elements can be prescribed in terms of the angle of shear strain $d\theta$. Figure 5.4 indicates a thin sheet of fluid element ABCD placed between two plates' distances dy apart. The length and the width of the plates are much larger than the thickness dy so that the edge effects can be neglected. When force F is applied to the upper plate, it causes it to move at a small speed du relative to the bottom plate. Velocity gradient sets up shear stress $\tau = F/A$ which makes the fluid element distort to position AB′C′D after a short time interval dt:

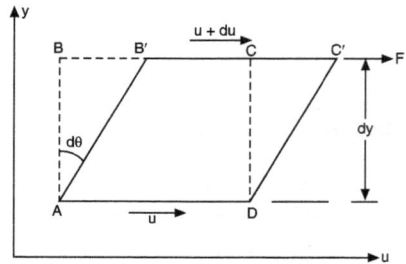

FIGURE 5.4 Shear stress and time rate of shear strain.

Distance BB′ = CC′ = speed × time = $du \times dt$

For small angular displacement $d\theta$, BB′

$$= dy \times d\theta$$

$$\therefore \qquad du \times dt = dy \times d\theta;$$

$$\frac{du}{dy} = \frac{d\theta}{dt}$$

Invoke Newton's law of viscosity, that is, express the shear stress in terms of velocity gradient:

$$\tau = \mu \frac{du}{dy}; \; \tau = \mu \frac{d\theta}{dt} \tag{5.6}$$

Apparently, the shear stress in fluids is dependent on the rate of fluid deformation $\frac{d\theta}{dt}$. This characteristic serves to distinguish a solid from a fluid.

While the shear stress in a solid material is generally proportional to shear strain, the shear stress in a viscous fluid is proportional to the time rate of strain.

Dimensional Formula and Units of Viscosity

The units of viscosity can be worked out from Newton's equation of viscosity; $\tau = \mu \; du/dy$. Solving for the viscosity μ and inserting dimensions F, L, and T for force, length, and time:

$$\mu = \frac{\tau}{du/dy} = \left[\frac{F}{L^2}\right] \div \left[\frac{L}{T} \times \frac{1}{L}\right] = \left[\frac{FT}{L^2}\right]$$

When the force dimension is expressed in terms of mass, $F = \left[\dfrac{ML}{T^2}\right]$, the dimensions for viscosity in terms of mass, length, and time become $\left[\dfrac{M}{LT}\right]$.

When appropriate units are inserted for force, length, and time, the dynamic viscosity will have the units:

$$\mu = \frac{\tau}{du/dy} = \frac{N/m^2}{\left(\dfrac{m}{s} \times \dfrac{1}{m}\right)} = \frac{Ns}{m^2} = Pa\; s$$

Sometimes, the coefficient of dynamic viscosity μ is distinguished by poise (P)

$$1 \; poise = \frac{1\; gm}{cm\; sec} = \frac{1\; dyne\; sec}{cm^2}$$

$$= \frac{10^{-5}}{\left(10^{-2}\right)^2} \frac{Ns}{m^2} = \frac{0.1\; N\; s}{m^2} = 0.1\; Pa\; s$$

A poise turns out to be a relatively large unit; hence, the unit centipoise (cP) is generally used: 1 cP = 0.01 P. Typical values of viscosity for water and air at 20°C and at standard atmospheric pressure are

$$\mu \text{ water} = 1.0 \text{ cP} = 10^{-3} \text{ N s/m}^2$$
$$\mu \text{ air} = 0.0181 \text{ cP} = 0.0181 \times 10^{-3} \text{ N s/m}^2$$

that is, water is nearly 55 times as viscous as air.

Specific viscosity is the ratio of the viscosity of fluid to the viscosity of water at 20°C. Since water has a viscosity of 1 cP at 20°C, the viscosity of any fluid expressed in centipoise units would be a measure of the viscosity relative to water.

Kinematic Viscosity

The ratio between the dynamic viscosity and density is defined as **kinematic viscosity** of fluid and is denoted by v (pronounced new):

$$\text{Kinematic viscosity} = \frac{\text{dynamic viscosity}}{\text{mass density}}; \ v = \frac{\mu}{\rho} \tag{5.7}$$

The dimensional formula for kinematic viscosity is

$$v = \left[\frac{M}{LT}\right] \div \left[\frac{M}{L^3}\right] = \left[\frac{L^2}{T}\right]$$

The kinematic viscosity does not involve force; its only dimensions are length and time as in kinematics of fluid flow. Typical units of v are m²/s or cm²/s, the latter being referred to as stoke (St) to perpetuate the name of the English physical Sir George Stokes. A centistoke (c St) is one-hundredth of a stoke: 1 c St = 0.01 St. Typical values of kinematics viscosity at 20°C and at standard atmospheric pressure are

$$v \text{ water} = 1.0 \text{ c St} = 1 \times 10^{-6} \text{ m}^2/\text{s}$$
$$v \text{ air} = 15.0 \text{ c St} = 15 \times 10^{-6} \text{ m}^2/\text{s}$$

that is, the kinematic viscosity of air is about 15 times greater than the corresponding value of water.

EXAMPLE 5.3

i. **A lubricating oil of viscosity μ undergoes steady shear between a fixed lower plate and an upper plate moving at speed V. The clearance between the plates is h. Show that a linear velocity profile results if the fluid does not slip at either plate.**

ii. **Two horizontal plates are placed 1.25 cm apart, the space between them being filled with the oil of viscosity 14 poise. Compute the** shear stress in the oil if the upper plate is moved with a velocity of 2.5 m/s.

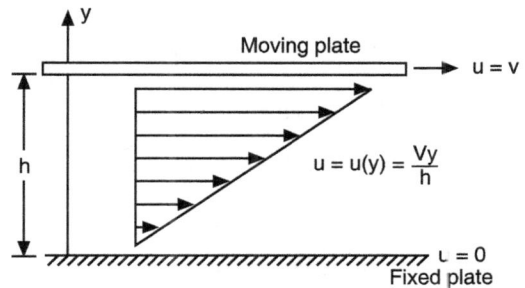

FIGURE 5.5

Solution: The shear stress τ is constant throughout the fluid for the given geometry and motion and, therefore, from Newton's law of viscosity:

$$\frac{du}{dy} = \frac{\tau}{\mu} = \text{constant}$$

or $\qquad u = a + by$

The constants a and b are evaluated from the no slip conditions at the upper and lower plates.

$u = 0$ at $y = 0$; $0 = a$

$u = V$ at $y = h$; $V = a + bh$

Hence, $a = 0$ and $b = V/h$. The velocity profile between the plates is than given by $u = \dfrac{V}{h}y$ and is linear as indicated in Figure 5.6.

(*ii*) Viscous shear stress is given by the Newton's law of viscosity:

$$\tau = \mu \, \frac{du}{dy}$$

Given $\quad \mu = 14$ poise $= 1.4$ N s/m^2

$du = 2.5$ m/s and $dy = 1.25 \times 10^{-2}$ m

$$\therefore \qquad \tau = 1.4 \times \frac{2.5}{1.24 \times 10^{-2}} = 280 \text{ N/m}^2 = \textbf{280 Pa}$$

Although oil is very viscous, this is a modest shear stress, about 360 times less than atmospheric pressure.

EXAMPLE 5.4

The clearance space between a shaft and a concentric sleeve has been filled with Newtonian fluid. The sleeve attains a speed of 60 cm/s when a force of 500 N is applied to it parallel to the shaft. What force is needed if it is desired to move the sleeve with a speed of 300 cm/s?

Solution: For a Newtonian fluid, $\tau = \mu \dfrac{du}{dy}$. Since the space between the shaft and the sleeve is very small, that is, the oil film is thin, it can be presumed that $\dfrac{du}{dy} = \dfrac{u}{t}$ where u is the sleeve speed and t is the oil film thickness. Further,

$$\text{Shear stress } \tau = \frac{\text{force}}{\text{area}} = \frac{F}{A}$$

$\therefore$
$$\frac{F}{A} = \mu \frac{u}{t} \text{ or } F = A \mu \frac{u}{t}$$

A, μ, and t are constant, and therefore, $F \propto u$ and accordingly $\dfrac{F_1}{u_1} = \dfrac{F_2}{u_2}$.

Inserting the appropriate values,

$$\frac{500}{60} = \frac{F_2}{300}; F_2 = \textbf{2500 N.}$$

EXAMPLE 5.5

Two horizontal flat plates are placed 0.15 mm apart and the space between them is filled with oil of viscosity 1 poise. The upper plate of area 1.5 m² is required to move with a speed of 0.5 m/s relative to the lower plate. Determine the necessary force and power required to maintain this speed.

Solution: Viscous shear stress $\tau = \mu \dfrac{du}{dy}$

$$\text{Given } \mu = 1 \text{ poise} = 0.1 \text{ N s/m}^2; du = 0.5 \text{ m/s}$$
$$dy = 0.15 \text{ mm} = 0.15 \times 10^{-3} \text{ m}$$

$\therefore$
$$\text{Shear stress } \tau = \frac{0.1 \times 0.5}{0.15 \times 10^{-3}} = 333.3 \text{ N/m}^2$$

i. Shear resistance or force,

$$F = \text{shear stress} \times \text{area}$$
$$= 333.3 \times 1.5 = \textbf{500 N}$$

ii. Power required to move the upper plate at a speed of 0.5 m/s.

$$= Fu = (500 \times 0.5) \text{ Nm/s} = 250 \text{ W} = \textbf{0.25 kW}$$

EXAMPLE 5.6

A dash pot 10 cm diameter and 12.5 cm long slide vertically down in a 10.05 cm diameter cylinder. The oil filling the annular space has a viscosity of 0.80 poise. Find the speed with which the piston slides down if load on the piston is 10 N.

Solution: Since the space between the dash pot and the cylinder is very small, that is, the oil film is thin, we can presume that $\dfrac{du}{dy} = \dfrac{u}{t}$ where u is the piston speed and t is the oil film thickness.

$$\text{Shear stress } \tau = \mu \frac{du}{dy} = \mu \frac{u}{t}$$

$$\text{Shear or viscous force = shear stress} \times \text{area} = \mu \frac{u}{t} (2\pi\, rl)$$

$$\text{Given } r = \frac{10}{2} = 5 \text{ cm} = 0.05 \text{ m}$$

$$u = 0.8 \text{ poise} = 0.08 \text{ N/sm}^2$$

$$t = \frac{10.05 - 10}{2} = 0.025 \text{ cm} = 0.00025 \text{ m}$$

Viscous force is equal to the load of 10 N

$$\therefore\ 10 = 0.08 \times \frac{u}{0.00025} \times (2\pi \times 0.05 \times 0.125)$$

Hence, piston speed $u =$ **0.796 m/s.**

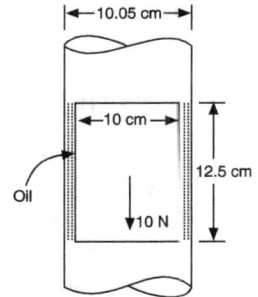

FIGURE 5.6

EXAMPLE 5.7

A cylinder of diameter 15 cm and weight 90 N slides, a distance of 12.5 cm in a lubricated pipe. The clearance between the cylinder and pipe is 2.5×10^{-3} cm. The cylinder is noted to decelerate at a rate of 0.6 m/s^2 when the speed is 6 m/s. Calculate the viscosity of the oil used for lubricating the pipe.

Solution: Viscous shear stress $\tau = \mu \dfrac{du}{dy} = \mu \dfrac{u}{t}$

$$\text{Viscous resistance or force = shear stress} \times \text{area}$$

$$= \mu \frac{u}{t} \times \pi\, dl = \frac{\mu \times 6}{2.5 \times 10^{-5}} \times \pi(0.15)(0.125)$$

$$= 14130\, \mu \text{ N}$$

Invoking Newton's second law: $\Sigma F = $ mass $\times$ acceleration

$$\therefore \quad 90 - 14130 \,\mu = \frac{90}{9.81}\,(-0.6)$$

$$90 - 14130 \,\mu = -5.5$$

$$\therefore \quad \mu = \frac{95.5}{14130} = \mathbf{6.76 \times 10^{-3} \ N \ s/m^2}$$

EXAMPLE 5.8

Find the kinematic viscosity of a liquid in stokes of which specific gravity is 0.95 and dynamic viscosity is 0.012 poise.

Solution: $\mu = 0.012$ poise $= 0.012 \times 0.1 = 1.2 \times 10^{-3}$ N s/m^2

Mass density of liquid = specific gravity $\times$ mass density of water

$$\rho = 0.95 \times 1000 = 950 \ \text{kg/m}^3$$

$$\therefore \ \text{Kinematic viscosity } \nu = \frac{m}{\rho} = \frac{1.2 \times 10^{-3}}{950} = 1.263 \times 10^{-6} \ \text{m}^2/\text{s}$$

$$= 1.263 \times 10^{-2} \ \text{cm}^2/\text{s} = \mathbf{1.263 \times 10^{-2} \ stokes}$$

EXAMPLE 5.9

A hydraulic lift used for lifting automobiles has a 20 cm diameter ram which slides in a 20.016 cm diameter cylinder. The annular space between the cylinder and ram is filled with an oil of kinematic viscosity 3.5 stokes and relative density 0.85. If the travel of 3.2 m long ram has a uniform rate of 15 cm/s, estimate the frictional resistance experienced by the ram.

Solution: Kinematic viscosity $\nu = 3.5$ stokes $= 3.5$ cm^2/s $= 3.5 \times 10^{-4}$ m^2/s

$$\text{Mass density } \rho = 0.85 \times 1000 = 850 \ \text{kg/m}^3$$

$$\therefore \ \text{Dynamic viscosity } \mu = \rho \, \nu$$

$$= 850 \times (3.5 \times 10^{-4}) = 0.2975 \ \text{N s/m}^2$$

$$\text{Thickness of oil film} = \frac{(20.016 - 20)}{2} \times 10^{-2} = 0.00008 \ \text{m}$$

$$\text{Shear stress } \tau = \mu \frac{du}{dy} = \mu \frac{V}{t} = \frac{0.2975 \times 0.15}{0.00008 \ \text{m}} = 557.81 \ \text{N/m}^2$$

Frictional resistance = shear stress $\times$ area

$$= 557.81 \times (\pi \times 0.20 \times 3.2) = 1121 \ \text{N} = \mathbf{1.12 \ kN}$$

EXAMPLE 5.10

Two square flat plates with each side 60 cm are spaced 12.5 mm apart. The lower plate is stationary and the upper plate requires a force of 100 N to keep it moving with a velocity of 2.5 m/s. The oil film between the plates has the same velocity as that of plates at the surface of contact. Assuming a linear velocity distribution determines

i. **the dynamic viscosity of the oil in poise, and**

ii. **the kinematic viscosity of the oil in stokes if the specific gravity of the oil is 0.95.**

Solution: Shearing stress $\tau = \mu \dfrac{du}{dy} = \mu \dfrac{u}{t} = \mu \times \dfrac{2.5}{0.0125} = 200 \, \mu$

Shearing area $= 0.6 \times 0.6 = 0.36 \text{ m}^2$

Shearing force $= \tau \times A = 200 \, \mu \times 0.36 = 72 \, \mu$

Equating it to the given force,

$$72 \, \mu = 100; \quad \mu = \frac{100}{72} = 1.39 \text{ Ns/m}^2 = \textbf{13.9 poise}$$

(*ii*) Kinematic viscosity of the oil,

$$\nu = \frac{\mu}{\rho} = \frac{1.39}{0.95 \times 1000} = 14.63 \times 10^{-4} \text{ m}^2/\text{s}$$

$$= 14.63 \text{ cm}^2/\text{s} = \textbf{14.63 stokes.}$$

EXAMPLE 5.11

A cubical block weighing 4.5 N and having a 40 cm edge is allowed to slide down an inclined plane surface making an angle of 30° with the horizontal on which there is a uniform layer of oil with 0.005 cm thickness. If the expected steady-state velocity of the block is 12.5 cm/s, determine the viscosity of the oil. Also, express the kinematic viscosity in stokes if the oil has a mass density of 800 kg/m^3.

Solution: Component of weight of block along the plane $= W \sin \alpha$

Assuming a linear velocity profile in the oil film,

$$\text{Shear stress } \tau = \mu \frac{du}{dy} = \mu \frac{u}{t}$$

Shearing force opposing motion $= \mu \dfrac{u}{t} A$

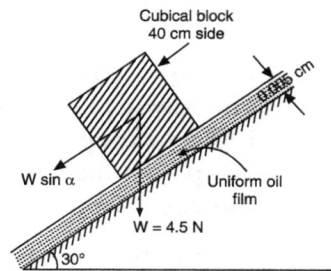

Under equilibrium conditions:

$$W \sin \alpha = \mu \frac{u}{t} A$$

$\therefore$ Dynamic viscosity $\mu = \dfrac{W \sin \alpha \times t}{uA}$

FIGURE 5.7.

$$= \frac{4.5 \sin 30° \times (0.005 \times 10^{-2})}{(12.5 \times 10^{-2})(0.4 \times 0.4)}$$

$$= 0.0056 \text{ N s/m}^2 = \textbf{0.056 poise}$$

Kinematic viscosity $\nu = \dfrac{\mu}{\rho} = \dfrac{0.0056}{800} = 7 \times 10^{-6} \text{ m}^2/\text{s}$

$$= 7 \times 10^{-2} \text{ cm}^2/\text{s} = \textbf{7} \times \textbf{10}^{-2} \textbf{ stokes.}$$

5.4.3. Surface tension and capillarity

Cohesion and Adhesion

Liquids have characteristic properties of cohesion and adhesion **cohesion** refers to the force with which the neighboring or adjacent fluid molecules are attracted toward each other. **Adhesion** represents the adhering or clinging of the fluid molecules to the solid surface with which they come in contact. In brief, forces between like molecules are cohesive, and the forces between unlike molecules are adhesive. When a liquid, like mercury, is spilled on a smooth horizontal surface, it tends to gather into droplets because the cohesive molecular forces are greater than the adhesive forces between the mercury molecules and the material of the surface. Mercury tends to stay away from the surface and is said to be a **not-wetting** liquid. In case of water, adhesive forces are greater than cohesive forces. Naturally, when water is poured on the same smooth horizontal surface, it would spread out and wet the horizontal surface. The wetting and non-wetting of the surface are dictated by the angle of contact between the liquid and the surface material.

Refer to Figure 5.8, which illustrates the liquid-gas interface with a solid surface. Liquid would wet the surface when $\theta < \pi/2$ and the degree of wetting increases as θ decreases to zero. For a non-wetting liquid, $\theta > \pi/2$. The contact angle is dependent on the nature and type of liquid, the solid surface, and its cleanliness. For pure water in contact with a clean glass surface, θ is essentially 0°. Even when the water is slightly contaminated, θ becomes as high as 25°. Mercury, a non-wetting liquid has θ between 130 and 150°.

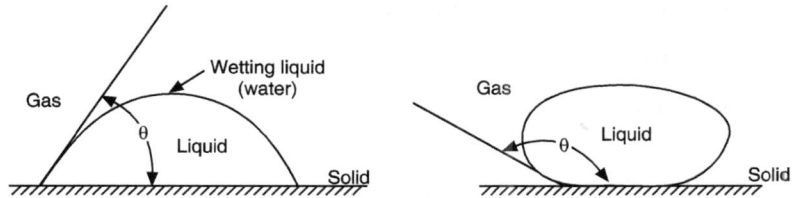

FIGURE 5.8 Wetting and non-wetting liquids.

Surface Tension

A liquid molecule lying well beneath the free surface of a liquid mass is surrounded by other molecules all around it. Consequently, the molecule is acted upon by the molecular forces of attraction (cohesion) that are equal in all directions. These equal and opposite forces cancel out; there is no resultant force acting upon the molecule within the fluid mass, and this aspect keeps the liquid mass in equilibrium. Nevertheless, a liquid molecule at the free surface has no liquid molecules above it to counteract the forces due to molecules below it. Consequently, as depicted in the molecular arrangement of Figure 5.9, the molecules lying at the surface have a net attraction tending to pull them into the interior of the liquid mass.

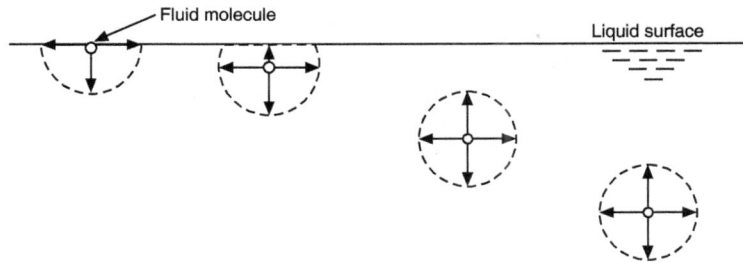

FIGURE 5.9 Forces of attraction on a liquid molecule.

A quantum of energy/work is, thus, extended to bring the molecule to the free liquid surface which then acts like an elastic or stretched membrane. Energy expended per unit area of the surface is called **surface tension**, designated by sigma σ. Surface tension occurs at the interface of a liquid and a gas or at the interface of two liquids and is essentially due to inter-molecular forces of cohesion.

It is primarily due to surface tension effects that

- an isolated drop of liquid takes nearly a spherical shape;

- birds can drink water from ponds;

- water can be poured into a clean glass tumbler to a level above the lip of tumbler;

- stretched water surface can support small objects like dust particles and a needle placed gently upon it; and

- capillary rise and depression in thin-bored glass tubes.

Surface tension forces are generally negligible in comparison with the pressure and gravitational forces but become quite significant when there is a free surface and the boundary dimensions are small, for example, in the small-scale models of hydraulic engineering structures.

The dimensional formula for surface tension is $\left[\dfrac{F}{L}\right]$ or $\left[\dfrac{M}{T^2}\right]$; it is usually expressed in N/m. The value of surface tension depends upon (*i*) nature of the liquid, (*ii*) nature of the surrounding matter which may be a solid, liquid, or a gas, (*iii*) kinetic energy and hence the temperature of liquid molecule. Growth in temperature results in a reduction of the inter-molecular cohesive forces and, hence, in a reduction of the surface tension force. At a critical point where the liquid and vapor phases become indistinguishable, the surface tension becomes zero. Surface tension values for liquids are generally quoted when in contact with air as the surrounding medium.

$$\sigma = 0.073 \text{ N/m} \qquad \text{for airwater interface}$$
$$\sigma = 0.480 \text{ N/m} \qquad \text{for air-mercury interface}$$

The surface tension values drop with a rise in temperature.

EXAMPLE 5.12

List some occurrences which can be attributed to the physics of surface tension. Why the concept of surface tension is not applied to gases?

Solution: The following occurrences can be attributed to the physics of surface tension:

i.   capillary rise or depression,

ii. break-up of liquid jets,

iii. formation of dew drops on grass early morning,

iv. dust particles collecting on water surface,

v.   formation of large soap bubbles with gentle blowing,

vi. floating of a greased needle of steel on water surface. Certain insects (fly, mosquitoes) can creep freely on the water surface,

vii. spherical shape of a droplet of liquid. When a molten metal is poured into water from a suitable height, the falling stream of molten metal breaks up and the detached portions acquire spherical shape. This technique is used for preparing lead shots and glass marbles.

Small drops of mercury are always spherical, but larger ones are somewhat flattened. Shape of the drop is governed by the combined influence of the surface tension and the gravitational force due to weight. For a tiny drop, the gravitational force is negligible and surface tension makes the drop spherical. However, in case of larger drops, force of gravity is appreciable and tries to flatten the drop so as to lower the position of the center of gravity. Eventually, the drop acquires an oval or elliptical shape.

Rain drops, tiny dew drops, and small water drops ejected from a burette are always spherical.

viii. On adding soap or detergent powder into water, its surface tension is lowered considerably. This soap-water solution can more easily seep into the pores of clothes and remove the dirt and grease, etc.

ix. Antiseptic creams are generally prepared in oil or greasy base of low surface tension so that even small amount of cream spreads on the whole of cut (wound) and prevents oozing out of the blood.

For gases, the inter-molecular distance among gas molecules is very large, and consequently, there is no appreciable force of cohesion, and as such the characteristic property of surface tension is not exhibited/manifested by gases even though a gas is also a fluid.

Capillary or Meniscus Effect

When a small diameter glass tube, called the capillary tube, is dipped into a water container, water rises in the tube to a level that stands higher than the level of water in the container. Conversely, the surface of mercury is depressed down in the capillary tubing when it is dipped in mercury. The phenomenon of liquid rise or fall in a capillary tube is called the **capillary** or **meniscus effect**. Capillary is a surface tension effect that depends upon the relative inter-molecular attraction between different substances; it is due to both cohesion and adhesion.

Adhesion between glass and water molecules is greater than cohesion between water molecules. Consequently, the water molecules spread over the glass surface and form a concave meniscus with small, angle of contact. Opposite conditions hold good for mercury, that is, cohesion between mercury molecules is greater than adhesion of mercury to glass. Mercury then displays a convex meniscus with the angle of contact greater than 90°.

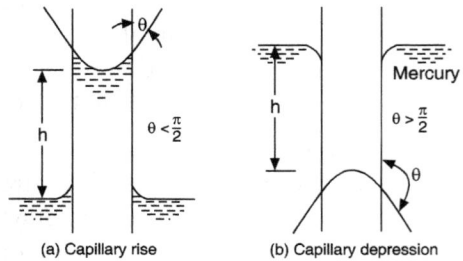

(a) Capillary rise   (b) Capillary depression

FIGURE 5.10 Capillary rise and depression.

Knowing the surface tension σ, angle of contact θ, tube diameter d, and specific weight of liquid w, the rise (for water) or depression (for mercury) of the liquid in the capillary tube can be worked out by the following analysis:

Weight of liquid raised or lowered in the capillary tube

$$= (\text{area of tube} \times \text{rise or fall}) \times \text{specific weight}$$

$$= \left(\frac{\pi}{4}d^2 h\right) w$$

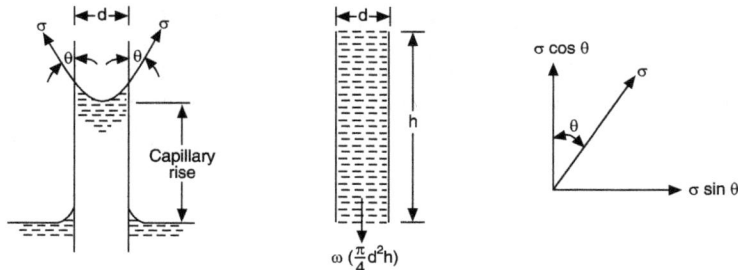

FIGURE 5.11 Rise or depression of liquid in a capillary tube.

Vertical component of surface tension force

$$= \sigma \cos\theta \times \text{circumference}$$
$$= \sigma \cos\theta \times \pi d = \pi d \sigma \cos\theta$$

When in equilibrium, the downward weight of the liquid column h is balanced by the vertical component of the force of surface tension.

Hence, $\quad \dfrac{\pi}{4} d^2 h w = \pi d \sigma \cos\theta$

or $\qquad h = \dfrac{4\sigma \cos\theta}{wd} \qquad\qquad (5.8)$

It is to be noticed that for $0 \le \theta < 90°$, h is positive (concave meniscus and capillary rise) and that for $90 \le \theta < 180°$, h is negative (convex meniscus and capillary depression).

Evidently, the capillary action is inversely proportional to the tube diameter. For precise work, the small diameter tubes are to be avoided; recommended minimum tube diameter for water and mercury is 6 mm. Further, since the presence of dirt affects the surface tension and, hence, the capillary rise or depression, the interior surface of the tube is to be kept clean.

EXAMPLE 5.13

Explain why in a capillary tube, the meniscus of water is concave upwards while the meniscus of mercury is convex upwards.

Calculate the capillary effects in millimeters in a glass tube of 4 mm diameter, when immersed in (*i*) water and (*ii*) in mercury. The temperature of the liquid is 20°C and the values of surface tension of water and mercury at 20°C in contact with air are 0.0735 N/m and 0.48 N/m, respectively. The contact angle of water $\theta = 0°$ and for mercury $\theta = 130°$.

Solution: The rise or depression h of a liquid in a capillary tube is given by

$$h = \dfrac{4\sigma \cos\theta}{wd}$$

Case (*i*): *Capillary effect in water:*

$$\sigma = 0.0735 \text{ N/m; angle of contact } \theta = 0°$$
$$w = 9800 \text{ N/m}^3 \text{ at } 20°C \text{ (say)}$$

$\therefore \qquad h = \dfrac{4 \times 0.0735 \times \cos 0°}{9800 \times 0.0004}$

$$= 7.50 \times 10^{-3} \text{ m} = \textbf{7.50 mm (rise)}$$

Case (ii): *Capillary effect in mercury:*

$$\sigma = 0.48 \text{ N/m; angle of contact } \theta = 130°$$
$$w = (9800 \times 13.6) \text{ N/m}^3$$

$$\therefore \qquad h = \frac{4 \times 0.48 \times \cos 130°}{(9800 \times 13.6) \times 0.004}$$

$$= -2.31 \times 10^{-3} \text{ m} = \textbf{2.31 mm} \text{ (depression)}.$$

EXAMPLE 5.14

Why should a mercury column in a thin glass tube be depressed while a water column is lifted up?

Calculate the size of glass tube if the capillary rise is limited to 2.2 mm of water. Assume suitable values of required data at 20°C and 1 atm pressure.

Solution: Capillary rise $h = 2.2$ mm $= 2.2 \times 10^{-3}$ m

The rise or depression h of a liquid in capillary tubing is given by

$$h = \frac{4\sigma \cos\theta}{wd}$$

The most appropriate values of desired data at 20°C and 1 atm pressure are

Surface tension $\sigma = 0.073$ N/m   for air–water interface
Angle of contact $\theta = 0°$ for water-clean glass surface
Density of water $\rho = 998$ kg/m^3

$$\therefore \quad 2.2 \times 10^{-3} = \frac{4 \times 0.073 \times \cos 0°}{(998 \times 9.81) \times d} ; d = 0.0136 \text{ m}$$

Thus, minimum diameter of tube should be **1.36 cm**.

EXAMPLE 5.15

Find the smallest diameter of a manometer tube such that the error due to capillary action in the measured gage pressure of 100 N/m^2 is less than 5 percent. The manometric liquid is water.

Solution: For water $\sigma = 0.073$ N/m, $\theta = 0°$ and $w = 9810$ N/m^3

From hydrostatic equation, $p = wH$, the gage pressure in terms of height of water column is

$$H = \frac{p}{w} = \frac{100}{9810} = 0.01019 \text{ m}$$

The rise of water in a capillary tube is given by

$$h = \frac{4\sigma \cos\theta}{wd}$$

$$= \frac{4 \times 0.073 \times \cos 0°}{9810 \times d} = \frac{2.976 \times 10^{-5}}{d}$$

Percentage error

$$= \frac{2.976 \times 10^{-5}}{d \times 0.1019} \times 100 \le 5$$

$$\therefore \qquad d = \frac{2.976 \times 10^{-5}}{5 \times 0.01019} \times 100 = 0.0584 \text{ m} \approx 5.84 \text{ cm}$$

The minimum diameter of the manometer tube should be **5.84 cm.**

EXAMPLE 5.16

a. **Water peeps up through sand columns. Why?**

Estimate the height to which water would rise in a clay soil of average grain diameter 0.06 mm. It may be assumed that surface tension at air–water interface is 0.0735 N/m and interspaces in clay are of size equal to one-fifth of mean diameter of clay grain. Take angle of contact $\theta = 0°$.

b. **What should be the average diameter of capillary tubes in a tree if the sap is carried to a height of 8 m? It may be assumed that sap in trees has the same characteristics as water, and it rises purely due to the capillary phenomenon.**

Solution: The rise of liquid in a capillary tubing is given by

$$h = \frac{4\sigma \cos\theta}{wd}$$

a. Size of interspaces in clay, $d = \dfrac{1}{5}$ of soil grain diameter

$$= \frac{1}{5} \times 0.06 = 0.012 \text{ mm} = 0.012 \times 10^{-3} \text{ m}$$

Surface tension $\sigma = 0.0735$ N/m
Angle of contact $\theta = 0$-degree

$$\therefore \qquad h = \frac{4 \times 0.0735 \times \cos 0°}{9810 \times (0.012 \times 10^{-3})} = 2.5 \text{ m}$$

b. Sap is stated to have the characteristics of water

$$\sigma = 0.0735 \text{ N/m}; \ \theta = 0\text{-degree}$$

$$\therefore \quad d = \frac{4\sigma \cos\theta}{wh} = \frac{4 \times 0.0735 \times \cos 0°}{9810 \times 8}$$

$$= 3.746 \times 10^{-6} \text{ m}$$

$$= \mathbf{3.746 \times 10^{-3}} \text{ mm}.$$

EXAMPLE 5.17

A U-tube is made up of two capillaries of bore 1 mm and 2 mm, respectively. The tube is held vertically and is partially filled with liquid of surface tension 0.05 N/m and zero contact angle. Calculate the mass density of the liquid if the estimated difference in the level of the two menisci is 1.25 cm.

Solution: Let h_1 and h_2 be the heights of liquid columns in the two limbs of bore d_1 and d_2, respectively. Then

$$h_1 = \frac{4\sigma \cos\theta}{wd_1} = \frac{4\sigma}{wd_1} \ ; h_2 = \frac{4\sigma}{wd_2}$$

$$h_1 - h_2 = \frac{4\sigma}{w}\left[\frac{1}{d_1} - \frac{1}{d_2}\right] = \frac{4\sigma}{\rho g}\left[\frac{1}{d_1} - \frac{1}{d_2}\right]$$

Substituting $h_1 - h_2 = 0.0125$ m
$d_1 = 0.001$m; $d_2 = 0.002$ m and $\sigma = 0.05$ N/m

$$0.0125 = \frac{4 \times 0.05}{\rho \times 9.807}\left[\frac{1}{0.001} - \frac{1}{0.002}\right];$$

$$\rho = 816 \text{ kg/m}^2$$

Hence, mass density $\rho = \mathbf{816 \text{ kg/m}^3}$.

EXAMPLE 5.18

A single-column U-tube manometer, made of glass tubing having a nominal inside diameter of 2.5 mm, has been used to measure pressure in a pipe or vessel containing air. If the limb opened to the atmosphere is 10% oversize, find the error in mm of mercury in the measurement of air pressure due to surface tension effects. It is stated that mercury is the manometric fluid for which surface tension $\sigma = 0.514$ N/m and angle of contact $\alpha = 140°$.

Solution: The surface tension manifests the phenomenon of capillary action due to which rise or depression of manometric liquid in a tube is given by

$$h = \frac{4\sigma \, \cos\theta}{wd}$$

For the given case,

$$d_1 = 2.5 \text{ mm}$$
$$d_2 = 2.5 \times 1.1 = 2.75 \text{ mm}$$

$$\therefore \qquad h_1 = \frac{4 \times 0.514 \times \cos 140°}{(9810 \times 13.6) \times \left(2.5 \times 10^{-3}\right)} = 4.72 \times 10^{-3} \text{ m}$$

$$h_2 = \frac{4 \times 0.514 \times \cos 140°}{(9810 \times 13.6) \times \left(2.75 \times 10^{-3}\right)}$$

$$= 4.29 \times 10^{-3} \text{ m}.$$

Hence, error in measurement due to surface tension effects

$$= (4.72 - 4.29) \times 10^{-3}$$
$$= 0.43 \times 10^{-3} \text{ m} = \textbf{0.43 mm.}$$

5.4.4. Newtonian and non-Newtonian fluids

The distinction between Newtonian and non-Newtonian fluids can be readily illustrated when the velocity gradient du/dy is plotted against the viscous shear stress τ.

Fluids for which the viscosity is independent of velocity gradient are called Newtonian fluids. For these fluids, the plot between shear stress and velocity gradient is a straight line passing through the origin. Slope of the line is equal to the coefficient of viscosity, $\mu = \tau/(du/dy)$. Fluids represented by curves (a) and (b) are Newtonian fluids; fluid represented by line (a) is more viscous than that represented by line (b). Fluids like air, water, kerosene, and thin lubricating oils are essentially Newtonian in the chapter under normal working conditions.

Fluids such as human blood, thick lubricating oils, and certain suspensions for which the viscosity coefficient depends upon velocity gradient are referred to as non-Newtonian fluids. The viscous behavior of a non-Newtonian fluid may be prescribed by the power-law equation $\tau = k \, (du/dy)^n$ where k is a consistency index and n is a flow behavior index. For a Newtonian fluid, the consistency index k becomes the dynamic viscosity coefficient μ and the flow behavior index n assumes a unity value.

Fluids for which the flow behavior index n is less than unity are called **pseudo-plastic**. Viscosity coefficient is smaller at greater rates of velocity gradient, and the curve becomes flattered as the shear rate (i.e., velocity gradient) increases (curve c). Examples of pseudo-plastic fluids are the milk, blood, clay, and liquid cement. Fluids for which the index n is greater than unity are called **dilatant**. Viscosity coefficient is more at greater rates of viscosity and the flow curve steeps with increasing shear rate (curve d). Concentrated solution of sugar and aqueous suspension of rice starch are examples of dilatant fluids.

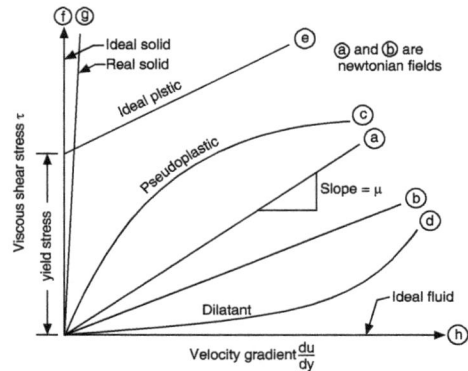

FIGURE 5.12 Variation of shear stress with velocity gradient (time rate of deformation).

An **ideal plastic** substance indicates no deformation when stressed up to a certain point (yield stress) and beyond that, it behaves like a Newtonian fluid and, hence, is represented by line (e). For certain substances, there is finite deformation for a given load, that is, the rate of deformation is zero. These materials plot as ordinate (curve f) and are called **elastic materials** or **ideal solids**. Actual solids deform slightly when subjected to shear stress of larger magnitude and, hence, plot as a straight line almost vertical (g). A fluid for which shear stress is zero (even if there is velocity gradient) is the **ideal fluid,** and it plots as abscissa (h). Fluids that show an apparent increase in viscosity with time are called **thixotropic**. Conversely, if the apparent viscosity decreases with time, the fluid is called **rheopectic**.

5.5. PRESSURE AND ITS RELATIONSHIP WITH HEIGHT

5.5.1. Pressure

A fluid element or mass is essentially acted upon by two categories of forces: body forces and surface forces. **Body forces** on fluids element are caused by agencies such as gravitational, electric, or magnetic fields. The magnitude of these forces is proportional to the mass of the fluid. **Surface forces** represent the action of the surrounding fluid on the element under consideration through direct contact. These forces are due to surface stresses like pressure (normal force) and shear (tangential force). In fluids at rest, there is no relative motion between the layers of the fluid. The velocity gradient is zero, and hence, there is no shear in the fluid. Consequently, there is no tangential component of force, and hence for a stationary fluid, the force exerted is normal to the surface of the containing vessel. This normal surface force is called the pressure force. The mathematical definition of ***intensity of pressure*** (or simply pressure), in the absence of shearing stress, is

$$p = \frac{dF}{dA}$$

where dF represents the resultant force acting normal to an infinitesimal area dA. If the total force F acts uniformly over the entire area A, then $p = F/A$. Pressure has the dimensions of $[FL^{-2}]$ and is usually expressed in N/m^2 (pascal), bar or atmosphere.

$$1 \text{ bar} = 10^5 \text{ N/m}^2 = 100 \text{ kPa}$$
$$1 \text{ atm} = 101.3 \text{ kPa}$$

5.5.2. Pascal's law

An important and unique property of hydrostatic pressure is reflected in Pascal's law, which states that

Intensity of pressure at a point in a fluid at rest is same in all directions

Consider a small wedge-shaped element of stationary fluid and assume that the element has a unit depth perpendicular to the plane of the paper (Figure 5.13). The element is acted upon by the normal pressure forces and the vertical forces due to weight. Let p_x, p_y, and p_θ be the pressure intensities on the faces AB, BC, and AC, respectively. Then,

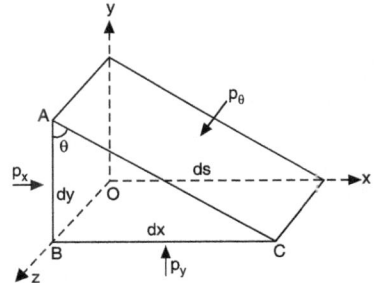

FIGURE 5.13

Force on face AB $= p_x \times$ area of face AB

$$= p_x (dy \times 1) = p_x \, dy$$

Likewise, Force on face BC $= p_y \, dx$

Force on face AC $= p_\theta \, ds$

The weight of fluid element is

$$= (\text{area of triangular element} \times \text{depth}) \times \text{specific weight}$$

$$= \left(\frac{1}{2} dx \, dy \times 1\right) \times w = \frac{1}{2} w \, dx \, dy,$$

and it acts through the center of gravity. Since the fluid element is in equilibrium, the forces in the horizontal and vertical directions must balance.

Resolving the forces in the x-direction:

$$p_x \, dy = p\theta \, ds \, \sin\theta$$

From Figure 5.13, $dy = ds \, \sin\theta$

$\therefore$
$$p_x \, dy = p_\theta \, dy; \, p_x = p_\theta$$

Resolving the forces in the y-direction,

$$p_y \, dx = \frac{1}{2} \, w \, dx \, dy + p_\theta \, ds \, \cos\theta$$

Let the size of the elemental system approach smaller and smaller dimensions; then the gravitation force (weight) which diminishes as the product of two dimensions (dx and dy) can be neglected in comparison with the pressure forces for which the diminishing effect is proportional to be reduction in single dimension (dx). Thus, in the limit,

$$p_y \, dx = p\theta \, ds \, \cos\theta$$

From Figure 5.13, $\quad dx = ds \, \cos\theta$

$\therefore$
$$p_y \, dx = p_\theta \, dx; \, p_y = p_\theta$$

From Equations 5.2 and 5.3, we have

$$p_x = p_y = p_\theta \tag{5.9}$$

This result is independent of the angle θ and, therefore, it follows that pressure acts equally in all directions in a stationary fluid. Pressure at a point has only one value regardless of the orientation of the area upon which it is determined. Independence of direction implies that pressure is a scalar quantity.

5.5.3. Hydrostatic law

Rate of increase of pressure in a vertical direction is equal to weight density (specific weight) of the fluid.

The fundamental equation relating pressure, density, and vertical distance can be established by considering the equilibrium of an imaginary cylindrical element in a body of fluid at rest. The cylindrical element is of cross-sectional area dA and height dy.

The pressure forces acting on the fluid element are

i.   Pressure force on bottom face AB $= p\,dA$ acting in the upward direction.

ii. Pressure force on top face CD

$$= \left(p + \frac{\partial p}{\partial y} dy \right) dA \text{ acting in the downward direction.}$$

iii. Weight of fluid element = specific weight × volume $= w\,dA\,dy$

iv. Pressure forces on surface AC and BD are equal and opposite and, hence, cancel out.

For the element to be in equilibrium, the sum of downward forces on the element must be equal to the upward forces. That is

$$p\,dA - \left(p + \frac{\partial p}{\partial y} dy \right) dA - w\,dA\,dy$$

or $\dfrac{\partial p}{\partial y} = -w$

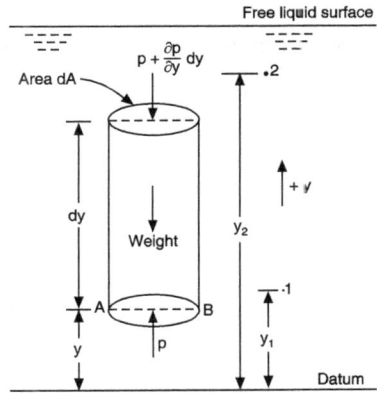

FIGURE 5.14 Pressure–density–height relationship.

Since we are considering the variation of pressure only in the y-direction, the partial differential $\dfrac{\partial p}{\partial y}$ can be replaced by exact differential $\dfrac{\partial p}{\partial y}$.

$$\frac{dp}{dy} = -w \text{ or } dp = -w\,dy \tag{5.10}$$

This is the fundamental equation of fluid statics and indicates that a negative pressure gradient exists upward along any vertical. Thus, the pressure decreases in the upward direction and increases in the downward direction with magnitude equal to specific weight.

Letting height dy to be zero, then $dp = 0$, that is, the pressure intensity remains the same if there is no change in elevation. Thus, the pressure will be constant everywhere over the same level surface in a continuous body of static fluid. A surface layer where pressure is the same at all points is called an **isobaric surface** or **equipotential surface**. The equipotential surfaces are horizontal planes; the free surface (a surface separating the liquid from the atmosphere) being one of them.

Equation 5.10 can be integrated directly to determine the difference in pressure between any two points in the fluid mass

$$\int_1^2 dp = -\int_1^2 w \, dy \tag{5.11}$$

Evaluation of this integral is possible only for a given dependence of specific weight w with elevation y.

5.5.4. Pressure variation for an incompressible fluid

For incompressible fluid, the specific weight is independent of pressure intensity, and it remains fairly constant with height. Then for a homogeneous fluid of constant specific weight, Equation 5.11 becomes

$$p_2 - p_1 = -w \, (y_2 - y_1)$$

Rearrangement gives

$$\frac{p_1}{w} + y_1 = \frac{p_2}{w} + y_2 \text{ or } \frac{p}{w} + y = \text{constant}$$

Each term has dimensions of length (meter). The coordinate y is called the **position** or **elevation head**, the $\dfrac{p}{w}$ term is called the **pressure head**, and the sum $\left(\dfrac{p}{w} + y\right)$ is called the **piezometric head**. Evidently, at every point in a homogeneous fluid at rest, the piezometric head is constant.

Representing the difference in elevation between two points by h and measuring it from the free liquid surface where atmospheric pressure p_{at} prevails, we have

$$p - p_{at} = wh; \; p = p_{at} + wh$$

Thus, the pressure p at any point in a static fluid is partly due to atmospheric pressure p_{at} at the free surface and partly due to the distance of that point beneath the free surface. If atmospheric pressure p_{at} is regarded as zero of pressure scale, then

$$p = wh \tag{5.12}$$

This pressure p at any point in a static fluid when expressed above atmospheric pressure is called the **positive** or **gage pressure**. Equation 5.10

demonstrates that pressure at any point submerged below a free liquid surface exposed to the atmosphere is equal to the product of the vertical distance below the surface and the specific weight of the liquid.

Apparently, the pressure intensity depends only upon the height of column and not at all upon the size of column. Accordingly, in vessels of different shapes and configurations, the same unit pressure would be exerted against the bottom of the container.

5.5.5. Atmospheric, absolute, and gage pressure

The following terms are generally associated with pressure and its measurement:

i. **Atmospheric pressure** (P_{at}): This is the pressure exerted by the envelope of air surrounding the earth's surface. Atmospheric pressure is usually determined by a mercury column barometer shown in Figure 5.15. A long, clean thick glass tube closed at one end is filled with pure mercury. The tube diameter is such that capillarity effects are minimal. The open end is stoppered and the tube is inserted into mercury container; the stoppered end kept well beneath the mercury surface. When the stopper is removed, mercury runs out of the tube into the container and eventually, mercury level in the tube settles at height h above mercury level in the container. The atmospheric pressure p_{at} acts at the mercury surface in the container, and mercury vapor pressure P_{vp} exists at the top of mercury column in the tube. From hydrostatic equation,

$$P_{at} - P_{vp} = wh$$

Mercury has a low vapor pressure (=0.17 N/m^2 at 20°C), and thus for all intents and purposes, it can be neglected in comparison with P_{at} which is about 10^5 N/m^2 at mean sea level. Then

$$P_{at} = wh$$

Mercury is preferred as other liquids possess sufficient vapor pressure at room temperature and that creates an error in pressure indication. Atmospheric pressure varies with altitude because the air nearer the earth's surface is compressed by the air above. At sea level, value of atmospheric pressure is close to 1.01325 bar or 760 mm of Hg column or 10.33 m of water column.

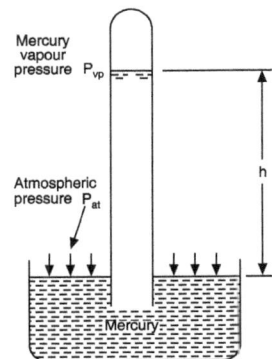

FIGURE 5.15 Simple mercury barometer.

ii. ***Absolute pressure (P$_{abs}$):*** Pressure has been defined as the force per unit area due to the interaction of fluid particles among themselves. Zero pressure intensity will occur when molecular momentum is zero. Such a situation can occur only when there is a perfect vacuum, that is, a vanishingly small population of gas molecules or of molecular velocity. Pressure intensity measured from this state of vacuum or zero pressure is called absolute pressure.

iii. ***Gage pressure (P$_g$) and vacuum (P$_{vac}$):*** Instruments and gages used to measure fluid pressure generally measure the difference between the unknown pressure P and the existing atmospheric pressure P_{at} as shown in Figure 5.16.

When the unknown pressure is more than atmospheric pressure, the pressure recorded by the instrument is called *gage pressure.* A pressure reading below the atmospheric pressure is known as *vacuum, rarefaction,* or *negative pressure.* Actual absolute pressure is then the algebraic sum of the gage indication and the atmospheric pressure

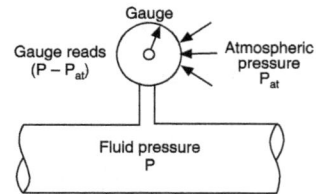

FIGURE 5.16 Gage pressure.

$$P_{abs} = P_{at} + P_g$$
$$P_{abs} = P_{at} - P_{vac}. \qquad (5.13)$$

Relation between these pressure terms is illustrated in Figure 5.17.

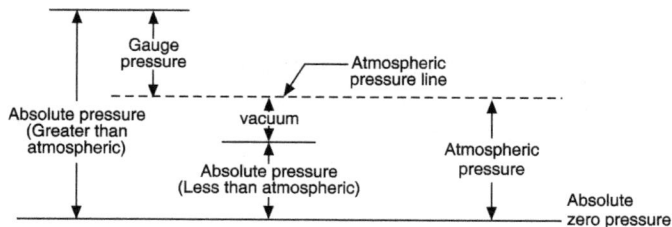

FIGURE 5.17 Relation between absolute, gage, and atmospheric pressure.

Some of the commonly used pressure units are

$1 \text{ bar} = 10^5 \text{ N/m}^2 = 100 \text{ kPa} = 750.06 \text{ mm of Hg}$
$1 \text{ micron} = 1 \mu = 10^{-3} \text{ mm of Hg}$
$1 \text{ torr} = 1 \text{ mm of Hg}$
$1 \mu \text{ bar} = 1 \text{ dyne/cm}^2$

Quite often pressure is expressed in the unit *atmosphere*. This unit simply uses the standard atmospheric value of 101.3 kPa and is defined as one atmosphere. Two atmospheres would be then 202.6 kPa. Sometimes, we assume atmospheric pressure equivalent to a rounded figure of 100 kPa and call it a *technical atmosphere*.

EXAMPLE 5.19

A piston of area 150 cm² exerts a force of 600 N. Find the intensity of pressure on the fluid in contact with the underside of piston if the piston is in equilibrium.

Solution: Under equilibrium conditions, pressure exerted by the fluid is equal to the piston force

$$\therefore \text{ Intensity of pressure} = \frac{\text{force}}{\text{area}}$$

$$= \frac{600}{\left(150 \times 10^{-4}\right)} = \mathbf{4 \times 10^4} \text{ N/m}^2$$

EXAMPLE 5.20

Explain the terms intensity of pressure and pressure head.

Convert a pressure of 1.5 bar to (*i*) meters of water (*ii*) cm of mercury (sp gr 13.6).

Solution: From hydrostatic equation, $p = wh$

i. $h = \dfrac{p}{w} = \dfrac{1.5 \times 10^5}{9810} = \mathbf{15.29}$ **m** of water

ii. $h = \dfrac{1.5 \times 10^5}{(13.6 \times 9810)}$

$= 1.124$ m of mercury
$= \mathbf{112.4}$ **cm** of mercury

EXAMPLE 5.21

At a certain location in the flow field, pressure is equal to 50 m of water column. Obtain the equivalent pressure head in terms of (*i*) specific gravity of kerosene is 0.8 (*ii*) specific gravity carbon tetrachloride is 1.5.

Solution: Invoking hydrostatic equation $p = wh$

∴ Pressure intensity corresponding to 50 m of water column

$$p = 9810 \times 50 = 490\,500 \text{ N/m}^2$$

Equivalent pressure head in terms of

i. specific gravity of kerosene: 0.8

$$= \frac{490\,500}{0.8 \times 9810} = \mathbf{62.5} \text{ m of kerosene}$$

ii. specific gravity of carbon tetrachloride: 1.5

$$= \frac{490\,500}{1.5 \times 9810} = \mathbf{33.33} \text{ m of carbon tetrachloride}$$

EXAMPLE 5.22

Measurements of pressure at the base and top of a mountain are 74 cm and 60 cm of mercury, respectively. Workout the height of mountain if air has a specific weight of 11.97 N/m³.

Solution:

$$p_{base} = 74 \text{ cm of mercury}$$

$$= \frac{74}{100} \times (13.6 \times 9810) \text{ N/m}^2$$

$$p_{top} = \frac{60}{100} \times (13.6 \times 9810) \text{ N/m}^2$$

Now, $\quad p_{base} - p_{top} = wh$

or $\quad \dfrac{(74-60)}{100} \times 13.6 \times 9810 = 11.97\,h$

Solution gives: $h = 1560.4$ m

∴ Mountain has a height of **1560.4 m**.

EXAMPLE 5.23

A gage on the suction side of a pump shows a negative pressure of 0.285 bar. Express this pressure in terms of (*i*) pressure intensity kPa (*ii*) N/m² absolute (*iii*) m of water gage (*iv*) m of oil (specific gravity 0.85) absolute and (*v*) cm of mercury gage. Take atmospheric pressure as 76 cm of mercury and relative density of mercury as 13.6.

Solution: (*i*) Gage reading = 0.285 bar
$$= 0.285 \times 10^5 \text{ N/m}^2$$
$$= 0.285 \times 10^5 \text{ Pa}$$
$$= 28.5 \text{ kPa gage (vacuum)}$$

(*ii*) Atmospheric pressure p_{at}
$$= 76 \text{ cm of mercury}$$
$$= (13.6 \times 9810) \times 0.76$$
$$= 101396 \text{ N/m}^2$$

Absolute pressure
$$= \text{atmospheric pressure} - \text{vacuum pressure}$$
$$p_{abs} = p_{at} - p_{vac}$$
$$= 101396 - 28500$$
$$= 72896 \text{ N/m}^2 \text{ absolute}$$

(*iii*) Equivalent heads of water, oil, and mercury can be worked out by applying the hydrostatic equation, $p = wh$

$$\text{Head of water (gage)} = \frac{0.285 \times 10^5}{9810} = 2.905 \text{ m of water (gage)}$$

$$\text{Head of oil (absolute)} = \frac{72896}{0.85 \times 9810} = 8.742 \text{ m of water (absolute)}$$

$$\text{Head of mercury (gage)} = \frac{0.285 \times 10^5}{13.6 \times 9810}$$
$$= 0.2136 \text{ m of mercury}$$
$$= 21.36 \text{ cm of mercury (gage)}.$$

EXAMPLE 5.24

A diver is working at a depth of 20 m below the surface of sea water (sp. wt. = 10 kN/m³). Calculate the pressure intensity at this depth. What would be the absolute pressure if barometer reads 760 mm of mercury column at the sea level?

Solution: From hydrostatic equation, $p = wh$

∴ Pressure intensity at the given depth,

$$p_g = 10000 \times 20 = 200000 \text{ N/m}^2 = \textbf{200 kPa}$$

Atmospheric pressure p_{at}
$$= 760 \text{ mm of mercury}$$
$$= (13.6 \times 9810) \times 0.76$$

$$= 101\,396 \text{ N/m}^2 \approx 101.4 \text{ kPa}$$

$$\therefore \text{ Absolute pressure } p_{abs} = p_{at} + p_g$$
$$= (101.4 + 200) = \textbf{301.4 kPa.}$$

EXAMPLE 5.25

The inlet to pump is 10 m above the bottom of sump from which it draws water through a suction pipe. If the pressure at the pump inlet is not to fall below 30 kN/m² absolute, workout the minimum depth of water in the tank. Take atmospheric pressure as 100 kN/m².

Solution: Let p_{vac} be the vacuum (suction) pressure at the pump inlet. Then

$$p_{vac} = p_{at} - p_{abs}$$
$$= (100 - 30) = 70 \text{ kN/m}^2 = 70000 \text{ N/m}^2$$

Further, let h represent the distance between the pump inlet and the free water surface in the sump.

Applying hydrostatic equation, $p = wh$, we have

$$70000 = 9810\,h; h = 7.136 \text{ m}$$

$\therefore$ Minimum depth of water in the sump

$$= (10 - 7.136) = \textbf{2.864 m.}$$

EXAMPLE 5.26

The cylindrical fuel tank of a motor car, 20 cm in diameter and with its horizontal axis is filled with petrol with mass density as 800 kg/m³. A 35 mm diameter filler pipe rises from the top of the tank to a height of 50 cm. Calculate the force on one end of the fuel tank when the filler pipe is full.

Solution: Pressure intensity at any point in a flow field varies directly with depth and is prescribed by the hydrostatic equation,

$$p = wh$$

i. Pressure intensity at the tank top is due to 50 cm head of petrol in the filler pipe and is it equal to

$$p_t = wh$$
$$= (800 \times 9.81) \times 0.5 = 3924 \text{ N/m}^2$$

ii. Pressure intensity at the tank bottom is due to 70 cm head of petrol in the filler pipe and the tank and is it equal to

$$p_b = (800 \times 9.81) \times 0.7 = 5494 \text{ N/m}^2$$

Average pressure intensity on end of the tank,

$$p_{av} = \frac{3924 + 5494}{2} = 4709 \text{ N/m}^2$$

Force on one end of the tank,

$$F = p_{av} \times \text{area}$$

$$= 4709 \times \frac{\pi}{4} (0.2)^2 = \textbf{147.86 N.}$$

EXAMPLE 5.27

What forces act on a fluid element in static equilibrium?

A cylindrical tank 5 m diameter × 10 m height is completely filled with water. Find (a) the intensity of pressure and total force on the bottom of the tank (b) minimum and maximum an average pressure intensities on the vertical surface (c) the total force on the vertical surface.

Solution: Pressure intensity at the bottom,

$$p = wh = 9810 \times 10 = 98100 \text{ N/m}^2$$

Total pressure force at the bottom,

$$F = pA$$

$$= 98100 \times \frac{\pi}{4} (5)^2 = 1925212 \text{ N}$$

(b) Pressure intensity varies directly with the depth.
Minimum pressure intensity occurs at the top and is equal to

$$p_{min} = 0 \qquad\qquad (\because h = 0 \text{ at the top})$$

Maximum pressure intensity occurs at the bottom and is equal to

$$p_{max} = wh = 9810 \times 10 = 98100 \text{ N/m}^2$$

Average pressure intensity is at the middle,

$$p_{av} = \frac{0 + 98100}{4} = 49050 \text{ N/m}^2$$

(c) Total force on the wall

$$= p_{av} \times \text{lateral surface area}$$

$$= 49050 \times (\pi \times 5 \times 10)$$

$$= 7704756 \text{ N} = \textbf{7705 kN.}$$

EXAMPLE 5.28

A cylindrical tank of 3 m height and 5 cm² cross-sectional area is filled with water up to a height of 2 m and remaining with oil of specific gravity 0.8. The vessel is open to atmosphere. Calculate
(*i*) pressure intensity at the interface, (*ii*) absolute and gage pressure on the base of the tank in terms of water head, oil head, and N/m².

Also workout the net force experienced by the base of the tank. Take atmospheric pressure as 1.0132 bar.

Solution: Applying hydrostatic equation, the pressure intensity at any point in the flow field is given by $p = wh$

i. Pressure at the interface between the oil and water is due to 1 m of oil and it is equal to

$$p_{inetrface} = (0.8 \times 9810) \times 1 = \textbf{7848 N/m}^2$$

ii. Pressure at the base of the tank is equal to the sum of pressures at the interface (due to 1 m of oil) and pressure due to 2 m of water.

$$p_{base} = 7848 + (9810 \times 2)$$
$$= 27468 \text{ N/m}^2 \text{ (gage)}$$
$$= \frac{27468}{9810} = 2.8 \text{ m of water (gage)}$$
$$= \frac{27468}{0.8 \times 981} = 3.5 \text{ m of oil (gage)}$$

Atmospheric pressure,
$$p_{at} = 1.0132 \text{ bar}$$
$$= 1.0132 \times 10^5 \text{ N/m}^2$$
$$= \frac{1.0132 \times 10^5}{9810} = 10.328 \text{ m of water}$$
$$= \frac{1.0132 \times 10^5}{0.8 \times 981} = 12.910 \text{ m of oil}$$

Absolute pressure = atmospheric pressure + gage pressure
$$\therefore p_{base} \text{ absolute} = 10.328 + 2.8$$

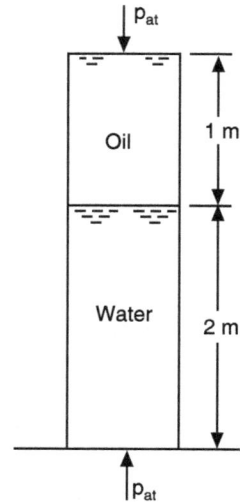

FIGURE 5.18.

= **13.128 m** of water

= 12.910 + 3.5

= **16.410 m** of oil

= 101320 + 27468

= **128788 N/m²**

The surface of the base exposed to water is acted upon by p_{base} absolute where as atmospheric pressure acts on the other surface. As such, the net force experienced by the base of the tank would be due to gage pressure at the base.

$$F = (p_{base})_{gage} \times \text{cross-sectional area}$$
$$= 27468 \times (5 \times 10^{-4})$$
$$= \textbf{13.734 N.}$$

EXAMPLE 5.29

State Pascal's law and discuss its applications.

A hydraulic press has a ram 30 cm diameter and plunger 5 cm diameter. Neglecting all losses due to friction etc., calculate the force to be applied at the plunger to lift a load of 20 kN. If the plunger has a stroke of 30 cm and if it executes 100 strokes per minute, calculate the distance through which the load is raised per minute. Also, make calculations for the power expended in driving the plunger.

Solution: Pascal's law states that *"intensity of pressure is transmitted equally in all directions through a mass of fluid at rest."*

This characteristic property of the fluid forms the working principle of hydraulic jack, hydraulic lift, and hydraulic crane.

Hydraulic press consists of two cylinders which are interconnected at the bottom through a pipeline and are filled with some liquid. The larger cylinder contains a ram of area A and a plunger of area a recip-

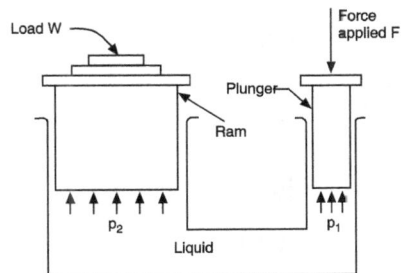

FIGURE 5.19 Working principle of hydraulic press.

rocates inside the smaller cylinder. A force F applied to the plunger produces an intensity of pressure p_1 which is transmitted in all directions through the liquid. If the plunger and the ram are at the same level and if their weights are neglected, then pressure intensity p_2 acting on the ram must equal to p_1.

Now, $p_1 = \dfrac{F}{a}$ and $p_2 = \dfrac{W}{A}$ where W is the weight to be lifted by the ram. If $p_1 = p_2$, then

$$\frac{F}{a} = \frac{W}{A} \; ; \; W = F\left(\frac{A}{a}\right)$$

The above expression indicates that by applying a small force F on the plunger, a larger load can be lifted by the ram.

Mechanical advantage of the system is equal to A/a, the ratio of the area of ram and plunger. By proper adjustment of plunger and ram areas, even a small force can be multiplied many times; the hydraulic press is often called a *machine of multiplying forces.*

Given: $W = 20 \times 10^3$ N;

$$a = \frac{\pi}{4}(5)^2 \text{ cm}^2 \text{ and } A = \frac{\pi}{4}(30)^2 \text{ cm}^2$$

Substituting these data in the expression,

$$\frac{F}{a} = \frac{W}{A}$$

$$\frac{F}{\dfrac{\pi}{4}(5)^2} = \frac{20 \times 10^3}{\dfrac{\pi}{4}(30)^2} \; ; \; F = \textbf{555.6 N}$$

(*b*) Stroke volume on the plunger side

$$= \text{stroke} \times \text{area} = 30 \times \frac{\pi}{4}(5)^2 \text{ cm}^2$$

If S_r donates the distance moved by the ram in one stroke, then stroke volume on ram side is equal to $S_r \times \dfrac{\pi}{4}(30)^2$. Stroke volumes on the plunger and ram side equal each other. Therefore:

$$S_r \times \frac{\pi}{4}(30)^2 = 30 \times \frac{\pi}{4}(5)^2 ;$$

$$S_r = 0.833 \text{ cm}$$

∴ Distance through which load is raised on the ram side

$$= \text{(distance moved by the ram in one stroke)} \times \text{number of strokes/minute}$$

$$= 0.833 \times 100 = \textbf{83.3 cm/min}$$

(c) Work done = load × distance moved

$$= (20 \times 10^3) \times 0.833 = 16660 \text{ Nm/min}$$

∴ Power required to operate the plunger

$$= \frac{16660}{60} = 277.67 \text{ Nm/s} = \mathbf{277.67 \text{ W.}}$$

5.6. EQUATIONS OF MOTION

5.6.1. Flow rate and continuity equation

Consider flow of an ideal fluid through a stream tube (Figure 5.20). Since no flow takes place across the streamlines, the fluid must enter and leave the tube only at the end sections. At the inlet section 1–1, the flow characteristics are tube cross-sectional area A, average fluid density ρ, and the mean flow velocity V. The corresponding parameters at the exit section 2–2 are $(A + dA)$, $(\rho + d\rho)$, and $(V + dV)$.

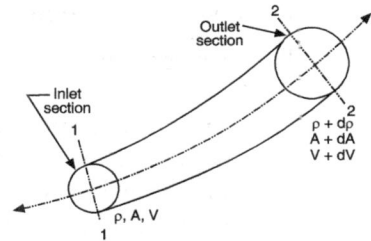

FIGURE 5.20 Steady flow through a stream tube.

The mass flow entering the stream tube at section 1–1 during time interval dt is given by $(AV \rho dt)$. During the same time interval, a quantity of mass flow $(A + dA)(V + dV)(\rho + d\rho) dt$ out flows from the exist section 2–2. The fluid mass that accumulates between the two sections is then given by

$dm = AV\rho\, dt - (A + dA)(V + dV)(\rho + d\rho)\, dt$

Simplification yields

$$\frac{dm}{dt} = -(AV\, d\rho + V \rho\, dA + A \rho\, dV)$$

For steady flow $\dfrac{dm}{dt} = 0$ and, therefore,

$$AV\, d\rho + V\rho dA + A \rho\, dV = 0$$

Dividing throughout by ρAV, one obtains

$$\frac{d\rho}{\rho} + \frac{dA}{A} + \frac{dV}{V} = 0$$

or

$$d(\rho AV) = 0$$

$$\rho AV = \text{constant} \qquad (5.14)$$

Evidently the mass of fluid per unit time passing through any section of a stream tube is constant.

For an incompressible fluid, the mass density ρ is constant and, therefore,

$$AV = \text{constant}$$

that is, $\qquad A_1 V_1 = A_2 V_2 \qquad (5.15)$

Equation 5.15 represents the continuity equation for the steady incompressible flow through an elementary stream tube. The continuity equation states that in a varying duct, the average velocity may change along the direction of flow but the product (area × velocity) remains constant. Further, the mean velocities are inversely proportional to the cross-sectional areas of the flow passage, that is, $\dfrac{V_1}{V_2} = \dfrac{A_2}{A_1}$.

Continuity relation $Q = A_1 V_1 = A_2 V_2$ works with the restrictions: constant density, no friction, steady flow, parallel stream tubes at each sections 1 and 2, and no branching of stream tubes.

The product AV is the quantity (volume measure) of fluid which passes a reference point per unit time and is called the *flow rate* or discharge. *Mass flow* rate is the quantity (mass measure) ρAV of fluid which passes per unit time; the units are expressed in kg/s. *Weight flow* rate is the quantity (weight measure) $w AV$ of fluid which passages the reference point per unit time; value is expressed in N/s.

For a two-dimensional flow, the stream tube can be considered to be a unit dimension in the z-direction. Then the area along the stream tube is numerically equal to the spacing t of the streamlines. Continuity equation for a stream tube in a two-dimensional, steady flow or incompressible fluid may then be prescribed as follows:

$$V_1 t_1 = V_2 t_2 = \text{constant} \qquad (5.16)$$

Apparently, the mean velocity is inversely proportional to the spacing of the streamlines.

EXAMPLE 5.30

From a flow net diagram, it was found that the distances between two consecutive streamlines at two successive sections are 1 cm and 0.6 cm, respectively. If the velocity at the first section is 1 m/s, find the velocity at the other section. Also, find the discharge between the two streamlines.

Solution: For a two-dimensional flow,

$$V_1 t_1 = V_2 t_2 = \text{constant}$$

Given

$$V_1 = 1 \text{ m/s}; t_1 = 1 \text{ cm} = 0.01 \text{ m}: t_2 = 0.6 \text{ cm} = 0.006 \text{ m}$$

$$\therefore \qquad \text{Velocity } V_2 = \frac{V_1 t_1}{t_2} = \frac{1 \times 0.01}{0.006} = \textbf{1.67 m/s}$$

Further, discharge $= V_1 t_1 = 1 \times 0.01 = \textbf{0.01 m}^2\textbf{/s unit depth.}$

EXAMPLE 5.31

Water is flowing through a pipe of 0.5 m diameter with an average velocity of 1 m/s. What is the rate of discharge of water? The same flow then passes through another section where the diameter is 1 m. What is the average flow velocity at this section?

Solution: Discharge $Q = \text{area} \times \text{velocity} = \dfrac{\pi}{4}(0.5)^2 \times 1 = \textbf{0.196 m}^3\textbf{/s}$

Let V_2 be the velocity at the section where diameter is 1 m. From continuity considerations,

$$Q = A_1 V_1 = A_2 V_2$$

$$V_2 = \frac{A_1}{A_2} V_1 = \frac{\dfrac{\pi}{4}(0.5)^2}{\dfrac{\pi}{4}(1)^2} \times 1 = \textbf{0.25 m/s.}$$

EXAMPLE 5.32

A pipe AB branches into two pipes C and D as shown in Figure 5.21. The pipe has a diameter of 45 cm at A, 30 cm at B, 20 cm at C, and 15 cm at D. Determine the discharge at A if the velocity at A is 2 m/s. Also determine the velocities at B and D, if the velocity at C is 4 m/s.

Solution: The quantity of liquid passing through section A is

$$Q_A = A_A \times V_A = \frac{\pi}{4}(0.45)^2 \times 2 = 0.318 \text{ m}^3\text{/s}$$

From continuity considerations,

$$Q = A_A \times V_A = A_B \times V_B;$$

$$\therefore \qquad V_B = \frac{Q}{A_B} = \frac{0.318}{\dfrac{\pi}{4}(0.3)^2} = \textbf{4.5 m/s}$$

From the geometry of branching pipes, it follows that

$$Q_A = Q_C + Q_D$$

$$0.318 = \frac{\pi}{4}(0.2)^2 \times 4 + \frac{\pi}{4}(0.15)^2 \times V_D$$

$$= 0.1256 + 0.0177\, V_D$$

∴ Velocity at section is

$$V_D = \frac{0.318 - 0.1256}{0.0177} = \textbf{10.6 m/s}$$

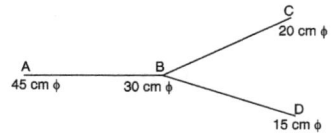

FIGURE 5.21.

5.6.2. Euler's equation along a streamline

Euler's equation of motion is established by applying Newton's second law of motion to a small element of fluid moving within a stream tube (Figure 5.22). The element has a mean cross-sectional area dA, length ds, and the centroid of the downstream face lies at a level dy higher than the centroid of the upstream face. Motion of the element is influenced by

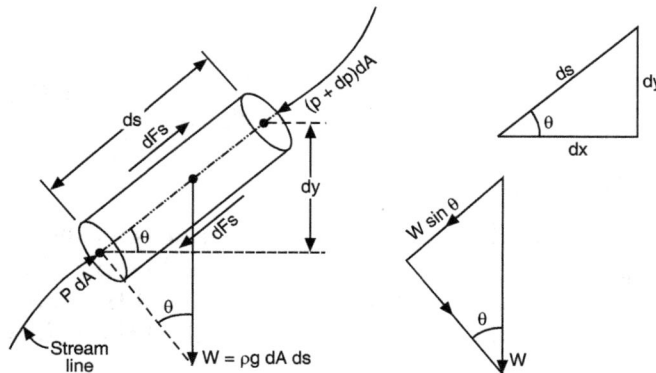

FIGURE 5.22 Fluid element with friction.

*Normal forces** due to pressure: Let p and $(p + dp)$ be the pressure intensities at the upstream and downstream face, respectively. Net pressure force acting on the element in the direction of motion is then given by

$$p\,dA - (p + dp)\,dA = -dp\,dA \tag{5.17}$$

*Tangential force** due to viscous shear: If the fluid element has a perimeter dP, then shear force on the element is

$$dFs = \tau\,dP\,ds$$

where τ is the frictional surface force per unit area acting on the walls of the stream tube. The sum of all the shearing forces is the measure of the energy lost due to friction.

*Body force such as gravity acting in the direction of gravitational field. If ρ is the density of the fluid mass, then the body force is equal to $\rho g \, dA \, ds$. Its component in the direction of motion is

$$= \rho g \, dA \, ds \, \sin \theta$$

$$= \rho g \, dA \, dy \qquad \left(\because \sin\theta = \frac{dy}{ds} \right)$$

The resultant force in the direction of motion must equal to the product of mass and acceleration in that direction. That is

$$-dp \, dA - \rho g \, dA \, dy - \tau \, dP \, ds = \rho \, dA \, ds \, a_s \qquad (5.18)$$

It may be recalled that the velocity of an elementary fluid particle along a streamline is a function of position and time

$$u = f(s, t)$$

$$du = \frac{\partial u}{\partial s} ds + \frac{\partial u}{\partial t} dt$$

$$\frac{du}{dt} = \frac{\partial u}{\partial s} \frac{ds}{dt} + \frac{\partial u}{\partial t}$$

or

$$a_s = u \frac{\partial u}{\partial s} + \frac{\partial u}{\partial t}$$

In a steady flow, the changes are with respect to position only; so $\frac{\partial u}{\partial t} = 0$ and the partial differentials become the total differentials. Evidently, for a steady flow, the acceleration of fluid element along a streamline is equal to, $a_s = u \frac{\partial u}{\partial s}$. Substituting this result in Equation 5.18, we obtain

$$-dp \, dA - \rho g \, dA \, dy - \tau \, dP \, ds = \rho \, dA \, u \, du$$

Dividing throughout by the fluid mass $\rho \, dA \, ds$ and rearranging,

$$u \frac{du}{ds} + \frac{1}{\rho} \frac{dp}{ds} + g \frac{dy}{ds} = -\frac{\tau}{\rho} \frac{dP}{dA} \qquad (5.19)$$

which is Euler's equation of motion. Here,

i. the term $u \frac{du}{ds}$ is a measure of convective acceleration experienced by the fluid as it moves from a region of one velocity to another region of a different velocity; evidently, it represents a change in kinetic energy.

ii. the term $\dfrac{1}{\rho}\dfrac{dp}{ds}$ represents the force per unit mass caused by the pressure distribution.

iii. the term $g\dfrac{dy}{ds}$ represents the force per unit mass resulting from gravitational pull.

iv. the term $-\dfrac{\tau}{\rho}\dfrac{dP}{dA}$ prescribes the force per unit mass caused by friction.

For ideal fluids, $\tau = 0$ and, therefore, Equation (5.15) reduces to

$$u\,du + \frac{dp}{\rho} + g\,dy = 0$$

$$\frac{1}{2}\,d(u^2) + \frac{dp}{\rho} + dy = 0 \qquad (5.20)$$

Euler's Equations 5.19 and 5.20 have been set up by considering the flow with a stream tube and as such apply to the flow with a stream tube or along a streamline because as dA goes to zero, the stream tube becomes a streamline.

5.6.3. Bernoulli's theorem: integration of Euler's equation for one-dimensional flow

Bernoulli's equation relates velocity, pressure, and elevation changes of a fluid in motion. The equation is obtained when Euler's equation is integrated along the streamline for a constant density (incompressible) fluid. Integration of Euler's equation,

$$\frac{1}{2}d\left(V^2\right) + \frac{dp}{\rho} + g\,dy = 0 \text{ gives}$$

$$\int \frac{1}{2}\,d(V^2) + \int \frac{dp}{\rho} + g\int dy = \text{constant.}$$

Assuming ρ to be constant,

$$\frac{V^2}{2} + \frac{p}{\rho} + gy = \text{constant} \qquad (5.21)$$

Equation 5.21 is one of the most useful tools of fluid mechanics and is known as Bernoulli's equation in honor of the Swiss mathematician Daniel Bernoulli (1700–82).

The constant of integration (called the Bernoulli constant) varies from one streamline to another but remains constant along a streamline in steady, frictionless, incompressible flow. Each term has the dimension $\left(\dfrac{L}{T}\right)^2$

$$\frac{m^2}{s^2} \equiv \frac{m\left(\text{kg m/s}^2\right)}{\text{kg}} \equiv \frac{\text{N m}}{\text{kg}} \qquad\qquad [\because 1\ \text{N} = 1\ \text{kg m/s}^2]$$

or units of $\dfrac{\text{N m}}{\text{kg}}$ and as such represents the energy per unit kilogram mass. Evidently, the energy per unit mass of fluid is constant along a streamline for steady, compressible flow of non-viscous fluid.

Dividing Equation 5.21 by g and using the relation $w = \rho g$, we obtain

$$\frac{V^2}{2g} + \frac{p}{\rho g} + y = \text{constant}$$

or
$$\frac{V^2}{2g} + \frac{p}{w} + y = \text{constant} \qquad\qquad (5.22)$$

This is the form of Bernoulli's equation commonly used by hydraulic engineers. Here, each term represents the energy per unit weight (mN/N) and has physical dimensions of length (meter). The quantities are called heads. Thus,

$$\frac{V^2}{2g} = \text{velocity head}$$

$$\frac{p}{w} = \text{pressure head or static head}$$

$$y = \text{elevated head, position head, potential head or geodetic head.}$$

The sum $\left(\dfrac{V^2}{2g} + \dfrac{p}{w} + y\right) = H$ is called the total or hydrodynamic head.

Evidently, the Bernoulli's equation may be restated as "the sum of the kinetic energy (velocity head), the pressure energy (static head) and the potential energy (elevation head) of a ideal, incompressible fluid constant along a streamline."

The different heads appearing in the Bernoulli equation (5.22) are shown diagrammatically in Figure 5.23. The **energy line** represents the sum of quantities $\left(\dfrac{V^2}{2g} + \dfrac{p}{w} + y\right)$ at each point along the flow, and the **hydraulic**

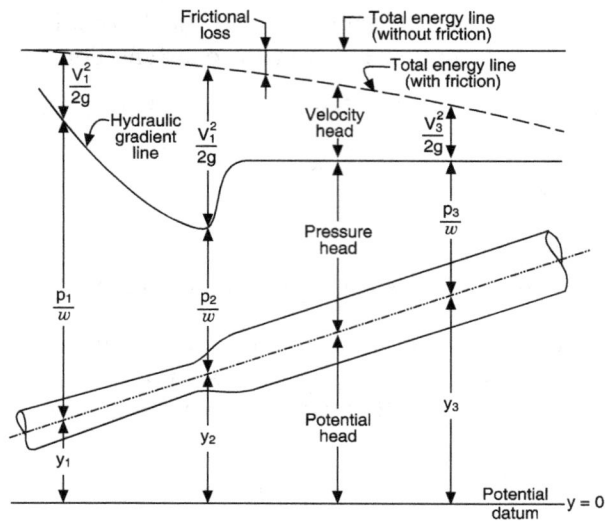

FIGURE 5.23 Head variations through a tube.

grade line represents the sum of quantities $\left(\dfrac{p}{w} + y\right)$ at each point along the flow. Thus, the hydraulic grade line lies everywhere at a distance $\dfrac{V^2}{2g}$ below the energy line. For an ideal frictionless flow, the total energy line is horizontal, parallel to the potential datum line. However, if the fluid is not ideal and exhibits friction effects, the energy line would drop thereby indicating a gradual loss of total energy.

Bernoulli's equation can also be recast as follows:

$$w \frac{V^2}{2g} + p + wy = \text{constant}$$

or
$$\rho \frac{V^2}{2g} + p + wy = \text{constant} \tag{5.23}$$

which is the customary from in which the Bernoulli's equation is applied to gas flow. Each term is in $\dfrac{\text{Nm}}{\text{m}^3}$, that is, energy per unit volume. Since elevation change is frequently very small, and the term wy may be dropped out.

The Bernoulli's equation forms the basis for solving a wide variety of fluid flow problems such as jets issuing from an orifice, jet trajectory, flow under a gate and over a weir, flow metering by obstruction meters, flow around submerged objects, and flows associated with pumps and turbines, etc.

While working with Bernoulli's equation, one must have a clear understanding of the assumptions involved in its derivation and the corresponding limitations of its applications.

- *flow is steady,* that is, at a given point, there is no variation of fluid properties with respect to time.

- *flow is ideal,* that is, it does not exhibit any frictional effects due to fluid viscosity.

- *flow is incompressible;* no variation in fluid density.

- *flow is essentially one-dimensional,* that is, along a streamline. However, Bernoulli's equation can be applied across streamlines if the flow is irrotational.

- *flow is continuous and velocity is uniform over a section.*

- *only gravity and pressure forces are present.* No energy in the form of heat or work is either added to or subtracted from the fluid.

5.6.4. Bernoulli's theorem: principle of energy conservation

Consider steady flow of incompressible liquid through a non-uniform pipe (a converging stream tube) lying entirely in the x-y plane. Flow can be assumed to be uniform and normal to the inlet and outlet area. The fluid mass has an area A_1, average velocity V_1, and pressure p_1 at the entrance; the corresponding values at exist are A_2, V_2, and p_2. The mass of the fluid in the region abcd is shifted to net position a'b'c'd' during an infinitely small internal of time.

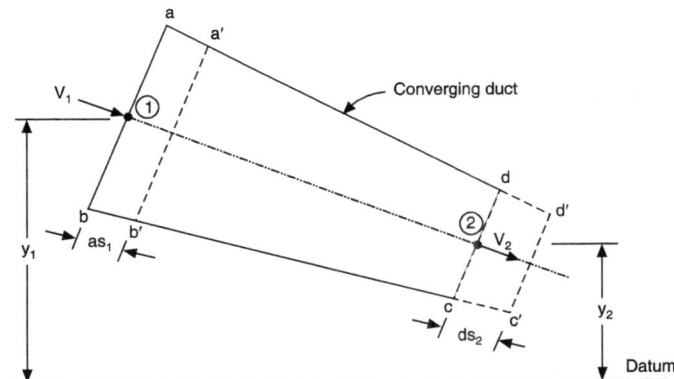

FIGURE 5.24 Energy conservation during fluid flow.

Since the area a'b'cd is common to both the regions abcd and a'b'c'd', it will not experience any energy change. Evidently, then the energy changes of the fluid masses in the section abb'a' and cc'd'd have to be considered.

i. Invoking the principle of mass conservation, the following continuity equation applies
    Fluid mass within the region abb'a' = fluid mass within the region cc'd'd

$$m = \rho \, A_1 \, ds_1 = \rho A_2 \, ds_2$$

ii. Work done during displacement of fluid mass from ab to a'b'

$$= \text{force} \times \text{displacement} = p_1 \, A_1 \, ds_1$$

The displacement ds_1 is so small that any variation in fluid properties can be neglected. Likewise, the work done at exist is equal to

$$= -p_2 \, A_2 \, ds_2$$

This work done is in opposite direction, that is, force and displacements are in opposite directions; hence, negative sign
Net flow work (work done by pressure)

$$= p_1 \, A_1 \, ds_1 - p_2 \, A_2 \, ds_2$$

$$= \frac{m}{\rho} (p_1 - p_2) \qquad \left(\because A_1 ds_1 = A_2 ds_2 = \frac{m}{\rho} \right)$$

iii. The fluid flows downwards and, therefore,

$$\text{Loss of potential energy} = mg \, (y_1 - y_2)$$

iv. The fluid mass accelerates from velocity V_1 to V_2 and, therefore,

$$\text{Gain of kinetic energy} = \frac{m}{2} \left(V_2^2 - V_1^2 \right)$$

From the principle of conservation of energy,
loss of potential energy + work done by pressure = gain in kinetic energy

$$mg(y_1 - y_2) + \frac{m}{\rho} (p_1 - p_2) = \frac{m}{2} \left(V_2^2 - V_1^2 \right)$$

Dividing throughout by mg and using the relation $w = \rho g$, we obtain

$$(y_1 - y_2) + \frac{(p_1 - p_2)}{\rho g} = \frac{\left(V_2^2 - V_1^2 \right)}{2g}$$

Rearrangement gives

$$\frac{V_1^2}{2g} + \frac{p_1}{w} + y_1 = \frac{V_2^2}{2g} + \frac{p_2}{w} + y_2$$

or $\qquad \dfrac{V^2}{2g} + \dfrac{p}{w} + y = \text{constant}$

which is the Bernoulli equation.

Thus, "In a steady flow system of frictionless incompressible fluid, the sum of velocity, pressure and elevation heads remains constant at every section." This is, however, on the assumption that energy is neither added to nor taken away by some external agency.

EXAMPLE 5.33

Water is flowing with a velocity of 15 m/s and under a pressure of 300 kPa. If the height above the datum is 30 m, calculate the total energy per unit weight of water.

Solution: Total energy per unit weight = kinetic energy + pressure energy + potential energy

$$= \frac{V^2}{2g} + \frac{p}{w} + y$$

$$= \frac{15^2}{2 \times 9.81} + \frac{300 \times 10^3}{9810} + 30 = 11.42 + 30.58 + 30 = \textbf{72 Nm/N}$$

EXAMPLE 5.34

A 7.5 cm diameter hose contains water flowing at a rate of 0.085 m^3/s. If the pressure within the pipe is 70 kPa gage, work out the maximum height to which the water may be sprayed?

Solution: The problem deals with the conversion of all the energy of water into potential energy. At maximum height, both the pressure and velocity will become zero.

$$\text{Flow velocity } V = \frac{\text{discharge}}{\text{sectional area}} = \frac{0.085}{\dfrac{\pi}{4}(0.075)^2} = 19.25 \text{ m/s}$$

$$\text{Kinetic energy of flow} = \frac{V^2}{2g} = \frac{(19.25)^2}{2 \times 9.81} = 18.88 \text{ m}$$

$$\text{Pressure energy} = \frac{p}{w} = \frac{70 \times 10^3}{9810} = 7.13$$

Considering level of hose as the potential energy of datum, the total energy is

$$18.88 + 7.13 + 0 = \mathbf{26.01 \ m}$$

EXAMPLE 5.35

A pipe 12.5 cm in diameter is used to transport oil of relative density 0.75 under a pressure of 1 bar. If the total energy relative to a datum plane 2.5 m below the center of pipe is 20 Nm/N (or J/N or m), work out the flow rate of oil.

Solution: Energy per Newton of oil

$$= \frac{V^2}{2g} + \frac{p}{w} + y$$

$$20 = \frac{V^2}{2 \times 9.81} + \frac{1 \times 10^5}{0.75 \times 9810} + 2.5 = \frac{V^2}{2 \times 9.81} + 13.59 + 2.5$$

$$\frac{V^2}{2 \times 9.81} = 3.91$$

Flow velocity $V = \sqrt{2 \times 9.81 \times 3.91} = 8.76 \ \text{m/s}$

Discharge $Q = AV = \dfrac{\pi}{4}(0.125)^2 \times 8.76 = \mathbf{0.1075 \ m^3/s}$

EXAMPLE 5.36

A horizontal water pipe of diameter 15 cm converges to 7.5 cm diameter. If the pressures at the two sections are 400 kPa and 150 kPa respectively, calculate the flow rate of water.

Solution: Apply the Bernoulli's equation at the entrance section 1–1 and exit section 2–2 of the horizontal pipe,

$$\frac{V_1^2}{2g} + \frac{p_1}{w} + y_1 = \frac{V_2^2}{2g} + \frac{p_2}{w} + y_2$$

Since the pipe is laid horizontal: $y_1 = y_2$

$$\therefore \qquad \frac{V_2^2 - V_1^2}{2g} = \frac{p_1 - p_2}{w} = \frac{(400 - 150) \times 10^3}{9810} = 25.48 \qquad (i)$$

From continuity considerations,

$$A_1 \, V_1 = A_2 \, V_2$$

$$V_2 = \frac{A_1}{A_2} V_1 = \frac{\dfrac{\pi}{4} \times (0.15)^2}{\dfrac{\pi}{4}(0.075)^2} \, V_1 = 4V_1 \qquad (ii)$$

From expressions (i) and (ii),

$$\frac{(4V_1)^2 - V_1^2}{2 \times 9.81} = 25.48$$

$$V_1 = \sqrt{\frac{2 \times 9.81 \times 25.48}{15}} = 5.77 \text{ m/s}$$

Flow rate $Q = A_1 \, V_1 = \dfrac{\pi}{4}(0.15)^2 \times 5.77 = \textbf{0.102 m}^3\textbf{/s}$.

EXAMPLE 5.37

A pipe 300 meters long has a slope of 1 in 100 and tapers from 1 m diameter at the high end to 0.5 meter at the low end. Quantity of water flowing is 5400 liters per minute. If the pressure at the high end is 70 kPa, find the pressure at the low end.

Solution: Flow area at the lower end, $A1 = \dfrac{\pi}{4}(0.5)^2 = 0.196 \text{ m}^2$

Flow area at the high end, $\qquad A2 = \dfrac{\pi}{4}(1.0)^2 = 0.785 \text{ m}^2$

Discharge $Q = 5400 \text{ liter/min} = 90 \text{ liters/s } 0.09 \text{ m}^3\text{/s}$

From continuity considerations,

$$Q = A_1 V_1 = A_2 V_2$$

$$\therefore \quad V_1 = \frac{0.09}{0.196} = 0.46 \text{ m/s}$$

and $\quad V_2 = \dfrac{0.09}{0.785} = 0.115 \text{ m/s}$

Applying Bernoulli's equation between the two pipe ends,

$$\frac{V_1^2}{2g} + \frac{p_1}{w} + y_1 = \frac{V_2^2}{2g} + \frac{p_2}{w} + y_2$$

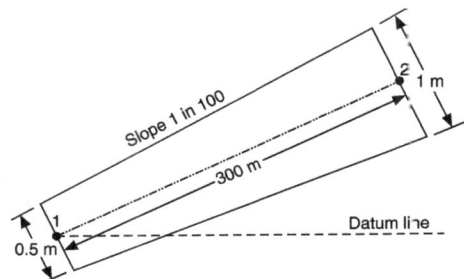

FIGURE 5.25.

With horizontal line through the lower end as the datum for elevation head,

$$y_1 = 0 \text{ and } y_2 = \frac{1}{100} \times 300 = 3\text{m}$$

$$\therefore \quad \frac{(0.46)^2}{2 \times 9.81} + \frac{p_1}{9810} + 0 = \frac{(0.115)^2}{2 \times 9.81} + \frac{70 \times 10^3}{9810} + 3$$

Solution gives: $p_1 = 99330 \text{ N/m}^2 = \textbf{99.33 kPa}$

EXAMPLE 5.38

A 2 m long pipeline tapers uniformly from 10 cm diameter to 20 cm diameter at its upper end. The pipe center line slopes upwards at an angle of 30° to the horizontal and the flow direction is from smaller to bigger cross section. If the pressure gages installed at the lower and upper ends of the pipeline read 200 kPa and 230 kPa respectively, determine the flow rate and the fluid pressure at the mid-length of the pipeline. Assume no energy losses.

Solution: Applying Bernoulli's equation between sections 1 and 2,

$$\frac{p_1}{w} + \frac{V_1^2}{2g} + y_1 = \frac{p_2}{w} + \frac{V_2^2}{2g} + y_2$$

With horizontal line through section 1 as the datum for elevation head,

$$y = 0 \text{ and } y_2 = 2\sin 30° = 1 \text{ m}$$

$$\therefore \quad \frac{200 \times 10^3}{9810} + \frac{V_1^2}{2g} + 0 = \frac{230 \times 10^3}{9810} + \frac{V_2^2}{2g} + 1$$

$$\frac{V_1^2 - V_2^2}{2g} = \frac{(230 - 200) \times 10^3}{9810} + 1$$

$$= 4.058$$

For continuity considerations: $A_1 V_1 = A_2 V_2$

$$V_2 = \frac{A_1}{A_2} V_1 = \left(\frac{D_1}{D_2}\right)^2 V_1$$

$$= \left(\frac{0.1}{0.2}\right)^2 V_1 = \frac{V_1}{4}$$

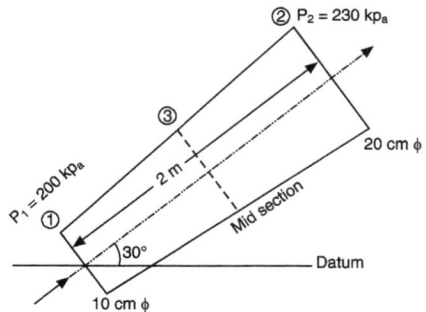

FIGURE 5.26.

$$\therefore \quad \frac{V_1^2 - \left(V_1/4\right)^2}{2 \times 9.81} = 4.058 \text{ and } V_1 = 9.215 \text{ m/s}$$

$$\text{Discharge } Q = A_1 \, V_1 = \frac{\pi}{4} \, (0.1)^2 \times 9.215 = 0.0723 \text{ m}^3/\text{s}$$

At the mid-length of pipeline, $D_3 = 15$ cm $= 0.15$ m

$$V_3 = \left(\frac{D_1}{D_3}\right)^2 V_1 = \left(\frac{0.1}{0.15}\right)^2 \times 9.215 = 4.095 \text{ m/s}$$

Applying Bernoulli's equation to sections 1 and 3,

$$\frac{p_1}{w} + \frac{V_1^2}{2g} + y_1 = \frac{p_2}{w} + \frac{V_2^2}{2g} + y_2$$

$$\frac{200 \times 10^3}{9810} + \frac{(9.215)^2}{2 \times 9.81} + 0 = \frac{p_3}{9810} + \frac{(4.095)^2}{2 \times 9.81} + 1 \times \sin 30°$$

Solution gives $p_3 = 2.29 \times 10^5$ N/m^2 = **2.29 bar**.

5.7. BERNOULLI'S EQUATION FOR REAL FLUID: FRICTION CONSIDERED

All liquids are more or less viscous and their flow is accompanied by frictional forces or resistances, which hinder motion. Therefore, from section to section, there must be a continual expenditure of energy by the liquid in overcoming the resistance. Obviously, energy at the downstream is less than that at the upstream section by an amount equivalent to frictional loss. Bernculli's equation for real fluids may then the written as follows:

$$\frac{V_1^2}{2g} + \frac{p_1}{w} + y_1 = \frac{V_2^2}{2g} + \frac{p_2}{w} + y_2 + h_f \qquad (5.24)$$

where h_f is the frictional loss of energy of head along the streamline between two sections.

Expressing the theorem in words: In steady flow, with friction present, the total head (or total energy per unit weight) at any section is equal to that at any subsequent section, plus the lost head (or lost energy per unit weight) occurring between the two sections.

EXAMPLE 5.39

A 60 cm diameter pipeline carries oil (sp. gr. = 0.85) at 82500 m^3 per day. The friction head loss is 8.5 m per 1000 m of pipe run. It is planned to place pumping stations every 20 km along the pipe. Make calculations for the pressure drop in kN/m^2 between pumping stations.

Solution: Applying Bernoulli's equation between two adjacent pumping stations

$$\frac{p_1}{w} + \frac{V_1^2}{2g} + y_1 = \frac{p_2}{w} + \frac{V_2^2}{2g} + y_2 + h_f$$

Since the pipeline is laid horizontal and is of uniform cross-sectional area,

$$y_1 = y_2 \text{ and } V_1 = V_2$$

∴
$$\frac{p_1 - p_2}{w} = h_f = \frac{8.5}{1000} \times (20 \times 1000) = 170 \text{ m}$$

$$p_1 - p_2 = 170 \times (9810 \times 0.85)$$
$$= 1417545 \text{ N/m}^2 \approx \mathbf{1417.55 \text{ kN/m}^2}$$

EXAMPLE 5.40

Water is flowing upwards through a pipeline having diameters of 15 cm and 30 cm at the bottom and upper ends, respectively. When a discharge of 50 liters/sec is passed through the pipeline, the pressure gages at the bottom and upper section read 30 kPa and –54 kPa, respectively. If the friction loss in the pipe is 2 m, determine the difference in elevation head. Take specific weight of water 10 kN/m^3.

Solution: Let suffixes 1 and 2 refer to bottom and upper ends of the pipeline.

$$V_1 = \frac{Q}{A_1} = \frac{0.05}{\frac{\pi}{4}(0.15)^2} = 2.83 \text{ m/s}$$

$$V_2 = \frac{Q}{A_1} = \frac{0.05}{\frac{\pi}{4}(0.3)^2} = 0.707 \text{ m/s}$$

Applying Bernoulli's equation to sections 1 and 2;

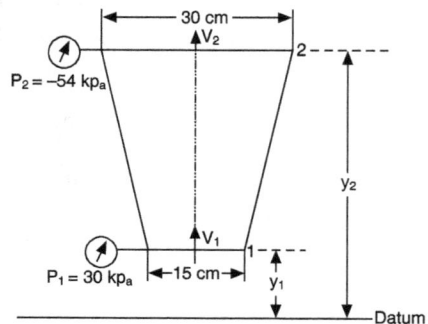

FIGURE 5.27.

$$\frac{p_1}{w} + \frac{V_1^2}{2g} + y_1 = \frac{p_2}{w} + \frac{V_2^2}{w} + \frac{V_2^2}{2g} + y_2 + h_f$$

$$\frac{30}{10} + \frac{(2.83)^2}{2 \times 9.81} + y_1 = \frac{-54}{10} + \frac{(0.707)^2}{2 \times 9.81} + y_2 + 2$$

Solution gives: $y_2 - y_1 = $ **6.78 m**

EXAMPLE 5.41

Enumerate the limitations that have to be born in mind while applying Bernoulli's theorem to engineering problems.

A conical tube 1.5 m long is fixed vertically with its smaller end upwards, and it forms a part of pipeline. Water flows down the tube and measurements indicate that velocity is 4.5 m/s at the smaller end, 1.5 m/s at the larger end, and the pressure head is 10 m of water at the upper end. Presuming that loss of head in the tube is expressed as $\dfrac{0.3(V_1 - V_2)^2}{2g}$ where V_1 and V_2 are the velocities at the upper and lower ends, make calculations for the pressure head at the lower end of the conical tube.

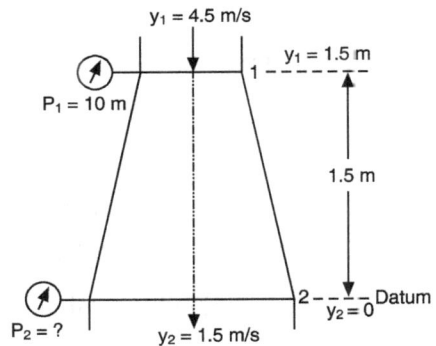

FIGURE 5.28.

Solution: Apply Bernoulli's equation to small end (1) and larger end (2) with horizontal line through section 2 as the datum for elevation head:

$$\frac{p_1}{w} + \frac{V_1^2}{2g} + y_1 = \frac{p_2}{w} + \frac{V_2^2}{2g} + y_2 + \text{losses}$$

or
$$\frac{p_2}{w} = \frac{p_1}{w} + \frac{V_1^2 - V_2^2}{2g} + (y_1 - y_2) - 0.3\frac{(V_1 - V_2)^2}{2g}$$

$$= 10 + \frac{4.5^2 - 1.5^2}{2 \times 9.81} + (1.5 - 0) - 0.3\frac{(4.5 - 1.5)^2}{2 \times 9.81}$$

$$= 10 + 0.917 + 1.5 - 0.1376$$

$$= 12.279 \text{ m of water}$$

$$= 12.279 \times 9.81$$

$$= \textbf{120.46 kN/m}^2.$$

EXAMPLE 5.42

A pipeline carrying oil of specific gravity 0.87 changes in diameter from 200 mm at a position *A* to 500 mm at another position *B* which is 4 meters at a higher level. If the pressures at A and B are 1 bar and 0.6 bar respectively and the discharge is 0.2 m³/s, determine the loss of head and the direction of flow.

Solution: Velocities of flow at sections A–A and B–B are

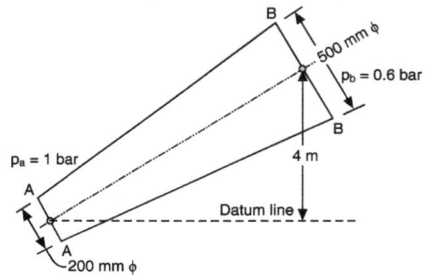

FIGURE 5.29.

$$V_a = \frac{0.2}{\frac{\pi}{4}(0.2)^2} = 6.37 \text{ m/s}$$

$$V_b = \frac{0.2}{\frac{\pi}{4}(0.5)^2} = 1.02 \text{ m/s}$$

Energy head available at section A–A with respect to elevation datum passing through section A–A

$$E_a = \frac{V_a^2}{2g} + \frac{p_a}{w} + y_a$$

$$= \frac{(6.37)^2}{2 \times 9.81} + \frac{1 \times 10^5}{0.87 \times 9810} + 0 = 13.785 \text{ m}$$

Energy head available at section B–B

$$= \frac{(1.02)^2}{2 \times 9.81} + \frac{0.6 \times 10^5}{0.87 \times 9810} + 4 = 11.083 \text{ m}$$

Since energy at section A–A is greater than at section B–B, the direction of flow is from A to B.

$$\text{Friction loss, } h_f = E_a - E_b$$
$$= 13.785 - 11.083$$
$$= \textbf{2.702 m} \text{ of oil}$$

REVIEW QUESTIONS

A. Conceptual and Conventional questions:

1. Define a fluid and distinguish between ideal and real fluids.

2. How does (*i*) a fluid differ from a solid, (*ii*) a liquid differ from a solid?

3. Cite some examples to illustrate the importance of fluid mechanics in the engineering field.

4. Define and distinguish between the following set of liquid properties:

    i. specific weight and mass density

    ii. cohesion and adhesion

    iii. surface tensity and capillarity

    iv. dynamic viscosity and kinematic viscosity

5. Define mass density, specific weight, specific volume, and specific gravity.

6. What is meant by viscosity of a liquid? How does it manifest and in what units it is measured?

7. Explain how the surface tension accounts for

    i. formation of a droplet, and

    ii. rise of liquid in a capillary.

8. State and prove Pascal's law.

9. What is meant by intensity of pressure and pressure head?

10. State and prove the hydrostatic law and adapt it for incompressible fluids.

11. Explain the concept of absolute, gage, and vacuum pressures.

12. Consider one-dimensional, frictionless steady flow of a fluid in an elementary stream tube. Show that

$$\frac{d\rho}{\rho} + \frac{dA}{A} + \frac{dV}{V} = 0$$

    Hence deduce the continuity equation $AV = $ constant for an incompressible fluid flow.

13. Mention the different forces that are included while setting the Euler equation.

14. What are the different forms of energy in a fluid? Explain each of them.

15. Establish Bernoulli's theorem from the Euler equation of motion through a stream tube. Mention the assumptions made.

16. Bernoulli's equation for steady flow may be expressed as follows:

$$\frac{V^2}{2g} + \frac{p}{w} + y = \text{constant}$$

Each term on the left side is an energy term of different type. What are they and how they represent energy form?

17. Water flows through a 10 cm diameter pipe with velocity 8 m/s. Compute the discharge rate. If the same flow now takes place through a 20 cm diameter pipe, evaluate the new flow velocity. [**Ans.** 0.628 m³/s, 2 m/s]

18. Air is flowing through a high-speed subsonic wind tunnel. At a point in the tunnel, the velocity and density are 30 m/s and 1.2 kg/m³, respectively. At another point in the tunnel, where the cross section is one-third of that at the first point, the density is 0.6 kg/m³. Estimate the velocity at the second point. [**Ans.** 180 m/s]

19. A 30 cm diameter pipe carrying 0.045 m³/s of water has a pressure of 3.5 bar at a certain flow section. Work out the total energy relative to a datum 6 m below the pipe. [**Ans.** 41.70 m of water]

20. A uniform tapering pipe is 20 cm diameter at one end A and 10 cm at the other end B. The pipe is 3 m long, is inclined to the horizontal at an angle $\alpha = \tan^{-1}(1/4)$ with end A above B. If the flow velocity at section B is 0.6 m/s, determine the difference of pressure between the two sections. [**Ans.** 6.96 kN/m²]

21. In a smooth inclined pipe of uniform diameter 250 mm, a pressure of 50 kPa was observed at section 1 which was at an elevation of 10 m. At another section 2 at elevation 12 m, the pressure was 20 kPa and the velocity was 1.25 m/s. Determine the direction of flow and the head loss between these two sections. The fluid flowing through the pipeline has a mass density of 998 kg/m³. [**Ans.** Flow is from section 1 to section 2, h_f = .064 m]

22. In a vertical pipe conveying water, pressure gages are inserted at A and B where the diameters are 15 cm and 7.5 cm, respectively. The point B is 2.4 cm below A and when the rate of flow down the pipe is 0.02 m³/s,

the pressure at B is 12 kPa greater than that at A. Assuming that losses between A and B are expressed as $\dfrac{KV_a^2}{2g}$ where V_a is the velocity at point B, find the value of K. [**Ans.** 3.49]

B. Fill in the blanks with appropriate word(s):

i. When subjected to .... force, a fluid deforms continuously no matter how small this force may be

ii. An ideal fluid is both .... and ....

iii. .... is a property of fluid by virtue of which it offers resistance to relative motion between adjacent layers.

iv. The ratio between dynamic viscosity and density is defined as ....

v. Stoke is the unit of ....

vi. According to .... law, the intensity of pressure at a point in a fluid at rest is same in all directions.

vii. The hydrostatic law states that the rate of increase of pressure in a vertical direction is equal to ....

viii. A pressure reading below the atmospheric pressure is known as ....

ix. The Bernoulli equation refers to the conservation of ....

x. Newton's law of viscosity is a relationship between shear stress and rate of ....

xi. At a liquid–air–solid surface, the contact angle θ, measured in the liquid is less than 90°. This indicates that the liquid is ....

Answers:
(*i*) shear; (*ii*) incompressible, non-viscous; (*iii*) viscosity; (*iv*) kinematic viscosity; (*v*) kinematic viscosity; (*vi*) Pascal's law; (*vii*) specific weight; (*viii*) vacuum; (*ix*) energy; (*x*) angular deformation; (*xi*) wetting.

C. Multiple choice questions:

1. A fluid is a substance that
  (a) is practically incompressible and non-viscous
  (b) obeys the Newton's law of viscosity
  (c) cannot remain at rest under the action of any shear force
  (d) always expands until it fills any container.

2. A fluid which obeys the relation $\mu = \dfrac{\tau}{du/dy}$ is called
   (a) real fluid   (b) perfect fluid
   (c) ideal fluid   (d) Newtonian fluid

3. Poise is a unit of
   (a) density   (b) surface tension
   (c) dynamic viscosity   (d) kinematic viscosity

4. All liquid surfaces tend to stretch. This phenomenon is called
   (a) cohesion   (b) adhesion
   (c) surface tension   (d) cavitation

5. All of the following statements are correct, except
   (a) For an ideal fluid $\mu = 0$ and ρ = constant
   (b) Capillary action is due to surface tension
   (c) The water is about 55 times more viscous than air
   (d) The dimension of surface tension is J/m^2.

6. Continuity equation deals with the law of conservation of
   (a) mass   (b) energy
   (c) momentum   (d) force

7. Each term of Bernoulli's equation stated in the form

   $$\frac{p}{w} + \frac{V^2}{2g} + y = \text{constant}$$

   has units of
   (a) N   (b) Nm/N
   (c) Nm/kg   (d) J/s

8. The Bernoulli's equation refers to conservation of
   (a) mass   (b) momentum
   (c) force   (d) energy

9. Stoke is a unit of
   (a) surface tension   (b) kinematic viscosity
   (c) rate of strain   (d) velocity gradient

10. What are the dimensions of kinematic viscosity of a fluid?
    (a) $L\,T^{-2}$ (b) $L\,T^{-1}$
    (c) $L^2\,T^{-1}$ (d) $M\,L^{-1}\,T^{-1}$

11. For a liquid having specific gravity 0.95 and dynamic viscosity 0.012 poise, the kinematic viscosity in stroke will be about
(**a**) 1.26   (**b**) 12.6   (**c**) 0.126   (**d**) 0.0126

12. Consider the following aspects for liquid flow:
1. steady flow
2. conservation of energy
3. viscous flow
4. incompressible flow

Which of these aspects pertain to Bernoulli's theorem?
(**a**) 1, 2 and 4   (**b**) 2, 3 and 4   (**c**) 1 and 2   (**d**) 1, 2 and 3

13. The Euler equation of motion is a statement of
(**a**) energy balance
(**b**) conservation of momentum for an inviscid fluid
(**c**) conservation of momentum for an incompressible flow
(**d**) conservation of momentum for a real fluid

Answers:
1. (*c*)   **2.** (*d*)   **3.** (*c*)   **4.** (*c*)   **5.** (*b*)   **6.** (*a*)   **7.** (*b*)
8. (*d*)   **9.** (*b*)   **10.** (*c*)   **11.** (*d*)   **12.** (*a*)   **13.** (*b*)

CHAPTER 6

HYDRAULIC MACHINES

Power engineers are primarily concerned with the design, fabrication, installation, operation, and maintenance of fluid machines that function to affect an exchange of energy between a fluid medium and a mechanical system. In the fluid-mechanical system either the machine does work on the fluid or conversely, the fluid imparts work to the machine. The wide variety of fluid machines that have been built up to affect a transfer of energy to or from a fluid medium can be broadly classified into two main types; the dynamic or kinetic type; and the displacement or static type.

1. Dynamic or **kinetic machine** in which mechanical work is done on or by the fluid entirely as a result of dynamic or kinetic action between a fluid and a rotating system of blades. The energy transfer is accompanied both by pressure and momentum changes. Examples of dynamic machines, also called the turbomachines are:

- water, steam, and gas turbines,

- pumps, fans, blowers, and compressors,

- propellers, windmills, and unshrouded fans, and

- fluid coupling and torque converters.

In these bladed machines, an impervious boundary containing the fluid does not exist at any time. Further, the operation of these machines is both steady and continuous.

Turbomachines can be identified as power generating, power absorbing, and power transmitting machines. In a power generating turbomachine, the energy of the incoming fluid is transferred to the rotating element which in turn may drive another machine. These machines produce power by expanding fluid to low pressure or head; there is a conversion from static pressure to kinetic energy. Examples of power generating machines are water, steam, and gas turbines. In a power-absorbing turbomachine, the rotating element imparts energy and thereby increases the pressure or head of the outgoing fluid; there is a transformation from kinetic energy into static pressure. Pumps, ducted fans, blowers, and compressors are notable examples of power-absorbing machines. In a power transmitting turbomachine, the energy of one rotating shaft is first transferred to the fluid which in turn transfers the energy to another shaft; this may result in a change of speed and torque. The best-known examples of power transmitting machines are fluid couplings and torque convertors. These machines are used for power transmission in automobiles and trucks, and industrial machines.

The path of fluid in the rotating element may be axial, radial, or in some cases a combination of both. Naturally, the turbomachines can be classified into axial flow, radial flow, and mixed flow machines.

2. Positive displacement or **static machines** wherein the static pressure is developed by a displacement action rather than by a velocity or kinetic energy change. During certain stages of operation, the fluid is fully contained within solid boundaries and it cannot escape except through leakage. A positive displacement machine can be reciprocating (i.e., petrol, diesel, and steam engines; reciprocating pumps and compressors) or a rotary (gear and screw pumps, root blowers, and vane 199 compressors) device while a turbomachine is always a rotary machine. Further, in a positive displacement machine, an interaction between the moving part and the fluid involves a change in the volume and/or displacement of the fluid. When the fluid volume increases, there is a transfer of energy from the fluid to the mechanical system. Conversely, when the fluid volume diminishes, energy is transferred to the fluid system.

Compared to the positive displacement machine, a turbomachine unit has the advantages of few balancing problems due to the absence of reciprocating and rubbing parts, exceptionally low consumption of lubricating oil, no conversion loss of power from rectilinear motion into rotary motion, and high reliability.

6.1. HYDRAULIC TURBINES

Hydraulic turbines are required to transform fluid energy into usable mechanical energy as efficiently as possible. Further, depending on the site, the available fluid energy may vary in its quantum of potential and kinetic energy.

Accordingly, a suitable type of turbine needs to be selected to perform the required job.

Depending upon the basic operating principle, hydraulic turbines are categorized into impulse and reaction turbines depending on whether the pressure head available is fully or partially converted into kinetic energy in the nozzle.

- *Impulse turbine* wherein the available hydraulic energy is first converted into kinetic energy by means of an efficient nozzle. The high-velocity jet issuing from the nozzle then strikes a series of suitably shaped buckets fixed around the rim of a wheel (Figure 6.1). The buckets change the direction of jet without changing its pressure. The resulting change in momentum sets buckets and wheels into rotary motion and thus mechanical

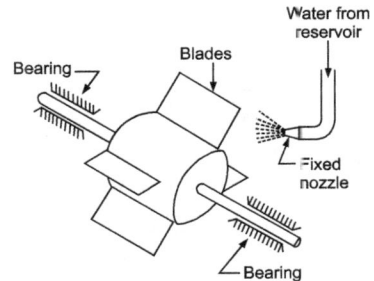

FIGURE 6.1 Principle of an impulse turbine.

energy is made available at the turbine shaft. The fluid jet leaves the runner with reduced energy. An impulse turbine operates under atmospheric pressure: there is no change of static pressure across the turbine runner and the unit is often referred to as a free-jet turbine. Important impulse turbines are Pelton wheel, Turgo-impulse wheel, Girad turbine, Banki turbine, Jonval turbine, etc.; Pelton wheel is predominantly used at present.

- *Reaction turbine* wherein a part of the total available hydraulic energy is transformed into kinetic energy before the water is taken to the turbine runner. A substantial part remains in the form of pressure energy. Subsequently, both the velocity and pressure change simultaneously as water glides along with the turbine runner. The flow from inlet to outlet of the turbine is under pressure and, therefore, blades of a reaction turbine are closed passages sealed from atmospheric conditions.

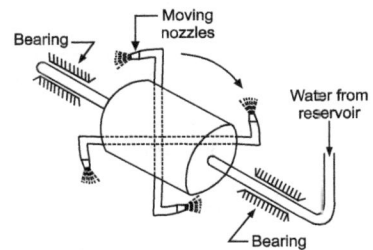

FIGURE 6.2 Principle of a reaction turbine.

Figure 6.2 illustrates the working principle of a reaction turbine in which water from the reservoir is taken to the hollow disc through a hollow shaft. The disc has four radial openings, through tubes that are shaped as nozzles. When the water escapes 4 through these tubes its pressure energy decreases and there is an increase in kinetic energy relative to the rotating disc. The resulting reaction force sets the disc in rotation. The disc and shaft rotate in a direction opposite to the direction of the water jet.

Important reaction turbines are: Fourneyron, Thomson, Francis, Kaplan, and Propellor turbines; Francis and Kaplan turbines are widely used at present.

The following table lists salient points of difference between the impulse and reaction turbines with regard to their operation and application.

TABLE 6.1 Impulse versus Reaction Turbines.

Impulse Turbine	Reaction Turbine
• All the available energy of the fluid is converted into kinetic energy by an efficient nozzle that forms a free jet.	Only a portion of the fluid energy is transformed into kinetic energy before the fluid enters the turbine runner.
• The jet is unconfined and at atmospheric pressure throughout the action of water on the runner and during its subsequent flow to the tail race.	Water enters the runner with an excess pressure, and then both the velocity and pressure change as water passes through the runner.
• Blades are only in action when they are in front of the nozzle.	Blades are in action all the time.
• Water may be allowed to enter a part or whole of the wheel circumference.	Water is admitted over the circumference of the wheel.
• The wheel does not run full and air has free access to the buckets.	Water completely fills the vane passages throughout the operation of the turbine.
• Casing has no hydraulic function to perform; it only serves to prevent splashing and to guide the water to the tail race.	Pressure at the inlet to the turbine is much higher than the pressure at the outlet; the unit has to be sealed from atmospheric conditions and, therefore, the casing is absolutely essential.
• Unit is installed above the tail race.	Unit is kept entirely submerged in water below the tail race.
• Flow regulation is possible without loss.	Flow regulation is always accompanied by loss.
• When water glides over the moving blades, its relative velocity either remains constant or reduces slightly due to friction.	Since there is a continuous drop in pressure during flow through the blade passages, the relative velocity does increase.

In addition to the concept of impulse and reaction, hydraulic turbines may be further classified into various kinds according to:

i. *Direction of water flow through runner:* Classification of turbines based on consideration of the direction of flow is given below:

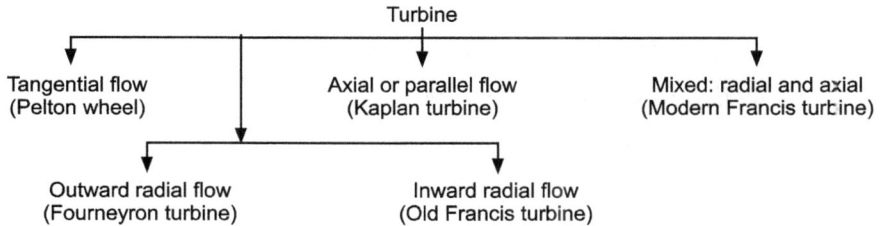

```
                          Turbine
        ┌────────────────────┼────────────────────────┐
        ▼                    │                         ▼
Tangential flow              │               Mixed: radial and axial
(Pelton wheel)               ▼               (Modern Francis turbine)
                    Axial or parallel flow
                      (Kaplan turbine)
              ┌──────────────┴──────────────┐
              ▼                             ▼
      Outward radial flow           Inward radial flow
      (Fourneyron turbine)          (Old Francis turbine)
```

Flow path in different types of runners has been illustrated in Figure 6.3.

- Pelton wheel is the tangential flow turbine; here the centerline of a jet is tangential to the path of rotation of the runner.

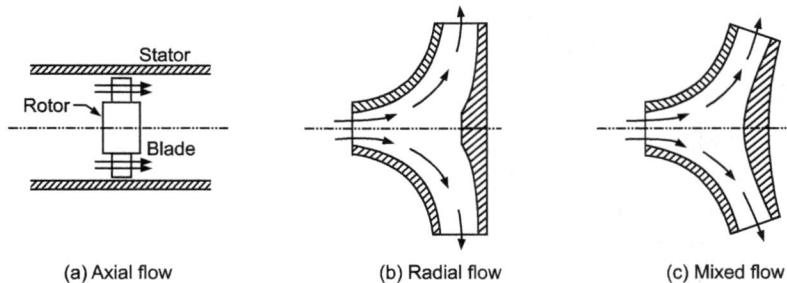

FIGURE 6.3 Flow path in different types of rotors.

- Propellor and Kaplan turbines are axial flow turbines; here water enters and leaves the runner along a direction parallel to the axis of the shaft.

- Radial flow turbines wherein the fluid passes through the runner in a plane practically perpendicular to the axis of rotation; water flows radially through the turbine. Further, the flow of water may be radially inward or radially outwards. In a radially inward flow, turbine water enters at the outer periphery, glides over the moving blades, and then flows radially inwards toward the center of the runner. The old Francis turbine and the Thomson turbine are the inward flow turbines. In a radially outward flow turbine water enters at the inner periphery, glides over the blades, and then moves radially outward/ toward the outer periphery of the runner. Fourneyron turbine is an example of an outward radial flow turbine.

- Mixed flow turbines where water enters the runner at the outer periphery in the radial direction and leaves it at the center in the direction parallel to the axis of rotation of the runner. The modern Francis turbine is a mixed flow machine.

ii. *Available head and discharge:*

- High head turbines operate under high head (above 250 m) and require relatively small rates of flow. Pelton wheel is a high head turbine.

- Medium head turbines operate under medium heads (60–250 m) and require medium flow rates. Modern Francis turbine belongs to this category.

- Low head turbines operate under heads up to 30 m and require very large volumetric rates of flow. Units of axial flow turbine (Propeller and Kaplan) are examples of a low head turbine.

iii. *Specific speed:*

Refers to the speed of a geometrically similar turbine (i.e., a turbine identical in shape, blade angles, gate openings, etc.) which would develop unit power when working under a unit head. The turbine specific speed is prescribed by the relation $N_s = N \sqrt{P}/H^{5/4}$ where P is the power in kW, H is the net available head in m and N is the speed in rpm.

Specific speed is a characteristic index that serves to identify the types of hydraulic turbine.

* For Pelton wheel: N_s = 9–17 for a slow runner
   = 17–25 for a normal runner
   = 25–30 for a fast runner = 40 for a double jet

* Francis turbine: N_s = 50–100 for a slow runner
   = 100–150 for a normal run
   = 150–250 for a fast runner

* Kaplan turbine N_s = 250–850

iv. *Disposition of shaft:* Impulse turbines have usually a horizontal shaft and vertical runner arrangement. Reaction turbines may be either of vertical or horizontal shaft type.

6.2. PELTON TURBINE

The oldest form of water turbine is the water wheel. The natural head (difference in water level) of a stream is utilized to drive it. In its conventional form, the water wheel is made of wood and is provided with buckets or vanes around the periphery. The water thrusts against these, causing the wheel to rotate. The latter drives the millstones and sometimes other machinery. In the case of an *overshot wheel* the water pours onto the buckets from above. If the water thrusts against the vanes on the underside of a water wheel, it is called an *undershot wheel.* The principle of the old water wheel is embodied in the modern Pelton wheel.

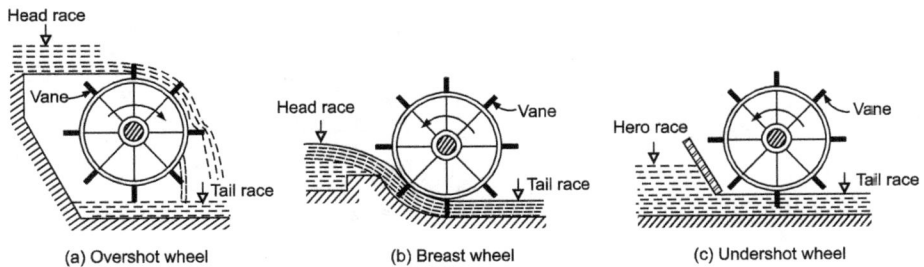

FIGURE 6.4 Water wheels.

A Pelton wheel is a free-jet impulse turbine named after the American engineer Lesser Pelton (1829–1908) who contributed much to its development. It is simple, robust and the only hydraulic turbine which operates efficiently and is invariably used for heads in excess of 450 m. Smooth running and good performance are other common features of this unit.

Component Parts: Construction and Operation

i. *Penstock:* It is a large-sized conduit that conveys water from the high-level reservoir to the turbine. Depending upon low head or high head installations, a penstock may be made of wood, concrete, or steel. Further, the penstock may be of any length depending upon the distance between the reservoir and powerhouse. For the regulation of water flow from the reservoir to the turbine, the penstock is provided with control valves. Again screens called trash racks are provided at the inlet of the penstock to prevent the debris from entering into it.

ii. *Spear and nozzle:* At its downstream end, the penstock is fitted with an efficient nozzle that converts the whole of hydraulic energy into a high-speed jet. To regulate the water flow through the nozzle and to obtain a good jet of water at all loads, a spear or needle is so arranged that it can move forward or backward thereby decreasing or increasing the annular area of the nozzle flow passage. The movement of the spear is controlled either manually by a hand wheel (in case of very small units) or automatically by a governing mechanism (in case of almost all the bigger units).

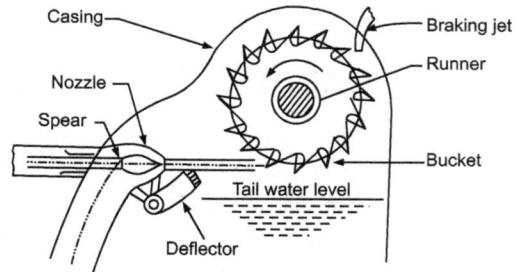

FIGURE 6.5 Elements of a Pelton wheel.

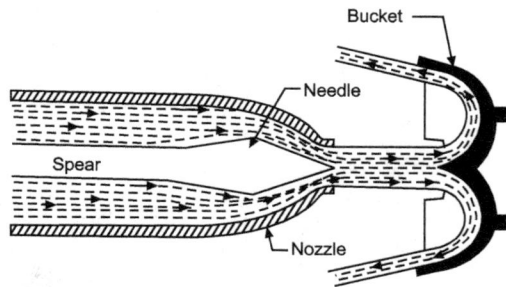

FIGURE 6.6 Spear and bucket of a Pelton wheel.

iii. *Runner with buckets:* The turbine rotor, called the runner, is a circular disk carrying a number (seldom less than 15) of cup-shaped buckets which are arranged equidistantly around the periphery of the disk. The runner is generally mounted on a horizontal shaft supported in small thrust bearings, and the buckets are either cast integrally with the disk or fastened separately. The bolt fastening facilitates the easy replacement of buckets when necessary. For low heads, the buckets are made of cast iron, but for higher heads, they are made of bronze, cast steel, or stainless steel. Further, the inner surface of the buckets is polished to reduce frictional resistance to the water jet.

Each bucket has a ridge or splitter which distributes the striking jet equally into two halves of the hemispherical bucket. Again there is a cut (notch) in the outer rim of each bucket; this notch is provided to make the jet face the bucket only when it has come into proper position with respect to the jet. This position occurs when the face of the bucket and axis of the jet is approximately 90° to each other. Maximum driving force will be exerted on the disk when the jet gets deflected through 180°, that is, when the bucket is exactly hemispherical. However, in practice, the angular deflection of a jet in the bucket is limited to about 165–170°. This is to ensure that the water jet while leaving one bucket does not strike the back of the succeeding bucket. This avoids the splashing of water and unnecessary interference which could impair the overall efficiency of the turbine.

Since the two hemispherical cups are joined together and water is directed at the junction, the side thrusts produced by the fluid in each half balance each other. The arrangement has thus the advantage that bearings supporting the wheel shaft are not subjected to any axial or end thrust.

iv. *Casing:* Outflow from the runner buckets is in the form of a strong splash that scatters in all directions. To prevent this and to guide the water to the tail race, a casing is provided all around the runner. The casing also acts as a safeguard against accidents. Evidently, the casing has no hydraulic function to perform. A baffle is arranged in the casing to prevent the discharged water from being carried along the runner direction.

v. *Governing mechanism:* Speed to the turbine runner is required to be maintained constant so that the electric generator coupled directly to the turbine shaft runs at a constant speed under varying load conditions. The task is accomplished by a governing mechanism that automatically regulates the quantity of water flowing through the runner in accordance with any variations in the load.

6.3. FRANCIS TURBINE

The Francis turbine is an inward flow reaction turbine that was designed and developed by the American engineer James B. Francis (1815–1892). In the earlier stages of its development, the Francis turbine had a purely radial flow runner; the flow passing through the runner had a velocity component only in a plane normal to the axis of the runner. The modern Francis turbine is, however, a mixed flow unit in which the water enters the runner radially at its outer periphery and leaves axially at its center. This arrangement provides a large discharge area with the prescribed diameter of the runner. Francis turbine with its full peripheral admission enjoys a great superiority and is well adopted in the hydroelectric power plants where a large quantity of water is available at low and medium heads.

Component Parts: Construction and Operation

The main features of the Francis turbine are illustrated schematically in Figure 6.7.

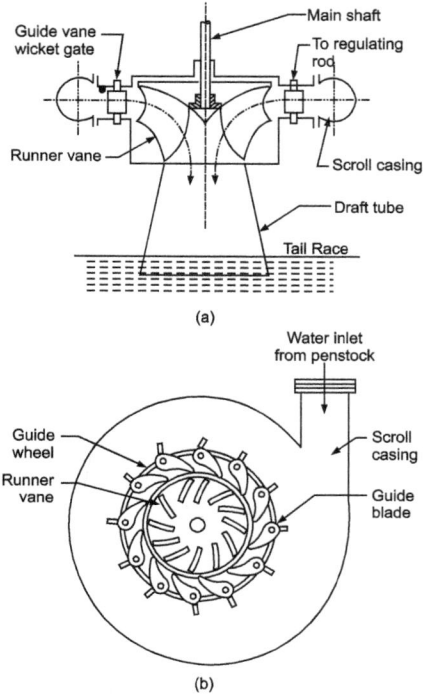

i. Penstock: It is a large-sized conduit that conveys water from the upstream of the dam to the turbine runner. Because of the large volume of water flow, size of the penstock required for a Francis turbine is larger than that of a Pelton wheel. The penstock is invariably made of steel and is embedded inside the dam. Trashrack is provided at inlet of the penstock in order to obstruct the entry of debris and other foreign matter.

ii. Scroll casing: Penstock is connected to and feeds water directly into an annular channel surrounding the turbine runner. The channel is spiral in its layout and is known as the spiral or scroll casing. Casing constitutes a closed passage whose cross-section area gradually decreases along the flow direction; area is maximum at inlet and nearly zero at exit. The decrease in area is in proportion to the decreasing volume of water to be handled and that ensures that the velocity of water is constant along its path. After entry into the casing, the water starts distributing itself into the guide blades which are arranged inside the casing. At the turn of 360°, the entire water has passed to the guide blades.

FIGURE 6.7 Elements of a Francis turbine.

The casing is made of cast steel, plate steel, and concrete depending upon the pressure/head to which the casing is subjected. Further, in the case of bigger units, stay vanes are usually provided inside the casing to support it and to direct the water from the casing to the guide vanes.

iii. Guide vanes or wicket gates: A series of airfoil-shaped vanes, called the guide vanes or wicket gates, are arranged inside the casing to form a number of flow passages between the casing and the runner blades. The guide vanes direct the water onto the runner at an angle appropriate to the design. They direct the flow just as the nozzle of the Pelton wheel. The configuration and arrangement of the guide vanes are such that the energy of water is not consumed by eddies and other undesirable flow phenomena causing energy losses.

Guide vanes are fixed in position, that is, they do not rotate with the rotating runner. However, they can swing around their own axes and that helps to bring about a change in the flow area between two consecutive runner blades. This provides a degree of adaptability to the quantity of water to be admitted to the runner in the wake of load variations. Motion is given to the guide vanes either by means of a handwheel or automatically by a governor.

iv. *Guide wheel and governing mechanism:* The governing mechanism changes the position of guide blades to affect a variation in the water flow rate in the wake of changing load conditions on the turbine. The system consists of a centrifugal governing mechanism, linkages, servomotor with its oil pressure governor, and the guide wheel. When the load changes, the governing mechanism rotates all the guide blades about their axes through the same angle so that the water flow rate to the runner and its direction essentially remain the same at all the passages between any two consecutive guide vanes. The penstock pipe feeding the turbine is often fitted with a relief valve, also known as the pressure regulator. When the guide vanes are suddenly closed, the relief valve opens and diverts the water directly to tail race. The simultaneous operation of guide vanes and relief valves is termed as double regulation.

v. *Runner and runner blades:* Runner of the Francis turbine is a rotor that has passages formed between crown and shroud in one direction and two consecutive blades on the other. These passages take water in at the outer periphery in the radially inward direction and discharge it in a direction parallel to the axis of the rotor. The driving force on the runner is both due to impulse (deviation in the direction of flow) and reaction (change in pressure and velocity energy) effects.

The number of runner blades usually varies between 16 and 24. With small units, the runner is made of cast iron while the bigger units have runner essentially made of stainless steel or a non-ferrous metal like bronze when the water is chemically impure and there is danger of corrosion. The runner is keyed to the shaft which may be of vertical or horizontal disposition; mostly vertical.

vi. *Draft tube:* After passing through the runner, the water is discharged to the tail race through a gradually expanding tube called the draft tube. The free end of the draft tube is submerged deep into the tail race. Evidently then the entire water passage from the head race to the tail race is totally closed; does not communicate with the surrounding atmospheric pressure.

Because of its gradually increasing cross-section, the discharge velocity from the turbine runner is not all wasted; it is partly converted into a useful pressure head and the water discharges at a relatively low velocity to the tail water.

6.4. PROPELLER AND KAPLAN TURBINES

The propeller turbine is a reaction turbine that is particularly suited for low head (up to 30 m) and high flow rate installations, that is, at barrages in rivers. The unit is like the propeller of a ship operating in reverse. The ship propeller rotates, thrusts the water away behind it, and thus causes the ship to move forward. In a propeller turbine, the water flows through the propeller and sets it in motion. Water enters the turbine laterally, gets deflected by the guide vanes, and then flows through the propeller. For this reason, these machines are referred to as axial flow units.

Component Parts: Construction and Operation

The main features of a propeller turbine are illustrated schematically in Figures 6.8 and 6.9. Except the runner, all other parts such as the scroll casing, stay ring, guide mechanism (arrangement of guide vanes), and the draft tube of a propeller turbine are similar to those of a Francis turbine. Between the guide vanes and the runner, the water turns through a right angle and subsequently, flows parallel to the

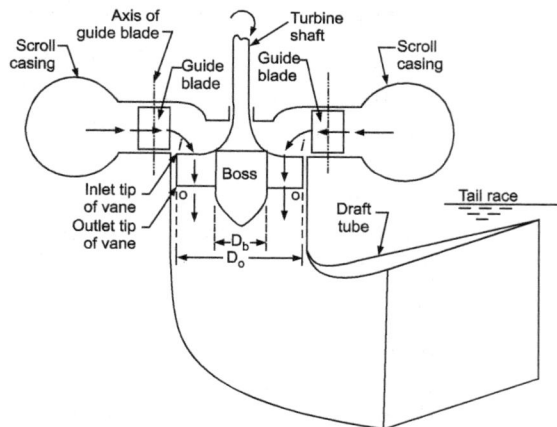

FIGURE 6.8 Elements of a Kaplan turbine.

shaft. This purely axial flow arrangement provides the largest flow area; even at larger flow rates, the flow velocities are not too large.

The runner is in the form of a boss which is nothing but the extersion of bottom end of the shaft into a bigger diameter. On the periphery of the boss are mounted equidistantly 3 to 6 vanes made of stainless steel. Thus compared to the Francis turbine which has 16–24 blades, a propeller turbine with only 3–6 vanes will have less contact surface with water and as such a low value of frictional resistance. Furthermore, the runner blades are directly attached to the hub and this feature eliminates the frictional losses which are caused by the bend provided in a Francis turbine.

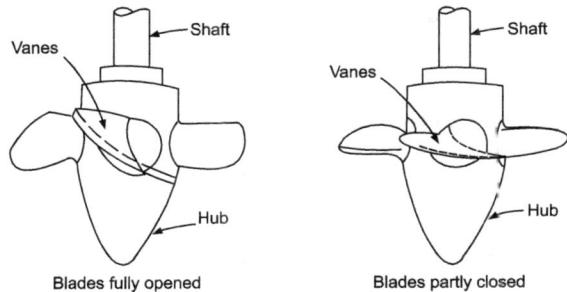

FIGURE 6.9 Kaplan turbine runner.

The fixed-blade propeller turbine is installed only at the sites where the head and load are constant. At part load, the power efficiency curve of such a unit is very much peaked, that is, poor performance is indicated. This problem of poor efficiency at part load was successfully solved by the Australian engineer Victor Kaplan who introduced the concept of adjusting the runner vanes in the face of changing load conditions on the turbine. Hence the name *variable pitch propeller turbine* is often given to the Kaplan turbine. With proper adjustment of blades during its running, the Kaplan turbine is capable of giving a high efficiency for a wide range of load conditions. The pitch of the runner blades is automatically adjusted by the governor through the action of a servomotor.

The Kaplan turbine has double regulation which comprises the movement of guide vanes and rotation of runners blades (Figure 6.10). The mechanism employs two servomotors; one controls the guide vanes and the second operates on the runner vanes. The governing is done by the governors (servomotors) from the inside of the hollow shaft of the turbine runner and the movement of the piston is employed to twist the blades through suitable linkages. The double regulation ensures a balanced and most satisfactory relationship between the relative positions of the guide and working vanes. Both the servomotors are synchronized; they are actuated simultaneously and high efficiency is maintained at all loads.

Kaplan turbines are capable of taking overloads from 15 to 20% and give a very high efficiency at all the gate openings while working at full load and part load conditions (speed and the head remain constant). The velocity plots do change with the flow rate variations caused by changes in load on the turbine. However, the blade angles also get simultaneously adjusted and as such under all working conditions, water enters and flows through the runner blades without shock. As such the eddy losses, which are inevitable in the Francis and the fixed-blade propeller turbine, get entirely eliminated in the Kaplan turbine.

Quite often the electric generator coupled to the Kaplan turbine is enclosed and works inside a straight passage having the shape of a bulb. The water tight bulb is submerged directly into the stream of water, and the bends at inlet to casing, draft tube, etc., which are responsible for the loss of head are dispensed with. The unit then needs less installation space with a consequent reduction in excavation and other civil engineering works. These turbines are referred to as *bulb* or *tubular* turbines and the power stations using such turbines are called *under water power stations.* The tubular turbines have become very popular and are invariable employed for very low head (as low as 4 m) installations such as those in tidal power plants and in rivers at very modest falls.

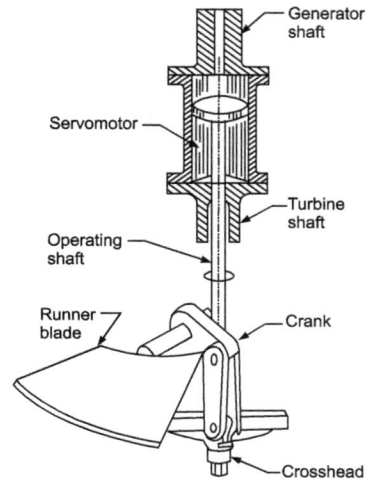

FIGURE 6.10 Kaplan runner blade mechanism.

The salient points of difference between the Francis and Kaplan turbines are enumerated below:

Francis turbine	Kaplan turbine
1. Radially inward or mixed flow turbine.	Purely axial flow turbine.
2. Horizontal or vertical disposition of shaft.	Only vertical shaft disposition.
3. Runner vanes are not adjustable.	Runner vanes are adjustable.
4. Large number of vanes; 16–24 blades.	Small number of vanes; 3–8 blades.

(Continued)

Francis turbine	Kaplan turbine
5. Large resistance needs to be overcome owing to more vanes and greater area of contact with water.	Less resistance as there are fewer vanes and less wetted area.
6. Medium head turbine (60 m to 250 m) works under medium flow rate.	Low head turbine (up to 30 m) and requires very large volumetric flow rates.
7. Specific speed ranges from 50–250.	Specific duty ranges from 250–850
8. Ordinary governor is sufficient for speed control as the servomotor is of larger size.	Heavy-duty governor is essential for speed control due to smaller sizes of the servomotors.

6.5. HYDRAULIC PUMPS

Man's economic and technical progress through the ages might be measured in terms of development from the primitive pumping devices operated either by man or animal to the positive displacement and dynamic pumps he fashions today. Typical progress in the development of pumps is found in municipal water works, power plants, agriculture, transport, and many other utility services and industries.

A *pump* has been defined differently by different investigators; the different definitions are, however, all are similar and nearly equivalent:

• a device that raises or transfers liquids at the expense of power input

• a machine designed to elevate, deliver, and move various liquids

• a unit that transfers the mechanical energy of a motor or an engine into potential and kinetic energy of a liquid.

By their action, the pumps require that energy must be expended and as such, they belong to the category of power-absorbing machines. Further, since the temperature gradients are minimal, pumps are non-thermal machines. The expended energy enables the pump to overcome the hydraulic resistance and make the fluid rise through a geodetic elevation.

6.6. PUMP CLASSIFICATION AND SELECTION CRITERION

According to design and principal of operation, pumps may be placed in one of the two general categories:

a. dynamic pumps, and

b. positive displacement pumps.

These two categories are further subdivided as depicted below:

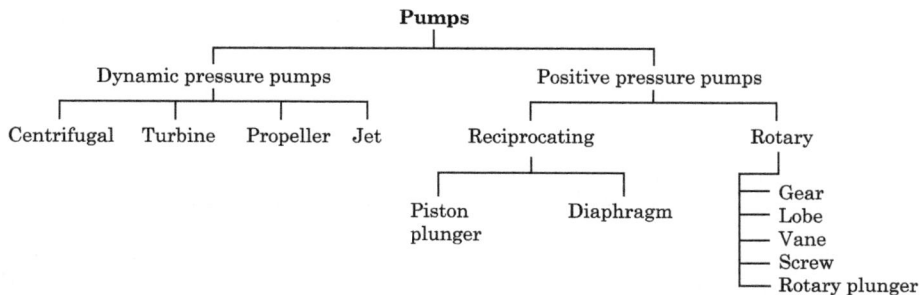

Essential data for the selection of a pump include:

• pressure and capacity of the liquid being handled,

• properties such as viscosity, temperature, corrosiveness, grittiness, etc., of the flowing liquid,

• initial and maintenance cost,

• pump duty, that is, whether the pump is to transfer the liquid or to meter it also,

• availability of space, size, and position of locating the pump,

• speed of rotation and power required,

• standardization with respect to the types and makes of pumps already available at the site, and

• scale-up problems.

Each pump has its own operating characteristics that limit its practical applications. For example in a centrifugal pump, a small change in pressure differential causes a relatively large change inflow. A positive displacement pump, on the other hand, delivers an almost constant quantity regardless of pressure fluctuations. Thus if finite pressure differences are known to exist in a particular application, then the demand of a constant supply of liquid would be met by installing a positive displacement type of pump. Likewise, a centrifugal pump would be the obvious choice if in a particular pump application it is necessary to maintain a constant head/pressure on the mains despite fluctuations in capacity/discharge.

Centrifugal pumps have high output and high efficiency. Their simple design and convenient operation have resulted in their widespread use. In general, it is always advantageous to go for a centrifugal pump unless:

i. viscosity of the liquid is greater than 1000 centipoise,

ii. low capacity and high heads are in demand, and

iii. percentage volume of dissolved gases is greater than 5%.

Reciprocating pumps are best in the field of high pressure and moderate capacity pumping. Rotary positive displacement pumps are employed in oil conduits, hydraulic devices, etc.

6.7. PUMP APPLICATIONS

A pump adds to the pressure, existing on a liquid, and increment sufficient to do the required service. This service may be increasing the pressure, imparting kinetic energy, lifting and circulating, exhausting or extracting liquids, etc., Some notable applications of pump installation are in the fields of:

- agriculture and irrigation works,

- municipal water works and drainage system (sewage disposal),

- fire protection systems,

- condensing water, condensate, boiler feed, sump drain, and such other services in a steam power plant,

- hydraulic control systems,

- circulation of water in compressor and diesel engine cooling systems,

- oil pumping, and

- transfer of raw materials, materials in manufacture, and the finished products in industry.

6.8. CENTRIFUGAL PUMPS

Centrifugal pumps belong to the category of dynamic pressure pumps wherein the pumping of liquids or generation of head is affected by rotary motion of one or more rotating wheels called the impellers.

A centrifugal pump consists essentially of the following elements:

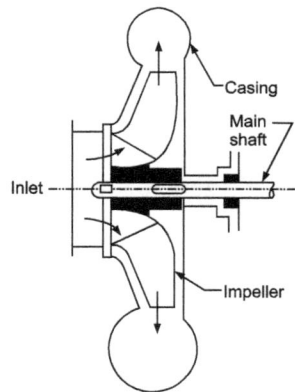

FIGURE 6.11.

i. *Rotating element* consisting of shaft and a vaned rotor called impeller. The vanes are curved, cylindrical, or have more complex surfaces. The unit has a finite number of vanes; the number is selected to assure motion of the liquid in the desired direction and varies with diameter of the impeller eye and the radial depth of the vanes. The number usually ranges between six and twelve.

The impeller is mounted on a shaft coupled to the driving unit which may be an internal combustion engine or an electric motor. By virtue of force interaction between the vanes and the liquid, the mechanical energy of the driver is transformed into the energy of flow.

ii. *Stationary element* consisting of casing, stuffing box, and bearings. The casing is an airtight chamber surrounding the pump impeller; it collects liquid from the impeller and leads it away under high pressure to the delivery side. Packings, labyrinths, and glands are needed to reduce the shaft leakage, both internal and external.

iii. *Suction pipe, strainer, and foot valve:* Suction pipe connects the center (eye) of the impeller to the sump from which the liquid is to be lifted. The pipe is laid airtight so that there is no possibility of the formation of air pockets.

Suction pipe is provided with a strainer at its lower end so as to

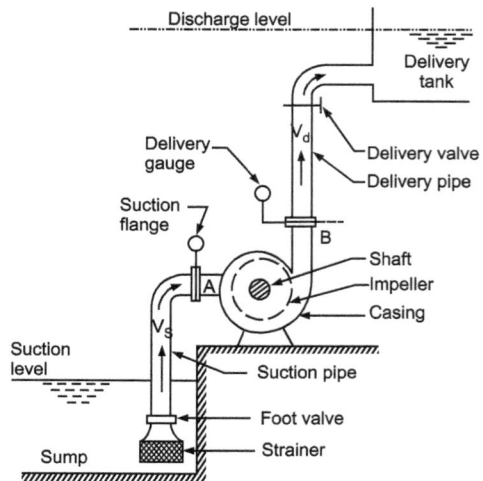

FIGURE 6.12 Typical installation of a centrifugal pump.

prevent the entry of solid particles, debris, etc., into the pump. These foreign materials, if carried into the pump would adversely affect its performance. The foot valve is a one-way valve located above the strainer into the suction pipe. It serves to fill the pump with liquid before it is started, and prevents backflow when the pump is stopped.

iv. *Delivery pipe and delivery valve:* Delivery pipe leads the liquid from the pump outlet to the point of use. A regulating valve provided just near the pump outlet serves to control the flow of liquid into the delivery pipe.

Working

The pump is initially primed wherein the suction pipe, casing, and portion of the delivery pipe up to the delivery valve are completely filled with the liquid to be pumped. Rapid motion imparted to impeller then builds up the centrifugal force which throws the liquid toward the impeller periphery. This causes a pressure gradient in the suction pipe, that is, a partial vacuum exists at the impeller eye while the liquid in the sump is at atmospheric pressure. Consequently, liquid from the sump is sucked in toward the impeller eye. When the liquid passes through the impeller, it receives energy and that results in the growth of both pressure and velocity. The casing collects the liquid from the impeller and guides it to the delivery pipe. Since the casing increases in a cross-sectional area toward the delivery, kinetic head represented by the high discharge velocity is partially transferred into pressure head before the liquid leaves the pump. The process is continuous as long as motion is given to the impeller and there is a supply of liquid to draw upon.

6.9. CLASSIFICATION OF CENTRIFUGAL PUMPS

Based on their utility, design, and constructional features, centrifugal pumps can be classified with respect to the following characteristics:

Shape and Type of Casing

Liquid leaving the impeller has an appreciable high velocity. This necessitates some arrangement to bring about the desired conversion of kinetic energy to pressure energy before the liquid reaches the discharge end of the pump. This conversion has to be accomplished with minimum energy loss. The task is accomplished by providing a casing around the pump impeller.

In general, there are three casing arrangements, and the pump is named after the casing arrangement it uses.

- *Volute* or *spiral casing* wherein the cross-section of the moving stream gradually increases from the tongue toward the discharge pipe. This increase in the area results in a gradual decrease in velocity with a corresponding increase in pressure. Most of the single-stage pumps are built with volute casings. However, the volute casing has greater eddy losses and hence lower overall efficiency.

- *Vortex* or *whirlpool chamber* wherein an annular space is provided between the volute and impeller. This arrangement arrests the formation of eddies and gives an improved performance.

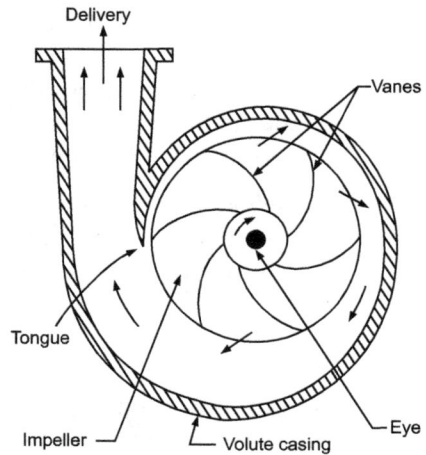

FIGURE 6.13 Pump with volute casing.

- *Volute casing with guide blades* wherein fixed guide blades is provided around the impeller periphery. When the liquid flows through diverging passages formed between the guide vanes, conversion of dynamic into static head occurs. Liquid leaving the vanes is then collected in a volute chamber where further diffusion occurs before the liquid is discharged to the delivery pipe. Pumps fitted with guide vanes are called "diffuser pumps" or "turbine pumps" as distinct from "volute pumps." Diffuser type casing is adopted when the pump impellers are to be connected in series, that is, in multistage deep well pumps. Machines with diffuser blades have rather a maximum efficiency but are less satisfactory when a wide range of operating conditions is required. This may be attributed to the losses caused by the change of blade incidence angle with flow rate.

With respect to the mechanical construction of casing, we have:

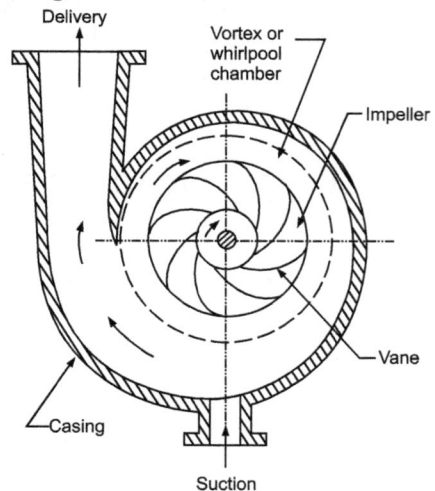

FIGURE 6.14 Pump with vortex chamber.

i. integral casing pumps: pumps equipped with a casing made in a single piece,

ii. horizontally split casing pumps: pumps equipped with a casing split on the horizontal centerline,

iii. vertically split casing pumps: pumps equipped with a casing split on the vertical centerline,

iv. diagonally split casing pumps: pumps equipped with a casing split diagonally, and

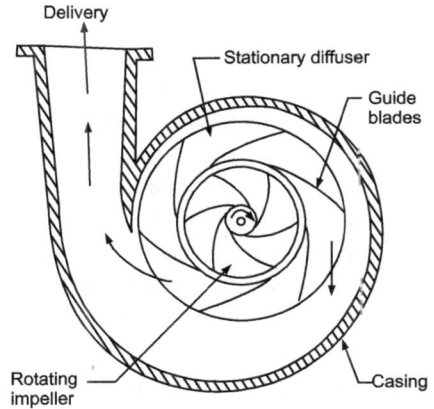

FIGURE 6.15 Pump with guide vanes.

v. segmented casing pumps: pumps equipped with a casing made up of segments. These may either be of the band type for multi-purpose pumps or of the bowl type for turbine pumps.

Closed, Semi-closed, and Open Impellers

In the closed or shrouded impellers the vanes are covered with shrouds (side-plates) on both sides. The back shroud is mounted into the shaft and the front shroud is coupled to the former by the vanes. The arrangement provides a smooth passage for the liquid; wear is reduced to a minimum. This ensures full capacity operation with high efficiency for a prolonged running period. This type is, however, meant to pump only clear liquids of low viscosity; liquids may be ordinary water, hot water, and acids.

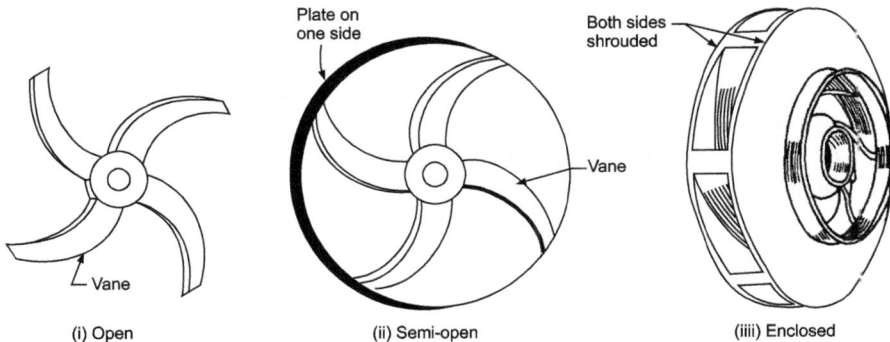

FIGURE 6.16 Closed, semi-closed, and open impellers.

The semi-open impeller has a plate (shroud) only on the backside. The design is adapted to industrial pump problems which require a rugged pump to handle liquids containing fibrous material such as paper pulp, sugar molasses, sewage water, etc.

In an open impeller, no shroud or plate is provided on either side. That is the vanes are open on both sides. Such pumps are used where the pump has a very rough duty to perform, that is, to handle abrasive liquids such as a mixture of water sand, pebbles, and clay. The presence of these foreign materials is liable to clog between the impeller and stationary side plates of a closed or semi-closed type impeller.

Axial, Radial, and Mixed Flow Impellers

In the axial flow pumps, the head is developed by the propelling or lift action of the vanes on the liquid which enters the impeller axially and discharges axially. The action is similar to the generation of lift by the wings of an airplane. Axial flow pumps have a very large discharge and are best suited for irrigation purposes.

In radial flow impellers, the head is developed by the action of centrifugal force upon the liquid which enters the impeller axially at the center and flows radially to the periphery. Flow through a mixed flow impeller is a combination of axial and radial flows. The head is developed partly by the

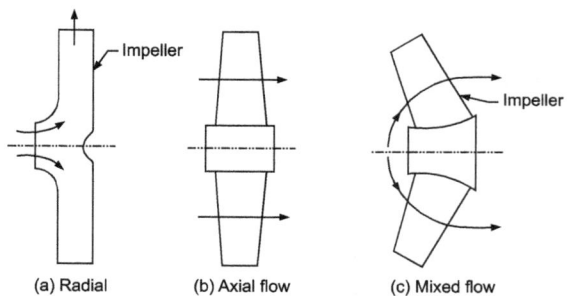

(a) Radial   (b) Axial flow   (c) Mixed flow

FIGURE 6.17 Axial, radial, and mixed flow impellers.

action of centrifugal force and partly by axial propulsion as a result of which the fluid entering the impeller axially at the center is discharged in an angular direction. Mixed flow impellers resemble the shape of a screw and are sometimes called screw impellers. Centrifugal pumps with mixed flow impellers are best suited for irrigation purposes where large quantities of water at low head are required.

Shape and Number of Vanes

Impeller of a centrifugal pump has a finite number of vanes; usually from 6 to 12. These vanes may be curved, cylindrical, or of more complex surfaces.

Working Head and Number of Stages

Based upon the range of working head, centrifugal pumps are called low head (up to 15 m), medium head (15–40 m), and high head (over 40 m) pumps. Maximum head built up by a single a stage centrifugal pump seldom exceeds 40 m of water. Greater heads are achieved by having pumps with several stages; the number of stages is indicated by the number of impellers in series. In a multistage pump, the liquid discharging from one impeller and its volute enters the eye of the succeeding impeller, and so forth, thereby increasing the head. The total head added by a multistage pump equals the sum of the heads built up by each impeller.

Single Suction and Double Suction

With respect to how the liquid enters the impeller, the pumps may be with one-sided suction (admission) or with two-sided suction. The one-side feed arrangement has the liquid entering through one side of the impeller. In a two-sided

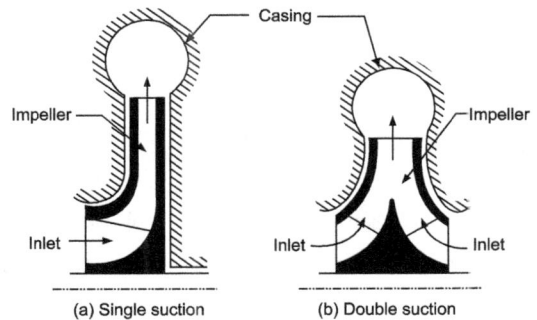

FIGURE 6.18 Single suction and double suction pumps.

feed, the liquid enters on both sides thereby increasing the discharge of the pump. Further, this arrangement eliminates the axial thrust. A single-sided impeller would, however, experience an axial thrust toward the inlet end.

Specific Speed

Specific speed is a term used for classifying pumps on the basis of their performance and dimensional proportions regardless of their actual size or the speed at which they operate. It is defined as the speed of an imaginary pump geometrically similar in every respect to the actual pump and capable of delivering unit quantity against a unit head. Mathematically, the specific speed N_s for a pump is given by,

$$N_s = \frac{N\sqrt{Q}}{H^{3/4}}$$

where N = pump speed in rev/min; Q = discharge in m³/s of a single suction impeller, and H = head per stage in meters.

Representative values of specific speed for different type of pump impellers are tabulated below:

Pump	Speed	Specific speed
Radial flow	Slow	10–30
	Medium	30–35
	High	50–80
Mixed flow		80–160
Axial flow		100–450

Shaft Position

Most of the centrifugal pumps are of horizontal shaft disposition. However, to affect the economy in space, the pumps may be designed with vertical shafts as is done for deep well and mine pumps.

Classification of pumps can also be made on the basis of:

- type of the liquid to be handled such as water, solids in suspension, and viscous liquids,

- application such as irrigation, boiler feed, and condensate circulation, and

- power used such as I.C. engine or electric motor.

6.10. HYDRAULIC SYSTEMS

There exist numerous hydraulic systems and devices in which force and energy are transmitted through an incompressible fluid, generally an oil. Notable examples are hydraulic accumulators, intensifiers, lifts and cranes, fluid coupling, and torque converters.

Figure 6.19 illustrates one typical layout of a hydraulically operated machinery (press, crane, or lift) incorporating a pump (power source to provide hydraulic energy), an accumulator, and an intensifier unit. The operation of these devices is essentially based on the principles of hydrostatics and hydro-kinetics.

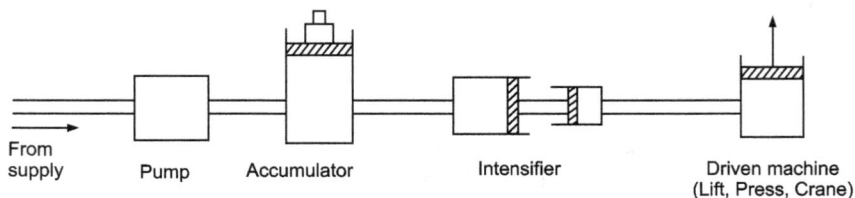

FIGURE 6.19 Typical layout of a hydraulic device.

6.10.1. Hydraulic Accumulator

Function: To store the energy of fluid under pressure and make this energy readily available as a quick secondary source of power to fluid machines such as presses, lifts, and cranes. This function is analogous to that of an electric storage battery, and the flywheel of a reciprocating engine. An accumulator also functions as a pressure regulator; it serves to damp out pressure surges and shocks in the hydraulic system.

Construction and operation: The accumulator is a simple mechanical device that essentially consists of a long vertical ram mounted within a honed or ground cylinder. The ram is weighted at the top to create pressure in the cylinder chamber (Figure 6.20).

During idle periods of the driven machine (crane or lift), high-pressure liquid discharged from the continuously working pump is admitted into the hollow space of the cylinder. This creates an upward force on the lower end of the ram. Eventually, the ram is lifted upwards when the upward force overcomes the dead weight on the ram and its packing friction. The flow of more liquid continues and fluid energy under pressure is stored in the cylinder. This high-pressure liquid stored in the accumulator is later discharged to the driven machine during its working stroke, that is, when it has to do the maximum amount of work.

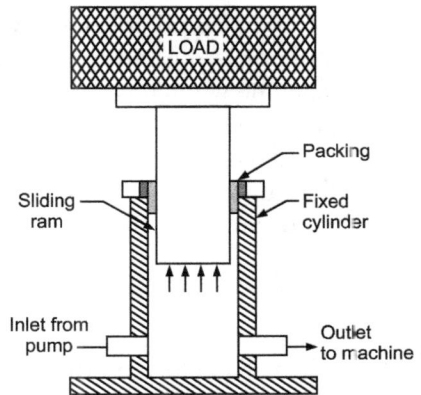

FIGURE 6.20 Simple hydraulic accumulator.

Energy stored in the accumulator refers to the work done in lifting the ram.

Work done = force × distance moved
$$= (p \times a)\,h$$
$$= p \times (ah)$$
$$= p \times \text{cylinder volume} \tag{6.1}$$

where p is the pressure of the liquid supplied by the pump, a is the cross-sectional area of ram, and h is the stroke, that is, height or fall of ram.

Evidently, the capacity of a simple accumulator is the product of the intensity of pressure and volume of the accumulator.

In its differential arrangement (Figure 6.21), the accumulator consists of a fixed vertical plunger inside which is provided a central liquid passage of small diameter. This fixed cylindrical plunger is encased within a brass bush or sleeve which itself is enclosed within an inverted cylinder. The inverted cylinder is loaded but it can slide up and down under the influence of liquid pressure. Passages for liquid to enter and leave the unit are provided in the fixed ram, and connected to the inlet and outlet pipes.

Liquid under pressure from the pump main is taken to the bottom of the fixed ram, it passes through the central cavity and exerts an upward force on the annular roof of the weighted cylinder. Load on the cylinder is sufficient to overcome the pressure exerted by the liquid. During the downward movement of the loaded cylinder, liquid collected in the accumulator is pushed toward the driven machine when it is executing its working stroke.

If D is the external diameter of the bush and d is the diameter of the fixed ram, then the annular area

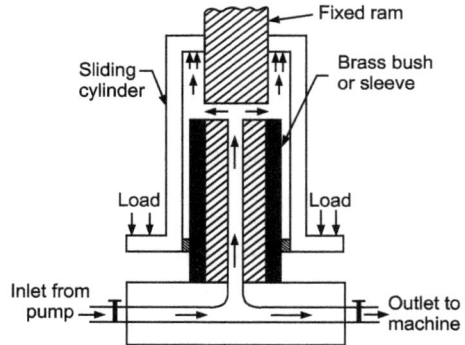

FIGURE 6.21 Differential hydraulic accumulator.

$$a = \frac{\pi}{4}\left(D^2 - d^2\right)$$

∴ Total load on the cylinder,

$$W = p \times \frac{\pi}{4}(D^2 - d^2)$$

where p is the pressure intensity of liquid. Again if h denotes the rise or fall of the sliding cylinder, then

Capacity of the accumulator $= W \times h$

Further, if lift h of the cylinder is for time t in the seconds, then

Work done or power supplied by the accumulator

$$= \frac{Wh}{t} = p\frac{\pi}{4} \times (D^2 - d^2) \times \frac{h}{t}. \tag{6.2}$$

EXAMPLE 6.1

a. **Explain the working principle of a hydraulic accumulator with the help of a neat sketch. Illustrate its function with suitable analogies from the field of electrical and mechanical systems.**

b. **An accumulator has a ram of 20 cm diameter and a lift of 7 m. If the liquid is supplied at a pressure of 5 MPa, find the load on the ram and the capacity of the accumulator in kW-h.**

Solution: (*i*) Load on ram,

$$W = p \times a$$
$$= (5 \times 10^6) \times \frac{\pi}{4} (0.2)^2 = \mathbf{0.157 \times 10^6 \ N}$$

(*ii*) Capacity of the accumulator

$$= W \times h$$
$$= (0.157 \times 10^6) \times 7 = \mathbf{1.099 \times 10^6 \ Nm}$$

Further, 1 kW-h

$$= (1000 \times 60 \times 60) = 36 \times 10^5 \ Nm$$
$$\therefore \text{ Capacity in kW-h} = \frac{1.099 \times 10^6}{35 \times 10^5} = \mathbf{0.314}.$$

EXAMPLE 6.2

The sliding ram of a hydraulic accumulator is 25 cm in diameter and it carries a load of 10 W kN. When the loaded ram moves down with a uniform velocity, the intensity of pressure inside the accumulator is 7.5 MPa. Assuming that the packing friction of the ram is equivalent to 4% of the load on the ram; work out the intensity of pressure when the load on the ram ascends up with a uniform velocity.

Solution: Load required to overcome friction

$$= 0.04 \times 10 \ W = 0.4 \ W \ kN$$

i. When the ram descends, the frictional force acts vertically upwards and hence the net load on the ram

$$= \text{weight of load the ram} - \text{load due to friction}$$
$$= (10 \ W - 0.4 \ W) = 9.6 \ W \ kN$$

Now, pressure intensity $= \dfrac{\text{net load}}{\text{area of the ram}}$

$$\text{or } 7.5 \times 10^6 = \frac{9.6W \times 1000}{\pi / 4 \times (0.25)^2}$$

Solution gives: $W = 38.3$ kN

∴ Load on the sliding ram

$$= 10\, W = 10 \times 38.3 = \mathbf{383\ kN}$$

ii. When the ram ascends up with a uniform velocity, the net load on the ram

$$= \text{weight of load} + \text{load due to packing friction}$$
$$= 383 + 0.04 \times 383 = 39.32\ \text{kN}$$

∴ Intensity of pressure $= \dfrac{39.32 \times 10^{3}}{\pi / 4 \times (0.25)^{2}} = \mathbf{81.19 \times 10^{5}\ N/m^{2}}.$

6.10.2. Hydraulic Intensifier

Function: To increase the liquid pressure above that available from a pump. The task is accomplished by utilizing the energy of a large quantity of liquid at low pressure. The intensifier is located between the pump and the machine (press, crane, and lift) that needs high-pressure liquid for its operation.

Construction and operation: The intensifier consists of a fixed ram surrounded by a sliding cylinder which itself is encased within a bigger and fixed cylinder. The fixed ram and the fixed cylinder are provided with valves V_1 and V_3

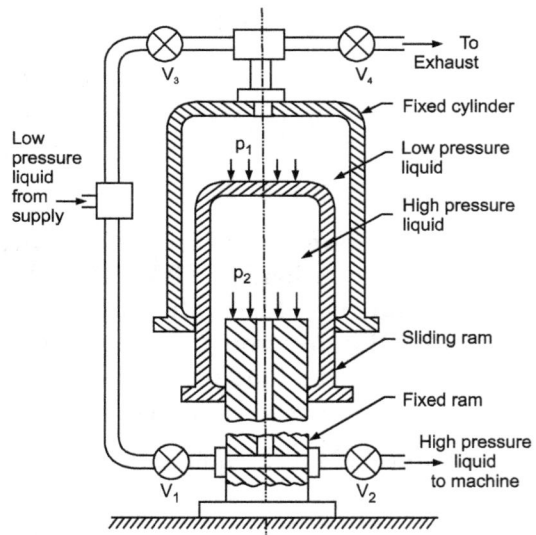

FIGURE 6.22 Hydraulic intensifier.

to admit low-pressure liquid from the main supply. Further, valve V_4 is for exhaust and valve V_2 allows high-pressure liquid to be supplied to the driven machine.

Initially, when the sliding cylinder lies at the bottom of its stroke, the fixed cylinder is full of low-pressure liquid. Keeping the valves V_2 and V_3 closed, the valve of V_1 is opened. That permits low-pressure liquid to enter the inside of the sliding cylinder. Meanwhile, exhaust valve V_4 is opened; the low-pressure liquid from the fixed cylinder is discharged to exhaust and the sliding ram moves upwards. Eventually, the sliding ram reaches its topmost position and the inside of the sliding ram gets completely filled up with the low-pressure liquid. At this stage valves V_2 and V_3 are opened and the valves V_1 and V_4 are closed. Low-pressure liquid entering the fixed cylinder through valve V_3 exerts a downward force on the ram. The sliding ram moves down, the pressure of liquid beneath it is raised and the high-pressure liquid is supplied to the driven machine.

Let A_1 and A_2 be the respective cross-sectional areas of the sliding cylinder and the fixed ram, p_1 be the pressure intensity of low-pressure liquid in the fixed cylinder and p_2 be the intensity of high-pressure liquid in the ram. Further, if k is the percentage of friction loss at each of the packings of ram, then for the equilibrium of sliding cylinder at any position:

$$p_1 A_1 \left(1 - \frac{k}{100} \right) = \frac{p_2 A_2}{\left(1 - \dfrac{k}{100} \right)}. \tag{6.3}$$

Thus, intensity of high-pressure liquid,

$$p_2 = p_1 \frac{A_1}{A_2} \left(1 - \frac{k}{100} \right)^2 \tag{6.4}$$

$$= \frac{p_1 A_1}{A_2} \text{ if the friction effects are neglected.}$$

The single-acting intensifier as described above supplies high-pressure liquid to be machine only during the downward motion of the sliding ram. For a continuous supply of high-pressure liquid, use is made of the double-acting intensifiers. Further depending upon the fluid used, an intensifier may be classified as:

- Hydro-pneumatic intensifier in which air is supplied to the fixed cylinder instead of low-pressure liquid.

- Steam intensifier wherein steam under pressure is supplied to the fixed cylinder instead of low-pressure liquid.

EXAMPLE 6.3

Describe the construction, working, and utility of a hydraulic intensifier.

A hydraulic intensifier has a ram diameter of 12 cm and a sliding cylinder diameter of 60 cm. Calculate the pressure at the outlet of the intensifier if the supply pressure is 1 MPa. It may be assumed that loss due to friction at each of the packings of the intensifier is 5% of the total force on each of the packings.

Solution: Considering equilibrium of the sliding for any position:

$$p_1 A_1 \left(1 - \frac{k}{100}\right) = \frac{p_2 A_2}{\left(1 - \dfrac{k}{100}\right)}$$

∴ Intensity of high-pressure liquid,

$$p_2 = p_1 \frac{A_1}{A_2} \left(1 - \frac{k}{100}\right)^2$$

$$= (1 \times 10^6) \times \frac{\dfrac{\pi}{4} \times 0.6^2}{\dfrac{\pi}{4} \times 0.12^2} \times 0.95^2$$

$$= 22.56 \times 10^6 \text{ N/m}^2$$

$$= 22.56 \text{ MPa.}$$

EXAMPLE 6.4

Mention some of the systems which have intensifiers as one of their basic elements.

A hydraulic intensifier gets the low-pressure liquid at a pressure of 50×10^5 N/m^2 and delivers it to a machine at a pressure of 200×10^5 N/m^2. If the intensifier has a capacity of 0.025 m^3 and stroke 1.25 m, calculate the diameters of the fixed ram and the sliding cylinder to be used for this intensifier.

Solution: Capacity of the intensifier

$$= \text{area of fixed ram} \times \text{stroke length}$$

∴ $$0.025 = A_2 \times 1.25; \; A_2 = 0.02 \text{ m}^2$$

Thus, diameter of the fixed ram

$$= \sqrt{0.02 \times 4 / \pi} = \textbf{0.1556 m}$$

(*ii*) Considering equilibrium of sliding cylinder,

$$p_1 A_1 = p_2 A_2 \qquad \text{(neglecting friction effects)}$$

∴ Area of cross-section of the sliding cylinder,

$$A_1 = \frac{p_2 A_2}{p_1}$$

$$= \frac{\left(200 \times 10^5\right) \times 0.02}{50 \times 10^5} = 0.08 \text{ m}$$

Hence diameter of sliding cylinder

$$= \sqrt{0.08 \times 4/\pi} = \textbf{0.319 m.}$$

6.10.3. Hydraulic Lift

Function: To lift or bring down load and passengers from one floor to another in a multi-storeyed building.

Construction and operation: A hydraulic lift consists essentially of a ram and cylinder arrangement with a cage or platform fitted to the top end of the ram. When fluid under pressure is forced into the cylinder, the ram gets a push vertically upwards. The platform carries loads or passengers and moves between the guides. At the requisite height, it can be made to stay in

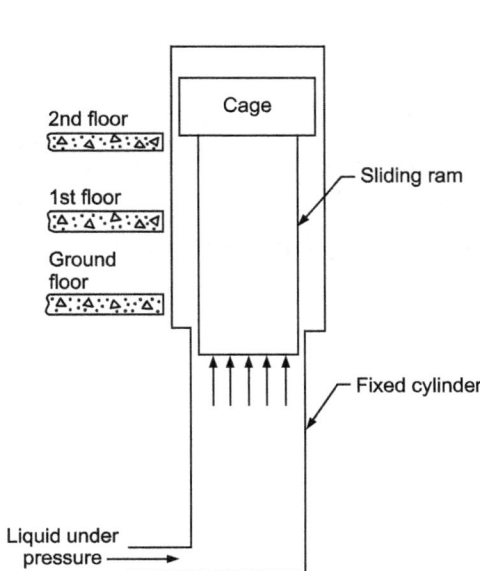

FIGURE 6.23 Direct-acting hydraulic lift.

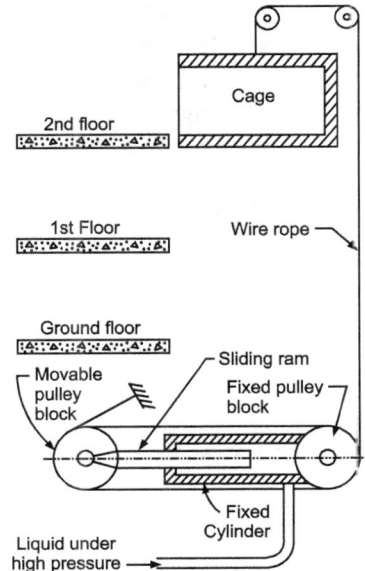

FIGURE 6.24 Suspended hydraulic lift.

level with each floor so that the goods/passengers can be transferred. In these direct-acting lifts, the stroke of the ram is equal to the lift of the cage.

In the suspended hydraulic lifts (Figure 6.24) motion of the platform or cage is obtained by the cylinder and ram arrangement of a hydraulic jigger. Modern lifts are generally of suspended type and these have lifting speeds of 150 m/min or even more.

Hydraulic lifts have, in general, been superseded by electric lifts. Hydraulic lifts then find applications as stand by units to electric lifts or in places where there is danger due to fire or explosion.

REVIEW QUESTIONS

A. Conceptual and conventional questions

1. How hydraulic turbines are classified?

2. Describe, with a sketch, the construction and working of
   (*a*) Pelton wheel;   (*b*) Francis turbine;   (*c*) Kaplan turbine

3. Sketch the Pelton turbine/Francis turbine/Kaplan turbine. Name the various components and state their function.

4. Distinguish between:
   (*a*)   Impulse and reaction turbine
   (*b*)   Kaplan and propeller turbine
   (*c*)   Inward and outward flow reaction turbine.

5. Draw the schematic arrangement of a centrifugal pump installation and state the function of different components.

6. Explain with a neat sketch the construction, operation, and utility of the following hydraulic devices:
   (*a*)   simple and differential accumulator
   (*b*)   hydraulic intensifier
   (*c*)   hydraulic lift

B. Fill in the blanks with appropriate word/words:

i.   Impulse machines operate under the _____ pressure whereas the reaction machines operate under _____ pressure.

ii. A Francis turbine is _____ flow _____ turbine.

iii. Draft tubes are provided only for _____ turbines and not for _____ turbines.

iv. The specific speeds of Kaplan, Francis, and Pelton turbines are in _____ order.

v. A _____ turbine can adjust both guide vane and blade angles according to the rate of discharge.

vi. In a centrifugal pump, the pressure energy of water is increased because of _____.

vii. The function of _____ is analogous to that of the flywheel of a reciprocating engine and an electric storage battery.

Answers:
(*i*) atmospheric, varying;   (*ii*) inward, reaction;   (*iii*) reaction, impulse;
(*iv*) decreasing;   (*v*) Kaplan;   (*vi*) centrifugal force;
(*vii*) hydraulic accumulator.

C. Multiple choice questions

1. An impulse turbine

    (*a*) is most suited for low head installation
    (*b*) makes use of draft tube
    (*c*) is not exposed to atmosphere
    (*d*) operates with initial complete conversion of pressure head to velocity head.

2. A Pelton turbine is ideally suited for

    (*a*) high head and low discharge   (*b*) high head and high discharge
    (*c*) low head and low discharge   (*d*) medium head and medium discharge

3. The modern Francis turbine is essentially a _____ turbine

    (*a*) tangent flow   (*b*) mixed flow   (*c*) axial flow   (*d*) radial flow

4. The movable wicket gates of a reaction turbine are used to

   (a) control the flow of water passing through the turbine
   (b) control the pressure under which the turbine is working
   (c) reduce the size of turbine
   (d) strengthen the casing of the turbine

5. The installation of a draft tube in a reaction turbine helps to

   (a) increase the flow rate
   (b) prevent air from entering
   (c) transport water downstream without eddies
   (d) convert the kinetic energy to pressure head

6. Which one of the following is a false statement about the centrifugal pump?

   (a) the vanes are generally curved backward
   (b) the impeller is of open type when dealing with muds
   (c) the discharge is pulsating
   (d) the value casing outside the rotating impeller prevents eddies and converts velocity head to pressure

7. The capacity of a hydraulic accumulator is defined as

   (a) maximum amount of energy it can store
   (b) maximum discharge it can deliver
   (c) total volume of liquid it can store
   (d) maximum height to which it is capable of lifting the liquid

Answers:
1. (d) **2.** (a) **3.** (b) **4.** (b) **5.** (d) **6.** (c) **7.** (a)

CHAPTER 7

MEASUREMENT AND INSTRUMENTATION

Engineering is a creative and learning profession. If engineers are to create, they must experiment and open the new frontiers of information. Experimentation is vital for progress in any field where information is lacking. There would thus be a need to measure the physical entities such as displacement, velocity, pressure, force, elapsed time, etc., in the operating devices and machines. Experimentation is considered to be the cornerstone in the field of engineering design, research, and development projects. In industry too, there is a need for the measurement and control of the physical conditions required for mass production and high-quality products. Similarly in commercial organizations, the measurement of water and electricity supplied to a consumer is a must.

The instruments for measurement, control, and transmission find such a wide and varied use that they have become an essential feature of technological operations and modern day-to-day life. It would be difficult to think of any man-made article whose manufacture did not at some stage involve measurement. There are instruments to control the flight of man-made satellites, probe the mysteries of outer space, and transmit related information. Nearer at home, we use instruments to control the temperature of our homes and to preserve food in refrigerators and cold storage. Our automobiles are equipped with instruments to measure speed, condition of the battery, and the amount of gasoline in the fuel tank. The national security devices and the sophisticated war weapons to utilize instruments for their functioning. The division of engineering science that deals with measuring techniques, devices, and their associated problems is called ***instrumentation***.

This chapter on measurement and instrumentation is intended to describe the measuring instruments and devices with regard to their construction and operation.

7.1. MEASUREMENT AND INSTRUMENT

The word ***measurement*** is used to tell us the length, the weight, the temperature, the color, or the change in one of these physical entities of a material. Measurement is the result of an opinion formed by one more observer about the relative size or intensity of some physical quantity. The opinion is formed by the observer after comparing the object with a quantity of the same kind chosen as a unit, called ***standard***. The result of measurement is expressed by a number representing the ratio of the unknown quantity to the adopted standard. This number gives the value of the measured quantity. For example, 10 cm length of an object implies that the object is ten times as large as 1 cm; the unit employed in expressing length.

The measurement standard is the physical embodiment of the unit of measurement. This places a sizable responsibility on the observer, he may be an engineer or a technician, to be certain that the standard used by him is accurately known and commonly accepted. Further, the procedure and apparatus employed for obtaining the comparison must be provable, that is, accuracy can be reproduced anywhere in the world. This is essential so that measurements obtained by him can be accepted with confidence. For consistency and quantitative comparison of physical parameters, certain standards of mass, length, time, temperature, and electrical quantities have been established. These standards are internationally accepted and well-preserved under controlled environmental conditions.

The physical quantity or the characteristic condition which is the object of measurement in an instrumentation system is variously termed as *measurand, measurement variable, instrumentation variable, and process variable*. The measurand may be a fundamental quantity (length, mass, and time), a derived quantity (speed, velocity, acceleration, power, etc.), or a quality (pressure, temperature, etc.).

The human senses cannot provide exact quantitative information about the knowledge of events occurring in our environments. The stringent requirements of precise and accurate measurements in the technological fields have, therefore, led to the development of mechanical aids called ***instruments***. Scientific instruments allow humans to observe and measure aspects of the physical universe beyond the range and precision of the unaided human senses. Instruments are the essential extensions of human sensing and perception without which scientific exploration of nature would be impossible. The instrument would sense a physical parameter (pressure, temperature, velocity, etc.), process and translate it into a format and range which can be interpreted by the observer.

FIGURE 7.1 Basic measuring process.

The man-made instruments are not only accurate and sensitive in their response but also retain their characteristics for extended periods of time. Instruments may be quite simple, such as a liquid-in-glass thermometer, or extremely complex such as the device to sense the physiological reactions of a man during space flight.

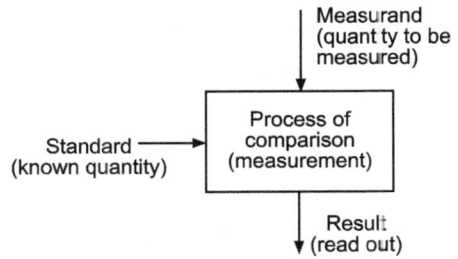

7.2. MEASUREMENT METHODS

7.2.1. Direct and Indirect Measurements

Measurement is a process of comparison of the physical quantity with a reference standard. Depending upon the requirement and based upon the standards employed, there are two basic methods of measurement:

1. Direct Measurement: The value of the physical parameter (measurand) is determined by comparing it directly with reference standards. The physical quantities like mass, length, and time are measured by direct comparison.

Direct measurements are not to be preferred because they involve human factors, are less accurate and are also less sensitive. Further, the direct methods may not always be possible, feasible, and practicable.

2. Indirect Measurement: The value of the physical parameter (measurand) is more generally determined by indirect comparison with secondary standards through calibration. The measurand is converted into an analogous signal which is subsequently processed and fed to the end device that presents the result of measurement. The indirect technique saves the primary or secondary standards from frequent and direct handling.

The accuracy of each approach is apparently traceable to the primary standard via secondary standard and the calibration.

7.2.2. Primary, Secondary, and Tertiary Measurements

The complexity of an instrument system depends upon the measurement being made and upon the accuracy level to which the measurement is needed. Based upon the complexity of the measurement system, the measurements are generally grouped into three categories, namely, the primary, secondary and tertiary measurements.

In the **primary mode**, the sought value of a physical parameter is determined by comparing it directly with reference standards. The requisite information is obtainable through senses of sight and touch. Examples are:

i. matching of two lengths while determining the length of an object with a ruler,

ii. matching of two colors while judging the temperature of red hot steel,

iii. estimating the temperature difference between the contents of containers by inserting fingers,

iv. use of beam balance to measure (actually compare) masses, and

v. measurement of time by counting the number of strokes of a clock.

The primary measurements provide subjective information only. That is, the observer can indicate only that the contents of one container are hotter than the contents of the other; one rod is longer than the other rod; one object contains more or less mass than the other.

In many technological activities, it is often difficult to make direct observation of the quantity being measured. The human senses are not equipped to make direct comparison of all the quantities with equal facility. Further, frequent measurements are extremely time-consuming and tedious if taken directly. Accordingly, we use *indirect* methods in which the measurand is converted into some directly measurable effect. The indirect methods make a comparison with a standard through the use of a calibrated system, that is, an empirical relation is established between the measurement actually made and the results that are desired. For example, an indirect method may consist of developing an electrical voltage proportional to a physical variable to be measured, measuring that voltage, and then converting the measured voltage back to the corresponding value of the original measurand. Electrical methods are preferred in the indirect methods due to their high speed of operation and simpler processing of the measured variable.

The indirect measurements involving one translation are called *secondary* measurements and those involving two conversions are called *tertiary* measurements.

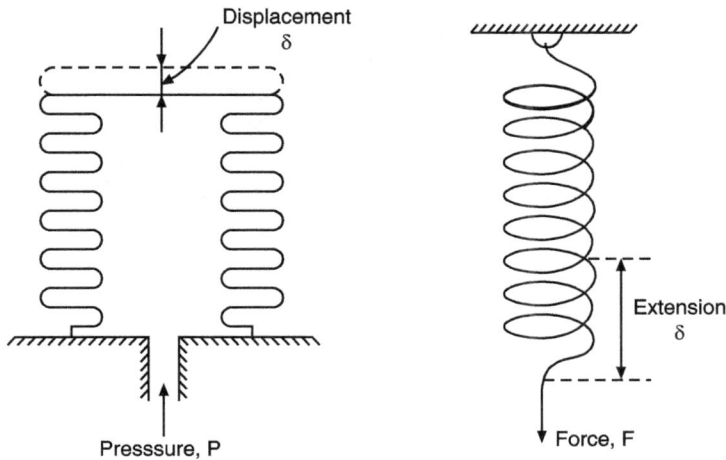

FIGURE 7.2 Secondary measurements: (a) bellows convert pressure into displacement, (b) springs convert force into displacement.

The conversion of pressure into displacement by means of bellows (Figure 7.2*a*) and the conversion of force into displacement by means of springs (Figure 7.2*b*) are simple examples of secondary measurements. When a pressure above that of the atmosphere is applied to the open end of the bellows, these expand and the resulting displacement is a measure of applied pressure. The displacement varies linearly with applied pressure provided that the range of pressure variation is small. Likewise, a spring stretches when a vertical force is applied at its free end. Different forces give rise to different displacements and so a measure of the spring deflection gives a unique indication of the force.

The pressure measurement by manometers and the temperature measurement by mercury-in-glass thermometers are other examples of secondary measurements. In these instruments, the primary signal (pressure or temperature) is first transmitted to a transducer where its effect is translated into a length change. The secondary signal of length change is subsequently converted into equivalent pressure or temperature change through a calibration process.

The measurement of static pressure by a bourdon tube pressure gauge (Figure 7.3) is a typical example of tertiary measurement.

When the static pressure (input signal) is applied to the bourdon tube, its free end deflects. The deflection which constitutes the secondary signal is very small and needs to be made larger for display and reading. The task is accomplished by an arrangement of lever, quadrant, gearing, and pointer.

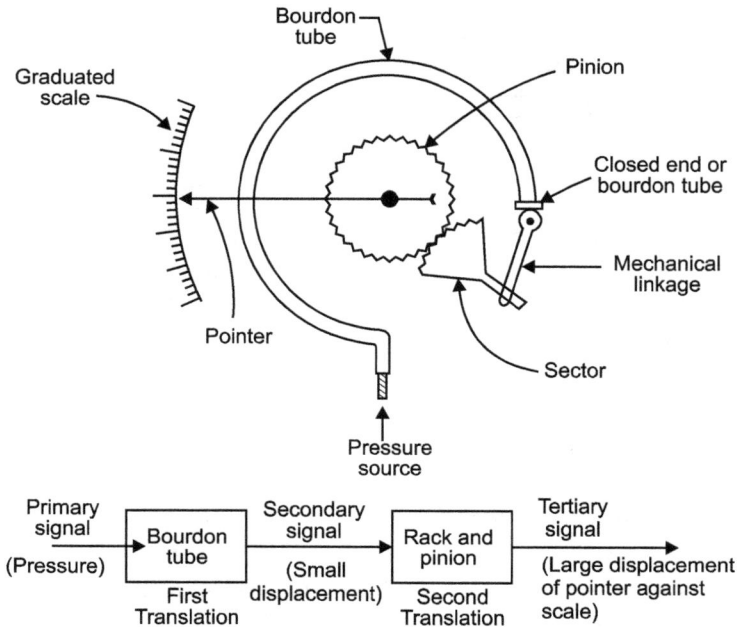

FIGURE 7.3 Tertiary measurement: measurement of pressure by a bourdon tube pressure gauge.

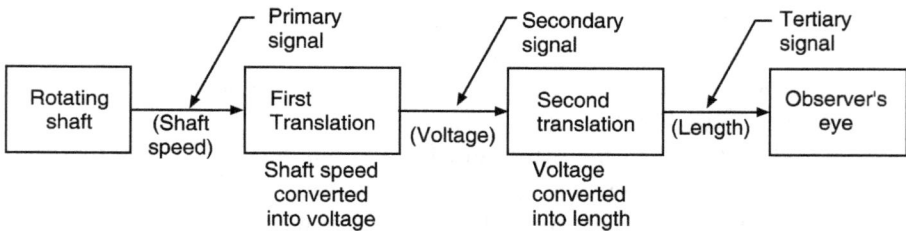

FIGURE 7.4 Tertiary measurement: measurement of angular speed by an electric tachometer.

The amplified displacement constitutes the tertiary signal, and it is indicated by the movement of the pointer against a graduated scale.

The measurement of the speed of a rotating shaft by means of an electric tachometer (Figure 7.4) is another typical example of tertiary measurement. The angular speed of the rotating shaft is first translated into an electrical voltage which is transmitted by a pair of wires to a voltmeter. In the voltmeter, the voltage moves a pointer on a scale, that is, voltage is translated into a length change. The tertiary signal of length change is a measure of the speed of the shaft and is transmitted to the observer.

The unit of a measuring system where the translation of a measurand takes place is called the *transducer* or *translator*. The term is usually applied to an electromechanical device that converts the measurand into a proportional electrical output. The electrical, mechanical, or any other variable which is actually measured is called the *measured signal*. In the example of a tertiary measurement cited above, the measured signal is the voltage which is an electrical analog of the speed of rotation of the unit coupled to the tachometer. In a thermocouple thermometer, the measured signal is an electromotive force which is the electrical analog of the temperature applied to the thermocouple. Likewise in a differential flowmeter, the measured signal would be the differential pressure which is the analog of the rate of flow through an orifice plate.

Needless to say, the majority of measurement systems are tertiary systems and they include the whole range of mechanical, electrical, pneumatic, electromechanical, and electro-pneumatic instruments.

Whereas the input to a measuring system is known as **measurand**, the output is called **measurement.**

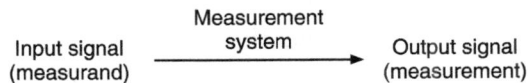

Input signal (measurand) ──── Measurement system ────▶ Output signal (measurement)

For example, in a bourdon tube gauge, the applied pressure (input to the measurement system) is the measurand. The output from the system is the movement of the pointer against a calibrated scale, and this pointer movement becomes the measurement. Likewise in an electric tachometer, the angular speed is the measurand and the movement of the pointer (length change) is the measurement.

7.2.3. Contact and Non-contact Type Measurements

Measurements may also be described as (*i*) contact type where the sensing element of the measuring device has contact with the medium whose characteristics are being measured and (*ii*) non-contact type where the sensor does not communicate physically with the medium. The optical, radioactive and some of the electrical/electronic measurements belong to this category.

7.3. STATIC TERMS AND CHARACTERISTICS

Range and Span

The region between the limits within which an instrument is designed to operate for measuring, indicating, or recording a physical quantity is called the *range* of the instrument. The range is expressed by stating the lower and upper values. *Span* represents the algebraic differences between the upper and lower range values of the instrument. For example,

Range −10° C to 80° C;   Span 90°C

Range 5 bar to 100 bar;   Span 95 bar

Range 0 V to 75 V;   Span 75 V

Accuracy, Error, and Correction

No instrument gives an exact value of what is being measured. There is always some uncertainty in the measured value. This uncertainty is expressed in terms of accuracy and error. Accuracy of an indicated (measured) value may be defined as conformity with or closeness to an accepted standard value (true value). The accuracy of the measured signal depends upon the intrinsic accuracy of the instrument itself, variation of the signal being measured, accuracy of the observer, and whether or not the quantity is being truly impressed upon the instrument. For example, the accuracy of a micrometer depends upon factors like an error in screw, anvil shape, temperature difference, applied torque variations, etc.

In general, the result of any measurement differs somewhat from the true value of the quantity being measured. The difference between the measured value (V_m) and the true value (V_t) of the quantity represents *static error* or absolute error of measurement (E_s).

$$E_s = V_m - V_t \tag{7.1}$$

The error may be either positive or negative. For positive static errors, the instrument reads high and for negative static errors, the instrument reads low.

From the experimentalist's viewpoint, *static correction* or simply correction (C_s) is more important than the static error. The static correction is defined as the difference between the true value and the measured value of a quantity:

$$C_s = V_t - V_m \tag{7.2}$$

The correction of the instrument reading is of the same magnitude as the error, but opposite in sign, that is, $C_s = -E_s$.

EXAMPLE 7.1.
A thermometer reads 73.5°C and the true value of the temperature is 73.15°C. Determine the error and the correction for the given thermometer.

Solution: Error E_s = measured value V_m – true value V_t
$$= 73.5 - 73.15 = 0.35°C$$
Correction $C_s = -E_s = \mathbf{-0.35°C}$

EXAMPLE 7.2.
A temperature transducer has a range of 0°C to 100°C and accuracy of ±0.5% of full-scale value. Find the error in a reading of 55°C.

Solution: Error $E_s = \pm\, 0.5°C$ percent of full-scale value

$$= \pm \frac{0.5}{100} \times 100 = \pm 0.5°C$$

Thus a nominal reading of 55°C actually indicates a temperature in the range of 54.5°C–55.5°C.

EXAMPLE 7.3.
  a. **The accuracy of the instrument has been specified as "accurate to within ± x for the prescribed or full range of the instrument." How do you interpret it?**
  b. **A thermometer is quoted as having the following specification:**

Range and subdivision °C	Maximum error
–0.75 to + 37.5 × 0.1	0.25°C

    How will you interpret this catalog?

Solution: a. The statement means that the instrument is accurate to within ±x at all points on the scale unless specified otherwise. This implies that irrespective of the indicated value, the error remains the same. For example, a given thermometer may be stated to read within ±0.5°C between 100°C and 230°C. Likewise, a scale of length may be read within ±0.025 cm.

    b. The given specification implies that a thermometer can be used for temperature measurement between –0.75°C and +37.5°C and has a scale that is subdivided into 0.1°C intervals. Further, the error has a temperature within a region bounded by plus or minus 0.25°C of the indicated value. Thus if the meniscus of the mercury-in-glass thermometer were read at 28.5°C, the actual temperature would lie between (28.5 ± 0.25)°C.

Hysteresis and Dead Zone

The magnitude of output for a given input depends upon the direction of the change of input. This dependence upon previous inputs is called *hysteresis*. Hysteresis is the maximum difference for the same measured quantity (input signal) between the upscale and downscale readings during a full range traverse in each direction. The maximum difference is frequently specified as a percentage of full scale. Hysteresis results from the presence of irreversible phenomenon such as mechanical friction, slack motion in bearings and gears, elastic deformation, magnetic and thermal effects. Hysteresis may also occur in electronic systems due to heating and cooling effects which occur differentially under conditions of rising and falling input.

Dead zone is the largest range through which an input signal can be varied without initiating any response from the indicating instrument. Friction or play is the direct cause of dead zone or band.

Drift

It is an undesired gradual departure of instrument output over a period of time that is unrelated to changes in input, operating conditions, or load. Wear and tear, high stress developing at some parts, and contamination of primary sensing elements cause drift. It may occur in obstruction flow meters because of wear and erosion of the orifice plate, nozzle, or Venturi meter. Drift occurs in thermocouples and resistance thermometers due to the contamination of the metal and a change in its atomic or metallurgical structure. Drift occurs very slowly and can be checked only by periodic inspection and maintenance of the instrument.

Sensitivity

Sensitivity of an instrument or an instrumentation system is the ratio of the magnitude of the response (output signal) to the magnitude of the quantity being measured (input signal), that is,

$$\text{Static sensitivity, } K = \frac{\text{change of output signal}}{\text{change of input signal}} \tag{7.3}$$

Sensitivity has a wide range of units, and these depend upon the instrument or measurement system being investigated. For example, the operation of a resistance thermometer depends upon a change in resistance (output) to change in temperature (input), and as such its sensitivity will have units of ohms/°C. Sensitivity of an instrument system is usually required to be as high as possible because then it becomes easier to take the measurement (read the output).

EXAMPLE 7.4
A spring scale requires a change of 150 N in the applied weight to produce a 2 cm change in the deflection of the spring scale. Determine the static sensitivity.

Solution:
$$k = \frac{\text{change of output signal}}{\text{change of input signal}} = \frac{2}{150} = 0.0133 \text{ cm/N}$$

EXAMPLE 7.5
Explain the following statements:

i. **A galvanometer has sensitivity specified as 15 mm/mA**

ii. **An automatic balance has a quoted sensitivity of 1 vernier division/0.1 mg**

Solution:

i. This means that for 1 mA input the display (which is the light spot moving across a scale) shows a movement of an index of 15 mm.

ii. This means that the index moves through one division when the mass changes by 0.1 mg.

Threshold and Resolution

The smallest increment of the quantity being measured which can be detected with certainty by an instrument represents the threshold and resolution of the instrument.

When the input signal to an instrument is gradually increased from zero, there will be some minimum value input before which the instrument will not detect any output change. This minimum value is called the threshold of the instrument. Thus *threshold* defines the minimum value of input which is necessary to cause a detectable change from zero output. Threshold may be caused by backlash or internal noise.

When the input signal is increased from non-zero value, one observes that the instrument output does not change until a certain input increment is exceeded. This increment is termed *resolution* or *discrimination*. Thus resolution defines the smallest change of input for which there will be a change of output. With analog instruments, the resolution is determined by the ability of the observer to judge the position of a pointer on a scale, for example, the level of mercury in a glass tube.

Threshold and resolution may be expressed as an actual value or as a fraction or percentage of full-scale value.

EXAMPLE 7.6

How resolution is reckoned for the analog and digital read-out devices?

A force transducer measures a range of 0-150 N with a resolution of 0.1% of full scale. Find the smallest change which can be measured.

Solution: Resolution = 0.1% of full-scale value

$$= 0.1/100 \times 150 = 0.15 \text{ N}$$

and this represents the smallest measurable change in force.

EXAMPLE 7.7

Distinguish between threshold and resolution (or discrimination).

The pointer scale of a thermometer has 100 uniform divisions, full-scale reading is 200°C and 1/10th of a scale division can be estimated with a fair degree of accuracy. Determine the resolution of the instrument.

Solution: 1 scale division = 200/100 = 2°C

$$\text{Resolution} = 1/10\text{th of scale division}$$

$$= 1/10 \times 2 = 0.2°C$$

Precision Repeatability and Reproducibility

These terms refer to the closeness of agreement among several measurements of the same true value under the same operating conditions.

Let us differentiate between accuracy and precision as applied to the realms of measurements. Accuracy refers to the closeness or conformity to the true value of the quantity under measurement. Precision refers to the degree of agreement within a group of measurements, that is, it prescribes the ability of the instrument to reproduce its reading over and over again for a constant input signal. This distinction can be elaborated by considering the following two examples:

i. Consider a micrometer normal in every respect but with its anvil displaced from its true position. The readings taken with this micrometer would be clearly defined and consistent, that is, a negligible scatter among different readings for the same dimension. We would say that the micrometer is as precise as ever. The readings, however, do not conform to the truth as the anvil is not placed at its correct position. The readings of the dimension with this micrometer are thus not accurate.

ii. Consider two voltmeters of the same model, make, and range. Further, let both have knife-edge pointers, carefully ruled and mirror-backed scale

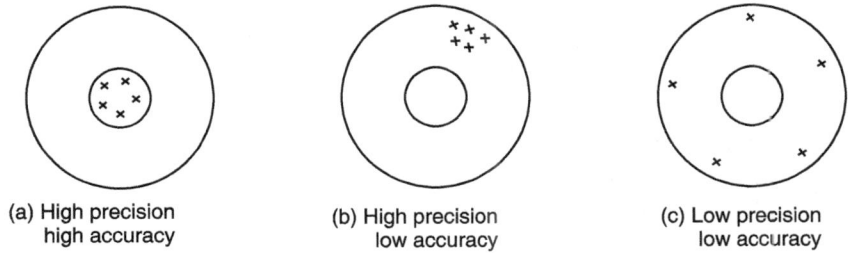

(a) High precision
high accuracy

(b) High precision
low accuracy

(c) Low precision
low accuracy

FIGURE 7.5 Difference between accuracy and precision.

to help avoid parallax errors. Both the voltmeters can be read to the same precision. In case the series resistance of one of the voltmeters is defective, its readings would be subjected to an error. The accuracy of the two instruments would then be different.

The difference between accuracy and precision has been illustrated in Figure 7.5. The arrangement may be thought to correspond to the game of darts where one is asked to strike a target represented by a center circle. The center circle then represents the true value, and the result achieved by the striker has been indicated by the mark X.

Two further terms used to define reproducibility are:

- *Stability* refers to the reproducibility of the mean reading of an instrument, repeated on different occasions separated by intervals of time which are long compared with the time of taking a reading. The conditions of use of the instrument remain unchanged.
- *Constancy* refers to the reproducibility of the mean reading of an instrument when a constant input is presented continuously and the conditions of the test are allowed to vary within specified limits. This variation may be due to some change in the external environmental conditions.

The above discussion also points out that it is possible to obtain high precision with poor accuracy, but not high accuracy with low precision. In other words, precision is a necessary prerequisite to accuracy but it does not guarantee accuracy.

Linearity

The working range of most of the instruments provides a linear relationship between the output (reading taken from the scale of the instrument) and input (measurand, signal presented to the measuring system). Linearity is defined as the ability to reproduce the input characteristics symmetrically, and this can be expressed by the straight line equation.

$$y = mx + c$$

where y is the output, x is the input, m is the slope, and c is the intercept. Apparently, the closeness of the calibration curve to a specified straight line is the linearity of the instrument.

Any departure from the straight line relationship is *non-linearity*. The non-linearity may be due to non-linear elements in the measurement device, mechanical hysteresis, viscous flow or creep, and elastic after-effects in the mechanical system.

Some other terms associated with the static performance of an instrument are:

- **Tolerance:** Range of inaccuracy which can be tolerated in measurements: it is the maximum permissible error. For example, the tolerance would be ±1% when an inaccuracy of ±1 bar can be tolerated for 100 bar value of pressure.

- **Readability and least count:** The term readability indicates the closeness with which the scale of the instrument may be read. The term least count represents the smallest difference that can be detected on the instrument scale. Both readability and least count are dependent on length scale, spacing of graduations, size of the pointer, and parallax effect.

- **Backlash:** The maximum distance or angle through which any part of a mechanical system may be moved in one direction without applying appreciable force or motion to the next part in a mechanical system.

- **Zero stability:** A measure of the ability of the instrument to restore to zero reading after the measurand has returned to zero, and other variations (temperature, pressure, humidity, vibration, etc.) have been removed.

- **Stiction (Static friction):** Force or torque that is necessary just to initiate motion from rest.

Calibration

The magnitude of the error and consequently the correction to be applied is determined by making a periodic comparison of the instrument with standards that are known to be constant. The entire procedure laid down for making, adjusting, or checking a scale so that readings of an instrument or measurement system conform to an accepted standard is called *calibration*. The graphical representation of the calibration record is called the *calibration curve* and this curve relates standard values of input or measurand to actual values of output throughout the operating range of the instrument.

A comparison of the instrument reading may be made with

i. a primary standard,

ii. a secondary standard of accuracy greater than the instrument to be calibrated, and

iii. a known input source.

For example, we may calibrate a flowmeter by comparing it with a standard flow measurement facility at the National Bureau of Standards; by comparing it with another flow meter (a secondary standard) that has already been compared with a primary standard; or by direct comparison with a primary measurement such as weighing a certain amount of water in a tank and recording the time elapsed for this quantity to flow through the meter.

The following points and observations need consideration while calibrating an instrument:

i. Calibration of the instrument is carried out with the instrument in the same position (upright, horizontal, etc.) and subjected to the same temperature and other environmental conditions under which it is to operate while in service.

ii. The instrument is calibrated with values of the measurand impressed both in the increasing and the decreasing order. The results are then expressed graphically; typically the output is plotted as the ordinate and the input or measurand as the abscissa.

iii. Output readings for a series of impressed values going up the scale may not agree with the output readings for the same input values when going down.

iv. Lines or curves plotted in the graphs may not close to form a loop.

In a typical calibration curve (Figure 7.6) ABC represents the readings obtained while ascending the scale; DER represents the readings during descent; KLM represents the median and is commonly accepted as the calibration curve. The term "median" refers to the mean of a series of up and down readings.

Quite often, the indicated values are plotted as abscissa and the ordinate represents the variation of the median from the true values (Figure 7.7).

A faired curve through the experimental points then represents the correction curve. This type of deviation presentation facilitates a rapid visual assessment of the accuracy of the instrument. The user looks along the abscissa for the value indicated by the instrument and then reads the correction to be applied.

A properly prepared calibration correction curve gives information about the absolute static errors of the measuring device, the extent of the instrument's linearity or conformity, and the hysteresis and repeatability of the instrument.

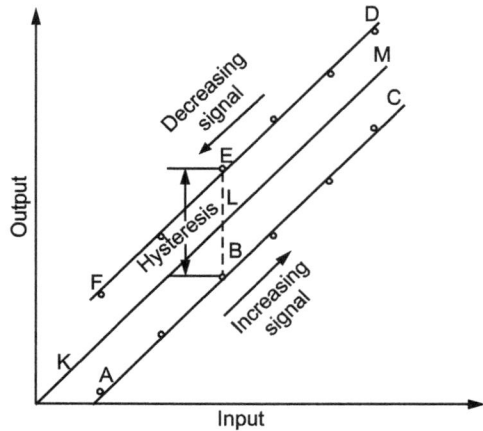

FIGURE 7.6 Calibration curve.

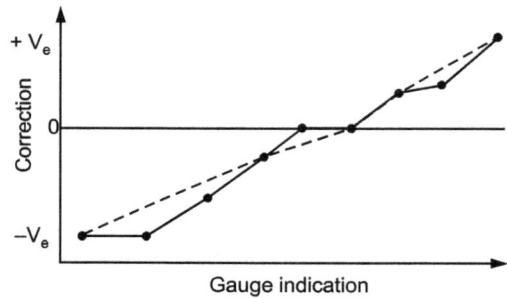

FIGURE 7.7 Corrective curve.

7.4. MEASUREMENT ERRORS

Despite utmost care and precautions an experimenter may take to eliminate all possible errors, the happy goal is seldom attained and certain errors are bound to creep in. For example, even in an apparently simple measurement of flow velocity with a pitot tube, any misalignment of the probe, leaks in the pressure tubing, changes in the bore and surface conditions of the manometer, any fluctuations in the atmospheric and stream pressure are likely to affect the probe readings and give rise to uncertainties. Errors and uncertainties are inherent in the process of making any measurement and in the instrument with which the measurements are made.

Errors may originate in a variety of ways and the following sources need examination.

Instruments Errors

There are many factors in the design and construction of instruments that limit the accuracy attainable. Instruments and standards possess inherent inaccuracies and certain additional inaccuracies develop with use and time. Example are:

- Improper selection and poor maintenance of the instrument.

- Faults of construction resulting from finite width of knife edges; lost motion due to necessary clearance in gear teeth and bearing; excessive friction at the mating parts, etc.

- Mechanical friction and wear, backlash, yielding of supports, pen or pointer drag, and hysteresis of elastic members due to aging.

- Unavoidable physical phenomenon due to friction, capillary attraction, and imperfect rarefaction.

- Assembly errors resulting from the incorrect fitting of the scale zero with respect to the actual zero position of the pointer, non-uniform division of the scale, and bent or distorted pointers.

The assembly errors do not alter with time and can be easily discovered and corrected. Uncertainty in measurement due to friction at the mating parts, and the pen and pointer drag is frequently reduced by gently tapping of the instruments; a vigorous tapping would however lead to delicate bearing being injured and thus increasing friction all the more.

Environmental Errors

The instrument location and the environment errors are introduced by using an instrument in conditions different for which it has been designed, assembled, and calibrated. The different conditions of use may be temperature, pressure, humidity, altitude, etc.; the effect of temperature being more predominant. A change in the temperature may alter the elastic constant of spring, may change the dimensions of a measuring element or linkage in the system, may alter the resistance values and flux densities of magnetic elements.

Consider a mercury-in-glass thermometer being used for the measurement of air temperature. The instrument will be located wrongly if during measurements the sun happens to be shining on the thermometer bulb. Similarly, the bulb would indicate an effect of heat radiation if the thermometer is placed too close to a window. Likewise, high air pressure would tend to compress the walls of the bulb and force the mercury to rise within the capillary and thus give a spurious temperature reading.

Environmental errors alter with time in an unpredictable manner. The following methods have been suggested to eliminate or at least reduce environmental errors.

- Use the instruments under the conditions for which it was originally assembled and calibrated. This may involve control of temperature, pressure, and humidity conditions.
- Measure deviations in the local conditions from the calibrated ones and then apply suitable corrections to the instrument readings.
- Automatic compensation for the departures from the calibrated conditions by using sophisticated devices.
- Make a completely new calibration under the local conditions.

The method chosen would depend on the local assessment of the problem.

Translation and Signal Transmission Errors

The instrument may not sense or translate the measured effect with complete fidelity. The error also includes the non-capability of the instrument to follow rapid changes in the measured quantity due to inertia and hysteresis effects. The transmission errors creep in when the transmitted signal is rendered faulty due to its distortion by resonance, attenuation, loss leakage, or being absorbed or otherwise consumed within the communication channel. The error may also result from unwanted disturbances such as noise, line pick-up, hum, ripple, etc. The errors are remedied by calibration and by monitoring the signal at one or more points along its transmission path.

Observation Errors

There goes a saying that "instruments are better than the people who use them." Even when an instrument has been properly selected, carefully installed, and faithfully calibrated, shortcomings in the measurement occur due to certain failings on the part of the observer. The observation errors may be due to:

- Parallax, that is apparent displacement when the line of vision is not normal to the scale.
- Inaccurate estimates of average reading, lack of ability to interpolate properly between graduations.
- Incorrect conversion of units in between consecutive readings, and non-simultaneous observation of interdependent quantities.

- Personal bias, that is, a tendency to read high or low, or anticipate a signal and read too soon.
- Wrong scale reading and wrong recording of data.

The poor mistakes resulting from the inexperience and carelessness of the observer are obviously remedied with careful training, and by taking independent readings of each item by two or more observers.

Operational Errors

A prerequisite to precise and meticulous measurements is that the instruments should be properly used. Quite often, errors are caused by poor operational techniques. Examples are:

- A differential type of flowmeter will read inaccurately if it is placed immediately after a valve or a bend.

- A thermometer will not read accurately if the sensitive portion is insufficiently immersed or is radiating heat to a colder portion of the installation.

- A pressure gauge will correctly indicate pressure when it is exposed only to the pressure which is to be measured.

- A steam calorimeter will not give the indication of the dryness fraction of steam unless the sample drawn correctly represents the condition of steam.

System Interaction Errors

The act of measurement may affect the condition of the measurements and thus lead to uncertainties in measurements. Examples are:

- Introduction of a thermometer alters the thermal capacity of the system and provides an extra path for heat leakage.
- A ruler pressed against a body results in a differential deformation of the body relative to the ruler.
- An obstruction-type flowmeter may partially block or disturb the flow conditions. Consequently, the flow rate shown by the meter may not be the same as before the meter installation.
- Reading shown by a hand tachometer would vary with the pressure with which it is pressed against the shaft.
- A milliammeter would introduce additional resistance in the circuit and thereby alter the flow current by a significant amount.

The job of an instrument designer is to see whether the alteration due to system interference is minimal. Many of the most precise, expensive, and elaborate measuring instruments owe their cost and complexity solely to the means adopted to eliminate, or at least reduce interaction between the instrument and the physical state being measured.

The errors discussed above may be grouped into systematic errors and random errors.

Systematic errors are repeated consistently with the repetition of the experiment and have the same magnitude and sign for a given set of conditions. They alter the instrument reading by a fixed magnitude and with the same sign from one reading to another. Because of the same algebraic sign, systematic errors tend to accumulate and hence are often called cumulative errors. Instrument bias is another term for systematic errors. These errors are caused by such effects as sensitivity shifts, zero offsets, and known non-linearity. Systematic errors cannot be determined by direct and repetitive observation of the measurand made each time with the same technique. The only way to locate these errors is to have repeated measurements under different conditions or with different equipment and where possible by an entirely different method. Some factors leading to systematic errors are:

i.   pointer offset

ii. change in ambient temperature

iii. poor design and construction of the instrument

iv. buoyant effect of the wind and the weights of a chemical balance

v.   inequality of the arms of a beam balance

vi. change in the original state of the system due to interaction between the instrument and the system.

Random errors are accidental, small, and independent and are mainly due to inconsistent factors such as spring hysteresis, stickiness, friction, noise, and threshold limitations. Since these errors vary both in magnitude and sign (are positive or negative based on chance alone), they tend to compensate one another and are referred to as chance/accidental/compensating errors. The random errors are detected by lack of consistency in the measured value when the same input is imposed repeatedly on the instrument (measured values are not precise and show a considerable scatter). The magnitude and direction of random errors cannot be predicted from a knowledge of the measurement system; however, these errors are assumed to follow the law of probabilities. Some factors leading to random errors are:

i. stickiness and friction,

ii. line voltage fluctuations,

iii. vibration of instrument supports,

iv. large dimensional tolerances between the mating parts,

v. spring hysteresis and elastic deformation, and

vi. inconsistencies associated with accurate measurement of small quantities.

7.5. PRESSURE MEASUREMENTS

Pressure measurement is undoubtedly one of the most common of all the measurements made on systems. In a company with temperature and flow, pressure measurements are extensively used in industry, laboratories, and many other fields for a wide variety of reasons. Pressure measurements are concerned not only with the determination of force per unit area exerted by a fluid at a point but are also involved in many liquid levels, density, flow, and temperature measurements. Measurement of pressure is also needed to maintain safe operating conditions, to help control a process, and to provide test data.

Pressure measurement by any technique is essentially based on the following well-known propositions:

i. Pressure at any point in a body of liquid at rest is proportional to the depth of the point below the free surface of the liquid, and increases in the downward direction at a rate equivalent to the density of the liquid. Further, within a continuous expanse of the same fluid, pressure is the same at any two points which lie in a horizontal plane.

ii. There is pressure equality throughout a fluid, that is, a pressure applied to a confined fluid via a movable surface would be transferred undiminished to all the boundary surfaces.

iii. Pressure is unaffected by the shape of the confining boundaries.

Manometers measure pressure by balancing a column of liquid against the pressure to be measured. Height of the column so balanced is measured and then converted to the desired pressure units. Manometers may be vertical, inclined, open, differential, or compound. Choice of any type depends on its sensitivity of the measurement, ease of operation, and the magnitude of pressure being measured. Manometers can be used to measure gauge, differential atmospheric, and absolute pressures.

7.5.1. Piezometer

It is a vertical transparent glass tube, the upper end of which is open to the atmosphere and the lower end is in communication with the gauge point; a point in the fluid container at which pressure is to be measured. The rise of fluid in the tube above a certain gauge point is a measure of the pressure at that point.

Fluid pressure at gauge point A
= atmospheric pressure p_a at the free surface + pressure due to a liquid column of height h_1

$$p_1 = p_a + wh_1$$

where w is the specific weight of the liquid.

Similarly for the gauge point B,

$$p_2 = p_a + wh_2$$

Pressures are generally prescribed with atmospheric pressure taken as the zero of the pressure scale. Evidently then, $p_1 = wh_1$ and $p_2 = wh_2$ and the pressures thus evaluated are the *gauge pressures*.

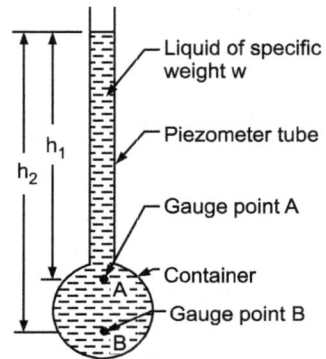

FIGURE 7.8. Piezometer

When using a piezometer to measure the pressure of a moving fluid, the axis of the tube should be absolutely normal to the direction of flow and its bottom end must flush smoothly with the pipe surface. Any burr or projection would cause obstruction resulting in a change in the pressure head. Further, to reduce the surface tension and capillary effects, a diameter of the tube must be kept at least 6 mm.

Piezometers cannot be used to measure pressures that are considerably excess of atmospheric pressure. Use of a very long glass tube would be unsafe, it being both fragile and unmanageable. Further, gas pressure cannot be measured as gas does not form any free surface with the atmosphere. Again measurement of negative pressure is not possible due to the flow of atmospheric air into the container through the tube. These difficulties are overcome by modifying the piezometer into a U-tube manometer, also called the double-column manometer.

7.5.2. U-Tube Double-Column Manometer

This simplest and most useful pressure measure device consists of a transparent tube bent in the form of letter U and filled with a manometric liquid whose density is known. The choice of a particular manometric liquid depends

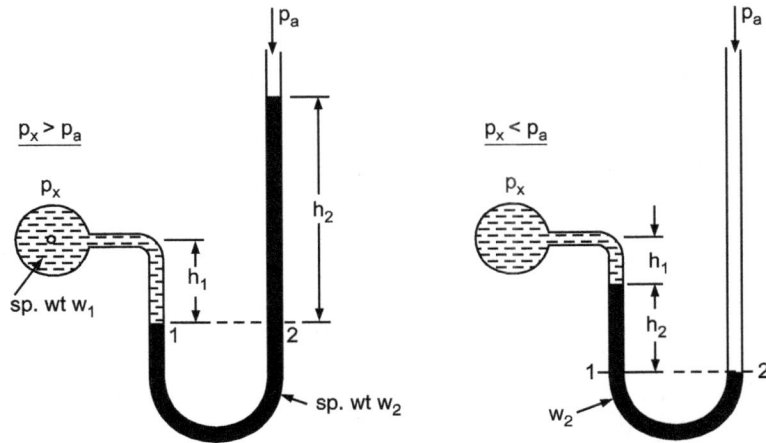

FIGURE 7.9 U-tube manometers

upon the pressure range and nature of the fluid whose pressure is sought. For high ranges, mercury (specific gravity 13.6) is the manometric/balancing liquid. For low-pressure ranges, liquid-like carbon tetrachloride (specific gravity 1.59) or acetylene tetrabromide (specific gravity 2.59) is employed. Quite often, some colors are added to the balancing liquid so as to get clear readings.

When both the limbs are open to atmosphere, manometric liquid stands at even height. Under the application of pressure p_x to one limb, the manometric liquid is forced down on the side with a corresponding rise on the other side until the column of liquid between the two levels balances the difference between the unknown pressure p_x and the atmospheric pressure p_a. Figure 7.9 shows the schematics of the U-tube being employed for measurement of positive and negative pressures.

Arrangement (*a*) *Measurement of pressure greater than atmospheric pressure*

Due to greater pressure p_x in the container, the manometric liquid is forced downward in the left limb of the U-tube and there is a corresponding rise of manometric liquid in the right limb.

For the right limb, the gauge pressure at point 2 is

p_2 = atmospheric pressure, that is, zero gauge pressure at the free surface
+ pressure due to head h_2 of manometric liquid of specific weight w_2
= $0 + w_2 h_2$

For the left limb, the gauge pressure at point 1 is

p_1 = gauge *pressure* p_x + pressure due to height h_1 of the liquid of specific weight w_1

$$= p_x + w_1 h_1$$

Points 1 and 2 are at the same horizontal plane; $p_1 = p_2$ and therefore

$$p_x + w_1 h_1 = w_2 h_2$$

∴ Gauge pressure in the container,

$$p_x = w_2 h_1 - w_1 h_1$$

or in terms of head of water column,

$$\frac{p_x}{w} = \left(\frac{w_2}{w} h_2 - \frac{w_1}{w} h_1 \right) = (s_2 h_2 - s_1 h_1) \qquad (7.4)$$

where w is the specific weight of water and symbol s denotes the specific gravity of a liquid.

Arrangement (*b*) *Measurement of pressure less than atmospheric pressure*

Due to negative pressure p_x in the container, the manometric liquid is sucked upwards in the left limb of the U-tube and there is a corresponding fall of manometric liquid in the right limb.

Pressure in the two legs at the same levels 1 and 2 are equal; $p_1 = p_2$ and therefore,

$$p_x + w_1 h_1 + w_2 h_2 = 0$$

∴ Gauge pressure in the container,

$$p_x = -(w_1 h_1 + w_2 h_2)$$

or in terms of head of water column,

$$\frac{p_x}{w} = -(s_1 h_1 + s_2 h_2) \qquad (7.5)$$

U-tube manometer necessitates two readings, h_1 and h_2 and that is likely to increase the chance of error. The difficulty is circumvented by adopting a single-column manometer.

EXAMPLE 7.8

The right limb of a U-tube manometer containing mercury is open to the atmosphere while the left limb is connected to a pipe through which flows a fluid of specific gravity 0.85. The center of the pipe lines 15 cm below the level of mercury in the right limb. If the difference of mercury level in the two limbs is 25 cm, determine the pressure of fluid of the pipe.

Solution: Let p_x be the gauge pressure of the fluid in the pipeline.

Consider pressure balance in the horizontal plane $0 - 0$; pressures in the left and right limbs at this plane are equal. That is:

$$p_x + w_1\, h_1 = w_m\, h_2$$

or

$$p_x = w_m\, h_2 - w_1\, h_1$$
$$= (9810 \times 13.6) \times 0.25$$
$$- (9810 \times 0.85) \times 0.1$$
$$= 33354 - 833.85$$
$$\approx \mathbf{32520\ N/m^2}$$

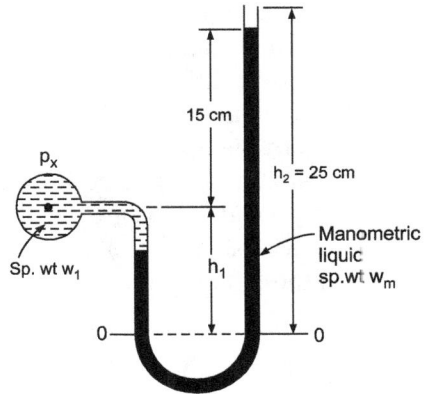

EXAMPLE 7.9

The right limb of a simple U-tube manometer containing mercury is open to the atmosphere and the left limb is connected to a pipe through which flows a fluid of specific gravity 0.8. Make calculations for the vacuum pressure in the pipe if the difference of mercury level in the two limbs is 30 cm and the level of fluid in the left limb is 10 cm below the center of pipe.

Solution: Let p_x be the gauge pressure of fluid in the pipeline.

Consider pressure balance in the horizontal plane $0 - 0$; the pressures in the left and right limbs at this level are equal. That is:

$$p_x + w_1\, h_1 + w_m\, h_2 = 0$$
$$p_x = -(w_1\, h_1 + w_m\, h_2)$$
$$= -(9810 \times 0.8 \times 0.1 + 9810 \times 13.6 \times 0.3)$$
$$= -(784.8 + 40024.8) = \mathbf{-40809.6\ N/m^2}$$

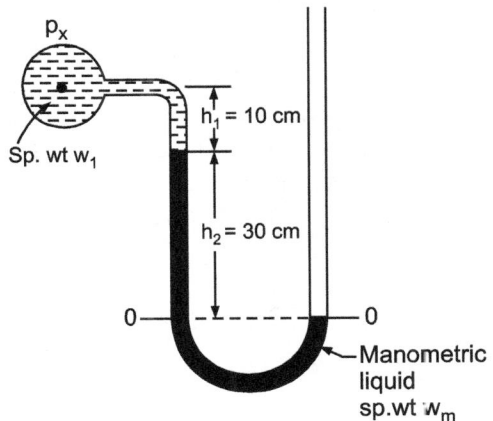

7.5.3. U-Tube Differential Manometer

A differential manometer is a device used to find the difference in pressure between two points in a pipeline or in two different pipes or containers. In general, a differential manometer consists of a U-tube filled with a manometric liquid and with its ends connected to the points between which the pressure difference is to be measured. Figure 7.10 shows the two common arrangements of a differential manometer.

In the *upright configuration* of the U-tube differential manometer, the manometric liquid contained in the U-tube is a heavier liquid, that is, its specific weight w_m is greater than that of the liquids in the containers.

Consider the pressure balance in the horizontal plane 0 – 0; the pressure in the left and right limbs at this plane are equal. That is:

$$p_A + w_1(h_1 + h_m) = p_B + w_2 h_2 + w_m h_m$$

or
$$p_A - p_B = h_m(w_2 - w_1) + w_2 h_2 - w_1 h_1$$

For the special and frequently encountered case of pipes A and B at the same level ($h_1 = h_2$) and carrying the same fluid ($w_1 = w_2$):

$$p_A - p_B = h_m(w_m - w_1)$$

In terms of head of water column

$$\frac{p_A - p_B}{w} = h_m\left[\frac{w_m}{w} - \frac{w_1}{w}\right] = h_m(s_m - s_1)$$

If water is the fluid in the two pipes and mercury is the manometric liquid, then $s_1 = 1$ and $s_m = 13.6$ and therefore, for the mercury-water differential manometer,

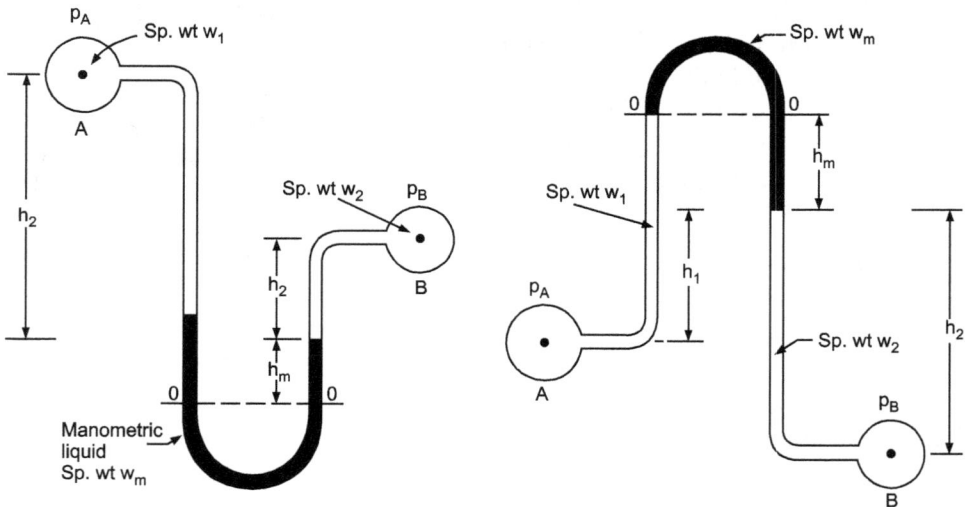

FIGURE 7.10.

$$\frac{p_A - p_B}{w} = 12.6\, h_m$$

That is the pressure difference measured as a head of the water column is 12.6 times the difference in height of the mercury column. The sensitivity of such a gauge may be defined as the ratio of the observed difference in levels h_m to the difference of pressure head $\dfrac{(p_A - p_B)}{w}$ of water being measured:

$$\text{Sensitivity} = \frac{h_m}{\dfrac{(p_A - p_B)}{w}} = \frac{h_m}{h_m(s_m - 1)} = \frac{1}{s_m - 1}$$

With mercury (relative density 13.6) sensitivity is $\dfrac{1}{12.6}$ and with paraffin (relative density 0.85) the sensitivity is $-\dfrac{1}{0.15}$. Negative sensitivity implies that a paraffin-water differential manometer must be used in the inverted position.

In the ***inverted differential manometer***, the manometric liquid is lighter than the fluid whose pressure difference is to be ascertained.

Consider the pressure balance in the horizontal place $0 - 0$; the pressures in the left and right limbs of the inverted U-tube at this place are equal. That is:

$$p_A - w_1 h_1 - w_1 h_m = p_B - w_2 h_2 - w_m h_m$$

or

$$p_A - p_B = w_1 h_1 - w_2 h_2 + h_m (w_1 - w_m)$$

If the pipes A and B are at the same level $(h_1 = h_2)$ and carry the same fluid $(w_1 = w_2)$ $p_A - p_B = h_m(w_1 - w_m)$

In terms of head of water column,

$$\frac{p_A - p_B}{w} = h_m \left(\frac{w_1}{w} - \frac{w_m}{w} \right) = h_m(s_1 - s_m)$$

where s_m and s_1 are the specific gravities of the manometric liquid and the fluid in the pipelines of the flow system.

The following points need to be noted:

i. If the manometric liquid is very light, that is, $s_m \lll s_1$, then

$$\frac{p_A - p_B}{w} = h_m$$

ii. If the manometric liquid is so chosen that its relative density is very nearly equal to that of fluids in the pipelines $(s_m \approx s_1)$ and that the fluids do not intermix, the manometer will become very sensitive. A sensitive manometer gives a large value of h_m for a small pressure difference.

EXAMPLE 7.10

A U-tube differential manometer containing mercury is connected on one side to pipe A containing carbon tetrachloride (sp. gr. 1.6) under a pressure of 120 kPa, and on the other side to pipe B con-taining oil (sp. gr. 0.8) under a pressure of 200 kPa. Pipe A lies 2.5 m above pipe B and the mercury level in the limb communicating with pipe A lies 4 m below pipe A. Determine the difference in the levels of mercury in the two limbs of the manometer.

Take the specific weight of water = 9.81 kN/m^3.

Solution: Consider pres-sure balance in the horizontal plane $0 - 0$; the pressures in the left and right limbs at this plane are equal. That is:

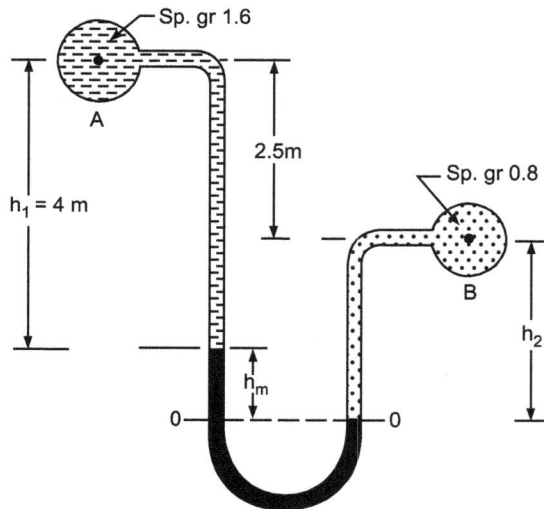

FIGURE 7.11.

$$p_A + w_c\, h_1 + w_m\, h_m = p_B + w_0\, h_2$$

$$120 + 9.81 \times 1.6 \times 4 + 9.81 \times 13.6\, h_m$$

$$= 200 + 9.81 \times 0.8(1.5 + h_m)$$

$$120 + 62.78 + 133.42\, h_m = 200 + 11.77 + 7.85\, h_m$$

$$h_m = \frac{200 + 11.77 - 120 - 62.78}{133.42 - 7.85}$$

$$= 0.231 \text{ m} = \mathbf{23.1 \text{ cm}}$$

EXAMPLE 7.11

Determine the difference of pressure between pipes A and B when connected to an inverted U-tube differential manometer containing oil of specific gravity 0.8 as the manometric liquid. Pipe A conveys water and fluid of sp. gr. 0.9 flows through pipe B. The position of

manometric liquid in the manometer limbs is as indicated in Figure 7.12. If $p_B = 5 \times 10^4$ N/m^2 and the barometer reading is 730 mm of mercury, find the pressure in pipe A in meters of water absolute.

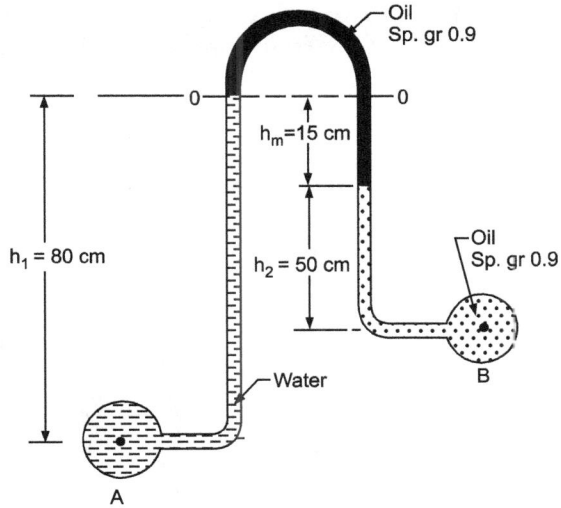

Solution: Consider pressure balance in the horizontal plane 0 – 0; the pressure in the left and right limbs at this plane is equal. That is:

FIGURE 7.12.

$$p_A - w_1\,h_1 = p_B - w_2\,h_2 - w_m\,h_m$$

$$p_A - 9810 \times 0.8 = p_B - 9810 \times 0.9 \times 0.5 - 9810 \times 0.8 \times 0.15$$

$$p_A - p_B = 9810(0.8 - 0.9 \times 0.5 - 0.8 \times 0.15)$$

$$= 9810(0.8 - 0.45 - 0.12)$$

$$= 2256 \text{ N/m}^2$$

Given: $p_A = 5 \times 10^4$ N/m^2

and $p_{at} = 9810 \times 13.6 \times 0.73 = 9.74 \times 10^4$ N/m^2

Pressure intensity in pipe A

$$p_A = p_B + 2256 = 5 \times 10^4 + 2256 = 52256 \text{ N/m}^2 \text{ (gauge)}$$

Absolute pressure = gauge pressure + atmospheric pressure

$$p_A \text{ (absolute)} = 52256 + 9.74 \times 10^4 = 149656 \text{ N/m}^2$$

$$= \frac{149656}{9810} = \textbf{15.25 m of water absolute}$$

EXAMPLE 7.12

When the pressure at a point is so large that the manometric fluid cannot be contained within the height of a single U-tube manometer, use is made of a compound U-tube manometer which essentially consists of a number of simple U-tube manometers arranged in series. For one such unit illustrated in Figure 7.13, calculate the pressure difference

between the points A and B. Take w_w = 10 kN/m³ for water w_m = 136 kN/m³ for mercury and w_0 = 8.5 kN/m³ for oil.

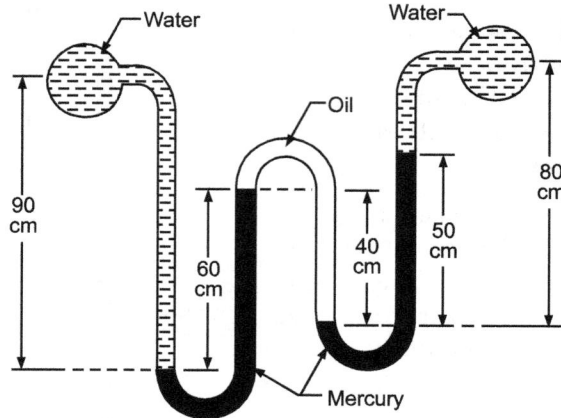

FIGURE 7.13.

Solution: Starting from point A, the governing manometric equation is:

$$p_A + 10 \times 0.9 - 136 \times 0.6 + 8.5 \times 0.4 - 136 \times 0.5 - 10\,(0.8 - 0.5) = p_B$$
$$p_A - p_B = -9 + 81.6 - 3.4 + 68 + 3 = \textbf{140.2 kN/m}^2$$

7.5.4. Single-Column Manometer

In the industrial well-manometers, one of the legs of the U-tube manometer is replaced by a large diameter well; the widened limb is made about 100 times greater than the cross-sectional area of the other limb. A change in the level of manometric liquid occurring in the wider limb due to pressure changes would then be so small that it can be neglected. The pressure difference would then be indicated only by the height of the liquid column in the narrow limb.

To start with, let both limbs of the manometer be exposed to atmospheric pressure. The liquid level in the wider limb (also called a reservoir, well, basin) and narrow limb will correspond to position 0–0. When the wider limb is connected to a vessel containing fluid at pressure p_x (which is greater than the atmospheric pressure p_a), manometric liquid level in the reservoir will fall down by δh and there will be a corresponding level rise h_2 in the narrow limb. By conservation of volume and applying manometric equations, the following expression can be set up for the pressure p_x of the fluid contained in the vessel,

$$p_x = w_2\, h_2 \left[1 + \frac{a}{A}\right] \qquad (7.13)$$

If the area ratio (a/A) is made so small that it can be neglected, then

$$p_x = w_2 h_2 \qquad (7.14)$$

Single-column manometers are used as primary standards for calibrating other pressure gauges and are more sensitive than simple U-tube manometers.

To expand the scale and thereby increase sensitivity, the narrow limb of the single-column manometer is not set vertically but is kept inclined to the horizontal axis by an angle θ as shown in Figure 7.15. Gauge pressure p_x is then given by:

$$p_x = w_2 l \left(\sin\theta + \frac{a}{A} \right) \qquad (7.15)$$

where l is the rise of liquid in the inclined tube.

The scale of the instrument is obviously expanded due to the presence of $\sin\theta$. By making θ quite small, l can be increased such that $l\sin\theta$ remains constant. Any desired value of sensitivity (normally up to about 25 times that of a U-tube) may be obtained by incorporating a swivel mechanism for the inclined limb. A minimum value of $\theta = 5°$. With inclination angles less than this, the exact position of the meniscus is difficult to determine.

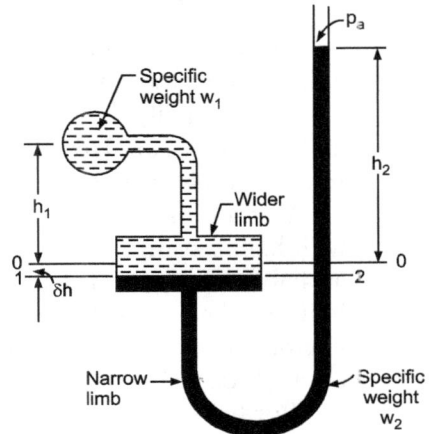

FIGURE 7.14.

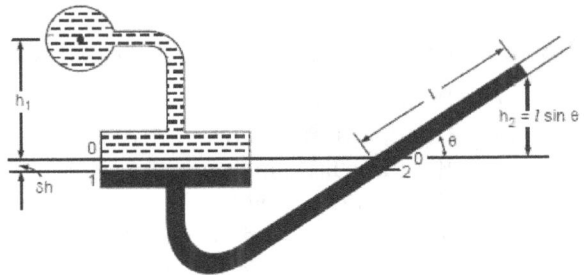

FIGURE 7.15 Inclined manometer.

This type of manometer is frequently called a ***draft gauge*** because it is so generally used for determining the draft in a steam generator setting and for measuring small pressure changes in low-velocity gas flows.

7.5.5. Manometric Liquids

Some of the desirable characteristics of a manometric liquid are:

- low viscosity, that is, *the* capability of quick adjustment with pressure changes,

- low coefficient of thermal expansion, that is, minimum density changes with temperature,
- low vapor pressure, that is, little or no evaporation at ambient conditions,
- negligible surface tension and capillary effects,
- non-corrosive, non-poisonous, non-sticky, and stable nature.

Mercury is usually used for measuring vacuum and moderate pressure of gas, vapor, or water where moderate sensitivity is required. Mercury does not evaporate readily, has a reasonably stable density, forms a sharp meniscus, and is clearly seen. However, it amalgamates or corrodes many metals, is poisonous, and is expensive.

Water is used for measuring small vacuums and small pressure differences with high sensitivity. Water has a fairly sharp meniscus, is cheap and readily available. However, it has a tendency to evaporate and dissolve some gases in it. Further, its transparent nature renders it difficult to be seen within the manometer tube. A dye added to give it a distinctive color would be deposited on the tube walls when water evaporates.

For high multiplication in two-liquid manometers, alcohol and kerosene are often used. Kerosene is, however, not satisfactory because of fractional vaporization and consequent changes in density. Alcohol is also apt to change in density by taking up water.

7.5.6. Advantages and Limitations of Manometers

- Relatively inexpensive and easy to fabricate.
- Good accuracy and sensitivity.
- Requires little maintenance; is not affected by vibrations.
- Particularly suitable to low pressure and low differential pressures.
- Sensitivity can be altered easily by affecting a change in the quantity of manometric liquid in the manometer.
- Generally large and bulky, fragile and gets easily broken.
- Measured medium has to be compatible with the manometric fluid used.
- Readings are affected by a change in gravity, temperature, and altitude.
- Surface tension of manometric fluid creates a capillary effect.
- Meniscus has to be measured by accurate means to ensure improved accuracy.

7.5.7. Mechanical Gauges: Elastic Pressure Transducers

The range of pressures that can be measured with a manometer depends upon the manometric fluid used, the minimum displacement which can be sensed, and the tube length. Manometers are employed to measure pressure as low as 0.35 N/m^2, and also the pressure difference in the range of

1.4×10^5 to 2.1×10^5 N/m^2. Pressures higher than two or three atmospheres are invariably measured with mechanical gauges of the Bourdon tube or diaphragm type. In these gauges, the fluid pressure is applied to a hollow tube, movable diaphragm, or bellows. The deflection thus obtained is transmitted through a suitable mechanism to a needle which indicates pressure on a pre-calibrated dial.

Bourdon Gauge: The pressure responsive element of a bourdon gauge consists essentially of a metal tube (called bourdon tube or spring), oval in cross-section, and bent to form a circular segment of approximately 200°–300°. The tube is fixed but open at one end and it is through this fixed end that the pressure to be measured is applied. The other end is closed but free to allow displacement under deforming action of the pressure difference across the tube walls. When pressure (greater than atmosphere) is applied to the inside of the tube, its cross-section tends to become circular. This makes the tube straighten itself out with a consequent increase in its radius of curvature, that is, the free end would collapse and curve.

The free end of the tube is connected to a spring-loaded linkage which amplifies the displacement and transmits it to the angular rotation of a pointer over a calibrated scale to give a mechanical indication of pressure (Figure 7.16). The linkage is so designed that the mechanism may be adjusted for optimum linearity and minimum hysteresis as well as to compensate for wear that may develop over a period of time. A hairspring is sometimes used to fasten the spindle to the frame of the instrument to provide necessary tension for proper meshing of the gear teeth and thereby freeing the system from backlash (lost motion). After prolonged use, the tooth gearing of the pinion and sector type linkage wears out and this impairs the accuracy of the gauge.

The reference pressure in the casing containing the bourdon tube is usually atmospheric and so the pointer indicates gauge pressure.

Bourdon tube shapes and configurations: The C-type bourdon tube has a small tip travel and this necessitates amplification by a lever, quadrant, pinion, and pointer arrangement. Increased sensitivity can be obtained by using a very long length of tubing in the form of a helix, and a flat spiral as indicated in Figure 7.17.

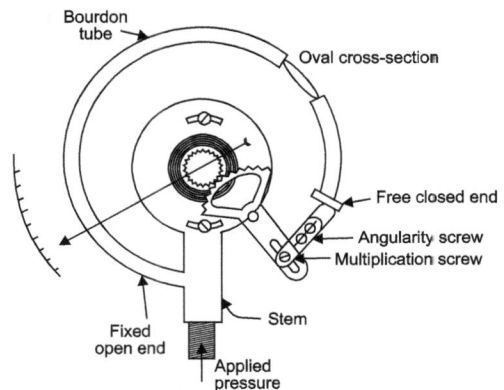

FIGURE 7.16. Bourdon tube pressure transducer

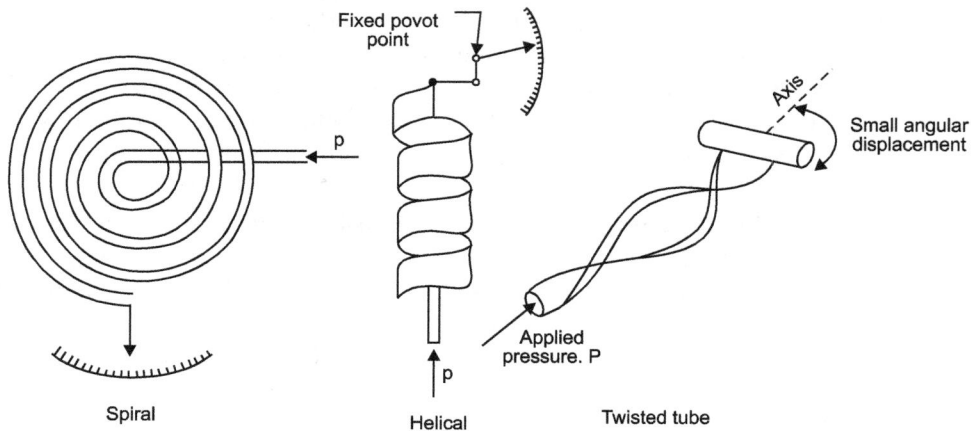

FIGURE 7.17 Bourdon tube configuration.

The spiral tubing produces the same effect as would be given by a number of C-tubes in series. The tip travel is of amount sufficient enough to indicate directly against a calibrated dial. Likewise, the increased number of turns of a helical bourdon makes it possible to obtain a greater angle of uncoil. Spiral and helical tubes frequently are used where it is desirable to eliminate the multiplication linkages between the pressure element and the indicating or recording arm. The absence of amplification linkage makes the system more robust; wear friction and the internal effects of the linkage are eliminated.

The twisted tube has a cross wise stability which reduces spurious output motion due to shock and vibration.

Advantages

- Low cost and simple construction.
- Capability to measure gauge, absolute and differential pressures.
- Simple and straightforward calibration with dead weight tester.
- Availability in several ranges and years of experience in application.
- Easily adapted to strain, capacitance, magnetic, and other electrical transducers.

Limitations

- Inherent hysteresis and slow response to pressure changes.
- Usually require geared movement for amplification.
- Susceptibility to shock and vibration.

7.6. VELOCITY MEASUREMENT

Flow velocity constitutes an important parameter in kinematics and dynamics of fluid flow and, therefore, its measurement is quite essential. Velocity measurement may be made with a view to:

- determine the volumetric flow rate,
- forecast the weather from wind velocity measurements, and
- locate the separation points from knowledge of velocity distribution.

Flow velocity measurements may be classified as:

i. Instruments like the pitot tube and hot wire anemometers measure the local velocity at a point in the channel or duct through which the fluid is flowing.

ii. Instruments like cup and vane anemometers, the current and turbine meters which measure the average velocity of fluid flow.

A very important class of velocity measuring devices consists of constriction meters. These involve Venturi meters, orifice plates, nozzles, and weirs; all of which measure the average velocity of flow from which the volumetric flow rate can be computed.

7.6.1. Pitot Tubes

In its elementary form, a pitot tube consists of an L-shaped tube; a tube bent through 90° and with ends unsealed. One limb called the body is inserted into the flow stream and aligned with the direction of flow while the other limb, called the stem, is vertical and open to the atmosphere (Figure 7.18).

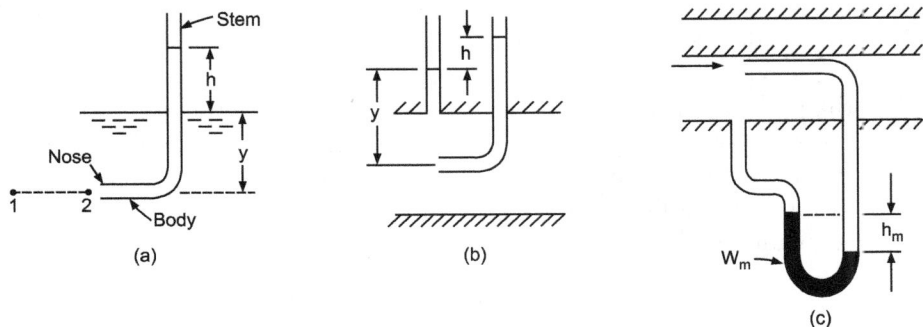

FIGURE 7.18 Pitot-static probes.

Applying Bernoulli's equation to point 1 (a point upstream from the submerged end of the tube and point 2 (a point at the tip or nose of the tube itself)

$$\frac{V_1^2}{2g} + \frac{p_1}{w} + y_1 = \frac{V_2^2}{2g} + \frac{p_2}{w} + y_2$$

The flowing fluid is brought to a state of zero velocity a the nose or tip of the tube and therefore, $V_2 = 0$. Further points 1 and 2 lie at the same elevation datum, that is, $y_1 = y_2$. That gives:

$$\frac{V_1^2}{2g} + \frac{p_1}{w} = \frac{p_2}{w}; \quad \frac{V_1^2}{2g} = \frac{p_2 - p_1}{w}$$

The pressure p_1 in the undisturbed free stream flow is called the **static pressure** p_s and the pressure p_2 at the stagnation point 2 where the velocity is zero is referred to as the **total pressure** p_t:

$$\therefore \qquad \frac{V_1^2}{2g} = \frac{p_t - p_s}{w} \qquad (7.6)$$

For Figure 7.18(b):

$$p_s = wy \text{ and } p_t = w(y + h)$$

where h is the rise of liquid level in the stem above the free surface:

$$\therefore \qquad \frac{V_1^2}{2g} = \frac{w(y + h) - wy}{w} = h$$

Hence, free stream velocity $V = \sqrt{2gh}$ $\qquad (7.7)$

Equations (7.16) and (7.17) are equally valid for determining the flow velocity at a point in the pipeline. However, the static pressure p_s has to be measured in addition to the stagnation pressure p_t indicated by the pitot tube. The static pressure p_s can be measured by means of a static pressure opening in the pipe wall (Figure 7.18 b) provided the static pressure distribution across the pipe is constant. However, in case of high static pressure, readings of the piezometer height y and $(y + h)$ become difficult and unmanageable. The difficulty is circumvented by connecting the pressure tappings to the two limbs of a U-tube differential manometer (Figure 7.18c). The difference between stagnation and static pressure is:

$$p_t - p_s = h_m(w_m - w_f) \qquad (7.8)$$

where h_m is the manometer deflection, and w_m and w_f are the specific weight of the manometric liquid and that of the liquid flowing through the pipeline.

The system depicted in Figure 7.18(c), though simple in construction, causes inconvenience because of the necessity of two connections to the pipe and the difficulty of obtaining correct static pressure by a single piezometer opening. Modern practice favors a combined pitot-static tube illustrated in Figure 7.19.

The unit consists of two concentrically arranged tubes bent to form a right-angled bend. The inner tube is open-ended and senses the fluid stagnation pressure. The

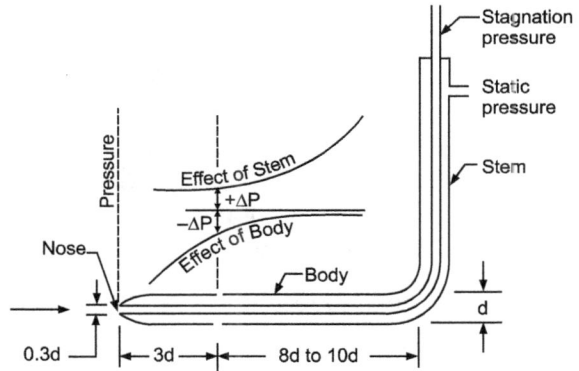

FIGURE 7.19 Combined pitot-static probe.

outer tube is sealed and has a streamlined shape. A number of holes (6 to 8) are drilled through the surface of the outer tube between the nose and stem; at a distance of $3d$ from the tip of the tube. The right-angled bent is at a further distance of $8d$ to $10d$. The presence of the body constricts the streamlines in its proximity, thereby causing an increase in the velocity or a fall in the static pressure. The stem produces a damping effect which tends to produce an error of the opposite sign. By careful placement of side tappings relative to the nose, a satisfactory reading of the fluid static pressure can be obtained.

The pressure differential $(p_t - p_s) = h_m(w_m - w_f)$ can be measured by taking leads from the two limbs to a differential manometer, and velocity computation made from the relation

$$V = \sqrt{2g \times \frac{(p_t - p_s)}{w_f}} = \sqrt{2gh_m \times \frac{(w_m - w_f)}{w_f}} = \sqrt{2gh_m(s_m - 1)} \qquad (7.9)$$

where s_m is the specific gravity of the manometric liquid relative to the liquid flowing through the pipeline.

The whole of the stream impinging on tip of the may not be brought to a state of rest; some may get deflected around the tip. Value of flow velocity as computed from Equation (7.9) may, therefore differ slightly from the true velocity value. This aspect is accounted for by introducing a factor C_v, called the *coefficient of velocity*,

$$V = C_v \sqrt{2gh_m(s_m - 1)} \qquad (7.10)$$

The instrument requires calibration to find the value of C_v; for a well-designed pitot tube, its value lies between 0.95 and 1.0.

EXAMPLE 7.13
Describe, with the help of a neat diagram, the construction, and operation of a pitot-static probe.
    Air flows through a duct, and the pitot-static tube measuring the velocity is attached to a differential manometer which shows a difference of head of 10 cm of water. The density of air is 1.22 kg/m³ and that of water is 1000 kg/m³. Workout the air velocity presuming that the coefficient for the pitot tube is 0.98.

Solution: The difference in mercury level is equivalent to head of air equal to,

$$h = (s_m - 1)\, h_m = \left(\frac{1000}{1.22} - 1\right) \times 0.1 = 8.1.87 \text{ m of air}$$

Flow velocity, $V = C_v \sqrt{2gh} = 0.98 \sqrt{2 \times 9.81 \times 81.87} = \textbf{39.28 m/s}$

EXAMPLE 7.14
A submarine moves horizontally in the sea with its axis much below the surface of the water. A pitot tube properly placed just in front of the submarine and along its axis is connected to the two limbs of a U-tube containing mercury. The difference in mercury level is found to be 17 cm. Find the speed of the submarine knowing that the density of mercury is 13.6 and that of sea water is 1.026 with respect to fresh water.

Solution: The difference in mercury level is equivalent to head of sea water equal to,

$$h = (s_m - 1)\, h_m = \left(\frac{13.6}{1.026} - 1\right) \times 0.17$$
$$= 2.083 \text{ m of sea water}$$

Flow velocity, $V = C_v \sqrt{2gh}$

$$= 0.98\sqrt{2 \times 9.81 \times 2.083} = \textbf{6.265 m/s}$$

The velocity coefficient for the pitot tube has been assumed equal to 0.98.

EXAMPLE 7.15
Explain how the velocity measurement at a point in fluid flow is made by a pitot tube.
    A pitot tube was used to measure the velocity of water at the center of a 25 cm diameter pipe. The difference between the stagnation

and static pressure heads was indicated to be 8.5 cm of water. What is the velocity? Take coefficient of velocity $C_v = 1.0$.

If the mean velocity of flow is presumed to be 0.67 of the velocity at the center, compute the discharge of water through the pipe in m³/s.

Solution: Velocity at center of pipe, $V = C_v \sqrt{2gh} = 1.0 \sqrt{2 \times 9.81 \times 0.85}$

Mean velocity $= 0.67 \times 1.291 = 0.865$ m/s

Area of pipe $= \dfrac{\pi}{4} (0.85)^2 = 0.049$ m^2

∴ Discharge = mean velocity × area = $0.865 \times 0.049 = $ **0.0424 m³/s**

7.6.2. Cup and Vane Anemometers

Cup and vane anemometers are the devices that measure the speed of air movement. The device consists essentially of a rotating element whose speed of rotation varies with the velocity of flow; the relationship between these variables is determined by appropriate calibration data.

In a cup, anemometer cups are attached to radial arms mounted on a shaft. Drag forces are set up on these cups when a flow stream in the plane of rotation approaches the unit from any direction. For the arrangement depicted in Figure 7.20(a), the drag on cup A (cup with its open end facing toward the stream) is greater than that on cup B (cup with rounded face toward the stream). The resultant torque rotates the assembly in the counterclockwise direction. The number of revolutions is read from a dial for a given period of time, and the frequency of rotation gives a measure of the average speed of air in the region traversed by the air.

In a vane anemometer, vanes of the windmill type are mounted in support so that the fluid flow is parallel to the axis of rotation. The rotor drives a low-friction gear train that in turn drives a pointer that indicates the wind speed on a dial.

The cup anemometer is usually mounted on a rigid shaft, and the vane type is held in the hand while readings are being taken. The cup type unit is best for relatively low speed whereas the vane type measures large

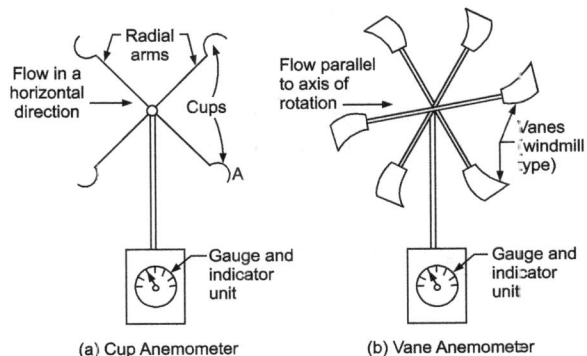

(a) Cup Anemometer   (b) Vane Anemometer

FIGURE 7.20 Cup and vane anemometers.

speeds more accurately. Experiments indicate that provided the wind speed is not too large, the relation between wind speed and angular velocity of cups/vanes is linear.

7.6.3. Current and Turbine Meters

Current and turbine meters are the devices that measure the speed of water movement.

A current meter consists of a horizontal wheel on which are fixed the conical buckets of V-shaped vanes. The unit is suspended into the flow stream by a suspension cable which is held taut (in tension) by a streamlined weight. Horizontal positioning (placement of the unit along the flow direction) is ensured by a streamlined tail vane. When the unit is held in a flowing stream, the liquid strikes the buckets, and that sets the wheel in rotation. At every revolution, signals are transmitted to the observer or to a revolution counter through electrical contacts. The frequency of rotation is directly related to the velocity of flow by appropriate calibration data.

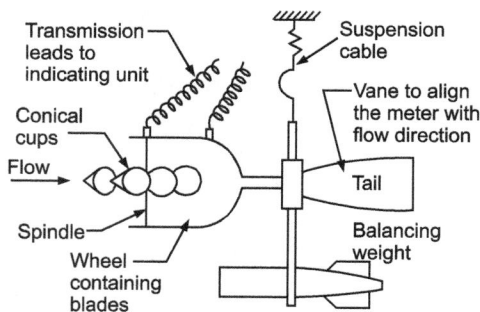

FIGURE 7.21 Current meter.

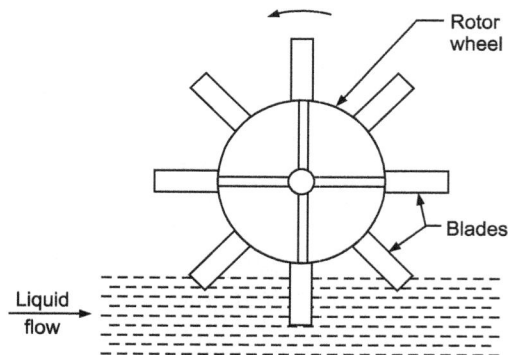

FIGURE 7.22 Turbine meter.

A turbine meter is similar in operation to a cup anemometer and depends on the application of an eccentric force applied by partial immersion of a bladed rotor in the fluid flow stream.

7.7. FLOW MEASUREMENT

The head meter in the form of Venturi meter, orifice, and flow nozzle is by far the most common flow meter for closed conduits. When the fluid flows through these obstruction meters, the fluid accelerates and a reduction in pressure occurs. The difference in pressure before and after the obstruction is measured by means of the differential pressure sensor and is related to the flow rate.

Venturi flow meter: The Venturi meter was invented by Clemens Hershel in 1887 and has been named in the honor of an Italian engineer Venturi. This simple and reliable device finds extensive use for water flow measurement, particularly in large-sized pipes and for large flow rates.

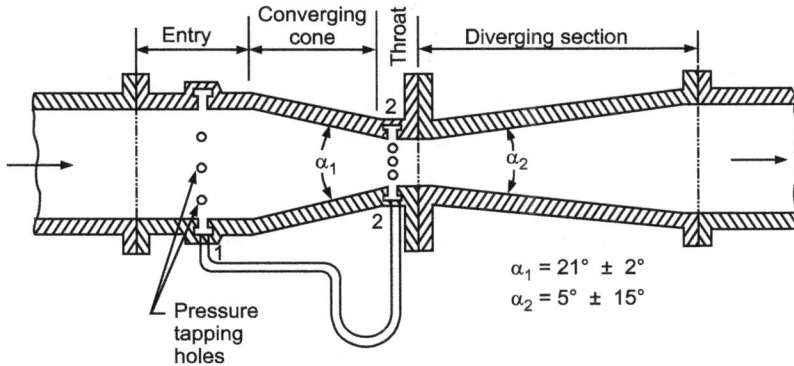

FIGURE 7.23 Venturi flow meter.

The important constructional features of a Venturi meter are shown in Figure 7.23. The meter consists of:

i.   *Cylindrical entrance section:* This section has the size of pipe to which it is attached. For satisfactory operation, the Venturi meter should be preceded by a straight pipe of not less than 5–10 pipe diameters and be free from fittings, misalignment, and other sources of large-scale turbulence. If these conditions cannot be met, straightening vanes should be placed upstream from the meter for reduction of rotational motion in flow.

ii. *Converging conical section:* The converging takes place at an angle of $21° \pm 2°$. The velocity of fluid increases as it passes through the converging section and correspondingly the static pressure falls.

iii. *Throat:* This is a cylindrical section of minimum area. At this section, the velocity is maximum and the pressure is minimum. The throat diameter is usually between 1/2 and 1/4 of the inlet diameter. The length of the throat equals its diameter.

iv. *Diverging section* in which there is a change of stream area back to the entrance area. The recovery of kinetic energy by its conversion to pressure is nearly complete and so the overall pressure loss is small. To accomplish a maximum recovery of kinetic energy, the diffuser section is made with an included angle of 5°–7°. This angle has to be kept less so that the flowing fluid has the least tendency to separate out from the boundary of the section. However, with small angles, the length and hence the cost of the meter would increase. So where pressure recovery is not of much importance, the angle of the diverging cone may be kept as high as 14°.

The pressure taps are made at the throat and at the entrance where the Venturi meter has a diameter equal to that of pipe. The pressure taps may be made either from a single hole or at piezometer rings thereby giving average values at the two sections.

The small-sized Venturi meter, suitable for pipelines less than 5 cm in diameter, are usually made of brass or bronze. The inside surface is smoothly finished to reduce friction. Large venturies are usually made of cast iron; the throat is, however, lined with brass or bronze and machined to a smooth finish. Very large venturies up to 6 m pipe diameter have been made of smooth surface concrete; only the throat is made of machined bronze.

Advantages and Limitations

- High-pressure recovery is attainable, that is, loss of head due to installation in the pipelines is small. Due to the low value of losses, the coefficient of discharge is high and it may approach unity under favorable conditions.
- Because of the smooth surface, the meter is not much affected by wear and tear. Less likelihood of becoming clogged with sediments.
- Well-established characteristics; years of application experience.
- Ideally suited for a large flow of water, process fluids, wastes, gases, and suspended solids.
- Long laying length; space requirements are more. Quite expensive in installation and replacement.

Venturi meters are not standardized yet to an extent that permits discharge coefficient from one meter to be used with another. From dimensional analysis and dynamic similarity conditions, the discharge coefficient for a Venturi meter is found to be a function of Reynolds number and meter size. These meters are not generally useful below 7.5 cm pipe diameter.

Flow nozzle: Nozzles are used in engineering practice for the creation of jets and streams for all purposes as well as for fluid metering. When placed in or at the end of a pipeline as metering devices, they are called flow nozzles.

The flow nozzle comprises a smooth, gradual contraction to the throat by a free uncontrolled expansion back to the pipe flow area. Because the device contains no provision for an orderly transformation of velocity into static pressure, the nozzle has a pressure loss from 80 to 90% of the differential pressure obtained. The discharge coefficient of a flow nozzle is dependent upon the smoothness of the approach to tangency, the length of the cylindrical portion of the nozzle, and the location of the pressure taps. Pipe wall taps located at one pipe diameter upstream and half pipe downstream from the inlet face of the nozzle give the best results. The flow nozzles are usually made of gun metal, stainless steel, or Monel metal.

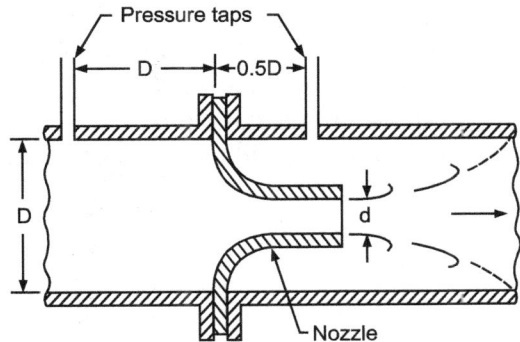

FIGURE 7.24 Flow nozzle.

Advantages and Limitations

- Cheaper than a standard Venturi meter; can be installed in an existing main without great difficulty.
- Increased coefficient of discharge when compared to orifice, and less physical length compared to Venturi meter.
- Widely accepted for high pressure/temperature steam flow; further good for fluids containing solids that settle.
- Pressure recovery is poor and so cannot be used where the available pressure head is small or where pressure recovery is a must. The pressure loss is, however, less than that of an orifice plate.
- Compared to the orifice meter, it is expensive and difficult to install.
- Limited to moderate pipe size; not available above 120 cm.

Orifice flow meter: The orifice meter consists of a thin, circular metal plate with a hole in it. The plate is held in the pipeline between two flanges called orifice flanges (Figure 7.25). The flow characteristics of the orifice differ from those of a nozzle in that the minimum section of the streamtube occurs not within the orifice but downstream from the orifice edge. Section of the minimum area is called vena contracta and minimum pressure exists at this section. Location of vena contracta depends on Reynolds number, area ratio between orifice and pipe, the roughness of pipe and, the compressibility of the flowing medium. It is also sensitive to the upstream velocity profiles and sharpness of the upstream edge of the orifice plate. Evidently, the location of downstream pressure connection at vena contracta is not feasible and as such, it is taken at a fixed proportion of the pipe diameter and a correction for vena contracta is made. Ideally, it should be placed where disturbing factors are least. For accuracy of results, there should be uniform flow conditions upstream of the orifice and for this a straight pipeline of at least 10 pipe diameter should precede the orifice. The discharge coefficient varies with the type of orifice, the pipe size, ratio of orifice diameter to pipe diameter, Reynolds number, and location of pressure connections.

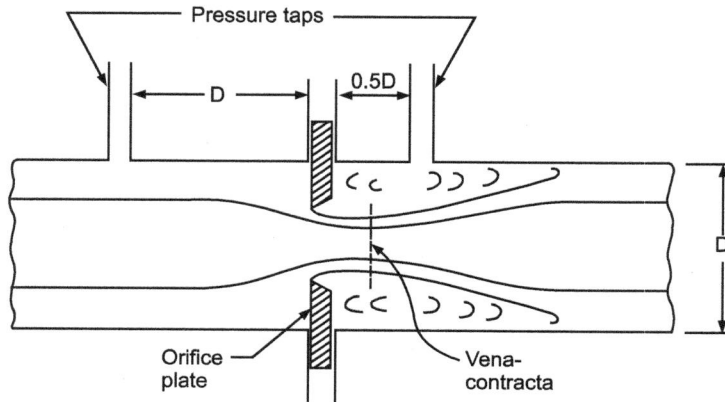

FIGURE 7.25 Orifice flow meter.

Orifice plates are made from steel, stainless steel, phosphor bronze, and other such materials that can withstand the corrosive effect of the flowing medium. Its thickness is only sufficient to withstand the buckling forces caused by the pressure differential. The circular hole is carefully made with a 90° square sharp edge. Wear and abrasion of this sharp edge greatly affect the accuracy of the orifice flow measurement. For this reason, an orifice should not be used to measure flow containing abrasives or other materials which would damage the edge. In some cases, it is advisable to replace the orifice plate frequently to maintain accuracy.

Advantages and Limitations

- Low initial cost, ease of installation, and replacement. Requires less space as compared to Venturi meter.
- Can be used in a wide range of pipe sizes (1.25–150 cm).
- Pressure recovery is poor; the overall pressure loss varies from 40 to 90% of the differential pressure.
- Coefficient of discharge has a low value.
- Necessity of providing straightening vanes upstream.
- Susceptible to inaccuracies resulting from erosion, corrosion, and scaling. Tends to clog and as such not suitable for slurries or entrained particles.

Theory of variable head meters: The equations convenient for practical use and for providing the basis of operation of variable head meters are derived by following three steps given below:

i. Apply Bernoulli's equation to the upstream and downstream pressure connections and modify it for the assumed conditions.

ii. Solve the modified equation for the downstream velocity.

iii. Express the flow rate by the product of area and velocity, that is, $Q = AV$.

Let subscript 1 refer to the pipe and the fluid at the upstream pressure connection and subscript 2 refer to the pipe and fluid at the section of minimum area. Bernoulli's equation gives:

$$\frac{p_1}{w} + \frac{V_1^2}{2g} + y_1 = \frac{p_2}{w} + \frac{V_2^2}{2g} + y_2 + \text{losses}$$

Neglecting losses and assuming the measuring device to be horizontal, that is, $y_1 = y_2$, we get

$$\frac{V_2^2 - V_1^2}{2g} = \frac{p_1 - p_2}{w} \qquad (7.11)$$

For incompressible fluids, the continuity relation for the situation is $A_1V_1 = A_2V_2$, and that yields

$$V_1 = \frac{A_2}{A_1} V_2 \qquad (7.12)$$

Solution of Equations (7.11) and (7.12) gives the outflow velocity V_2,

$$V_2 = \frac{A_1}{\sqrt{A_1^2 - A_2^2}} \times \sqrt{2g\frac{p_1 - p_2}{w}} \qquad (7.13)$$

The downstream fluid velocity as given by Equation (7.13) has been derived without considering any losses and so it is the ideal or theoretical velocity at the minimum section. The actual velocity can be obtained by multiplying the theoretical velocity by a factor C_v called the **coefficient of velocity**. The coefficient of velocity is the ratio of actual mean velocity which would occur without any friction loss

$$C_v = \frac{\text{actual mean velocity}}{\text{ideal mean velocity}}$$

For obtaining the volume flow rate, we apply the continuity equation:

discharge = area × velocity

$$\therefore \qquad Q = C_v \frac{A_1 A_2}{\sqrt{A_1^2 - A_2^2}} \times \sqrt{2g\frac{p_1 - p_2}{w}} \qquad (7.14)$$

During flow through an orifice meter, the fluid jet on leaving the orifice contracts to a minimum area at the vena contracta. Area of fluid jet at vena contracta is less than the area of the orifice and the two areas are related by the equation:

$$\text{area of jet at vena contracta} = C_c \times \text{orifice area}$$

where C_c is the coefficient of contraction. Thus if the orifice area is A_2 then the area at a minimum section that controls the flow rate and where the path of particles becomes parallel again would be $C_c A_2$

$$\therefore \qquad \text{Actual discharge } Q = C_v \frac{C_c A_1 A_2}{\sqrt{A_1^2 - A_2^2}} \times \sqrt{2g \frac{p_1 - p_2}{w}} \qquad (7.15)$$

For Venturi meter and flow nozzle, there is almost no formation of vena contracta and the coefficient of contraction can be taken as unity.

Combining C_v and C_c into single factor C_d called the coefficient of discharge, the volumetric flow rate through the meter can be written as:

$$Q = C_d \frac{A_1 A_2}{\sqrt{A_1^2 - A_2^2}} \times \sqrt{2g \frac{p_1 - p_2}{w}} \qquad (7.16)$$

Discharge coefficient C_d is not constant; it depends primarily on the flow Reynolds number and the channel geometry.

The quantity $\dfrac{A_1 A_2}{\sqrt{A_1^2 - A_2^2}} \sqrt{2g}$ is constant for a given meter; this quantity is generally designed by K and is known as the **meter constant**. Thus

$$Q = C_d K \sqrt{\frac{p_1 - p_2}{w}} \qquad (7.17)$$

The pressure differential $\dfrac{p_1 - p_2}{w}$, called the pressure head or piezometer head, is measured by a differential U-tube manometer. The manometer reading is the same for a given discharge irrespective of the inclination of the pipeline.

EXAMPLE 7.16

A Venturi meter with 200 mm diameter at inlet and 100 mm throat is laid with axis horizontal and is used for measuring the flow of oil of specific gravity 0.8. The difference of levels in the U-tube differential manometer reads 180 mm of mercury while 11.52×10^3 kg of oil is collected in 4 minutes. Calculate the discharge coefficient for the meter. Take specific gravity of mercury as 13.6.

Solution: $\qquad A_1 = \dfrac{\pi}{4}(0.2)^2 = 0.0314 \text{ m}^2;\ A_2 = \dfrac{\pi}{4}(0.1)^2 = 0.00785 \text{ m}^2$

$$\frac{A_1}{A_2} = \frac{0.0314}{0.00785} = 4$$

Piezometric head $P_h = h(s_m - 1)\ 0.18\left(\dfrac{13.6}{0.8} - 1\right) = 2.88$ m of oil

Mass of oil $= 11.52 \times 10^3$ kg in 4 minutes

∴ Discharge of oil, $Q = \dfrac{11.52 \times 10^3}{4 \times 60} \times \left(\dfrac{1}{800}\right) = 0.06 \text{ m}^3/\text{s}$ $\qquad (\because m = \rho Q)$

Substitute the given data in the discharge equation,

$$Q = C_d \frac{A_1 A_2}{\sqrt{A_1^2 - A_2^2}}\sqrt{2gP_h} = C_d \frac{A_1}{\sqrt{\left(\dfrac{A_1}{A_2}\right)^2 - 1}}\sqrt{2gP_h}$$

or $\qquad 0.06 = C_d \dfrac{0.0314}{\sqrt{4^2 - 1}} \times \sqrt{2 \times 9.81 \times 2.88} = 0.0609\ C_d$

∴ Discharge coefficient for the Venturi meter, $C_d = \dfrac{0.06}{0.0609} = \mathbf{0.985}$

7.7.1. Rotameter

The rotameter consists of a tapered metering glass tube, inside which is located the rotor or active element (float) of the rotor. This tapering tube is provided with suitable inlet and outlet connections. The float or bob material has specific gravity higher than that of the fluid to be metered.

With an increase in the flow rate, the float rises in the tube and there occurs an increase in the annular area between the float and the tube. The float adjusts its position in relation to discharge through the passage, that is, the float rises higher or lower depending on the flow rate. The discharge equation for flow through a rotameter is given by

$$Q = C_d A \left[2gV_f(\rho_f - \rho)/A_f\rho\right]^{\frac{1}{2}} \tag{7.18}$$

where Q is the volume flow rate, C_d is the discharge coefficient, V_f is the volume of float, ρ_f is the density of float material, ρ is the density of fluid flowing, A_f is the cross-sectional area of the float and A is the annular area between float and tube.

The glass tube is often made of high-strength borosilicate glass and the flow rate scale is engraved on the tube corresponding to a particular float material. However, where greater strength is required, the metallic tubes are used and the position of float is detected magnetically. Further, for remote indication, the position of float can be monitored electrically by using a suitable displacement transducer.

The rotameter is equally suitable for measuring both gas and liquid flows. When metering gas flows, a small sphere is used in a narrow tube and no guidance needs to be provided to its motion. While metering liquid fluids, spherical slots are cut on a part of the float and these slots cause it to rotate slowly about the axis of tube and keep it central. This spinning helps also to prevent the accumulation of any sediment on the top and sides of the float. Alternatively, stability of the bob is ensured by employing a guide along which the float would slide.

The float or bob material has a specific gravity higher than that of the fluid to be metered. The density difference $(\rho_f - \rho)$ required for metering a particular liquid or gas can be obtained by selecting different materials for the float. An examination of the discharge equation (7.28) would reveal that when $\rho_f >> \rho$, the volume flow rate becomes independent of the density r of the flowing fluid.

Rotameters are widely used for metering purge flows, pump-seal fluids, and coolants and lubricants for operating machinery. In these applications, flows are relatively small and accuracy requirements are not rigid.

Advantages and Limitations of a Rotameter

- Simplicity of operation, ease of reading, and installation.
- Relatively low cost.
- Handles a wide variety of corrosive fluids.
- Easily equipped with data transmission, indicating, and recording devices.
- Possibility of convenient and visible flow comparisons by mounting several rotameters side by side.
- Glass tube subject to breakage.
- Limited to small pipe sizes and capacities.
- Less accurate compared to Venturi and orifice meters.

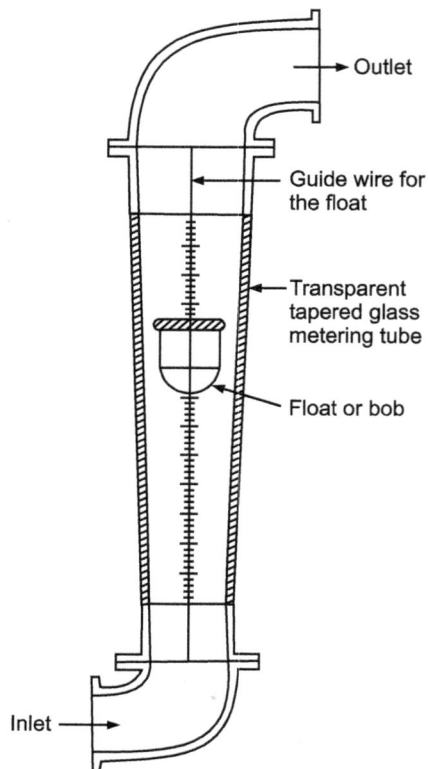

FIGURE 7.26 Rotameter.

- Must be mounted vertically.
- Subject to oscillations in pulsating flows.

7.8. TEMPERATURE MEASUREMENT

Temperature is probably the most widely measured and frequently controlled variable encountered in industrial processing of all kinds. Measurement of temperature potential is involved in thermodynamics, heat transfer, and many chemical operations. Basically, all the properties of matter such as size, color, electrical and magnetic characteristics, and the physical states (i.e., solid, liquid, and gas) change with changing temperatures The occurrence of physical and chemical changes is governed by the temperature at which a system is maintained.

Temperature measurement depends upon the establishment of thermodynamic equilibrium between the system and the device used to sense the temperature, for example, a thermometer bulb or thermocouple wires. The sensor has certain physical characteristics which change with temperature and this effect is taken as a measure of the temperature. The physical characteristics which are so used could be:

i. A change in dimension, that is, expansion or contraction of material in the form of solid, liquid, or gas.

ii. A change in electrical resistance of metals and semiconductors.

iii. A thermo-electric emf for two different metals and alloys joined together.

iv. A change in the intensity and color or radiation emitted by the hot body.

v. Fusion of materials when exposed to the temperature under investigation.

Calibration is then achieved through comparison with established standards. The international temperature scale serves to define temperature in terms of observable characteristic of materials.

7.8.1. Liquid-in-Glass Thermometers

The liquid-in-glass thermometer is one of the most common types of temperature measuring devices. The unit consists of a glass envelope, a responsive liquid, and an indicating scale. The envelope comprises a thick-walled glass tube with a capillary bore, and a spherical or cylindrical bulb filled with the liquid. The two parts are fused together and the top end of the capillary tube is sealed. The size of the capillary depends on the size of the sensing bulb, responsive liquid, and the desired temperature range of the instrument. Changes in the temperature will cause the fluid to expand and rise up the stem. Since the area of the stem is much less than the bulb, the relatively small change of fluid volume will result in a significant fluid rise in the stem. The length of the movement of the free surface of the fluid column serves, by a prior calibration to indicate the temperature of the bulb. The laboratory work thermometers have a scale engraved directly on the glass stem, while the industry types have separate scale located adjacent to the stem. Quite often the top of the capillary tube is also bulb-shaped to provide safety features in case the temperature range of the instrument is inadvertently exceeded.

The thermometer bulb is usually filled with mercury. It has the advantages of a broad temperature span between its freezing and boiling points, a nearly linear coefficient of expansion, relative ease of obtaining it in a very pure state, and its nonwetting glass characteristics. When measuring temperature above the boiling point of mercury (390°C atmospheric pressure), mercury may evaporate and condense at the top of the stem. This is prevented by filling the space above mercury with nitrogen or carbon dioxide under high pressure. This raises the boiling point and allows temperatures up to 610°C to be measured.

However, in many industrial applications, the escape of mercury through breakage causes considerable damage to the products. This may necessitate the use of other liquids such as alcohol, pentane, and toluene, which do not cause contamination. These liquids are also used for temperature measurements below the freezing point of mercury. These liquids have further advantages of superior readability to mercury when colored with inert dyes and of low cost. However, they have low boiling points, a greater tendency to separate in the capillary, and wetting glass characteristics.

The choice in the type of glass used is a matter of economics influenced by the range of the thermometer—the higher the range, the higher the cost. For temperature up to 450°C, normal glass is used. At high temperature up to about 520°C, borosilicate glass is used. Above this temperature, quartz thermometers have been used but they are not common.

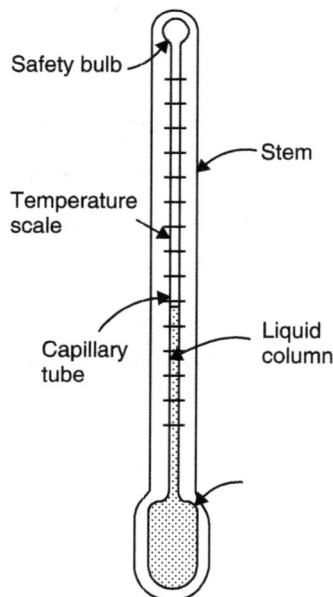

FIGURE 7.27 Liquid-in-glass thermometer.

Salient features/characteristics:

- simplicity of use and relatively low cost,
- easily portable,
- ease of checking for physical damage,
- absence of need for auxiliary power,
- no need for additional indicating instruments,
- fragile construction; range limited to about 600°C,
- lack of adaptability to remote reading, and
- time lag between change of temperature and thermometer response due to relatively high heat capacity of the bulb.

7.8.2. Solid Expansion or Bimetallic Thermometer

A bimetal strip consists of two pieces of different metals firmly bonded together by welding. For a bimetal in the form of a straight cantilever beam, temperature changes cause the free end to deflect because of the different expansion rates of the components. This deflection can be correlated quantitatively to the temperature change.

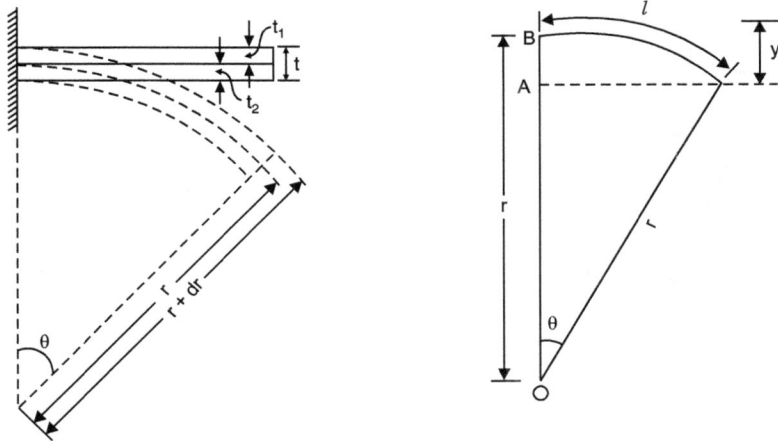

FIGURE 7.28 Bimetallic strip.

When a bimetal strip in the form of cantilever is assumed to bend through a circular arc, then

$$\frac{r+dr}{r} = \frac{\text{expanded length of strip having higher expansion coefficient}}{\text{expanded length of strip having lower expansion coefficient}}$$

$$= \frac{l\left[1+a_2\left(T-T_0\right)\right]}{l\left[1+a_1\left(T-T_0\right)\right]}$$

Simplification gives:

$$r = \frac{dr\left[1 + a_1\left(T - T_0\right)\right]}{\left(a_2 - a_1\right)\left(T_1 - T_0\right)}$$

The tip deflection can be increased with choice of materials that give a large value to the difference in their coefficient of expansion. Normally the low expansion material is invar (an iron-nickel alloy containing about 36% nickel) and the high expansion metal is brass.

Taking $\alpha_1 \approx 0$ and $dr = t/2$ (thickness of each metal strip), we get

$$r = \frac{t}{2a_2\left(T - T_0\right)}$$

The movement of free end of the cantilever in a perpendicular direction from the initial horizontal line is worked out as follows:

angular displacement $\theta = l/r$
vertical displacement $y = OB - OA = r - r \cos\theta$
$$= r\left(1 - \cos\theta\right)$$

Apparently, when one end of the bimetallic strip is fixed, the position is free and is a direct indication of the temperature of the strip.

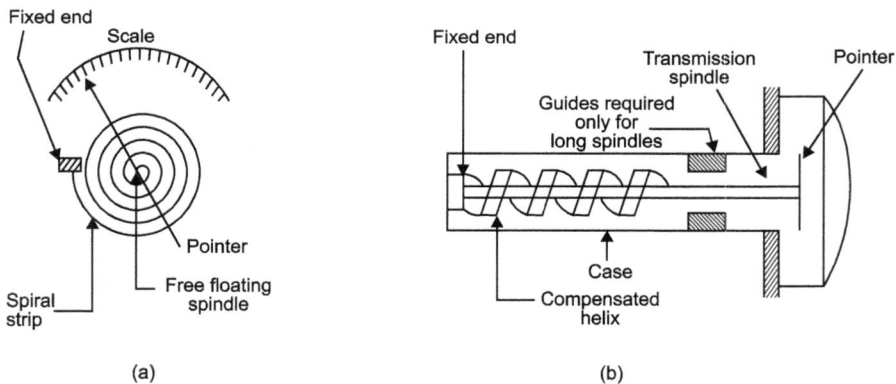

FIGURE 7.29 Bimetal strip thermometer.

Bimetalic elements can be arranged in flat, spiral, single helix, and multiple helix configuration. Figure 7.29 illustrates the functional principle of the usual industrial form of a bimetal thermometer. One end of the helix is anchored permanently to the casing and the other end is secured to a pointer that sweeps over a circular dial graduated in degree of temperature. In response to temperature change, the bimetal expands and the helical bimetal rotates at its free end, thus turning the stem and pointer to a new position on the dial. Likewise, the curvature of the bimetal spiral strip (Figure 7.29(b)) varies with temperature and causes a pointer to deflect. The continuous strip wound into helical or spiral form has the advantages of compactness while providing a long length of strip required for adequate indicator movement.

Bimetallic elements find wide application in simple thermometers in which the deflection of the elements is made to open or close electrical contacts in the electrical heat supply or to control a gas flow. Important applications include the switching devices used in domestic ovens, electric irons, car winker lamps, and refrigerators.

7.8.3. Thermocouples

When two conductors of dissimilar metals M_1 and M_2 are joined together to form a loop (a thermocouple) and two unequal temperatures T_1 and T_2 are imposed at the two interface connections, an electric current flows through the loop.

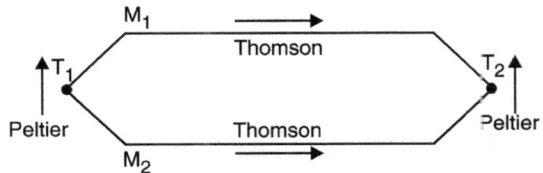

FIGURE 7.30 Basic thermocouple circuit.

Experimentally, it has been found that the magnitude of the current is directly related to the two materials M_1 and M_2 and the temperature difference (T_1-T_2). In the practical application of the effect, a suitable device is incorporated in the circuit to indicate any electromotive force or flow of current. For the convenience of measurement and standardization, one of the two junctions is usually maintained at some constant known temperature. The output voltage of the circuit then indicates the temperature difference relative to the reference temperature.

Thermo-electric effects arise in two ways:

- a potential difference always exists between two dissimilar metals in contact with each other (*Peltier effect*)
- a potential gradient exists even in a single conductor having a temperature gradient (*Thomson effect*)

In commercial instruments, the thermocouple materials are so chosen that the Peltier and Thomson emf's act in such a manner that the combined value is maximum and that varies directly with temperature.

Elements of a Thermo-electric Pyrometer

The essential elements of a thermo-electrical pyrometer are shown schematically in Figure 7.31.

- Two dissimilar conductors electrically insulated except at the hot junction, where the conductors may either be soldered or welded together, or may be completely separated from each other.

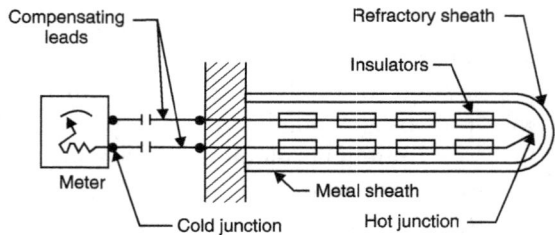

FIGURE 7.31 Element of a thermo-electric pyrometer.

- A refractory and a metal sheath to protect the thermocouple from injurious furnace gases and to prevent it from mechanical damage.
- Compensating leads which allow the measuring instrument to be placed at a considerable distance from the thermocouple without the necessity of using expensive thermocouple materials as extension leads.
- The cold or the reference junction provided by the instrument used for measuring the emf.

Thermocouple Materials

The desirable characteristics of thermocouple materials are:

i. The emf produced per degree of temperature change must be sufficient to facilitate detection and measurement.

ii. The temperature–emf relationship should be reasonably linear and reproducible. This will make the scale more easily read and also reduce the problem of reference junction compensation.

iii. The thermocouple should maintain its calibration without drift over a long period of time.

iv. The thermocouple should have a long life so that the cost of temperature measurement is not unnecessarily increased with frequent replacement of the thermocouple. For that, the thermocouple materials should be highly resistant to oxidation, corrosion, and contamination.

v. The material should physically be able to withstand high and rapidly fluctuating temperatures. Any phase change or other internal phenomenon will give rise to discontinuity in the temperature–emf relationship.

vi. The material must be such that successive batches can be produced with the same thermo-electric characteristics. This will allow the replacement of thermocouples without the necessity of recalibrating the temperature scale of the indicating instrument.

Depending upon the composition of metals used, the thermocouples are sometimes grouped into the following broad categories:

i. *Base metal thermocouples* use combinations of pure metal and alloys of iron, copper, and nickel, and are used in lower ranges of temperature up to 1375°C.

ii. Rare metal *thermocouples* use combinations of pure metals and alloys of

    a. platinum and rhodium for temperatures up to 1725°C, and

    b. tungsten, rhodium, and molybdenum for temperatures up to 2625°C.

7.8.4. Resistance Thermometers and Thermistors

The resistance R (ohms) of an electrical conductor of resistivity ρ (ohms Ω), length L (cm), and cross-sectional area A (cm^2) is given by

$$R = \rho \, L/A \qquad (7.19)$$

As temperature changes, the resistance of the conductor also changes. This is due to two factors: (i) dimensional change due to expansion or contraction and (ii) change in the current opposing properties of the material itself. For an unconstrained conductor, the latter is much more than 99% of the total change for copper. This change in resistance with temperature is used for measuring temperature.

Resistance Thermometers

Most metals become more resistant to the passage of electric current as they become hotter, that is, their resistance increases with growth in temperature. An adequate approximation of the resistance-temperature relationship is given by:

$$R_t = R_o \, (1 + \alpha t + \beta t^2) \qquad (7.20)$$

where R_t is resistance at any temperature $t°C$, R_o is resistance at 0°C, α and β are constants depending on the material. The constants R_o, α, and β are determined at the ice, steam, and sulfur points, respectively.

Over a limited temperature range around 0°C, the following linear relationship is equally valid:

$$R = R_o(1 + \alpha\theta)$$

where α is the temperature coefficient of resistance in $°C^{-1}$ and θ is temperature relative to 0°C. Some typical values for temperature coefficient are:

$\alpha = 0.0039°C^{-1}$ for platinum, $\alpha = 0.0043°C^{-1}$ for copper, $\alpha = 0.0068°C^{-1}$ for nickel

If a change in temperature from θ_1 to θ_2 is considered, then

$$R_1 = R_o(1 + \alpha\theta_1); R_2 = R_o(1 + \alpha\theta_2)$$

Rearrangement gives

$$\theta_2 = \theta_1 + \frac{R_1 - R_1}{\alpha R_o}; \frac{R_2 - R_1}{\theta_2 - \theta_1} = \alpha R_o \qquad (7.21)$$

Apparently, the linear relationship implies that changes in resistance are directly proportional to changes in temperature.

The thermometer comprises a resistance element or bulb, suitable electrical leads, and an indicating—recording or resistance measuring instrument. The resistance element is usually in the form of a coil of very fine platinum, nickel or copper wound non-conductively onto an insulting ceramic former which is protected externally by a metal sheath. A laboratory-type resistance thermometer is often wound on a crossed mica former and enclosed in a pyrx tube. The tube may be evacuated or filled with an inert gas to protect the metal wire. Care is to be taken to ensure that the resistance wire is free from mechanical stresses. A metal that has been strained will suffer a change in the resistance characteristic; the metal is therefore usually annealed at a temperature higher than that at which it is to operate.

Leads are taken out of the thermometer for the measurement of changes in resistance in order to determine the value of temperature. The change in resistance is usually measured by a wheat stone bridge which may be used either in the null (balanced) condition or in the deflection out of balance condition. For steady-state measurement, null conditions suffice whereas transient conditions usually require the use of the deflection mode.

A metal used for the fabrication of sensing elements is required to satisfy the following characteristics:

- Linearity of resistance—temperature relationship for convenience in measurement.
- Relatively large change in resistance with temperature in order to produce a resistance thermometer with good sensitivity.

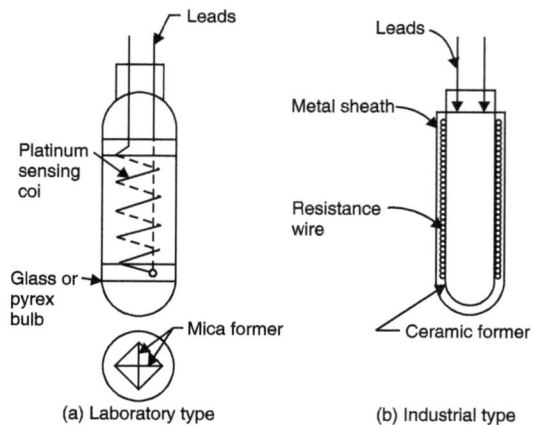

FIGURE 7.32 Resistance thermometers.

- No change of phase or state within a reasonable temperature change.
- Resistant to corrosion and absorption under conditions of use.
- Availability in a reproducible condition, that is, consistent resistance—temperature relationship to provide reliable uniformity.
- High resistivity so that the unit can be fabricated in a compact and convenient size.

Industrial resistance thermometers, often referred to as resistance-temperature detectors are usually made with elements of platinum (which shows little vocalization below 1000°C), nickel (up to 600°C), and copper (up to 250°C). For precise temperature measurements, platinum is preferred because it is physically stable (i.e., relatively indifferent to its environment, resists corrosion and chemical attack, and is not readily oxidized) and has high electrical resistance characteristics. It is stated that with careful and in scientific hands, the accuracy attainable with a platinum resistance thermometer is of the order of ±0.01°C up to 500°C, and with ±0.1°C up to 1200°C. Because of accuracy, stability, and sensitivity, the platinum resistance thermometer has been used to define international temperature scale from the boiling point of oxygen (−182.9°C) to the freezing point of antimony (630.5°C).

Thermistors

Thermistor is a contraction of the term "Thermal Resistor." They are essentially semiconductors that behave as resistors with a high negative temperature coefficient. As the temperature increases, the resistance goes down, and as the temperature decreases, the resistance goes up. This is just opposite to the effect of temperature changes on metals. A high sensitivity to temperature changes (decrease in resistance as much as 6% for each 1°C rise in temperature in some cases) makes the thermistors extremely useful for precision temperature measurement, control, and compensation in the temperature range of −100°C to 300°C.

Thermistors are composed of sintered mixture of metallic oxides such as manganese, nickel, cobalt, copper, iron, and uranium. These metallic oxides are milled, mixed in appropriate proportions, are pressed into the desired shape with appropriate binders and

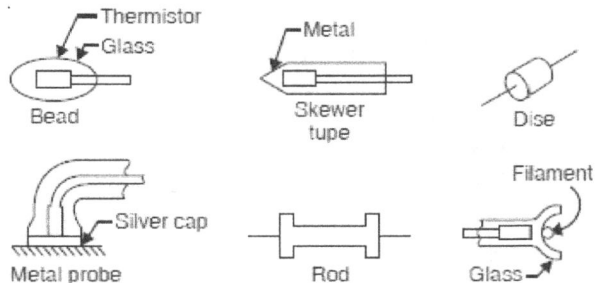

FIGURE 7.33 Typical thermistor forms.

finally sintered. The electrical terminals are either embedded before sintering or backed afterward. The electrical characteristics of thermistors are controlled by varying the type of oxide used and physical size and configuration of the thermistor. Thermistors may be shaped in the form of beads, disks, washers, rods and these standard forms are shown in Figure 7.33. Disks and rods are used more as time delay elements, temperature compensators, and for voltage and power control in electrical circuits. Glass and metal probes less than 2 mm diameter are used for temperature measurements metal surface, gases, and liquids.

Thermistors may be used bare but are usually glass coated or positioned under a thin metal cap. The change in resistance is measured by using circuitry similar to that of metal conductors.

7.8.5. Pyrometers

Total Radiation Pyrometers

The total radiation pyrometer is designed to collect the radiations from the radiating object (furnace) and focus it by means of mirrors or lenses onto a detector (say hot junction of a thermocouple). The emf developed by the thermocouple circuit is measured by a suitable milli-voltmeter or potentiometer, which after suitable calibration becomes a measure of the temperature of the radiating object. Figure 7.34 is a schematic drawing of typical radiation pyrometers.

Pyrometer consists of a blackened tube T open at one end to receive radiations from the object whose temperature is desired. The other end of the tube carries the sighting hole E which is essentially an adjustable eyepiece. The thermal radiations imping on a concave mirror M whose position can be adjusted by a rack and pinion. The mirror is centrally pierced to allow light to reach the eyepiece. The mirror provides maximum reflection of the incoming radiations onto a thermocouple C which is shielded from the incoming radiations and carries a blackened

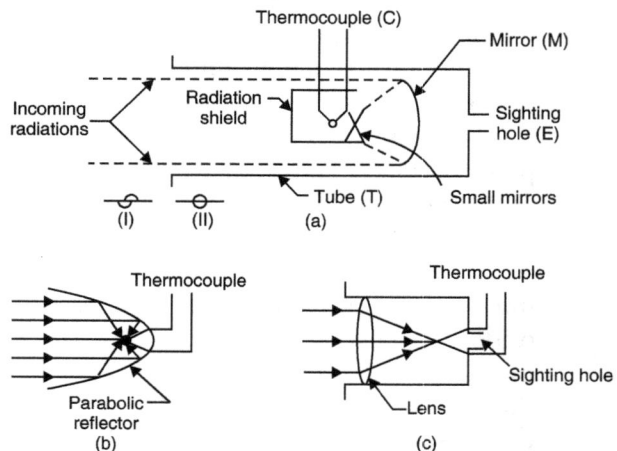

FIGURE 7.34 Typical radiation pyrometers.

copper target disk. There are two small semicircular flat mirrors that are inclined at a slight angle from the vertical plane. The resulting hole is smaller than the target and this allows radiation from the concave mirror to reach the thermocouple. The eyepiece and the concave mirror are adjusted to focus the radiation from the furnace onto the target. Small mirrors help in the focusing process. These mirrors appear as shown at (*i*) when the radiation is not focused onto the target and when focusing is achieved they appear as at (*ii*).

The object of directing radiations from the measured surface onto the temperature sensing element can also be achieved by a parabolic reflector [Figure 7.34(*b*)], or by a lens system [Figure 7.34(*c*)].

Characteristic of radiation pyrometers

1. High speed of response (0.01–0.02 min; fast response is due to small thermal capacitance of the detector. Accuracy ±2% of the scale range.

2. No direct contact is necessary with the object whose temperature is to be measured. This fact allows its use in situations where it is impossible or undesirable to bring the measuring instrument in contact with the object under consideration.

3. Primarily used to measure temperatures in the range 700–2000°C where thermocouple and resistance thermometers cannot be employed.

4. Capable of measuring the temperature of an object which may be either stationary or moving, and so adaptable to continuous industrial processing.

5. Suitable for measuring temperatures where the atmospheric or other environmental conditions prevent satisfactory operation of other temperature sensing devices.

6. Relatively independent of the distance between the measuring element and the heated body. However, for optimum working the distance from target to receiver should not be greater than 10 or 20 times the maximum useful diameter of the target. Further, with an increase in the distance, there will be greater opportunity for gases, smoke, etc., to intervene and absorb some of the radiant energy. This would tend to reduce the indicated temperature.

7. The effect of dust and dirt on the mirrors or lens is to cause the instrument to read too low.

8. Cooling is required to protect the instrument from overheating where the temperature may be high because of operating conditions.

Optical Pyrometers

A metallic surface is usually dark and dull-colored at room temperature. When the surface is heated, it emits radiations of different wavelengths; these radiations are, however, not visible at low temperatures. As the temperature is progressively increased beyond 540°C, the surface becomes dark red, orange, and finally white in color. A color variation with temperature growth may thus be taken as an index of the probable temperature.

This principle of temperature measurement by color or brightness compari-

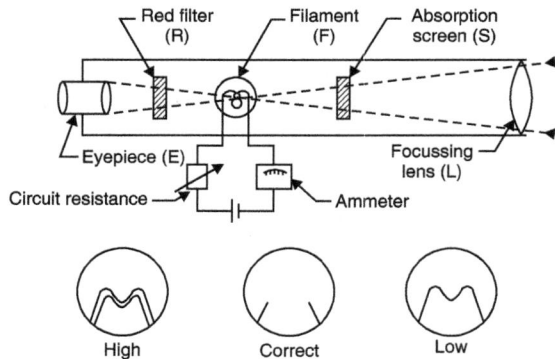

FIGURE 7.35 Disappearing—filament optical pyrometer.

son is utilized in optical pyrometers designed to measure temperatures in the range 700–3000°C. These pyrometers compare the energy emitted by a body at a given wavelength with that of a black body calibrated lamp. A sketch of one of the several types of optical pyrometers is shown in Figure 7.35.

Radiations from the target surface are focused by an objective lens (L) upon the plane filament (F) of an incandescent electric light bulb. The eye price (E) is also adjusted until the filament is in sharp focus and under these conditions, the filament is seen superimposed on the image of the target surface. A red filter (R) is placed between the eyepiece and filament, and it allows only a narrow band of wavelength 0.65 μ to pass through it. Matching the brightness of the lamp filament with that of the target surface is achieved by adjusting the current through the standard lamp by changing the value of circuit resistance. The variable resistance or the magnitude of milliammeter reading (a measure of current through the lamp) may then be calibrated in terms of the target temperature.

When the filament is indistinguishable, in terms of brightness, from the image of the target surface, then it is radiating at the same intensity as the target surface. Three different conditions of the filament as sighted through the eyepiece are also shown in Figure 7.35. When the filament is colder than the target surface, it appears as a dark wire against a light-colored background. Filament brightness is then increased by causing more current to pass through the filament. A filament hotter than the object would appear brighter than the target surface. The current through the filament is then reduced to provide the correct merging of the filament and the object.

In an alternative approach, the current through the lamp filament is maintained constant. An optical wedge of absorbing material is moved up and down and its variable thickness accentuates the incoming energy to match the filament. The wedge position is then calibrated for temperature. The pyrometer is calibrated by sighting it upon a black body at various known temperatures.

The notable characteristics of an optical pyrometer are:

1. No direct contact is necessary with the object whose temperature is to be measured. This aspect allows their use in situations where the measuring target is remote and inaccessible such as molten metals, furnace interiors, etc.

2. Excellent accuracy; the temperature in the useful operating range (700–1000°C) can be determined within ±5°C.

3. Measurement is independent of the distance between the target and the measuring instrument. The image of the target, however, should be sufficiently large to make it possible to secure a definite brightness match with the filament of the test spot.

4. The skill in operating the thermometer can be acquired readily. However, the skill of the operator has more effect upon the resulting temperature measurements when an optical pyrometer is used than when a radiation pyrometer is used.

5. Because of its manual null-balance operation, this pyrometer is not suitable for continuous recording or automatic control applications.

6. The lower measuring temperature is limited to 700°C. Below this temperature, the eye is insensitive to wavelength characteristics.

7.9. STRAIN MEASUREMENT

A strain gauge is a device for measuring dimensional change on the surface of a structural member under test. Measurement of strain is indispensable in a variety of applications due to the utility of strain measurement as a means of determining maximum stress values or in specialized transducers to measure force, pressure, accelerations, torque, etc.

The operation of an electrical resistance strain gauge is based on the fact that when a conductor is subjected to mechanical deformation, its length and diameter are altered and a change in its resistance occurs. The resistance change is measured by the Wheatstone bridge circuit and correlated to strain or the physical effect causing the strain. For many metals used as strain gauge material, the following correlation is applicable.

$$F = 1 + 2\mu$$

where μ = Poisson's ratio = $\dfrac{\text{diametral strain}}{\text{longitudinal strain}} = \dfrac{\delta d/d}{\delta d/l}$

and F = Gauge factor = $\dfrac{\text{fractional change in resistance}}{\text{longitudinal strain}} = \dfrac{\delta R/R}{\delta l/l}$

It is to be noted that

i. For most metals $\mu = 0.3$, and as such the guage factor has a value of around 1.6.

ii. For any given value of resistance R for the gauge element and strain, the change in resistance varies directly with the gauge factor.

iii. A high gauge factor is desirable because that would give a large change in resistance for a given strain input, thereby needing a less sensitive circuit for measuring the change in resistance.

Commercial solid strain gauges using doped crystal structures (semiconductors) have gauge factors from 100 to 5000. These gauges are becoming very popular in a modern instrumentation system.

7.10. FORCE MEASUREMENT

A measure of the unknown force may be accomplished by the methods incorporating the following principles:

i. Balancing the force against a known gravitational force on a standard mass (scales end balances).

ii. Translating the force to a fluid pressure and then measuring the resulting pressure (hydraulic and pneumatic load cells).

iii. Applying the force to some elastic member and then measuring the resulting deflection (proving ring).

iv. Applying the force to a known mass and then measuring the resulting acceleration.

v. Balancing the force against a magnetic force developed by interaction of a magnet and a current-carrying coil.

Scales and Balances

Force or weight is indicated by making a comparison between the force due to gravity acting on a standard mass and the force due to gravity acting on the unknown mass.

An *equal-arm beam balance* (Figure 7.36) consists of a beam pivoted on a knife-edge fulcrum at the center. Attached to the center of the beam is a pointer that points vertically downwards when the beam is in equilibrium. The equilibrium conditions exist when the clockwise rotating moment equals the counterclockwise rotating moment, that is, $m_1 l_1 = m_2 l_2$. Since the two arms of the beam are equal; the beam would be in equilibrium again when $m_1 = m_2$. Further for a given location, the earth's attraction acts equally on both the masses and therefore at the equilibrium conditions $W_1 = W_2$, that is, the unknown force or weights equal the known force or weights.

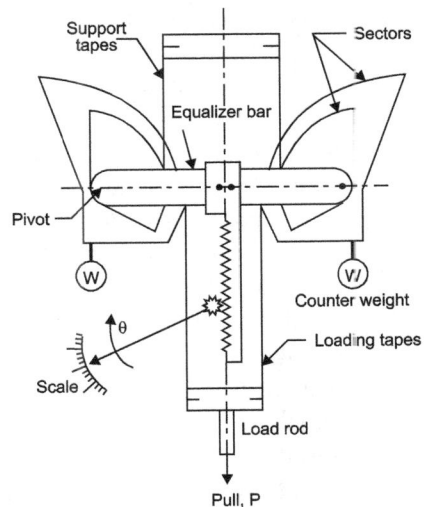

FIGURE 7.36 Equal-arm beam.

The *pendulum scale* (Figure 7.37) is a self-balancing and direct reading force measuring device of multiple lever tape. The weights are however mounted on bent levers, and the movement of the pendulum levers is magnified and transmitted to the indicator pointer.

When the unknown pull P is applied to the load rod, sectors tend to rotate due to unwinding of the loading tapes and consequently the counterweights W swing out. Equilibrium conditions are attained when the counterweight effective moment balances the load moment. The resulting linear movement of the equalizer bar is converted to indicator movement by a rack and pinion arrangement. An electrical signal proportional to the force can also be obtained by incorporating an angular displacement transducer that would measure the angular displacement θ.

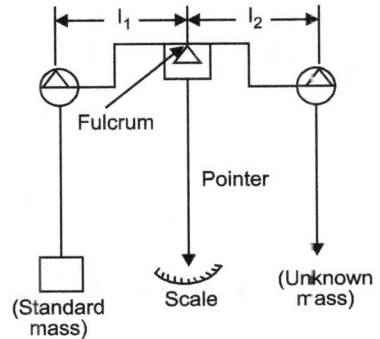

FIGURE 7.37 Essentials of a pendulum scale.

Elastic-Force Meters

These force measuring units measure the force by applying it to an elastic element and then measuring the elastic deformation. Within elastic range of the materials, the deflection of the element is exactly or nearly proportional to the force. Figure 7.38 illustrates the shapes of the more common elastic members used for force estimation.

- *Simple bar:* $x = \dfrac{FL}{AE}$
- *Simply supported beam:* $x = \dfrac{1}{48}\dfrac{FL^3}{EI}$
- *Cantilever:* $x = \dfrac{1}{3}\dfrac{FL^3}{EI}$
- *Spring:* $x = \dfrac{8FD_m^3 N}{E_3 D_w^4}$

where D_m is mean coil diameter, N is number of turns of the coil, D_w is wire diameter E_s is shear modulus.

The desirable properties of the materials used for constructing the elastic-force meters are (*i*) a large and proportional elastic range and (*ii*) freedom from hysteresis.

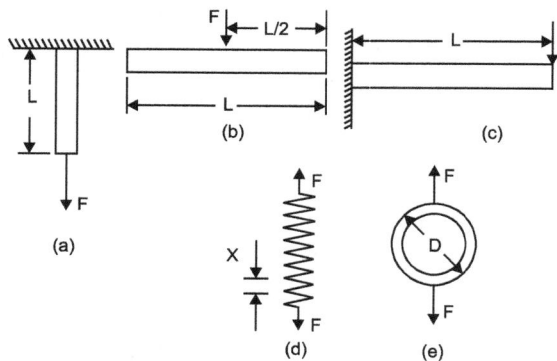

FIGURE 7.38 Elastic deflection elements.

The *proving* (stress) *ring* is a ring of known physical dimensions and mechanical properties. When an external compressive or tensile load is applied to the lugs or external bosses, the ring changes in its diameter; the change being proportional to the applied force. The amount of ring deflection is measured by means of a micrometer screw and a vibrating reed which are attached to the internal bosses. During use the micrometer tip is advanced and its contact with the reed is indicated by considerable damping of the reed vibration. The difference in the micrometer reading taken before and after the application of load is the measure of the amount of the elongation or compression of the ring. The proving ring deflection can also be picked by LVDT, resulting in a proportional voltage change. The device gives precise results when properly calibrated and corrected for temperature variations.

Instead of deflection, strain in an elastic member may be measured by a strain gauge, and then correlated to the applied force.

Mechanical Load Cells

The term "load cell" is used to describe a variety of force transducers which may utilize the deflection or strain of elastic member, or the increase in pressure of enclosed fluids. The resulting fluid pressure is transmitted to some form of pressure sensing device such as a manometer or a bourdon tube pressure gauge. The gauge reading is identified and calibrated in units of force.

In a *hydraulic load cell* (Figure 7.40) the force variable is impressed upon a diaphragm which deflects and thereby transmits the force to a liquid. The liquid medium, contained in a confined space, has a preload pressure of the order of 2 bar. Application of force increases the liquid pressure; it equals the force magnitude divided by the effective area of the diaphragm. The pressure is transmitted to and read on an accurate pressure gauge calibrated directly in force units. The system has a good dynamic response; the diaphram deflection being less than 0.05 mm under full load. This is because diaphragm has a low modulus and substantially all the force is transmitted to the liquid. These cells have been used to measure loads up to about 25×10^5 N with an accuracy of the order of 0.1% of full scale; resolution is about 0.02%.

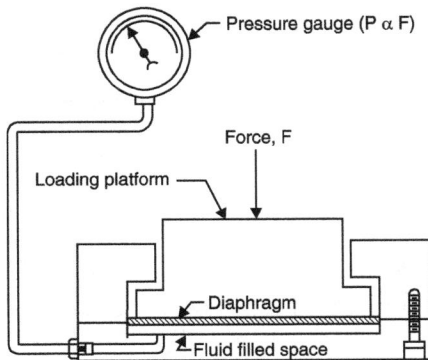

FIGURE 7.39 Proving ring.

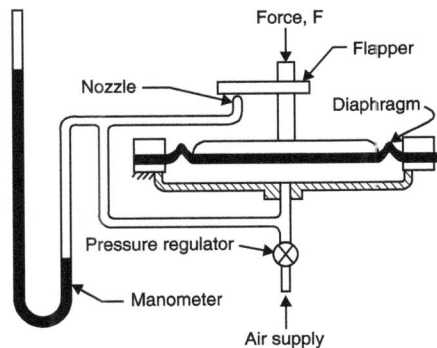

FIGURE 7.40 Hydraulic load cell.

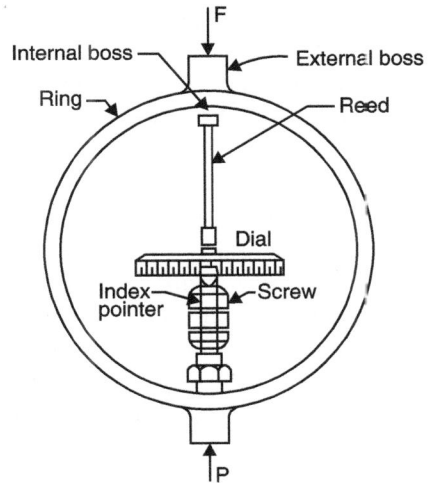

FIGURE 7.41 Pneumatic load cell.

A **pneumatic load cell** (Figure 7.41), operates on the force-balance principle and employs a nozzle-flapper transducer similar to the conventional relay system. A variable downward force is balanced by an upward force of air pressure against the effective area of a diaphragm. Application of force causes the flapper to come closer to the nozzle, and the diaphragm to deflect downwards. The nozzle opening is nearly shut off and this results in increased backpressure in the system. The increased pressure acts on the diaphragm produces an effective upward force that tends to return the diaphragm to its preload position. For any constant applied force, the system attains equilibrium at a specified nozzle opening and a corresponding pressure is indicated by the height of the mercury column in a manometer. Since the maximum pressure in the system is limited to the air supply pressure, the range of the unit can be extended only by using a larger diameter diaphragm. The commercially available load cells operating on this principle can measure loads up to 25×10^5 N with an accuracy of 0.5% of full scale. The air consumption is of the order of 0.17 m^3/h of free air.

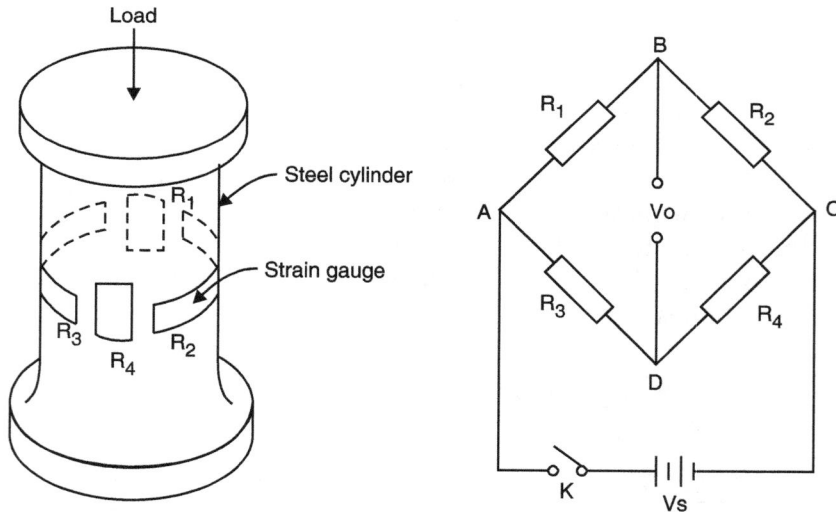

FIGURE 7.42 Strain gauge load cell.

The **strain gauge load** *cells* convert weight or force into electrical outputs which are provided by the strain gauges; these outputs can be connected to various measuring instruments for indicating, recording, and controlling the weight or force.

A simple load cell consists of a steel cylinder which has four identical strain gauges mounted upon it; the gauges R_1 and R_4 are along the direction of applied load and the gauges R_2 and R_3 are attached circumferentially at right angles to gauges R_1 and R_2. These four gauges are connected electrically to the four limbs of a Wheatstone bridge circuit.

The output voltage or the change in output voltage due to applied load is given by

$$dV_o = 2(1+m)\left(\frac{dR}{R} = \frac{V_s}{4}\right)$$

where R is the resistance of each gauge, m is the Poisson's ratio and V_s is the supply voltage.

Apparently, the output voltage is a measure of the applied load.

The strain gauge load cells are excellent force measuring devices, particularly when the force is not steady. They are generally stable, accurate and find extensive use in industrial applications such as drawbar and tool-force dynamometers, crane load monitoring, road vehicle weighing devices, etc.

7.11. TORQUE MEASUREMENT (TORSION METERS)

Measurement of torque may be necessitated for its own sake or as a part of power measurement for a rotating shaft.

In a **gravity balance method** (Figure 7.43), the known mass (m) is moved along the arm so that the value of torque ($F \times r$) equals the product (T) which is to be measured. Alternatively magnitude of the mass may be varied, keeping the radius constant. For the two arrangements we have:

$$r \alpha T \ (m \text{ and } g \text{ are constant})$$
$$m \alpha T \ (r \text{ and } g \text{ are constant})$$

Torque transmission through a shaft usually involves a power source, a power transmitter (shaft), and a power sink (also called the power absorber or dissipator). Torque measurement is accomplished by mounting either the source or the sink in bearing and measuring the reaction force F and the arm length L (Figure 7.44). This concept of bearing mounting is called *cradling* and this forms the basis of most shaft power dynamometers.

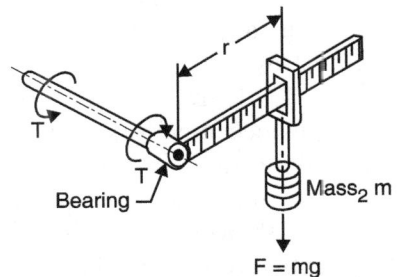

FIGURE 7.43 Gravity balance for torque measurement.

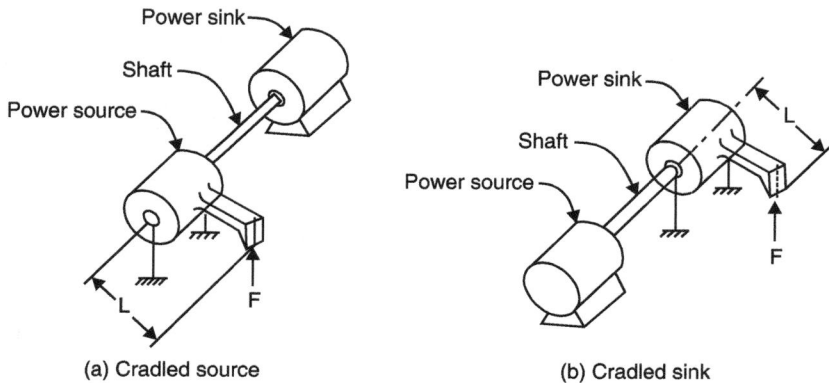

FIGURE 7.44 Torque measurement of rotating machines.

Further, it may be recalled that the following relation holds good for the angular deflection of a shaft subjected to torque within elastic limits:

$$\frac{T}{I_p} = \frac{f_s}{r} = \frac{C\theta}{l}$$

where T is the torque transmitted by the shaft, I_p is the polar moment of inertia of the shaft section, f_s is the maximum induced shear stress at the outside surface, r is the maximum radius at which the maximum shear stress occurs, C is the modulus of rigidity of the shaft material, θ is the angular twist, and l is the length of the shaft over which the twist is measured.

The shaft-twisting relation gives:

$$T = (I_p / I) \times f_s \text{, that is, } T = \text{constant} \times f_s$$

and

$$T = (I_p \, C/l) \times \theta \text{, that is, } T = \text{constant} \times \theta$$

Thus, torque for any given system can be calculated by measuring either the angle of twist or maximum shear stress.

Figure 7.45 shows the schematics of a mechanical torsion bar wherein angular deflection of a parallel length of the shaft is used to measure torque. The angular twist over a fixed length of the bar is observed on a calibrated disk (attached to the rotating shaft) by using the stroboscopic effect of intermittent viewing and

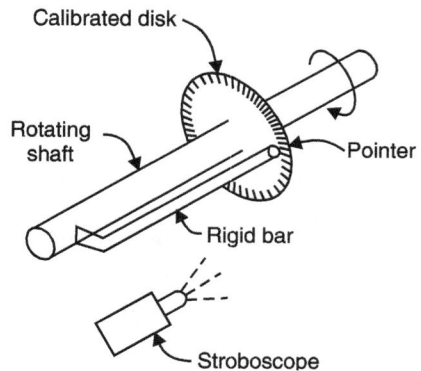

FIGURE 7.45 Mechanical torsion meter.

the persistence of vision. The system gives a varying angle of twist between the driving engine and the driven load as the torque changes.

Optical Torsion Meter

The meter uses an optical method to detect the angular twist of a rotating shaft.

The unit comprises two castings A and B which are fitted to the shaft at a known distance apart. These castings are attached to each other by a tension strip C which transmits torsion but has little resistance to bending. When the shaft is transmitting a torque, there occurs a relative movement between the castings which results in partial inclination between the two mirrors attached to the castings. The mirrors are made to reflect a light beam onto a graduated scale; angular deflection of the light ray is then proportional to the twist of, and hence the torque in the shaft.

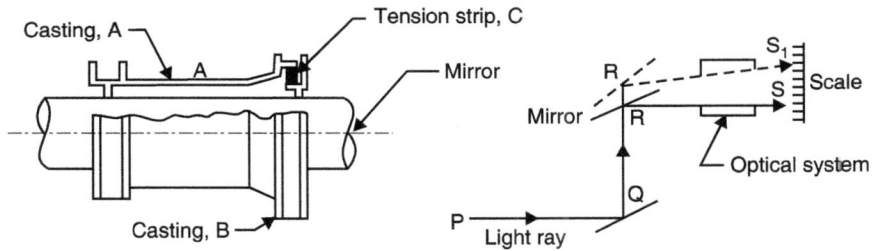

FIGURE 7.46 Optical torsion meter.

For constant torque measurements from a steam turbine, the two mirrors are arranged back to back and there occurs a reflection from each mirror during every half revolution. A second system of mirrors giving four reflections per revolution is desirable when used with a reciprocating engine whose torque varies during a revolution.

Electrical Torsion Meter

A system using two magnetic or photoelectric transducers, as shown in Figure 7.47, involves two sets of measurements.

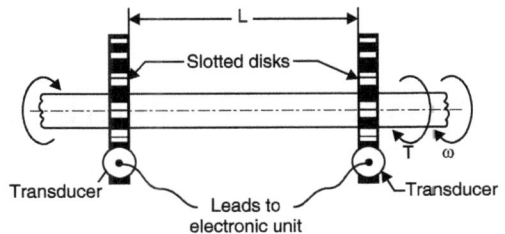

FIGURE 7.47 Electrical torsion meter.

i.   A count of the impulse from either slotted wheel. This count gives the frequency or shaft speed.

ii. A measure of the time between pulses from the two wheels. This signal is proportional to the twist θ of, and hence torque T in the shaft.

These two signals, T and ω, can be combined to estimate the power being transmitted by the shaft.

Strain Gauge Torsion Meter

A general configuration of a strain gauge bridge circuit widely employed for torque measurement from a rotating shaft is shown in Figure 7.48.

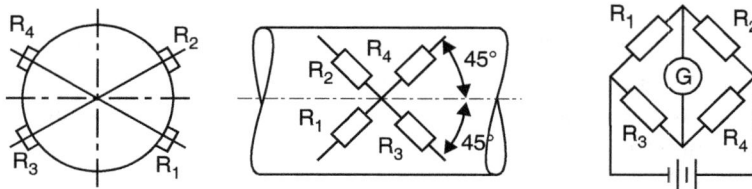

FIGURE 7.48 Strain gauge torsion meter.

Four bonded-wire strain gauges are mounted on a 45° helix with the axis of the rotation, and are placed in pairs diametrically opposite. If the gauges are accurately placed and have matched characteristics, the system is temperature compensated and insensitive to bending and thrust or pull effects. Any change in the gauge circuit then results only from torsional deflection. When the shaft is under torsion, gauges 1 and 4 will elongate as a result of the tensile component of pure shear stress on one diagonal axis, while gauges 2 and 3 will contract owing to compressive components on the other diagonal axis. These tensile and compressive principle strains can be measured, and the shaft torque can be calculated.

The main problem of the system is carrying connections from the strain gauges (mounted on the rotating shaft) to a bridge circuit which is stationary. For slow shaft rotations, the connecting wires are simply wrapped around the shaft. For continuous and fast shaft rotations, leads from the four junctions of the gauges are led along the shaft to the slip rings. Contact with the slip rings is made with the brushes through which connections can be made to the measuring instrument.

7.12. INTERCHANGEABILITY: LIMITS, FITS, AND TOLERANCES

In the design and manufacture of engineering products, attention is paid to be mating, assembly, and fitting of various components. The term ***interchangeability*** implies that the parts which go into the assembly can be selected at random from a large number erf identical parts that have been manufactured within the prescribed limits of dimensions. Then the selective fitting becomes unnecessary except when special allowances are encountered.

The dimension obtained by calculations for strength is known as basic size, basic dimension, or nominal dimension. The actual size of the actual dimension refers to the dimension as measured of a manufactured part. The algebraic difference between the actual size and the corresponding basic size is known as deviation.

Maximum (high) limit and minimum (low) limit are the dimensions with which the actual size of a part may vary.

i. *upper deviation:* Algebraic difference between the maximum limit of size and the corresponding basic size.

ii. *lower deviation*: Algebraic difference between the minimum limit of size and the corresponding basic size.

The terms "basic size," "deviations," and "tolerances" have been illustrated in Figure 7.49.

The degree of tightness or looseness between the two mating parts is known as the fit of the system. The two bases of limit size are the shaft basis system and the hole basis system. The term "shaft" refers to any external dimension of the component, and the term "hole" defines an internal dimension of the component.

In the shaft basis system, the size of the shaft is kept constant and the different fits are obtained by varying the hole size. The minimum limit of the

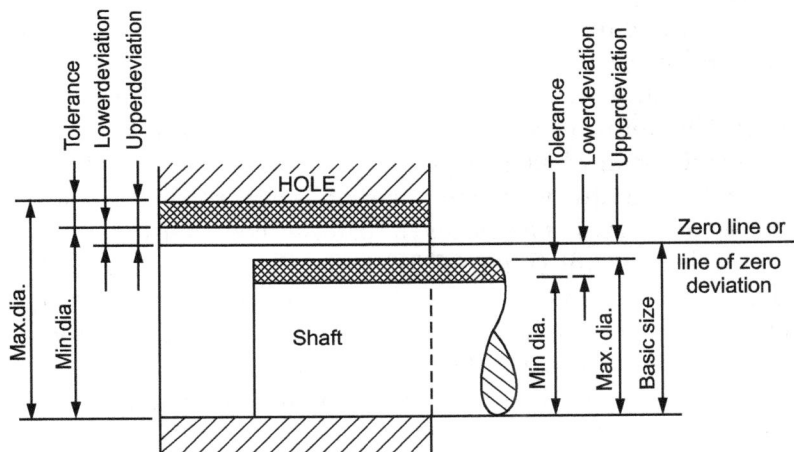

FIGURE 7.49.

shaft is the basic size and the upper deviation is zero. Figure 7.50 shows the representation of the basic shaft system.

In the hole basic system, the size of the hole is kept constant and different fits are obtained by varying the shaft size. The minimum limit of the hole size is the basic size and the lower deviation is zero. Figure 7.51 shows the representation of the basic hole system.

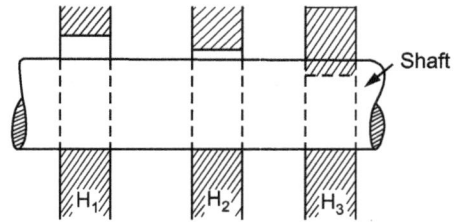

FIGURE 7.50.

According to IS specifications, there are 18 grades of fundamental tolerance, that is, grades of accuracy manufacture and 25 types of fundamental deviations. These deviations for holes are represented by capital letters *A* to *ZC* and for shaft by small letters *a* to *zc*.

The 18 grades of tolerances are designated by IT 01, IT 0, IT 1 to IT 16.

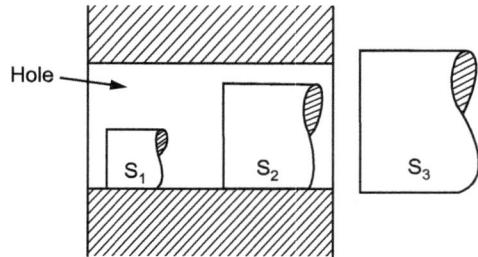

FIGURE 7.51.

From the production point of view, the basic hole system is economical and is preferred because a single drill or reamer of fixed size can produce a variety of fits by merely altering the shaft limits. Shafts are comparatively much easier to be accurately produced to size by turning and grinding. The shaft basic system is, however, used in industries where semifinished shafts are used as raw material.

The nature of fit is characterized by the size of clearance and interference. The term "clearance," refers to the difference between the sizes of a hole and a shaft which are assembled together when the shaft is smaller than the hole. In a situation where the shaft is larger than the hole, the difference between the sizes of a hole and a shaft is called interference.

A clearance fit (Figure 7.52) is one having limits of size so prescribed that a clearance always results when mating parts are assembled. The shaft is always smaller than the hole into which it fits. Clearance is the positive difference between the size of the hole and the shaft.

Typical applications of clearance fit are on rotating shafts, loose pulleys, bearings, cross-head sliders, etc.

An interference fit (Figure 7.53) is one having limits of size so prescribed that an interference always results when the mating parts are assembled. The shaft is always bigger than the hole into which it fits. Interference is the negative difference between the sizes of the hole and the shaft. Typical applications of the interference fit are crankpins, shrunk-on-couplings, iron types, railways wheels shrunk onto axles, pressed-in-bushes, etc.

A transition fit (Figure 7.54) is one having limits of sizes so prescribed that either a clearance or interference may result when mating parts are assembled. The shaft may be bigger, smaller, or of the same size as the hole into which it fits. Shaft

Typical applications of the transition fit are on bushes, spigots, fasteners, pins, keys, etc.

The tolerance or the error permitted in manufacturing a component of a specified dimension may be allowed to vary either on one side of the basic size or on either side of the basic size. In the unilateral system, tolerance is applied only in one direction, and it may be written as

$$2.500 + 0.001 - 0.000 \text{ or } 2.500^{+0.001}_{-0.000}$$

In the bilateral system, the tolerance is allowed on both sides of the nominal size, and it is written as

$$20.00 + 0.02 - 0.01 \text{ or } 20.00^{+0.02}_{-0.01}$$

Here 0.02 represents the upper limit and −0.01 represents the lower limit.

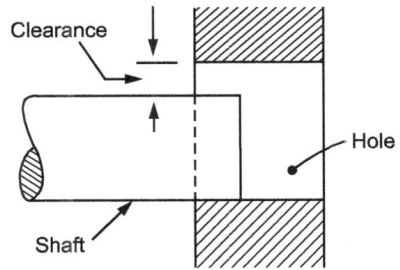

FIGURE 7.52.

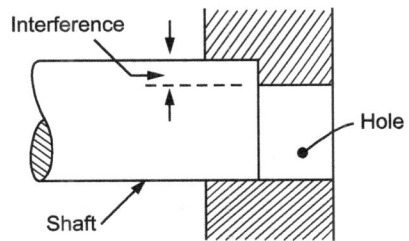

FIGURE 7.53.

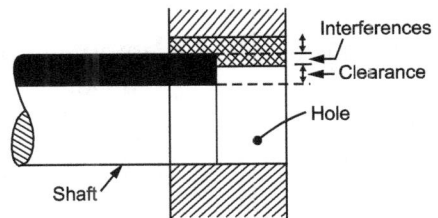

FIGURE 7.54.

REVIEW QUESTIONS

A. Conceptual and conventional questions:

1. (*a*) Define measurement.
   (*b*) What are direct and indirect methods of measurement?
   (*c*) Distinguish between primary, secondary, and tertiary measurements.

2. Distinguish between

    (*i*) range and span
    (*ii*) error and accuracy
    (*iii*) accuracy and precision
    (*iv*) threshold and resolution

3. Define sensitivity. Would you prefer sensitivity to be low or high for an instrument?

4. What is calibration and why is it necessary for an instrument?

5. Discuss the different sources of errors in the act of taking measurements.

6. Errors of measurement may be defined as systematic errors and random errors. Distinguish between these errors and illustrate by giving suitable examples of each type.

7. Explain absolute, gauge, and vacuum pressures.

8. State the principle employed in manometers used for the measurement of pressure.

9. Sketch the Bourdon tube pressure gauge and state the purpose of the details illustrated.

10. Explain the use of a pitot-static tube for velocity measurement.

11. Sketch a Venturi meter and manometer arrangement, apply the steady flow energy equation, and derive an expression for the flow rate of an incompressible fluid.

12. Compare and contrast the use of Venturi meter, flow nozzle, and orifice meter as a device for flow measurement.

13. Explain the principle of operation of rotameter for discharge measurement.

14. List any five physical properties of matter which are used to measure temperature and state clearly how each is used.

15. (a) Outline briefly the principle used in the act of temperature measurement by a thermocouple.
    (b) Sketch a thermocouple circuit showing the important details.

16. Sketch a typical radiation pyrometer and explain its working.

17. Describe, with a neat sketch, the working of a bimetallic thermometer.

18. What are thermistors? How do they differ from resistance thermistors?

19. (*a*) List the different principles on which force measurements are made.
    (*b*) Sketch the schematics of a pendulum scale. How it is used for force measurement?

20. Explain the working of a hydraulic or pneumatic load cell for the measurement of force.

B. Fill in the blanks with appropriate word/words:

(*i*) Measurement provides us with means of describing a natural phenomenon in _____ terms.

(*ii*) The term _____ refers to a piece of equipment having a known measure of the physical quantity.

(*iii*) The act of comparing the performance of the instrument against a precise standard is called _____.

(*iv*) _____ errors are indicated when repeated measurements of the same quantity result in differing values.

(*v*) Gearing, backlash, friction between moving parts, and scale inaccuracies are generally known as _____.

(*vi*) A barometer measures _____.

(*vii*) The stagnation (total) pressure at a point is measured by _____.

(*viii*) The _____ pressure differential that exists between entrance to the Venturi meter and the throat is a measure of _____ through the Venturi meter.

(*ix*) Semi-conductor resistors used for temperature measurement are known as _____

(*x*) _____ is the most suitable device for measuring the temperature of the furnace a boiler.

Answers:

(*i*) quantitative; (*ii*) standard; (*iii*) calibration; (*iv*) random; (*v*) instrument errors; (*vi*) atmospheric pressure; (*vii*) impact tube; (*viii*) static, flow rate; (*ix*) thermistors; (*x*) radiation pyrometer

C. Multiple choice questions:

1. In a measuring instrument, the ratio of output to input change for a given measuring system is referred to as

   (*a*) accuracy   (*b*) linearity
   (*c*) sensitivity   (*d*) range

2. The conformity (closeness, nearness) of the output (value indicated by an instrument) to the true value of the physical quantity is called

   (*a*) accuracy   (*b*) sensitivity
   (*c*) linearity   (*d*) precision

3. The minimum value of input that is necessary to cause a detectable change from zero output is called

   (*a*) least count   (*b*) resolution
   (*c*) drift   (*d*) threshold

4. The introduction of ammeter and voltmeter into an electrical circuit can alter the circuit current and voltage.
   The resulting error is known as

   (*a*) instrument error   (*b*) interaction error
   (*c*) operational error   (*d*) hysteresis error

5. All of the following statements are correct, except

   (*a*) zero error of a micrometer is a systematic error
   (*b*) random errors are accidental, small, and independent
   (*c*) loading error is the error caused by the act of measurement on the physical system being tested
   (*d*) backlash refers to the drift in the instrument output caused by hysteresis due to aging.

6. Turbine meters are used to measure

   (*a*) pressure   (*b*) flow
   (*c*) density   (*d*) viscosity

7. A pitot-static tube measures

   (*a*) undisturbed fluid pressure
   (*b*) dynamic pressure of a moving stream
   (*c*) pressure difference between two fluids
   (*d*) gauge pressure in a static mass of fluid

8. A Venturi meter is preferred to an orifice plate because

   (*a*) it is cheaper and easy to install
   (*b*) its coefficient of discharge has a constant value
   (*c*) its throat pressure is independent of the rate of flow
   (*d*) energy or head loss is less

9. These units are employed where it is not possible to have physical contact with the system whose temperature is to be measured.

   (**a**)   thermocouples
   (**b**)   resistance thermometers
   (**c**)   thermistors
   (**d**)   radiation pyrometers

10. Thermistors are:

   (**a**)   bimetallic strips used in a thermometers
   (**b**)   parallel combination of thermocouples
   (**c**)   semiconductors which have a negative coefficient of resistance
   (**d**)   transducers used for measuring the thermal properties of materials

11. Identify the device used for force measurement

   (**a**) bolometer   (**b**) proving ring
   (**c**) vane anemometer   (**d**) strain gauge

12. A strain gauge should have a high value of gauge factor to

   (**a**)   increase sensitivity
   (**b**)   reduce hysteresis effects
   (**c**)   reduce/eliminate the effect of variation in ambient temperature
   (**d**)   give a linear relationship between applied strain and resistance change

Answers:

   1. (a)   2. (c)   3. (d)   4. (b)   5. (d)   6. (b)
   7. (b)   8. (d)   9. (d)   10. (c)   11. (b)   12. (a)

8

CONTROL SYSTEMS

8.1. CONTROL SYSTEM: WHAT IS IT?

A system is an assemblage of devices and components connected or related by some form of regular interaction or interdependence to form an organized whole and perform specified tasks. The system produces an output corresponding to a given input. The thermometer and the mass-spring-damper system can be identified as systems.

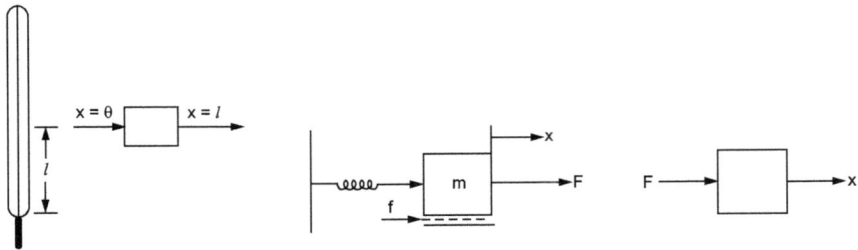

FIGURE 8.1. Systems comprising a thermometer and spring-mass

The thermometer has the input $x = \theta$ (temperature) and the output $y = l$ (length of the mercury column in the capillary). In the mass-spring arrangement, the force and the position of the mass constitute the input to and output from the system, respectively. In a rotational generator of electricity, the input would be the rotational speed of the prime-mover shaft and the output would either be the induced voltage at the terminals (with no load attached to the generator) or the unit of electrical power (with load attached to the generator).

The term *control* implies to regulate, direct, or command. A control system may thus be defined as:

i.   *means by which a set of variable quantities is field constant or caused to vary in a prescribed way, and*

ii. *an assemblage of devices and components connected or related so as to command, direct or regulate itself or another system.*

In a control system, deliberate guidance or manipulation is employed to maintain a system variable at a set point or to change it according to a preset program.

Control systems are intimately related to the concept of automation but have an ancient history. Romans maintained water levels in aqueducts by means of floating balls that opened and closed the valves at appropriate levels. Watt's flyball governor (1769) regulated steam flow to a steam engine to maintain constant engine speed despite a changing load. In World War II, control system theory was applied to anti-aircraft batteries and fire-control systems. The introduction of analog and digital computers has opened the way for much greater complexity in automatic control theory.

8.2. EXAMPLES OF CONTROL SYSTEMS

Illustrated below are some examples of control systems:

i.   An *electric switch* that serves to control the flow of electricity in a circuit. The input signal (command) is the flipping of the switch on or off, and the corresponding output (controlled) signal is the flow or non-flow of electric current.

ii. A *thermal system* where it is desired to maintain the temperature of hot water at a prescribed value.

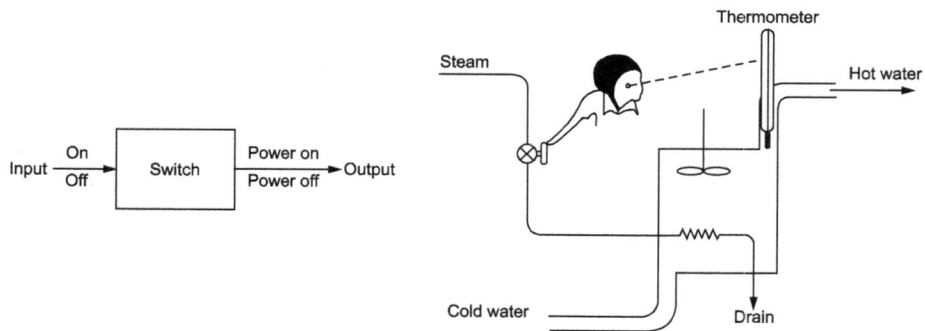

(a) Electrical on-off system (b) Thermal system: manual feedback control

FIGURE 8.2. An electric switch and a thermal control system

Before the operator can carry out this task satisfactorily, the following requirements must be met:

a. The operator must be told what temperature is required for the water. This temperature called the set point or desired value constitutes the input to the system.

b. The operator must be provided with some means of observing the temperature (sensing element). For that, a thermometer is installed in the hot water pipe and it measures the actual temperature of the water. This temperature is output from the system and is called the ***controlled variable***. The operator watches the thermometer and compares how the measured temperature compares with the desired value. This difference between the desired value and the actual measurement value is an error or ***actuating signal***:

$$e = r - c$$

where r refers to the set point or reference input and c denotes the controlled variable.

c. The operator must be provided with some means of influencing the temperature (***control element***) and must be instructed what to do to change the temperature in the desired direction (***control function***).

The sign of the error signal e indicates whether the controlled temperature is too high or too low, and this determines the direction of the corrective action required: whether to open up the valve or close it down. The size of the error signal determines the amount of correction action necessary. When the valve is turned in the correct direction by the correct amount the water will acquire the desired temperature value.

Hence the operator has been able to reduce the error signal to a minimum by changing the steam supply to water. The flow of steam constitutes the ***manipulated variable***.

iii. A *driving system* of an automobile (accelerator, carburettor, and an engine vehicle) where command signal is the force on the acceleration pedal and the automobile is the controlled variable. The desired change in engine speed can be obtained by controlling pressure on the accelerator pedal.

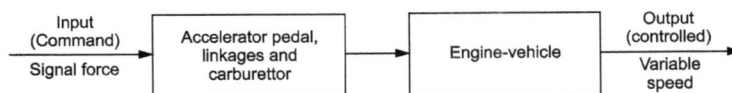

FIGURE 8.2. (c) Driving system of an automobile

iv. An *automobile steering system* where the driver is required to keep the automobile in the appropriate lane of the roadways. The eyes measure the output (heading of the automobile), the brain and hands react to any error existing between the input (applied lane) and the output signals and act to reduce the error to zero.

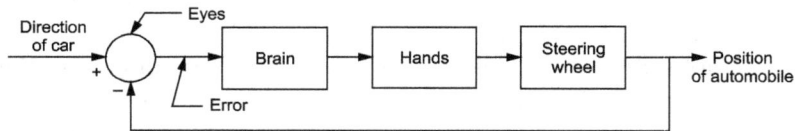

FIGURE 8.2. (d) Automobile steering system

v. A *biological control system* where a person moves his finger to point at an object. The command signal is the position of the object and the output is the pointed direction.

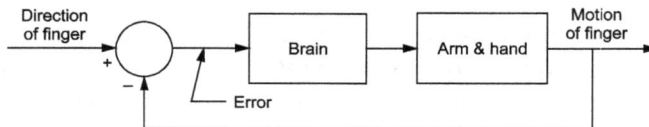

FIGURE 8.2. (e) Biological control system: a person pointing toward an object

Other well-known examples of control systems are electric frying pans, water pressure regulators, toilet-tank water levels, electric irons, refrigerators, and household furnaces with thermostatic control.

8.3. CLASSIFICATION OF CONTROL SYSTEMS

There are two basic types of control systems, open-loop, and closed-loop systems

8.3.1 Open-loop system (unmonitored control system)

The main features of an open-loop system are:

1. There is no comparison between the actual (controlled) and the desired values of a variable.

2. For each reference input, there corresponds to a fixed operating condition (output) and this output has no effect on the control action, that is, the control action is independent of output.

3. For the given set-input, there may be a big variation of the controlled variable depending upon the ambient conditions.

Since there is no comparison between actual output and the desired value, rapid changes can occur in the output if there occurs any change in the external load.

Some examples of open-loop systems are:

i. Trying to guide a car by setting the steering wheel, together with a pattern of subsequent changes of direction, at the beginning of a journey and making no alteration enroute as and when the car deviates from the desired path.

ii. Hitting the golf ball where the player knows his goal to get the ball into a particular hole. To achieve it, the player hits the ball correctly at the beginning of its flight. Once the moment of impact is passed, he loses his control on any further flight of the ball.

iii. A washing machine in which soaking, washing, and rinsing operations are carried out on a time basis. The machine does not measure the output signal, viz., the cleanliness of the clothes.

iv. An automatic toaster where the toasting time and temperature are preset quantities. The quality of the toast (darkness or lightness) is determined by the user and not by the toaster.

v. The automobile traffic control signals at roadways intersections are the open-loop systems. The red and green light times (input to the control action) are predetermined by a calibrated timing mechanism and are in no way influenced by the traffic (output).

The control systems depicted in Figure 8.2(a) and (c) are also the open-loop control system. In the electric switch control system, the flipping of the switch is independent of the flow of electric current through the circuit. Likewise, in the driving of the automobile, no correspondence is shown between the vehicle speed (controlled variable) and the force (command signal) on the pedal.

From the illustrations cited above, it may be noted that any control system which operates on a time basis is an open-loop system.

An open-loop system has the following advantages and limitations:

- simple construction and ease of maintenance,
- no stability problems,
- convenient when the controlled variable is either difficult to measure or it is economically not feasible,
- system affected by internal and external disturbances; the output may deviate from the desired value, and
- needs frequent and careful calibrations for an accurate result.

8.3.2. Closed-loop system (monitored control system)

The main features of the closed-loop system are:

1. There is a comparison between the actual (controlled) and the desired value of the variable. To accomplish it, the output signal is fed back and the loop is completed.

2. The error signal (deviation between the reference input and the feedback) actuates the control element to minimize the error and bring the system output to the desired valve.

3. The system operation is continually correcting any error that may exist. As the output does not coincide with the desired goal, there is likely to be some kind of error signal.

Evidently, the closed-loop systems correct the drift of the output away from the goal, which may be due to external disturbance or due to deterioration of the system.

Common phrases used to describe closed-loop are the feedback control or the monitored and automatic control systems. The performance of such a system is evaluated with reference to the following desirable characteristics:

- minimum deviation following a disturbance,
- minimum time interval before return to setpoint, and
- minimum offset due to change in operating conditions.

Examples of closed-loop systems are:

i. The control of the thermal system (Figure 8.2*b*) is a closed-loop system. When the operator detects that the output temperature is different from the desired or reference, he initiates an action to reduce the discrepancy by operating a valve that controls the steam supply to water.

ii. The automobile driving system (Figure 8.2*c*) would become a closed-loop system when the driver makes a visual observation of the speed indicated by a speedometer and compares this mentally with the desired speed.

Based on the deviation between the actual and the desired speed values, the driver would take the decision either to increase or decrease the speed. The decision is implemented by affecting a change in the pressure of his foot on the accelerator pedal. The driver's eye and the brain act as the error detectors.

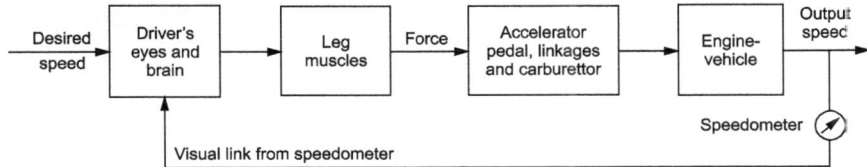

FIGURE 8.3. Driving system of an automobile: manually controlled

iii. The traffic control system at a roadway intersection is a closed-loop system when the traffic policeman allows a greater time interval to cope with a greater traffic volume coming from a particular direction.

The closed-loop systems listed above involve continuous manual control by human operators and are classified as *manual feedback* or *manual closed-loop systems*. In the many complex and fast-moving systems, the dull and time-consuming manual tasks are accomplished by incorporating in the system some equipment that would perform the desired functions more rapidly and consistently. A closed-loop system operating without human is called an *automatic control system*.

Examples of automatic control systems

i. In the automatic feedback control of a thermal system (Figure 8.4*a*), the human operator has been replaced by an automatic controller. The actual temperature of the hot water is measured by a thermometer and is fed to the controller for comparison with the reference temperature whose value has been specified by an appropriate setting of the thermostat/regulator. Based on the error signal, the controller generates an output (correcting signal) which is taken to the control valve in order to change the valve opening for the steam supply.

ii. The level control system depicted in Figure 8.4(*b*) is an automatic control system where the inflow of water to the tank is dependent on the water level in the tank. The automatic controller maintains the liquid level by comparing the actual level with the desired level correcting any error by adjusting the opening of the control valve.

iii. A pressure control system where the pressure inside the furnace is automatically tolled by adjusting the opening of the control valve (Figure 8.4*d*).

iv. The anti-aircraft radar tracking system (Figure 8.4*e*) incorporates a rotating antenna that senses the presence of the target plane. The detected signal (velocity and position) of the plane is transmitted to the computer which determines the firing angle. The firing angle then becomes the command signal which is passed on to the firing system through the power amplifiers. The angular position of the gun is fed back to the computer. The gun is triggered when the error between the command signal and the firing angle becomes zero.

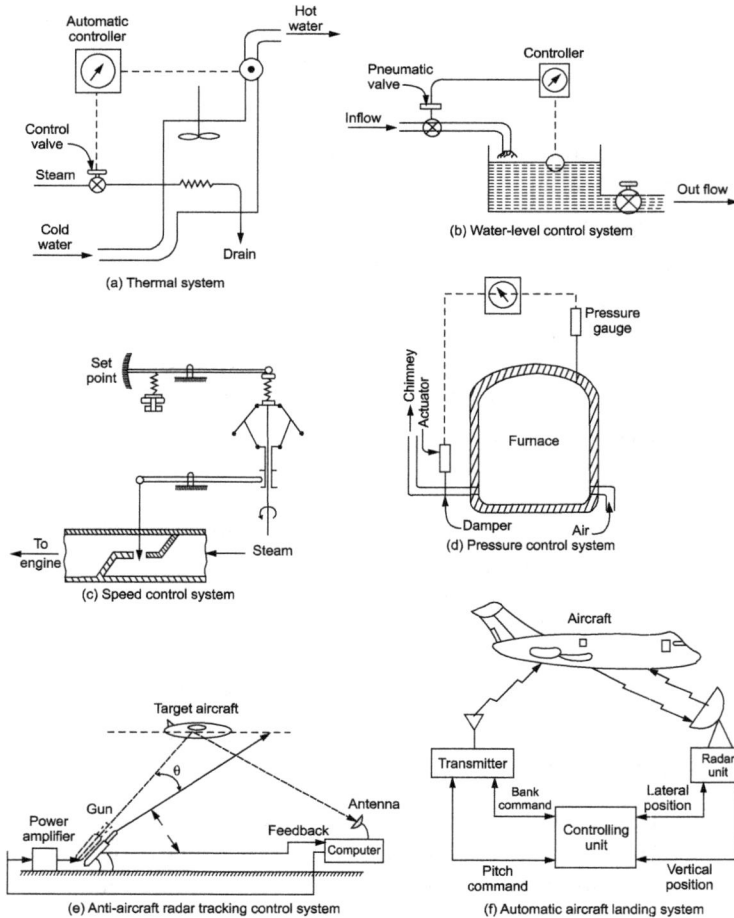

FIGURE 8.4. Automatic control systems

v. The automatic aircraft landing system (Figure 8.4*f*) has three basic parts, viz., the aircraft, the radar unit, and the controlling unit. The approximate vertical and lateral positions of the aircraft as measured by the radar unit are transmitted to the controlling unit. The controlling unit then determines the appropriate pitch and bank commands. These commands are later transmitted to the aircraft autopilots and that makes the aircraft respond.

vi. *Advantages and limitations of automatic control systems*

- suitability and desirability in the complex and fast-acting systems which are beyond the physical abilities of a man,
- relief to human beings from hard physical work, boredom and drudgery which normally result from a continuous repetitive job,
- economy in the operating cost due to the elimination of the continuous employment of a human operator,
- increased output or productivity,
- improvement in the quality and quantity of the products,
- economy in the plant equipment, power requirement, and processing material. The feedback permits to initiate precise control by using relatively inexpensive components,
- reduced effect of non-linearities and distortions, and
- satisfactory response over a wide range of input frequencies.

The system has however a tendency to over-correct errors and this may cause oscillations of constant or changing amplitude.

Comparison of open- and closed-loop control systems

Open-loop control system	Closed-loop control system
1. Less expensive and simple to construct.	1. Costly and complex construction.
2. Easy maintenance.	2. Maintenance is comparatively difficult.
3. Components incorporated in the system to be accurate.	3. Less accurate components can be used for the satisfactory operation of the system.
4. Generally stable.	4. Tends to become unstable under certain conditions.
5. No need to measure the output.	5. The output is necessarily to be measured.

Open-loop control system	*Closed-loop control system*
6. Slow in operation and there is no possibility of optimization.	6. Faster optimization is possible.
7. Feedback elements and error detectors are not needed.	7. Feedback elements and error detectors are essential components of the system.
8. Highly sensitive to disturbances and environmental changes.	8. Less sensitive to disturbance; the disturbances are taken care of by the feedback present in the system.
9. System needs to be calibrated and recalibrated for accuracy.	9. Calibration is not required; the error between the reference input and the output is measured through feedback and necessary correction is applied.

8.4. CONTROL SYSTEMS TERMINOLOGY

A closed-loop consists essentially of a process, error detector, and control elements. The source of the terms related to these basic components are defined below:

Process, plant, or controlled system (g_2): a body, process, or machine of which a particular quantity or condition is to be controlled, for example, a furnace, reactor, spacecraft, etc.

Controlled variable (c): the quality or condition (temperature, level, flow rate, etc.) characterizing a process whose value is held constant by the controller or is changed according to a certain law.

Controlled medium: the process material in the controlled system or flowing through it in which the variable is to be controlled.

Command: an input that is established or varied by some means which are external to and independent of the feedback control system.

Set point or reference input (r): a signal established as a standard of comparison for feedback control system by virtue of its relation to command. The setpoint either remains constant or changes with time according to a preset program.

Manipulated variable (m): the quality or condition that is varied as a function of the actuating signal so as to change the plant g_2 by the control element g_1.

Actuating signal (e): an algebraic sum of the reference input r and the primary feedback b. The actuating signal is also called the error or control action.

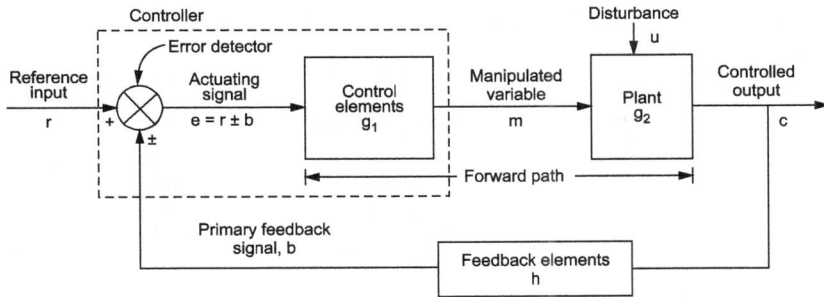

FIGURE 8.5. Elements of a control system

Primary feedback signal (b): a function of the controlled output c, which is compared with the reference input to obtain the actuating signal.

Error detector: an element that detects the feedback; essentially it is a summing point that gives the algebraic summation of two or more signals The detection of the flow of information is indicated by arrows and the algebraic nature of summation by plus or minus signs.

Negative feedback occurs when the feedback signal subtracts from the reference signal

$$e = r - b$$

If the feedback signal adds to the reference signal, the feedback is said to be positive

$$e = r + b$$

Negative feedback tries to reduce the error, whereas positive feedback makes the error large.

Disturbance (n): an undesired variable applied to the system which tends to affect adversely the value of the variable being controlled. The process disturbance may be due to changes in setpoint, supply, demand, environmental, and other associated variables.

Feedback element (h): an element of the feedback control system that establishes a functional relationship between the controlled variable c and the feedback signal b.

Control element (g_2): an element that is required to generate the appropriate control signal (manipulated variable) m applied to the plant.

Forward and backward paths: The transmission path from the actuating signal e to the controlled output c constitutes the forward path. The backward path is the transmission path from the controlled output c to the primary feedback signal b.

8.5. SERVOMECHANISM, PROCESS CONTROL, AND REGULATOR

A servomechanism is an automatic control system in which the controlled variable is mechanical position (displacement), or a time derivative of displacement such as velocity and acceleration. The output is designed to follow a continuously changing input or desired variable (command signal). The servomechanisms are inherently fast-acting (small time lag with response time in the order of milliseconds) systems and usually employ electric or hydraulic actuation. These systems are essentially used to control the position or speed of a mechanism that is either too heavy or too remote to be controlled manually. The complete automation of machine tools together with programmed instruction is another notable example of servomechanism.

Servomechanisms find utility in satellite-tracking antennas, automatic navigation systems on boats and planes, and anti-aircraft gun control systems. Other examples are fly-by-wire systems in aircraft which use servers to actuate the aircraft's control surfaces and radio-controlled models which use RC servers for the same purpose. Many autofocus cameras also use a servomechanism to accurately move the lens and then adjust the focus. A modern hard disk drive has a magnetic servo system with sub-micron positioning accuracy.

Process control refers to the control of such parameters as level, flow, pressure, temperature, and acidity of a process variable, A particular parameter has usually only one optimum desired value (setpoint) and the control system is required to ensure that the process output is maintained at this level inspite of changes in external conditions (disturbances) which affect the process. The load disturbance could be (*i*) a change in the boiler steam pressure affecting a temperature control system (*ii*) a change in raw materials affecting a mixing process. The process control systems are usually slow-acting (large time lags) and usually employ pneumatic actuation.

A ***regulator*** is a feedback control system in which the output (controlled variable) is maintained at a preset value irrespective of external load on the plant. The reference input or command signal, although adjustable, is held constant for long periods of time. The primary task is then to maintain the output at the desired value in the presence of disturbances (change in load on the system or changes in the environment or changes in the system itself). Examples of an automatic regulator are regulation of steam supply in steam engines by the fly ball governor; thermostat control of a home heating system; control of pressure and of electrical quantities such as voltage, current, and frequency.

In general, a control system that regulates a variable in response to a fixed command signal is known as a regulator system whereas a control system that accurately follows changes in the command signal is referred to as follow-up system.

8.6. SEQUENCE CONTROL

Sequence control is a special type of open-loop system which has the following main features:

i. the finish of one action initiates the start of the next,

ii. the acts take place in a certain fixed sequence, and

iii. there is no comparison of desired and actual value.

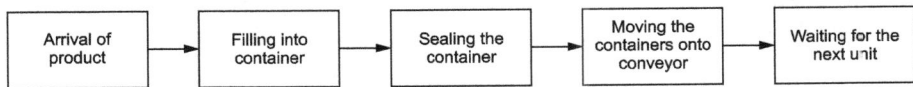

FIGURE 8.6. Block diagram of a sequential-control system

Figure 8.6 illustrates the block diagram of one such system involving mechanical handing packaging. The various actions are performed by pneumatic or hydraulic components and the completion of the operation is signaled by mechanical trip valves.

The cloth-washing machine too is a sequence control where the various operations, such as

a. filling of the tub   **b.** washing   **c.** draining the tub

d. rinsing and   **e.** spin-drying is controlled by timer switch

Likewise, the sequence of operations on a production machine may be

i. job in position   **ii.** guard in position

iii. tool in position   **iv.** tool motion

v. tool withdrawal and   **vi.** job withdrawal

Logic control devices are used to control each of these operations.

8.7. MANUAL AND AUTOMATIC SYSTEMS

Manual control systems involve a human operator who:

i. takes decision about the required output,

ii. ensures that the necessary input (the reference input) is applied to the system,

iii. observes the output and compares it with the desired value, and

iv. readjusts the control elements if the output is not what he wants.

In an automatic control system, the human operator determines the goal and sets up the system. Subsequently, the target output is achieved or maintained automatically. Only the reference input is provided and the necessary corrections are applied by mechanical (non-human) devices and that essentially forms the essence of automatic control.

Automatic control has become obligatory in a wide variety of engineering problems as is evident from the following few examples:

- control of temperature, pressure, humidity, viscosity, flow rate, etc., in the process industries like synthetic yarn production, oil refining and chemical plants,
- control of heat treatment, tooling, handling, and assembling of mechanical parts in the control of the manufacture of articles like refrigerators, radio, and automobile parts,
- control of position, speed and power in machine tools, pumps and compressors, electrical, and mechanical power supply units,
- speed regulation of devices like grinding wheel for precision grinding, tape recorders, strip rolling, and wire drawing, and
- transportation systems such as ship steering and rolling stabilization, aircraft flight control, the automatic landing of aircraft, etc. The positioning systems, radar travel systems and other military equipment are necessarily based on feed control systems.

Applications of control system engineering

Some of the common applications that involve the use of control systems are:

1. Range of human activities in the domestic domain such as picking a book from the table, eating meals from the plate, pointing a finger toward an object, walking from a starting point to a destination along a prescribed path.

2. On and off of electric supply to units such as washing machines, toasters, fans, air-conditioners, and other electrical appliances.

3. Speed and direction-control of transport vehicles.

4. Regulation of temperature and humidity of homes, offices, hospitals, and shopping malls for the comfort of human beings.

5. Control of temperature, pressure, water level and humidity, etc., in the process industry.

6. Quality control of manufactured products, automatic control of machine tools and assembly line in industry. Operation of computerized numerically controlled machines.

7. Regulation of voltage at electric power plants.

8. Space technology, missile launching, and guidance.

9. Military operations such as automatic positioning of guns, radar antennas, steering control of ships.

The control system engineering and its applications are not limited to engineering alone; it is applicable to all fields of knowledge pertaining to biological, biomedical, economic, and socioeconomic systems.

EXAMPLE 8.1
Name any three electrical devices used at home and which are equipped with suitable controls to achieve the desired purpose. Also mention the control category to which they belong.

Solution: Some common electrical devices used at home are:

i. Radio: one can adjust the volume, the tone, the station.

ii. Television set: one can adjust the volume, the channel, and the brightness.

iii. Oven: one can adjust the temperature.

In each of these arrangements, the human operator forms a part of the control loop. He provides the feedback path, makes adjustments with the controls and corrects for the errors so as to get the desired performance from the device. Evidently, these electrical devices constitute manual closed-loop control systems.

Electric fires and lights are open arrangements because their outputs cannot be adjusted if they deviate from the desired goal. If the electric fire does not give enough heat or if the lamp is not bright enough, these have to be discarded and replaced with new ones.

EXAMPLE 8.2
Identify the open-loop and closed-loop aspects of cooking.

Solution: Most of the cooking done in an oven by the novice cook is essentially an open-loop because:

i. the quantities of the ingredients are specified by the recipe,

ii. the mixing of the ingredients is as per instructions, and

iii. the oven settings and the cooking time are also according to the instructions.

If every act is in accordance with the instructions listed in the cookery book, the meal would emerge as required. Indeed, if the cook tries to add a little feedback to the process by opening the oven to take a look (presumably to make adjustments if things are not going right), he can make things worse.

However, the experienced cooks use their judgment and modify the open-loop instructions with feedback and achieve the required target rather more accurately. Using a frying pan is more of a feedback process. The quantities used, the temperature of the pan, the cooking time, etc., are all adjusted according to how closely the output (cooked meal) approaches the target (the desired quality of the meal).

EXAMPLE 8.3
Is the act of switching a light involve both manual and automatic controls?

Solution: The process of switching a light involves both manual and automatic controls.

i. The act of switching a light is a normal operation and is obviously a manual control. The person decides that he wants the light and accordingly the switch is turned on. Subsequently, the switch is turned off if the light is not needed.

ii. The bulb gives a predetermined brightness; this is achieved automatically by exercising careful control in the manufacture of the lamp and by keeping the constant voltage at the mains. Thus, the lamps either give the intended brightness or fail completely.

The operating system at the power station ensures that mains voltage stays close to 240 volts except during power cuts.

EXAMPLE 8.4
Identify the concept of plant, reference input, controlled output and feedback, etc., in the following control systems:

a. **control of temperature in the central heating system,**

b. **control of water level in a cistern,**

c. **control of the progress of an automobile vehicle,**

d. **control of water temperature for a shower bath.**

Solution: **a.** Refer to Figure 8.7(a) for the closed-loop central heating system:

   i. The *plant* refers to the room whose temperature is required to be controlled.

   ii. The thermostat is set to the specified temperature (*reference input*) and thus controls the fuel/oil input to the boiler.

   iii. The *controlled output* is the actual room temperature which is fed back to the thermostat.

   iv. The *controlled element* is the oil flow valve and the goal is the specified house temperature.

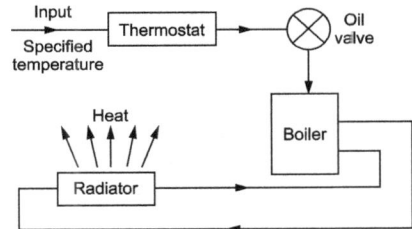

FIGURE 8.7. (a) A closed-loop central heating system

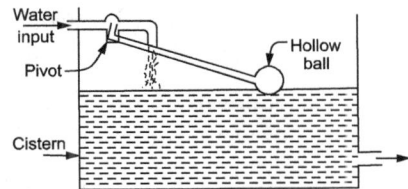

FIGURE 8.7. (b) Control of water level in a cistern

The thermostat switches on the oil flow to the boiler furnace when the room temperature drops below the specified value. This results in heat flow from die radiators in the room and the consequent rise in temperature, The supply of oil to the furnace is automatically shut off when the room temperature rises to the specified reference input.

b. Refer to Figure 8.7(b) for the control of water level in a cistern.

   i. The *plant* is the tank (cistern) wherein the level of water is to be controlled.

   ii. The *reference input* is prescribed by the initial setting of the ball lever.

   iii. The *output* is the constant level of water in the tank.

    iv. The *control element* is the pivot-ball arrangement and the feedback is the actual position of the floating ball.

When the ball is at a lower position (i.e., the water level is below the desired level), then will be the inflow of water into the cistern. With this inflow, the level of water would rise and with that, the ball would also rise. Eventually, the ball would move up to a position where the arm would cut off the supply of water.

c. Refer to Figure 8.3 for the control of the progress of an automobile vehicle,

    i. The *plant* is the automobile vehicle whose progress along a specified track is to be controlled.

    ii. The *input* includes turning off the steering wheel and the pressure of the foot on the acceleration pedal.

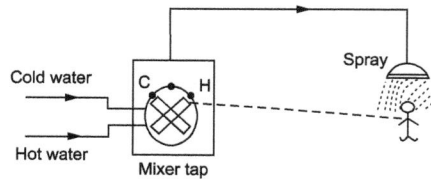

FIGURE 8.7. (c) Control of water temperature for shower bath

    iii. The engine, transmission links and steering are the *control elements*.

    iv. The *feedback* loop governing the motion of the vehicle along the road is the driver's observation. This feedback leads to a change in pressure on the accelerator on to an adjustment of the setting of the steering wheel (steering correction).

d. Refer to Figure 8.7(c) which depicts a simple arrangement for the control of water temperature for a shower bath.

    i. The *inputs* are the temperature and flow rates of hot and cold water.

    ii. The *plant* is the pipe taking the water up to the nozzle.

    iii. The *output* is the temperature of hot water flowing from the nozzle.

    iv. The *control element* is the mixing tap.

    v. The arms of the person taking the bath provide *feedback* from the output to the input and readjust the setting on the input side. If the water was too hot, he turns the mixer toward a cold position. If it was too cold, he would adjust the mixture toward hot. The adjustment is so made that difference between the actual output temperature and the desired output temperature is reduced to the minimum.

EXAMPLE 8.5
Draw the schematics and block diagram of a system representing a steam-generator set fitted with a speed governor.

Solution: Refer to Figure 8.8(a) for the schematics of a system for speed control of a turbo-governor.

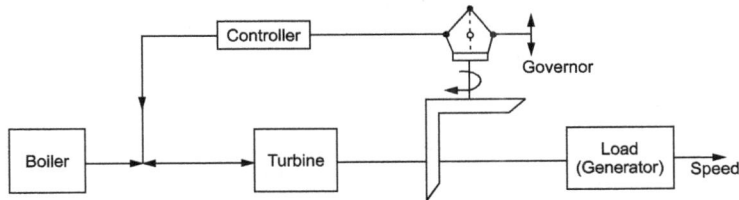

FIGURE 8.8. (a)

The system incorporates a centrifugal governor which uses the lift of centrifugal balls as speed monitor, it senses any speed change which may occur due to variation in load. The speed sensed by the governor is compared with the desired speed and an error or deviation signal is generated. A hydraulic amplifier serves as a controller that operates a control valve which moves by an amount proportional to the error. The valve then regulates the steam flow from the boiler to the turbine; which results in a change in speed until the output speed matches the desired speed.

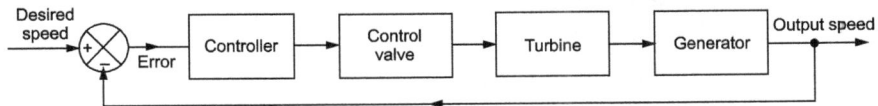

FIGURE 8.8. (b)

The system has been represented by a block diagram with various elements as shown in Figure 8.8(b).

REVIEW QUESTIONS

A. Conceptual and conventional questions:

1. Define the terms "system" and "control system."

2. Define and distinguish between open-loop and closed-loop control systems. Illustrate your answers with suitable examples.

3. Point out the merits and demerits in reference to open and closed-loop control systems.

4. Identify the following systems as open-loop or closed-loop control system. Give the reason thereof:
   (**a**) automatic electric toaster
   (**b**) home shower with separate valves for hot and cold water
   (**c**) a man walking in a prescribed direction
   (**d**) anti-aircraft radar tracking system

5. Draw the block diagram representation of a generalized feedback control system. Identify the various system components and state clearly the function performed by each component.

6. Define the following terms in reference to control system engineering:
   (**a**) command input and reference input
   (**b**) actual signal and feedback signal
   (**c**) controlled variable and controlled medium

7. The direction-control system for a guided missile operates as follows:

   "The relative directions of the missile and target are measured using a gyroscope in the missile and the error is fed to a controller which operates a servomotor to deflect the rocket thus altering the missile path."

8. Identify the components, input and output and describe the operation of a biological control system consisting of a human being reaching the push button of an electric bell.

9. Explain the operation and identify the pertinent quantities and components of an automatic domestic, refrigeration system. Which component or components comprise the plant, the controller and feedback?

10. Suggest a simple control system that automatically turns on a room lamp at dusk and turns it off in daylight. Draw the schematics and block diagram of the suggested control system.

11. Explain the operation of ordinary traffic signals which control automobile traffic at roadway intersections.

    Do they constitute an open-loop or closed-loop control system? How can traffic be controlled more effectively?

12. A control system contains an input transducer, output transducer, comparator, error amplifier, and actuator.

Draw a labeled block diagram showing these components arranged for the control of the load, and describe the function of each component.

13. Propose a control system to fill a container with water after it is emptied through a stop cock at the bottom.

   The system must automatically shut off the water when the container is filled. Draw the block diagram of the proposed system. Which component or components comprise the plant, the controller, and the feedback?

14. Describe a typical closed-loop control system that can be used in order to control the following processes:
   (*i*)   the speed of a steam engine,
   (*ii*) the pressure in a furnace,
   (*iii*) the temperature of water being heated by steam, and
   (*iv*) the speed of an automobile vehicle.

   Draw the block diagram of the arrangement and mention the use of feedback in the application.

15. Explain briefly the difference between:
   (*i*)   open-loop and closed-loop control systems,
   (*ii*) positive and negative feedback,
   (*iii*) servomechanism, process control, and regulator.

16. Distinguish between manual and automatic control systems and list some of the engineering situations where automatic control becomes obligatory.

17. Define a process control system. Sketch the schematics of a shower bath that operates by a manual closed-loop control arrangement. Identify the functions of each part of the system and explain how control is achieved.

18. "The description of a process control involves three acts identified as measurement, evaluation and control."

   Comment upon the validity of this statement.

B. Fill in the blanks with appropriate word(s)

   (*i*) In an open-loop system, there is no connection between the _____ from the system and _____ to the system.
   (*ii*) The operation of an automatic washing machine is _____ dependent and represents an _____ loop control system.

(*iii*) Closed-loop control systems are generally referred to as _____ control systems.

(*iv*) All systems operated by preset timing mechanism are _____ loop.

(*v*) The prime objective of any control system is to maintain the _____ variable at the reference or set point.

(*vi*) The particular variable to be regulated is known as _____ variable, and the desired condition of the controlled variable is the _____.

(*vii*) The controller acts on the _____ signal and produces an output which actuates the _____ element.

(*viii*) A controller system that has no connection between the output and the input is known as an _____ control system, whereas an error actuated control system is a _____ control system.

(*ix*) The _____ determines the difference between the reference or setpoint and the measured variable.
This difference is the _____ or _____ signal.

(*x*) The _____ is the channel of information that leads to a change in the _____ signal in order to bring the _____ closer to the required.

(*xi*) The systems having both analog and digital components are called _____ systems.

Answers:
(*i*) output, input; (*ii*) time, open; (*iii*) feedback; (*iv*) open; (*v*) controlled; (*vi*) controlled reference or set point; (*vii*) error, final control; (*viii*) open-loop, closed-loop; (*ix*) controller/error actuating; (*x*) feedback, actuating, output; (*xi*) hybrid.

C. Multiple choice questions:

1. All of the following statements about open-loop control system are correct except:
   (*a*) the control action is independent of the output
   (*b*) the system continuously corrects any error that may exist
   (*c*) the system is convenient when the controlled variable is difficult to measure or it is economically not feasible
   (*d*) the system is easier to build, is quite stable but requires careful calibration for accurate results

2. Which of the following aspects does not pertain to a closed-loop control system.
   (*a*) there is a comparison between the actual arid the desired value of the variable
   (*b*) the system continually corrects any error that may be caused by internal and external disturbances

(**c**)   the system is difficult to build, install, maintain but has a possibility of optimization

(**d**)   the system operates fast, is more reliable and stable.

3.   Examine the following statements and identify which of these are correct:

(**i**)   A servomechanism is a control system in which the controlled variable is a mechanical position or its time derivative.

(**ii**)   A regulator is a feedback control system in which the output is maintained at a preset value irrespective of external load on the plant.

(**iii**) A process control refers to the control of such parameters as level, flow, pressure, temperature, and humidity of a process variable.

Examine the above statements and identify which of these are correct:

(**a**)   (**i**) and (**ii**)   (**b**)   (**i**) and (**iii**)
(**c**)   (**ii**) and (**iii**)   (**d**)   (**i**), (**ii**) and (**iii**)

4.   Match the following

1. Process/Plant	(**a**) a unit that is required to generate the appropriate control signal applied to plant.
2. Command	(**b**) a signal established as a standard of comparison for feedback control system by virtue of its relation to command.
3. Control element	(**c**) a body of which a particular quantity or condition is to be controlled.
4. Setpoint	(**d**) an input that is established or varied by some means external to and independent of the feedback control system.

5.   Identify the false statement in the context of closed-loop control system

(**a**)   the command input is the input to the system and it is independent of the actual output of the system

(**b**)   the actuating signal means the results of the reference input signal and the feedback signal

(**c**)   the controller stands for a component that controls the output of the system

(**d**)   the term "servomechanism" is used for a system where the output is a position or its time derivative

6.   Identify the wrong statement

(**a**)   The traffic controlled by a policeman is an open-loop control system

(**b**)   The steering of a car by a person is a closed-loop feedback control system (**c**)   The radar system in an anti-craft radar tracking system, functions to sense the accurate position and velocity of the target plane

(**d**)   The term "servomechanism" refers to the control system when the output is position/velocity or acceleration

7.   The following features have been made in the context of a sequence control:

(**i**)   the finish of one action initiates the start of the next

(**ii**)   the acts take place in a certain fixed sequence

(**iii**) there is a comparison between the actual and desired value of the variable

Which of these aspects is correct?

(**a**)   (**i**) and (**ii**)   (**b**) (**ii**) and (**iii**)
(**c**)   (**i**) and (**iii**)   (**d**) (**i**), (**ii**) and (**ii**)

Answers:

1. (*b*)

2. (*d*) The closed-loop systems are less stable

3. (*d*)

4. $(1-c), (2-d), (3-a), (4-b)$

5. (*c*) The controller stands for a component that produces the control signal.

6. (*a*) it is a closed system.

7. (*a*) A sequence control is a special type of open-loop system; there is no comparison of desired and actual value.

CHAPTER 9

MECHATRONICS SYSTEMS: BASIC CONCEPTS AND APPLICATIONS

9.1. MECHATRONICS: WHAT IS IT?

Mechatronics → Mechanics + Electronics

Mechatronics is a fascinating branch of engineering science that has initially been a combination of mechanics and electronics. With the advancement of technology, Mechatronics became broad-based covering mechanical, electrical, electronics, software engineering, communication, control, and artificial intelligence. Essentially mechatronics is

- a broad term that integrates/unites principles of mechanics, electronics, and computing (frequently using microcontrollers to generate a simpler, economical, and reliable system).

- the synergistic integration of mechanical engineering with electronics and intelligent control algorithms in the design and manufacture of products processes.

The term "synergistic" implies the interaction of two or more disciplines to produce a combined effect greater than the sum of their separate effects. As such, a mechatronics engineer would study definitive portions of mechanical engineering, electrical/electronics engineering, computer engineering, and control engineering.

The different fields which make up mechatronics have been indicated in Figure 9.1. Mechatronics is treated as a modern buzzword synonymous with automation, robotics, and electromechanical systems.

The notable examples of mechatronic systems are:

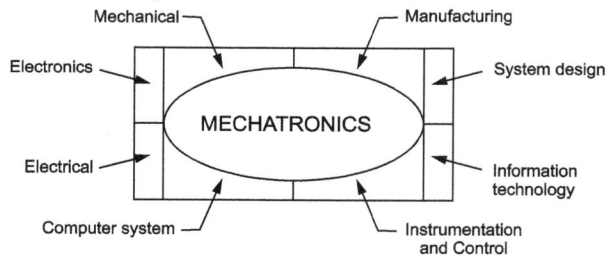

FIGURE 9.1 Constituent disciplines of a mechatronics system.

- digitally controlled combustion engines,

- machine tools with self-adaptive tools,

- contact-free magnetic bearings,

- automated guided vehicles, and

- robots.

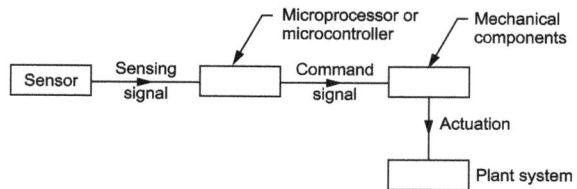

FIGURE 9.2 Components of mechatronics system.

Physically a mechatronics system consists of four prime components viz., sensors, actuators, controllers, and mechanical components arranged as shown in Figure 9.2.

A typical mechatronic system picks up signals from the environment, processes them to generate output signals, and transforms them into forces, motions and actions.

9.2. ORIGIN AND EVOLUTION

The term "mechatronics" was coined by Tetsure Mori and was used to describe a philosophy adopted in designing subsystems of electromechanical products. The term was a trademark of Yasakowa Electric Corporation from 1971 to 1982. Since those early days, there have been rapid advances in technology available to manufacturing industries, and the term "mechatronics" is now firmly established and is being freely used by industries around the world.

The different stages in the evolution of the discipline/field of mechatronics have been:

1. **Primary level:** The primary stage is the basic control level that covers input/output devices. There is the integration of electrical signaling with mechanical action using sensors and actuators; for example, relay switches and electrically controlled valves.

2. **Secondary level:** At this stage of development, there has been the integration of microelectronics into electrically controlled devices such as in a cassette tape player.

   The autofocus cameras and washing machines too are examples of such systems; often called stand-alone systems.

3. **Tertiary level:** The tertiary level mechatronics systems, often called smart systems, are of improved quality and sophistication. This results through the incorporation of advanced feedback functions into the control strategy that uses microelectronics, microprocessors, and other application-specific integrated circuits.

   Industrial robots are typical examples of tertiary mechatronics systems.

4. **Quaternary level:** There is an attempt in the quaternary level mechatronics systems to incorporate intelligent and fault detection isolation with the objective of enhancing smartness. There is a linkage of major subsystems such as machining centers, robots for part handling, automated inspection centers, etc., in the large factory systems. Further, intellectual capabilities of the human operator are captured through the concepts of artificial neural networks and fuzzy logic. A humanoid robot is one such mechatronics system.

9.3. AVIONICS, BIONICS, AND AUTOTRONICS

Avionics is a variant of a mechatronics system that is coined from a blend of **avi**ation and elect**ronics**. Every modern aircraft, spacecraft, and artificial satellite uses electronic systems of varying types to perform a range of functions pertinent to their purpose and mission. Such systems may include:

- Engine control and flight control systems in order to reduce pilot error and workload at landing or take off.

- Fuel control and monitoring system to report fuel remaining on board.

- Navigation and communication systems. Air navigation is the determination of position and direction on or above the surface of the earth.

- Weather and anti-collision systems. The transport aircraft uses a traffic alert and collision avoidance system which can detect the location of nearby aircraft and provide instructions for avoiding mid-air collision. The weather detectors give information on lightning and turbulence.

- Flight recorders (black boxes): These store flight information and audio from the cockpit. They provide information on control settings and other parameters when there is an unfortunate incident of a crash.

- Display and management of systems fitted to the aircraft to perform individual tasks.

There is also the integration of multiple functions to improve performance, simplify maintenance, and contain costs.

Bionics is a variant of the mechatronics system that has been coined by Jack Steel in 1958 from the subjects of **bio**logy and electro**nics**. This biologically inspired engineering pertains to the biological methods and systems and is made use of in the study and design of engineering systems. Bionics studies the mechanical and electronic systems that function like living organisms or parts of a living organism both internal and external.

Examples of bionics in engineering include:

- hulls of boat imitating the skin of dolphins,

- sonar, radar, and medical ultrasound imaging using sound waves and echoes to determine where the objects are in space. This is analogous to sending out sound waves (from the mouth or nose) by the bats to find food in the dark. When the sound waves hit the target (food), echoes are produced,

- producing artificial neurons, artificial neural networks, and swarm intelligence in the field of computer science,

- making artificial hands with sensors in the fingertips that monitor and adjust the strength of the hand's grip,

- development of dirt and water repellant paints/coatings from the observation that practically nothing sticks to the surface of the lotus flower plant,

- changing the shape of aircraft wings according to speed and duration of flight inspired by different bird species that have differently shaped wings according to the speed at which they fly,

- creation of new nanosensors to detect explosives inspired by wing structure of butterflies,

- development of smart clothing that adapts to changing temperatures. The smart fabric opens up when the weather is warm and sweating and shuts tight when cold. This development came from a study of pinecones (a type of plant), and

- application of the ways the animals move in the design of robots.

The above examples clearly indicate that bionics treats nature itself as a database of solutions that already exist. Further, it will be appropriate to mention how the subjects of bionics, cybernetics, and bioengineering differ from each other.

Bionics explores new ideas for building mechanical and electronics systems. Cybernetics focuses on seeking an explanation of the behavior of living organisms. Bioengineering uses living things to perform industrial tasks. For example, using bacteria in paper batteries to supply electrical energy would be an advancement in bioengineering and not bionics.

Autotronics is a variant of a mechatronics system that has been coined by blending automobile and electronics:

$$\textbf{auto}\text{mobile} + \text{elec}\textbf{tronics} \rightarrow \text{autotronics}$$

That makes autotronics flexible engineering that serves to develop and understand conversion principles in the design, construction, and working of mechanical systems and electronics systems combined with the advancement of sensors and microcontrollers. Analysts estimate that more than 80% of all automobile innovations now stem from electronics. Since 1970, there have been progressive changes in motor vehicle technology with many of the functions evolving from mechanical to becoming electronics and controlled by computers. Due to the embedding of electronics in automobile operations, there has been improvement in:

- fuel injection and engine ignition,
- steering, transmission, and suspension,
- antilock braking,
- navigation and general positioning system (GPS),
- audio and video entertainment system,
- safety control and security alarms,
- collision avoidance systems, and
- autolocking system and key-less entry.

All this has made driving more comfortable, more secure and more efficient, and has turned driving into a pleasurable experience.

9.4. APPLICATIONS OF MECHATRONICS

Mechatronics is one of the fastest developing fields with wide areas of application in marketing, design, and manufacturing. Marketing refers to information analysis related to the identification of uses needs and formulation of the product specification. The manufacturing domain looks into process development, production planning, material handling, inspection, and quality control. The current technological designs are highly complex and requires integration of knowledge from different interdisciplinary subjects.

Mechatronics finds application in:

- Industries where it is necessary to design and maintain automatic equipment.

- Large manufacturing units involved in high volume production. Automation and industrial robots perform consistently and quickly and that enables manufacturers to keep with demand while reducing costs.

- Many medical applications such as magnetic resonance, ultrasonic probes, and arthroscopic devices are mechatronics. Surgical robots have been developed for eye surgery, targeting lung cancer, knee surgery, and laparoscopy. Such examinations and treatments are less invasive and that leads to a fast recovery and low risk of infection.

- Computer machine tools like CNC milling machines, CNC water jets, and CNC plasma cutters.

- Computer-aided and integrated manufacturing systems.

- Consumer products, industrial goods packaging.

- Transportation and vehicular systems, automotive engineering equipment in the design of antilock brakes and stabilizers, and air-bag inflation. This has made driving safe and less accident-prone.

- Home applications such as automatic air-conditioning systems, security systems, washing machines, dishwashing, etc.

- Intelligent measuring devices like calibration, measuring, and testing of sensors.

- Automatically guided and unmanned aerial vehicles.

In recent times, a greater use of mechatronics is evident in manufacturing, mining, aviation, robotics, defence, and transport. Mechatronics engineers may program robots, design telecommunication systems, or develop nanotechnology.

9.5. ADVANTAGES AND DISADVANTAGES

Mechatronics is basically the application of various technical fields to have reliable product design and manufacturing solutions. The main advantages resulting from mechatronics systems are:

- enhancement of functionality and features,
- easy design of processes and products, improved design time and product size,
- rapid set up and cost-effective operation of manufacturing facilities,
- optimizing performance and quality,
- increased effectiveness and productivity,
- more user-friendly and safer to use,
- improved and less expensive controls,
- little interference from operators, and
- high level of integration.

There is better control of precision, position, speed, flow rate, and other variables due to the use of microcontrollers, software, and artificial intelligence in mechatronics.

However, the mechatronics systems have the following disadvantages too

- high initial cost,
- complicated design, and
- system complex and so difficult repair and maintenance.

9.6. SENSORS AND TRANSDUCERS

A generalized measurement system consists of two components, (*i*) sensing element which responds directly by reacting to the measurand and (*ii*) transducing element which is responsible for conversion of the measurand into the analog driving signal. The sensing element may also serve to transduce the measurand and put it into a more convenient form. The tie unit is then called *detector-transducer.*

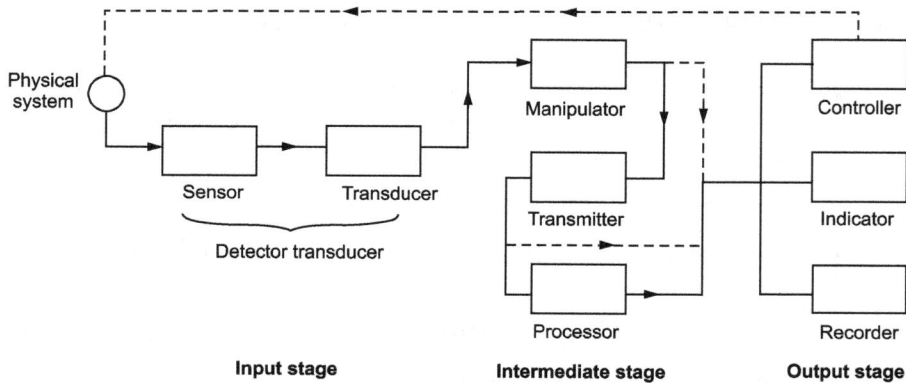

FIGURE 9.3 Generalized measurement system.

For instance:

- In the ordinary dial indicator the contacting spindle acts as a detector or sensing element for displacement. It simply performs the function of a detector and nothing else.

- The function of the bourdon tube of a pressure gauge is two-fold: firstly to sense the pressure and secondly to give the resulting effect or the output in the form of displacement. The tube then works as a detector-transducer.

- In a compressive load cell, the platform detects the force and gives an output in the form of deflection. This deflection may be further converted into an electrical output by-strain gauges mounted on the load cell. The strain gauge is called a secondary transducer because of the second translation.

Different types of sensors and transducers are available for the measurement of one particular quantity and the choice of a suitable unit depends upon the static and dynamic performance characteristics.

9.7. MECHANICAL DETECTOR-TRANSDUCER ELEMENTS

Transducers make use of different principles to convert the various quantities being measured into their analogs. Table 9.1 provides a list of the mechanical detector transducers, together with the functions they perform.

TABLE 9.1 Mechanical detector-transducer elements.

Detector-Transducer	Operation
Contacting spindle	Displacement to displacement
Elastic member	
• Bourdon tube (circumferential and twisted)	
• Bellows	Fluid pressure to displacement
• Diaphragms (flat, dished and corrugated)	
• Springs	Force to displacement
• Proving ring	
Concentrated mass	
• Seismic mass	Acceleration and vibration amplitude to displacement
• Pendulum	Force to displacement
• Manometric liquid	Fluid pressure to displacement
Thermal	
• Liquid-in-glass and filled system thermometers	Temperature to displacement
• Bimetallic rods	
• Temperature sticks	Temperature to phase change
• Thermocouples and thermopiles	Temperature to electric voltage
• Thermistors and resistance thermometers	Temperature to resistance change
Hydropneumatic	
(a) Static	
• Float	Liquid level to displacement
• Hydrometer	The specific gravity of liquid to displacement
(b) Dynamic	
• Orifice plate, venturi, and nozzle	Velocity to pressure
• Pitot tube	
• Vanes	Velocity to force
• Turbines	Linear to angular velocity

9.7.1. Elastic Elements

These units are frequently employed to furnish an indication of the magnitude of applied pressure/force through a displacement measurement. Operation of the elastic elements is based on one or a combination of the following three acts:

i.   compression that tends to force the molecules of the solid together,

ii.   tension that tends to force the molecules farther apart, and

iii. torsion that tends to twist the solids.

A force may be determined by applying it to an elastic element and measuring the resulting elastic deformation. The devices commonly used include springs, the proving ring, and the torsion rods.

- In a spring-type indicating scale, unknown weight applied to the free end of the spring causes displacement which is indicated by the pointer. A tape-and-drum movement can be employed to operate the pointer.

- The proving ring (stress ring) is a ring of known physical dimensions and mechanical properties. An external tensile or compressive force applied across the ring diameter causes distortion which is proportional to that force. The distortion is measured by means of a dial gauge, a sensitive micrometer, or a strain gauge. The proving ring is widely used as a calibration standard for large tensile testing machines.

- A torsion bar would twist in proportion to the applied torque and the resulting angular deformation can be used as a measure of the torque.

- Most pressure measuring devices use either a bourdon tube, a bellow, or a diaphragm. The action of these elements is based on elastic deformation brought about by the force resulting from pressure summation.

9.7.2. Mass Sensing Elements

The inertia of a concentrated mass provides another mechanical detector-transducer. The principle is employed in vibration pickups and accelerometers and serves to measure the characteristics of dynamic motion (displacement velocity, acceleration, and frequency) through the application of Newton's second law of motion.

- Any simple mechanical vibrating member such as a pendulum would serve as a time or frequency transducer, chopping the passage of time into discrete bits.

- The pressure measurement by manometers is also based on the principle of mass displacement.

9.7.3. Thermal Detectors

These units sense the temperature of a system by indicating some change in the property of a material which varies with temperature; properties that are so used include:

i. Expansion of solids and liquids: bimetallic thermometers, liquid-in-glass, and the filled system thermometers.

ii. Thermoelectric property of metals and alloys: thermocouple and thermopiles.

iii. Electrical resistance of metals and semiconductors: resistance thermometers and thermistors.

iv. Radiating ability: total radiation and optical pyrometers.

9.7.4. Hydropneumatic Elements

A simple float and a hydrometer are the two common examples of the hydropneumatic sensors applied to static conditions.

- The float converts the fluid level into displacement but makes no allowance for change of the density of the supporting liquid.

- A hydrometer senses specific gravity and uses the depth of immersion as a means for detecting variations in the specific gravity of the supporting liquid.

- The obstruction head flow meters (orifice plate, venturi, and nozzle) provide flow information in the form of pressure change as a result of energy transformation. The obstruction placed in the path of the fluid results in a change of fluid pressure which is dependent on the rate of flow. The difference in pressure before and after the obstruction is measured by means of a differential pressure gauge and is correlated to flow rate

- Aero-or-hydrodynamic principles are also applied to determine the fluid velocity. A point tube measures the pressures resulting from the total-flow rate rather than the change of rate. Flow rate is also sensed by vanes in the form of airfoil and turbine wheels. Flow rate a inferred from the vane displacement or the angular velocity of the turbine wheel.

9.8. ELECTRICAL TRANSDUCERS

Nowadays electrical/electronic techniques of measurement are being increasingly applied to measurements in many fields other than in electrical engineering. The advantages of such methods over others are:

- more compact instrumentation,

- good frequency and transient response,

- feasibility of remote indication and recording,

- possibility of mathematical processing of signals like summation, integration, etc.,

- minimum of friction and mass-inertia effects,

- possibility of non-contact measurements,

- less power consumption and less loading on the system to be measured, and

- amplification greater than that produced by a mechanical contrivance.

The different electrical phenomena employed in transduction elements of electrical transducers are listed in Table 9.2 along with their typical applications.

An examination of Tables 9.1 and 9.2 would reveal that whereas displacement is output from the mechanical and hydropneumatic devices, it is input to the electrical devices. This aspect results in a very workable combination with the mechanical device serving as detector-transducer and an electrical device serving as the electromechanical transducer (more often as transducer only) with the sole object of converting the linear or rotary displacement of the mechanical system into an electrical output. Transducers are also known as gauges, pickups, and signal generators. Most of the pickups have two basic elements in essential, viz., an actuating device and the transducing element. Some of the typical transducer actuating mechanisms are shown in, Figure 9.4.

TABLE 9.2 Electrical transducers.

Operating	Principle	Applications
Externally Powered (Passive) Transducers		
Resistance		
• Potentiometric device	Resistance in a potentiometric or a bridge circuit varies with change in the position of a slider by an externally applied load	Displacement and pressure
• Resistance strain gauge	Resistance of a wire or semiconductor is changed by elongation or compression due to an external load	Force, torque, and displacement
• Resistance thermometer	Resistance of a pure metal wire (with a positive temperature coefficient of resistance) varies with temperature	Temperature and radiation heat
• Thermistor	Resistance of certain metal oxides (with a negative tempera-true coefficient of resistance) varies with temperature	Temperature
• Pirani gauge	Resistance of a heating element changes due to convection cooling by the stream of gas flow	Gas flow and gas temperature
• Resistance hygrometer	Resistance of a conductive strip changes with moisture content	Relative humidity
• Photoconductive cell	Resistance of the cell as a circuit element varies with incident light	Photosensitive relay

(Continued)

Operating	Principle	Applications
Capacitive		
• Variable capacitance pressure gauge	Variation in capacitance due to change in distance between two parallel plates by an externally applied force	Displacement and pressure
• Dielectric gauge	Variation in capacitance by a change in the dielectric between the plates	Liquid level and thickness
• Capacitor microphone	Variation in capacitance between a fixed plate and a movable diaphragm due to sound pressure	Speech, music, and noise
Inductive		
• Magnetic circuit transducer	Variation in self-inductance or mutual inductance of an AC excited coil by changes in the magnetic circuit	Pressure and displacement
• Reluctance pickup	Variation in a reluctance of the magnetic circuit by changing the position of the iron core of a coil	Pressure, displacement, vibration, and position
• Differential transformer	Variation in the differential voltage of two secondary windings by positioning the magnetic core through an externally applied force	Pressure, force, displacement, and position
• Eddy current gauge	Variation in inductance of a coil by the proximity of an eddy current plate	Displacement and thickness
• Magnetostriction gauge	Variation in magnetic properties by the measurand	Force, pressure, and sound

(Continued)

Operating	Principle	Applications
Voltage and current		
• Photo emissive cell	Electron emission due to incident radiation upon the photoemissive surface	Light and radiation
• Photomultiplier tube	Secondary electron emission due to incident radiation on a photosensitive cathode	Light and radiation, photosensitive relays
• Ionization chamber	Electron flow induced by ionization of gas due to radioactive radiation	Particle counting and radiation
• Hall effect pickup	Setting up of potential difference across a semiconductor plate when there is an interaction of magnetic flux with an applied current	Magnetic flux and current
Self-generating (Active) Transducers (No External Power)		
• Thermocouple and thermopile	Generation of an emf across the junction of two dissimilar metals of semiconductors when one junction is heated	Temperature, heat flow, and radiation
• Piezoelectric pickup	Generation of an emf when an external force is applied to certain crystalline materials (e.g., quartz)	Pressure changes acceleration, vibration, and sound
• Photovoltaic cell	Generation of voltage in a semiconductor junction device when the cell is stimulated by the radiant energy	Solar cell and light meter
• Moving-coil generation	Generation of voltage due to motion of a coil in a magnetic field	Velocity and vibration

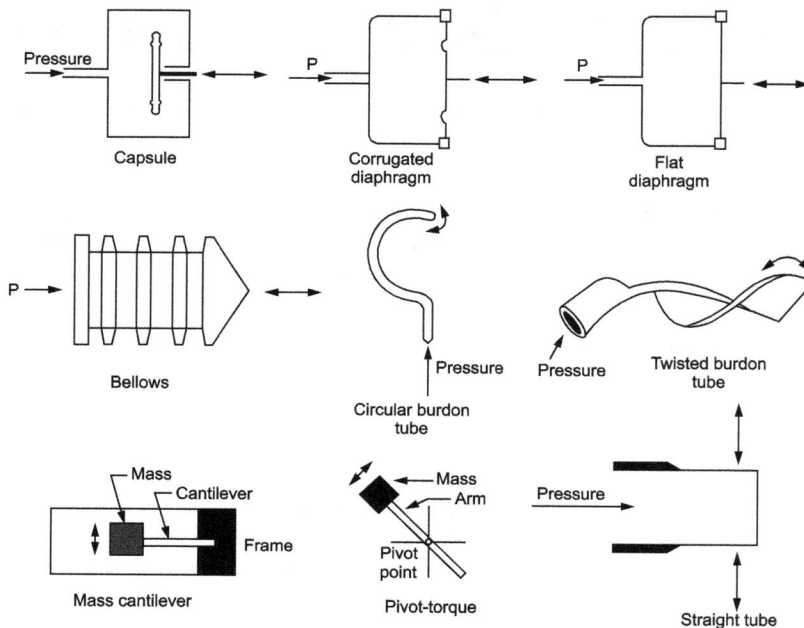

FIGURE 9.4 Transducer actuating mechanisms.

9.9. TRANSDUCER CLASSIFICATION AND DESCRIPTION

The transducers may be classified on the basis of their application (type and nature of measurand), method of energy conversion, nature of signal output, kind of sensing element mechanical or nonmechanical, and according to whether they are self-generating (active or externally powered (passive).

Self-Generating and Externally Powered Units

Self-generating transducers develop their own voltage or current. The energy required for this is absorbed from the physical quantity being measured. Examples are thermocouples and thermopiles, piezoelectric pickup, photovoltaic cell, etc.

Externally powered transducers derive the power required for energy conversion from an external power source. They may also absorb a little energy from the process variable being measured. Examples are resistance thermometers and thermistors, potentiometric devices, differential transformers, photoemissive cells, etc.

Input and Output Transducers

Input transducers convert a non-electrical quantity into an electrical signal (a strain gauge or photoelectric cell) and the output transducers convert the electrical signal back into a non-electrical quantity (movement of a pointer against a graduated scale). In between the input and output transducers, there is usually signal conditioning equipment (amplifier, filter, etc.).

With the fast-developing technology, there has been a rapid increase in the development and application of various types of transducers to convert all the measured quantities into their electrical analogs. The output electrical signal may be amplified, recorded, and processed in the instrumentation system.

Transducer Description

Information must be available about the following aspects while describing a particular transducer:

- the physical quantity or variable which is to be measured, that is, the measurand,

- the principle of operation of the transducer and where the output of the transducer originates,

- the sensing element which responds directly to the measurand,

- the built-in special features (if any), and

- the useful range, that is, the minimum and maximum values of the physical quantity the transducer can measure.

With regard to a DC tachometer (an instrument for measuring angular speed), the above-mentioned aspects are:

- angular speed in rpm is the measurand,

- principle of operation is electromagnetic,

- AC generator is the sensing element,

- commutator is the special built-in feature; it transforms AC voltage into DC voltage output, and

- minimum and maximum values of speed are 0 and 2000 rpm, that is, the useful range is 0–2000 rpm.

The DC tachometer would then be specified as: "0–2000 rpm, DC output, commutator type electromagnetic speed transducer."

9.10. VARIABLE RESISTANCE TRANSDUCERS

In terms of physical quantities, the equation for the electrical resistance of a metal conductor is

$$R = \rho \frac{1}{A}$$

where R is the resistance (ohms), ρ is the conductor resistivity or specific resistance (ohm-cm), l is the physical length (cm) and A is the uniform cross-sectional area of the resistor (cm^2). Any method of varying one of these quantities can be the design basis of an electrical transducer. In the variable resistance transducer, an indication of a measured physical quantity is given by a change in the resistance.

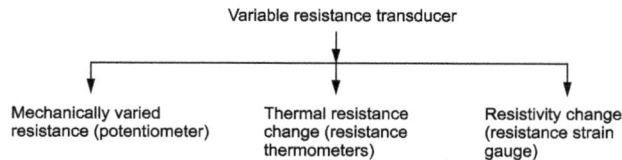

Variable resistance transducer

Mechanically varied resistance (potentiometer)   Thermal resistance change (resistance thermometers)   Resistivity change (resistance strain gauge)

Further, with some devices, resistance changes with light intensity (photoconductive effect) while with others, resistance changes on exposure to a magnetic field (magnetoresistive effect).

The variable resistance transducers are passive and they rely on an external excitation voltage for their operation. However, they are straightforward in design, simple, and easy to use.

9.10.1. Linear and Angular Motion Potentiometers

These potentiometers convert the linear motion (or the angular motion of a rotating shaft) into changes in resistance. Basically, a resistive potentiometer (or "pot") is a variable resistor whose resistance is varied by the movement of a slider over a resistance element (Figure 9.5a,b). Translatory devices have strokes from 2 mm to 50 cm, while rotational ones have a full scale ranging from 10° to as much 60 full turns.

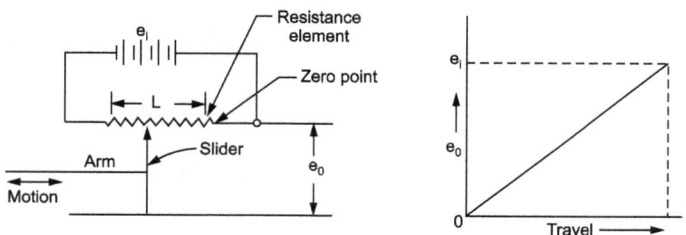

FIGURE 9.5 (a) Linear motion potentiometer schematics.

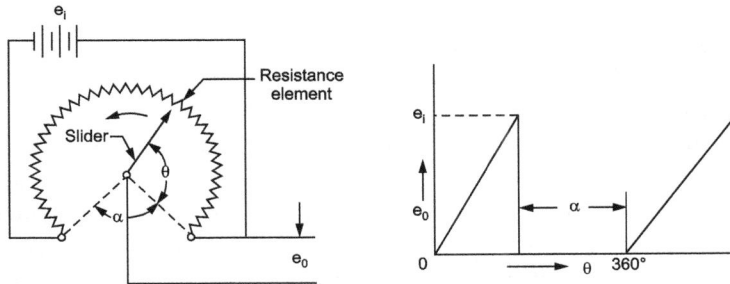

FIGURE 9.5 (b) Rotary motion potentiometer schematics.

The resistance elements in common use are wire wounds because they give sufficiently high resistance value in small space. The characteristics of the resistance wire are:

- precision drew wire with a diameter of about 25–50 μm, and wound over a cylindrical or a flat mandrel of ceramic, glass, anodized aluminum,

- resistivity of the wire ranges from 0.4 μΩ-m to 1–3 μΩ-m and temperature coefficient varies from 0.002%/°C to 0.001%/°C. With these values, the device operates with appreciable constant sensitivity over a wide temperature range, and

- the wire is strong, ductile, and protected from surface corrosion by enameling or oxidation.

The materials commonly employed are the alloys of copper-nickel, nickel-chromium, and silver-palladium.

9.10.2. Resistance Thermometers and Thermistors

Metals such as platinum, copper, tungsten, and nickel become more resistant to the passage of electric current as they become hotter. Their resistance increases with growth in temperature, that is, they have a positive temperature coefficient of resistance. For many practical purposes and within a narrow temperature range, the metal resistance thermometers depend upon the following relationship between metal resistance and temperature

$$R_1 = R_0 \left[1 + \alpha \left(t_1 - t_0\right)\right]$$

where R_0 is the resistance in ohms at the reference temperature (usually 0°C) and $\propto$ is the temperature coefficient of resistance in $°C_{-1}$. For precise temperature measurements, platinum is preferred because it is physically stable and has high electrical resistance characteristics because of accuracy, stability, and sensitivity, the platinum resistance thermometer has been used to define the international practical temperature scale from the boiling point of oxygen (−182.9°C) to the freezing point of antimony (630.5°C).

Thermistors are essentially semiconductors (sintered mixture of metallic oxides such as manganese, copper, iron, and uranium) which exhibit large non-linear resistance changes with temperature variation, that is, they have a high negative temperature coefficient, Thermistors are normally made in the form of beads, disks, washers, rods and can be made as small as 1 mm. Thermistors have the advantages of high sensitivity, very small size, fast thermal response, fairly low cost and, easy adaptability to electrical readout devices.

Thermistors and metal resistance thermometers find extensive application as temperature detecting elements for the purpose of measurement and control.

9.10.3. Resistance Strain Gauges

The operation of these gauges is based on the principle that the electrical resistance of a conductor changes when the resistance element is strained by an external force. Under no-load conditions, the gauge is bonded or cemented directly onto the surface of the body or structure which is being examined. The different forms of bonded strain gauges are:

i.   fine wire gauges cemented to a paper backing,

ii. photoetched grids of conducting foil on an epoxy-resin backing, and

iii. a single semiconductor filament mounted on an epoxy-resin backing with copper or nickel leads.

The wire grid participates in the subsequent deformations both in the specimen and the resistance element. A tensile or positive strain increases the resistance while compressive or negative strain decreases resistance. Resistance gauges made up of single elements measure strain in one direction only. A combination of elements, that is, rosettes will however permit simultaneous measurements in more than one direction.

The strain gauge is a versatile device and finds application in the measurement of different variables such as load, force, thrust, pressure, torque and displacement, etc.

9.11. THERMOELECTRIC TRANSDUCERS

When two dissimilar metal conductors are joined at the ends and the two junctions are kept at different temperatures, a small emf is produced in the circuit. The magnitude of this voltage depends upon the material of conductors and the temperature difference between the two junctions. This thermoelectric effect is used in thermocouples for the measurement of temperature. Any combination of metals may be used. Two commonly employed combinations are iron and constantan (an alloy of copper and nickel), chromel (an alloy of chromium and nickel), and alumel (an alloy of aluminum and nickel).

9.12. VARIABLE INDUCTANCE TRANSDUCERS

These transducers are based on a change in the magnetic characteristics of an electrical circuit in response to a measurand which may be displacement, velocity, acceleration, etc. Before discussing these transducers, it is pertinent to become familiar with the following terms and definitions:

- *Inductance or self-inductance:* When a varying current is made to pass through a coil, an induced counter emf results due to magnetic flux intersecting the turns of the coil. This effect causes resistance to flow of current and is called inductance or self-inductance.

- *Mutual inductance:* The term refers to the setup of an emf in a coil or in a circuit element due to varying flux fields in neighboring coil or circuit element.

- *Reluctance:* The term refers to that characteristic of a magnetic circuit that determines the total magnetic flux when a given magnetomotive force is applied. Reciprocal of reluctance is termed permeance.

- *Permeability:* It is defined as the ratio of the number of flux lines set up in a coil under given conditions to the number of magnetic flux lines that would occur if the path were air (other conditions remaining unchanged).

Variable inductance transducers have the advantages of freedom from mechanical hysteresis, good response to both static and dynamic measurements, continuous resolution, and high output. The performance is, however, adversely affected by the external magnetic fields. Variable inductance transducers can be classified into self-generating (active) and externally powered (passive) units.

9.12.1. Active Units

Active units in which the output signal is generated because of the relative motion between a conductor and magnetic field, and without the supply of energy from an external source.

The operation of self-generating inductance transducers depends upon the following well-known principles:

i. When a conductor is caused to move with a velocity through a magnetic field in a plane perpendicular to the magnetic field, an emf is generated along the conductor [Figure 9.6(a)].

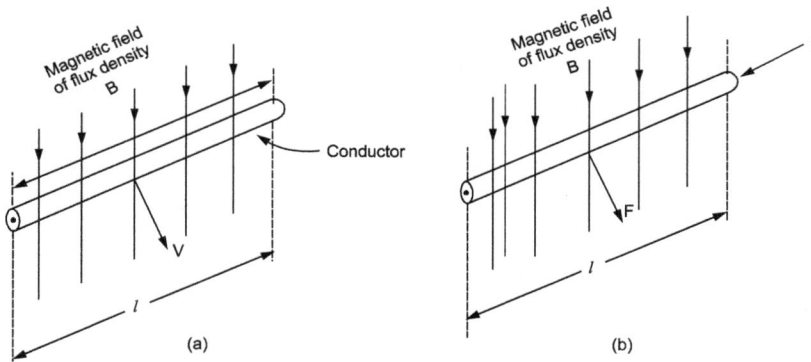

FIGURE 9.6 Flux cutting: operating principle of self-generating inductive transducers.

The relationship between the emf generated and the velocity is given by: $e = BlV$, where e is the generated emf, B is the flux density of the magnetic field, l is the conductor length and V is the conductor velocity. Evidently when B and l are maintained constant $e \propto V$. The emf generated along the conductor is then a measure of the velocity of the conductor. Transducers based upon this principle can be used for measuring velocities and are frequently used in the measurement of angular speed, vibration, and fluid flow.

ii. When a conductor is placed in a magnetic field with its longitudinal axis at right angles to the lines of flux and a current is allowed to flow through the conductor, a mechanical force is generated. This force acts on the conductor in a direction perpendicular to the lines of flux and to the conductor (Figure 9.6b). The relationship between the force generated and the current is given by

$$F = B \, i \, l \tag{9.1}$$

where F is the generated force, B is the flux density of the magnetic field, l is the conductor length and i is the conductor current. Evidently when B and l are maintained constant, then $F \propto i$. The force generated is then a measure of the current flowing through the conductor. This aspect forms the basis for the working of most of the moving-coil and moving-magnet type measuring instruments.

Some examples of self-generating inductive transducers are shown in Figure 9.7.

In the ***electromagnetic type***, a coil is wound directly on a permanent core. When a plate of iron or other ferromagnetic material is moved with respect to the magnet, the flux field expands or collapses and a voltage is induced in the coil. Practical application of the device lies in the angular speed indication. When the pickup is placed near the teeth of a rotating gear, speed measurements can be made with great accuracy.

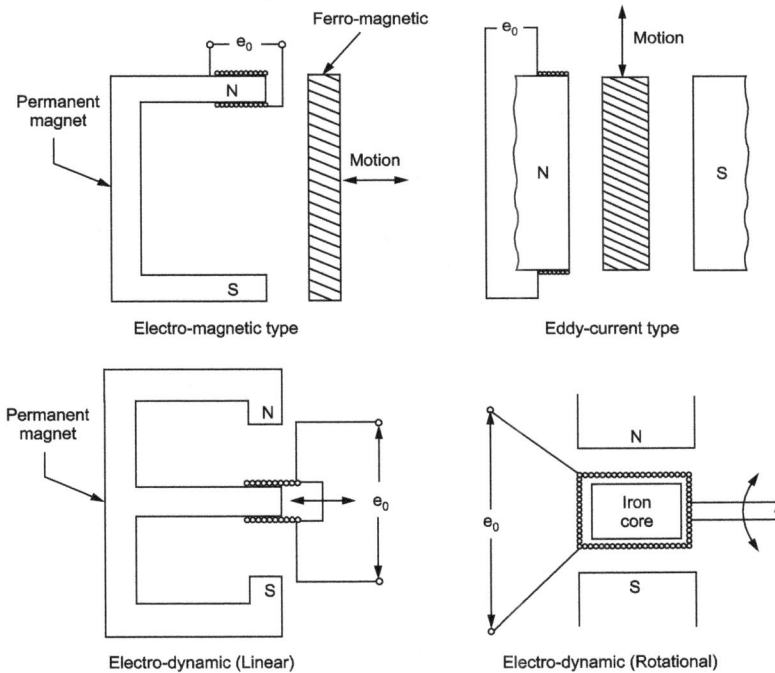

FIGURE 9.7 Self-generating variable inductance transducers.

In the ***electrodynamic type***, there is a movement of the coil or conductor within the field of a permanent magnet. The turns of the coil are perpendicular to the intersecting lines of force. The movement of the coil induces a voltage that at any moment is proportional to the velocity of the coil. The principle of electrodynamic transducers is used in the magnetic flow meters.

9.12.2. Passive Units

Passive units in which the motion of an object results in a change in the inductance of the coils of the transducer; energy is required to be supplied from an external source.

Passive type inductance transducers operate on the following aspects of flux linkage

i. When current i (amperes) passes through a coil having N turns and an air core (Figure 9.8), a magnetic flux ϕ (Weber), is generated

$$\phi \propto N i \text{ or } Ni = S\phi \qquad (9.2)$$

where S is called the reluctance of the coil. The reluctance is also prescribed by relation

$$S = \frac{1}{\mu_0 \mu_r A}$$

where A = cross-sectional area of magnetic circuit (m^2),
$\quad l$ = length of magnetic circuit (m),
$\quad \mu_0$ = permeability of free space = $4\pi \times 10^{-7}$ H/m, and
$\quad \mu_r$ = relatively permeability of the core of the coil. The value of μ_r depends on the core material, and for air $\mu_r = 1$.

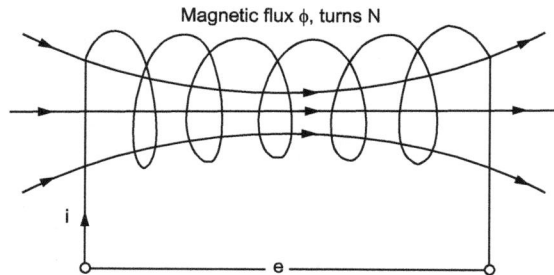

FIGURE 9.8 Flux linking: operating principle of passive inductive transducers.

ii. The coil inductance is a measure of the magnitude of magnetic flux and is defined as

$$L = \frac{N\phi}{i} \qquad (9.3)$$

where ϕ is the magnetic flux density.

Combining Eqs (9.1)–(9.3), we obtain

$$L = \frac{N^2 A \mu_0 \mu_r}{l} \text{ henrys} \qquad (9.4)$$

Evidently, the self-inductance of the coil is dependent upon the number of turns of coil, the geometrical configuration of the circuit, and the permeability of the core.

Variable inductance/reluctance transducers are constituted of magnetic field and core such that a gap exists between the core and the fixed coils. A change in the reluctance of the magnetic circuit by a mechanical input results in a similar change both in the inductance and inductive reactance of the coils. The change in inductance is then measured by suitable circuitry related to the value of mechanical input.

The reluctance of the magnetic circuit may be altered by affecting a change either in the air gap or in the amount/type of the core material. Transducers that make use of an air gap change are known as **reluctance type** and the transducers utilizing a variable core permeability change are referred to as **permeance type**.

Variable reluctance transducer: Figure 9.9 shows the variable reluctance transducer in which the variable air gap serves to alter the inductance of a single coil. The change in inductance may be calibrated in terms of the armature movement. The variable reluctance principle is particularly applicable to the measurement of dynamic quantities such as pressure, acceleration, force, displacement and angular position, etc.

Variable permeance transducer: Figure 9.10 illustrates the variable permeance transducer where the inductance of the coil is changed by varying the amount of core material.

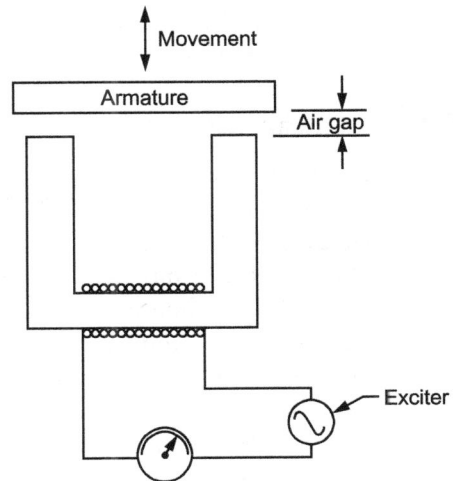

FIGURE 9.9 Self-inductance variable reluctance transducer.

The arrangement consists of a coil of many turns of wire wound on a tube of insulating material with a movable core of magnetic material.

As the coil is energized and the core enters the solenoid cell, the inductance of the coil increases in proportion to the amount of metal within the coil. A pickup of this type is used primarily for displacement, strain, and force measurement.

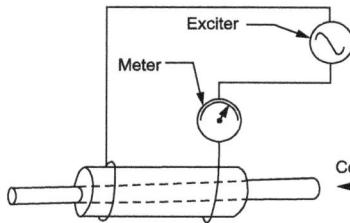

FIGURE 9.10 Self-inductance variable permeance transducer.

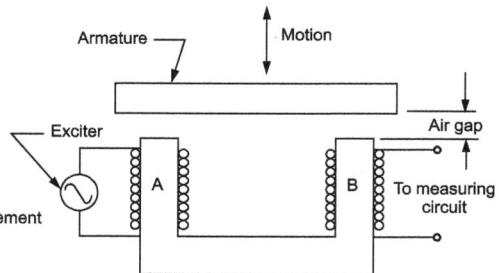

FIGURE 9.11 Mutual-inductance arrangement.

Figure 9.11 illustrates a form of a two-coil mutual inductance transducer. Coil A is the energizing coil, and coil B is the pickup coil. A change in the position of the armature by a mechanical input alters the air gap. This causes a change in the output from coil B which may be used as a measure of the armature displacement, that is, the mechanical input.

9.12.3. Linear Variable Differential Transformer (LVDT)

One of the most useful variable inductance transducers is the LVDT shown schematically in Figure 9.12. The device has one primary and two secondary windings with the magnetic core free to move inside the coils. The core is attached to the moving part on which the displacement measurements are to be made. When ac current is supplied to the primary winding, the magnetic flux generated by this coil is disturbed by the armature so that voltages are induced in the secondary coils. The secondary windings are symmetrically placed, are identical, and are connected in phase opposition so that emfs induced in them are opposite to each other. The net output from the transformer is then the difference between the voltages of the two secondary windings. The position of the magnetic core determines the flux linkages with each winding. When the core is placed centrally, equal but opposite emfs are induced in the secondary windings and zero output is recorded. This is termed as the balance point or null position.

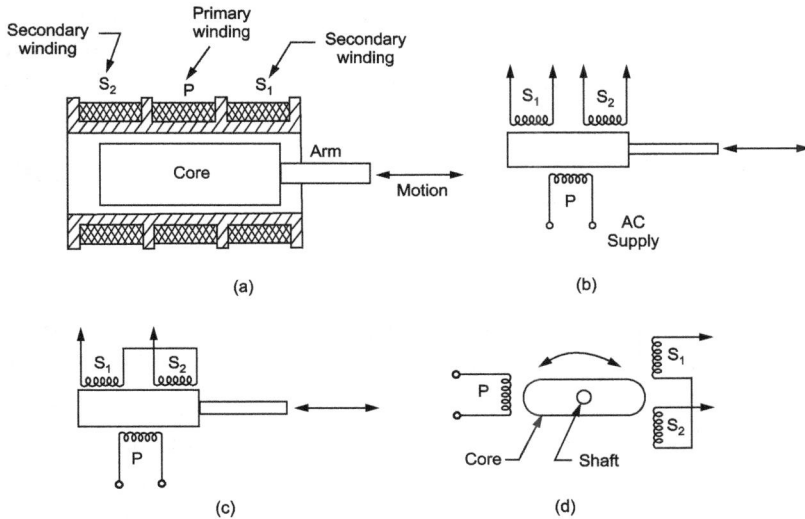

FIGURE 9.12 Variable differential transformer.

A variation in the position of the core from its null position produces an unbalance in the reactance of secondary windings to the primary windings. The voltage induced in the secondary winding toward which the core is displaced increases. A simultaneous decreased induced voltage results from the secondary coil. Thus, upon displacement of the armature, the result will be a voltage rise in one secondary coil and a decrease in the other. The asymmetry in the core position thus produces a differential voltage E_0 which varies linearly with change in the core position (Figure 9.13). Small residual voltages resulting from certain stray magnetic and capacitance effects may, however, not cancel and the output voltage may not necessarily become zero at the null position. Figure 9.13(*b*) represents an enlarged view of the residual voltage when the core is at the null position.

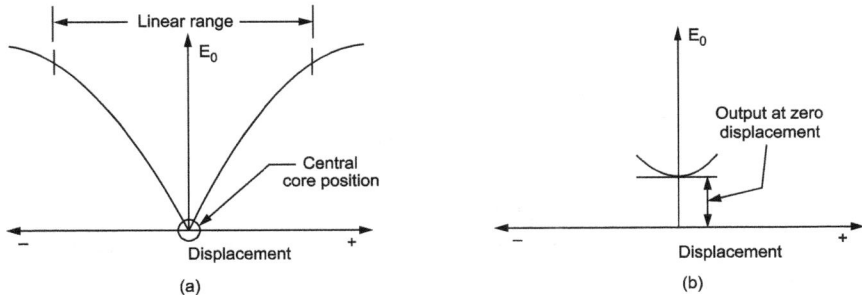

FIGURE 9.13 The output voltage of LVDT against the position of the core.

The differential transformers are available in a broad range of sizes and are widely used for displacement measurement in a variety of applications. Some important characteristics and features of the LVDT type of displacement sensors are:

- simplicity of design, ease of fabrication and installation, rugged and durable construction,

- wide range of displacement; displacement ranges available from 2×10^{-4} m to 0.5 m,

- high output, low hysteresis, continuous (infinite) resolution, and linear electrical response (linearity better than 0.5%) when actuated by linear mechanical motion, and

- negligible operating force and no wear of moving parts.

The device, however, is not particularly sensitive and must be excited with ac only; excitation frequency 50–20 kHz. Input voltage is limited by the current-carrying capacity of the primary coil.

Typical measurements are any quantities that can be transduced into displacement, for example, pressure, acceleration, vibration, force, and liquid level. The disadvantage lies in the area of dynamic measurements as its core is of appreciable mass in comparison to the strain gauge.

9.13. CAPACITIVE TRANSDUCERS

A capacitor comprises two or more metal plate conductors separated by an insulator. As voltage is applied across the plates, equal and opposite electric charges are generated on the plates. Capacitance is defined as the ratio of the charges to the applied voltage and for a parallel plate capacitor is given by:

$$C = \epsilon_0 \, \epsilon_r \frac{A}{t} (N - 1) \text{ farads}$$

where A = overlapping or effective area between plates (m^2),
$\quad t$ = distance between plates (m),
$\quad N$ = number of capacitor plates,
$\quad \epsilon_0$ = permittivity of free space = 8.854×10^{12} F/m, and
$\quad \epsilon_r$ = relative permittivity (or dielectric constant) of the material between the plates. The value of ϵ_r depends upon the insulator material and for air $\epsilon_r = 1$.

For a cylindrical capacitor, the capacitor is

$$C = \epsilon_0 \epsilon_r \frac{2\pi l}{\log_e \left(\dfrac{r_2}{r_1} \right)} \text{ farads} \qquad (9.5)$$

where l = length of overlapping part of cylinders (m),
r_1 = radius of inner cylindrical conductor (m), and
r_2 = radius of outer cylindrical conductor (m).

A capacitive pickup operates on the principle of a variation in capacitance produced by the physical quantity being measured. The capacitance can be made to vary by changing either the relative permittivity (dielectric constant) ϵ_r, the effective area A, or the distance between the plates t. The mechanical displacement is generally measured by noting the change in capacitance brought about by either change in area or by change in distance between the plates. The change in the dielectric is used to measure changes in liquid or gas levels.

Figure 9.14 represents the elementary diagram of the two arrangements of a capacitance transducer where capacitance change occurs because of change in the area of plates. Since capacitance is directly proportional to the effective area of the plates, the response of such a system is linear.

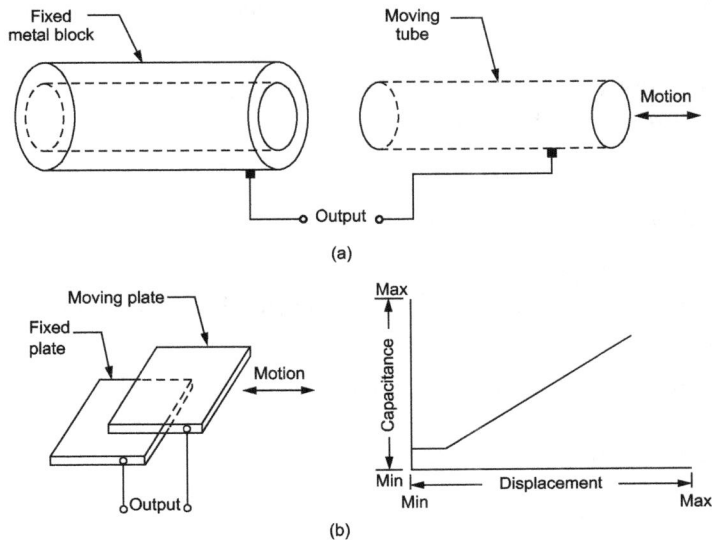

FIGURE 9.14 Capacitance transducer: area change.

Figure 9.15 represents the basic form of a capacitance transducer utilizing the effect of change of capacitance with changes in distance between the two plates. One is a fixed plate and the displacement to be measured is applied to the other plate which is moving. Since capacitance varies inversely as the distance between the plates, the response of this transducer is not linear.

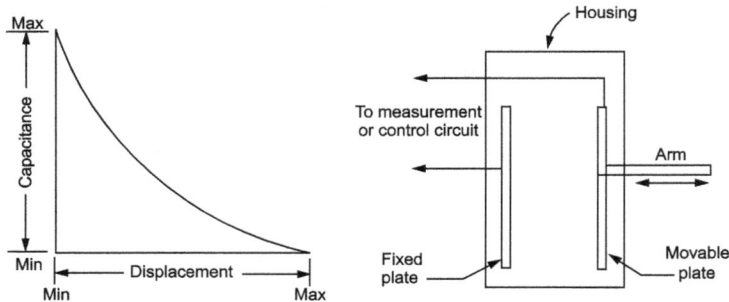

FIGURE 9.15 Capacitance transducer: change in distance between the plates.

Capacitance transducers can detect displacements as small as 2.5×10^{-6} m and produce a measurable signal. Parallel plate capacitance transducers have the advantages of: (*i*) easy fabrication, (*ii*) excellent high-frequency response, (*iii*) good linearity output, (*iv*) ability to measure static and dynamic quantities, and (*v*) a relatively low initial and maintenance cost.

EXAMPLE 9.1
The 3 cm region between the two plates of a parallel plate capacitor is filled by two dielectric layers:

i.   1 cm thick with dielectric constant 5, and

ii. 2 cm thick with dielectric constant 10.

What would be the relative permittivity (dielectric constant) of a material which gives the same capacitance if it completely fills the region between the plates?

Solution: For a multiple parallel plate capacitor with space between adjacent plates filled with different materials having dielectric constants $\in_{r1}$, $\in_{r2}$ .... $\in_{r3}$ and having respective distances $t_1, t_2, t_3, ...$, we have:

$$C = \frac{\in_0 A}{\dfrac{t_1}{\in_{r1}} + \dfrac{t_2}{\in_{r2}} + \dfrac{t_3}{\in_{r3}}}$$

Substituting the values from the given data,

$$C = \frac{\epsilon_0\, A}{\dfrac{0.01}{5} + \dfrac{0.02}{10}} = \frac{\epsilon_0\, A}{0.004} \qquad (i)$$

Let ϵ_r be the dielectric constant of the single medium which completely fills the gap of 3 cm between the two parallel plates of area A and gives the same capacitance. Then

$$C = \epsilon_0\, \epsilon_r\, \frac{A}{t} = \epsilon_0\, \epsilon_r\, \frac{A}{0.03} \qquad (ii)$$

Equating the capacitance values given by expressions (i) and (ii)

$$\frac{\epsilon_0\, A}{0.004} = \epsilon_0\, \epsilon_r\, \frac{A}{0.03}$$

That gives: $\quad \epsilon_r = \dfrac{0.03}{0.004} = 7.5$

EXAMPLE 9.2

A capacitive transducer using two quartz diaphragms of area 800 mm^2 and separated by a distance of 4 mm has a capacitance of 350 uF. When pressure of 1 MN/m^2 is applied to one of the diaphragms, a deflection of 0,75 mm is produced. Workout the change in the capacitance of the system.

Solution: Before application of pressure: $C_1 = \epsilon_0 \epsilon_r \dfrac{A}{t_1}$

After application of pressure: $C_2 = \epsilon_0 \epsilon_r \dfrac{A}{t_2}$

Given: $C_2 = 350\ \mu F$; $t_1 = 4$ mm and $t_2 = 4 - 0.75 = 3.25$ mm

$$C_2 = 350 \times \frac{4}{3.25} = 430.77\ \mu F$$

$\therefore$ Change in capacitance $\Delta C = C_2 \sim C_1$
$$= 430.77 - 350 = 80.77\ uF$$

9.14. PIEZOELECTRIC TRANSDUCERS

Piezoelectricity represents the property of a number of crystalline materials that cause the crystal to develop an electric charge or potential difference when subjected to mechanical forces or stresses along specific planes. Conversely, the crystal would undergo a change in thickness (and thus produce mechanical forces) when charged electrically by a potential difference applied to its proper axis. Elements exhibiting piezoelectric qualities are sometimes known as electro restrictive elements.

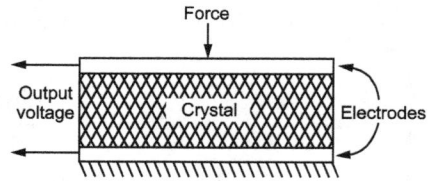

FIGURE 9.16 Piezoelectric transducer.

A typical mode of operation of a piezoelectric device for measuring the varying force applied in a simple plate is shown in Figure 9.16. Metal electrodes are attached to the selected faces of a crystal in order to detect the electrical charge developed. The magnitude and polarity in the induced charge on the crystal surface are proportional to the magnitude and direction of the applied force and is given by:

$$Q = KF \tag{9.6}$$

where Q is the charge in coulomb, F is the impressed force in N and K is the crystal sensitivity in C/N; it is constant for particular crystals and the manner in which they are out. The relationship between the force F and the change δt in the crystal thickness t is given by the stress-strain relationship.

$$\text{Young's modulus} = \frac{\text{stress}}{\text{strain}}; = Y = \frac{F/A}{\delta t / t} \tag{9.7}$$

$$F = AY \frac{\delta t}{t} \tag{9.8}$$

The charge at the electrode gives rise to voltage, such that

$$V_0 = \frac{Q}{C}$$

where C is the capacitance between electrodes. Furthermore

$$C = \epsilon_0 \epsilon_r \frac{A}{t} \text{ farads}$$

Combining the above equations, we obtain:

$$V_0 = \frac{K}{\epsilon_0 \epsilon_r} t \frac{F}{A} = gt \, P \tag{9.9}$$

where g is the crystal voltage sensitivity in Vm/N, and P is the applied pressure in N/m^2.

There are two main groups of piezoelectric crystals: (*i*) natural crystals such as quartz and tourmaline, (*ii*) synthetic crystals such as Rochelle salts, lithium sulfate, ammonia dihydrogen phosphate, ethylene diamine tartrate (EOT), dipotassium tartrate (DKT), etc. The advantages vary from crystal to crystal and one is chosen on the basis of a particular application. Tourmaline is the least active chemically while tartaric acid is most active electrically.

- *Natural crystals* have a very low electrical leakage when used with very high input impedance amplifiers and permit the measurement of a slowly varying parameter. They are, therefore, capable of withstanding higher temperatures; operating at low frequencies and sustaining shocks.

- *Synthetic crystals* exhibit a much high output for applied stress and are about a thousand times more sensitive than natural crystals. However, they are usually unable, to withstand high mechanical strain without fracture. Further, synthetic crystals have an accelerated rate of deterioration over natural ones.

The major advantages of piezoelectric transducers are:

- high-frequency response,

- high output,

- rugged construction,

- negligible phase shift, and

- small size. The small size of the transducer is especially useful for accelerometers where added mass will mechanically load a system.

The piezoelectric unit has the disadvantage in that it cannot measure static conditions and that its output is affected by changes in temperature. When an instrument is electrically connected to measure the electrical charge generated, it is slowly dissipated through the internal resistance of the crystal, that is, the charge decreases over a period of time. Because of this characteristic, the piezoelectric transducers have a poor steady-state response and as such are used mainly for measuring dynamic quantities (parameters varying rapidly with time). Special amplifiers with very high input impedance (10^{12} to 10^{14} ohms) can, however, be used to measure the static or quasi-static quantities, but that makes the measuring system increasingly expensive.

Applications: Piezoelectric transducers are most often used for accelerometers, pressure cells, and force cells in that order.

EXAMPLE 9.3

A quartz crystal having a thickness of 2 mm and a voltage sensitivity of 0.05 Volt-m/ newton is subjected to a pressure of 15×10^5 N/m². Calculate the voltage developed by the piezoelectric pickup and the charge sensitivity of the crystals. Take the permittivity of the quartz as 40.5×10^{-12} F/m.

Solution: The output voltage for a piezoelectric pickup is given by

$$= 0.05 \times 0.002 \times (15 \times 10^5) = \textbf{150 V}$$

(b) Charge sensitivity $= \epsilon_0 \epsilon_r g = \epsilon_g$

$$= (40.5 \times 10^{-12}) \times 0.05$$
$$= 2.025 \times 10^{-12} \text{ C/N} = 2.025 \text{ pc/N}$$

9.15. PHOTOELECTRIC TRANSDUCERS

These transducers operate on the principle that when light strikes a special combination of materials, a voltage may be generated, a resistance change may take place, or electrons may flow. Photoelectric cells are used for a wide variety of purposes in control engineering for precision measuring devices, in exposure meters used in photography. They are also used in solar batteries as sources of electrical power for rockets and satellites used in space research. Photoelectric transducers offer the advantage that they do not involve any contact being made with the system being measured; just interruption of a beam of light. Further, the light does not have to be visible; it can be selected to operate with infrared radiation. Photoelectric transducers can be grouped into photoemissive (phototube), photoconductive, and photovoltaic cells.

9.15.1. Photoemissive Cell

These transducers (Figure 9.17) operate on the photoemissive effect, that is, when certain types of materials are exposed to light, electrons are emitted and a current flow is produced.

$$\text{light information} \rightarrow \text{current information}$$

The light-sensitive photocathode may consist of a very thin film of cesium deposited by vaporization onto an oxidized silver base. Light strikes the cathode, causing the emission of electrons that are attracted toward the anode. This phenomenon produces the flow of electric current in the external circuit; the current being a function of radiant energy striking the cathode.

There exist three separate types of photoemissive cells; the high vacuum single cathode, the gas-filled and the multiplier tubes. The high vacuum and the gas-filled tubes are both diodes where the cathode and anode are enclosed in a glass or quartz envelope which is either evacuated or filled with an inert gas. The difference lies in the extent of the vacuum and the kind of inert gas. The photomultiplier tubes use the principle of secondary emission. The device consists of a series of reflecting electrodes, called dynodes, which amplify the original output current. The dynodes are so arranged that the electrons making a dynode produce further electron emission from the dynode. The number of emitted electrons can be increased and high gains are made possible by photomultipliers.

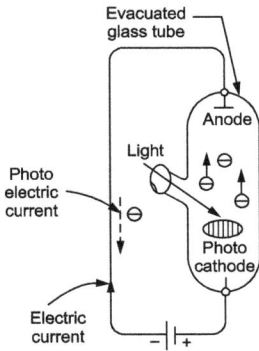

FIGURE 9.17 Photo tube.

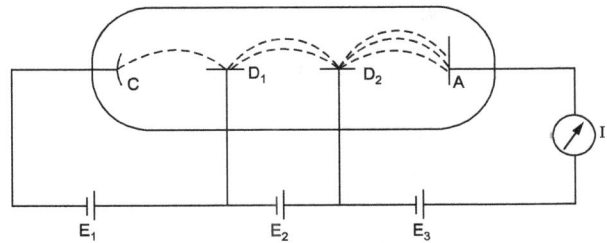

FIGURE 9.18 Photomultiplier tube.

With reference to Figure 9.18, voltage E_1 accelerates the electrons emitted by cathode C and these are focused onto the dynode D_1. Each incident electron causes emission of secondary electrons which subsequently get focused upon dynode D_3. Finally, these are attracted by the anode A leading to the generation of current l.

9.15.2. Photoconductive Cell

These are the variable resistance transducers. They operate on the principle of photoconductive effect, that is, some special type of semiconductor materials change their resistance when exposed to light.

light information → resistance information

Figure 9.19 shows schematically the construction and electrical circuit of a photoconductive cell. The sensitive material usually employed is cadmium sulfide, cadmium selenide, germanium, etc., in the form of thin coating between the two electrodes on a glass plate. Further, the cells are used in the circuit as a variable resistance and are put in series with an ammeter and a voltage source. When the light strikes the semiconductor material, there is a decrease in the cell resistance thereby producing an increase in the current indicated by the ammeter.

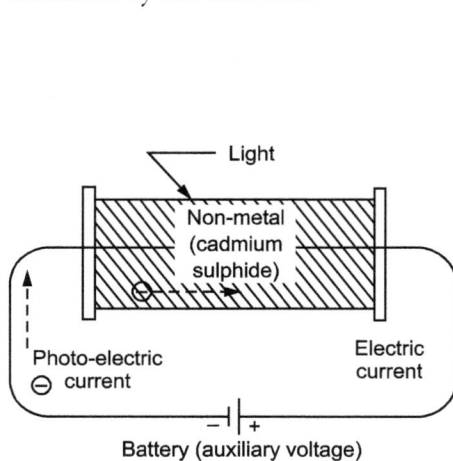

FIGURE 9.19 Photoconductive cell.

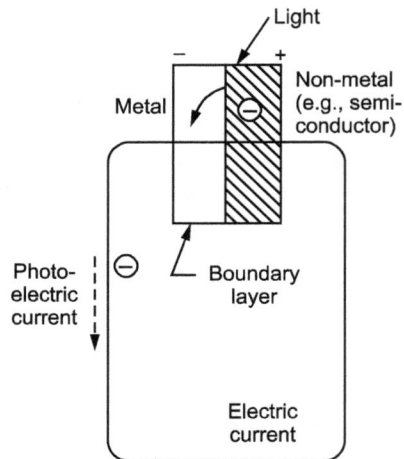

FIGURE 9.20 Photovoltaic cell.

9.15.3. Photovoltaic Cell

These transducers operate on the photovoltaic effect, that is, when light strikes a junction of certain dissimilar metals, a potential difference is built up

$$\text{light information} \rightarrow \text{emf information}$$

The cell consists of a metal base plate, a non-metal semiconductor, and a thin transparent metallic layer (Figure 9.20). Typical examples of the layers are the copper oxide on copper and iron oxide on an iron combination. The transparent layer may be in the form of a sprayed conducting lacquer. Light strikes the coating and generates an electric potential. The output is, however, low and is a non-linear function of the light intensity. In contrast to phototube and photoconductive cells, the photovoltaic unit is self-generated and requires no voltage source to operate it. Further, it need not be operated in a vacuum or gas-riled envelope. The most common application of photovoltaic cell is in light exposure meter in photographic work.

9.16. THE HALL EFFECT

The Hall effect relates to the generation of transverse voltage difference on a conductor which carries current and is subjected to magnetic field in a perpendicular direction. The current may be due to the movement of holes or that of face electrons.

Refer to Figure 9.21 which shows the schematic of a circuit which produces the Hall effect. Here

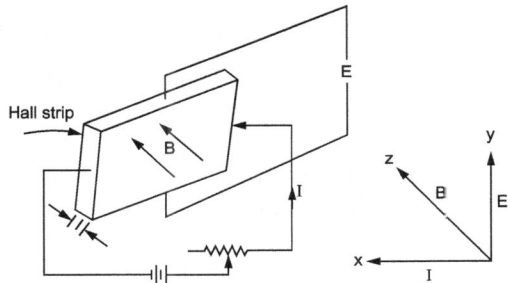

FIGURE 9.21 The Hall effect.

i. the Hall strip carries a current/in the x-direction and is subjected to magnetic field B in the z-direction,

ii. the thickness f of the strip is very small as compared to its length and width, and

iii. the voltage E is set up in the transverse or y-direction. This voltage is directly proportional to the current I, field strength B, and inversely proportional to thickness t of the strip. That is

$$E = K\frac{BI}{t}$$

The proportionality constant K is called the Hall effect coefficient. Taking the current in ampere, flux density in W_b/m^2, and the thickness of strip in meter, the units of K work out as $\dfrac{Vm^3}{AW_b}$

The notable aspects of transducers operating on the Hall effect are:

i. The Hall effect is present in metals and semiconductors in varying amounts depending upon the densities or mobiles of carriers. However, the Hall effect is more pronounced in semiconductors than in metals.

ii. The magnitude of current flow in the circuit is limited by heat dissipation and permissible temperature rise.

iii. The Hall effect transducers are of non-contact type and have small size and high resolution.

The Hall effect transducers are used:

a. to determine whether a semiconductor is of N-type or P-type,

b. to measure either the current or the strength of the magnetic field, and

c. to measure the displacement where it is possible to change the magnetic field strength by variation in the geometry of the magnetic structure.

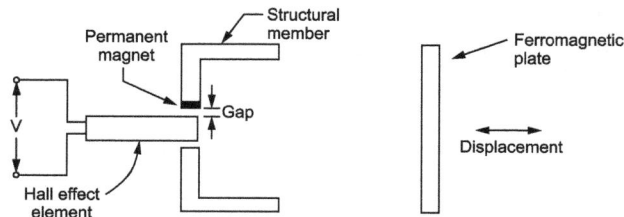

FIGURE 9.22 The Hall effect displacement transducer.

Figure 9.22 shows the arrangement of the Hall effect transducer as used for the measurement of linear displacement. This proximity pickup requires a ferrous target whose approach changes the reluctance of an internal magnetic circuit. The Hall effect element is located in the gap adjacent to the permanent magnet. When a plate of iron or other ferromagnetic material is moved with the suspect to the structure, the magnetic field in the gap due to the permanent magnet changes. That produces the output voltage which is a measure of the displacement of the ferromagnetic plate with respect to the structure.

This arrangement has been successfully employed for measuring displacement as small as 0.025 mm.

REVIEW QUESTIONS

A. Conceptual and conventional questions:

1. What is mechatronics? List the main technical areas under its domain.

2. Give a brief account of the different stages in the evolution of the discipline of mechatronics.

3. (*a*) How does a mechatronics system operate?

(*b*) Give a few examples of mechatronic systems.

4. Mention a few industrial applications of mechatronics.

5. "Mechatronics is a multi-disciplinary subject." Comment on the validity of this statement.

6. Define and explain the terms autotronics, bionics, and avionics.

7. Mention a few domains where the mechatronics-based design concept is considered fundamental to engineers.

8. Differentiate between a sensor and a transducer.

9. Distinguish between:
(*i*) active and passive transducers,
(*ii*) input and output transducers.
Illustrate your answer with suitable examples.

10. What is meant by transduction? List a few effects to which the principle of transduction can be attributed.

11. List the advantages and disadvantages of mechatronics systems.

12. Draw the schematic arrangement of the key elements of a typical mechatronics system.

13. What are transducers and how are they classified? Explain their importance in an instrumentation process.
Give some examples of mechanical transducers where there is transduction from (*i*) force to displacement (*ii*) velocity to pressure (*iii*) temperature to displacement (*iv*) fluid pressure to displacement.

14. In modern measurement systems, there is more reliance on electrical/electronic techniques of measurement.
List some advantages of electrical transducers over mechanical transducers. Suggest a suitable transducer to convert each of the following variables into electrical signals:
(*i*) pressure   (*ii*) force   (*iii*) acceleration
(*iv*) angular speed of a shaft, and   (*v*) liquid level.
Indicate in each case the measurements involved.

15. (*a*) Distinguish between:
    (*i*) active and passive transducers and
    (*ii*) input and output transducer. Illustrate your answer with suitable examples.
    (*b*) What information is needed to describe a transducer for a particular measurement?
    (*c*) Explain the major considerations which govern the selection of an instrument transducer.

16. Explain the use of wire wound potentiometers for the measurement of linear and rotary motions. Point out the advantages and limitations of such transducers.

17. (*a*) Explain the various physical principles involved in the operation of various categories of inductive transducers.
    (*b*) Give the essential features of inductive and capacitive transducers when used for the measurement of displacement.

18. (*a*) Describe the principle of operation of the linear variable differential transformer. Why it is necessary to connect the secondaries in a differential mode? Identify the input and output of the system and sketch the typical input-output graph.
    (*b*) Explain the use of a linear variable differential transformer (LVDT) for the measurement of pressure differential across an obstruction meter placed in fluid flow through a pipeline.

19. (*a*) Describe the principle of operation of a piezoelectric transducer. Identify the input and output of the system.
    (*b*) Mention some natural and synthetic materials that exhibit a piezoelectric effect.

20. Explain the difference in the principle of operation of a photoemissive cell, a photoconductive cell, and a photovoltaic cell. Give the applications of each of these cells.

21. Mention the different parameters of a parallel plate capacitor that may vary to bring about a change in the capacitance of the device. Point out the physical variable that is usually measured by employing a particular variation.

B. Fill in the blanks with appropriate word/words

(*i*)   A transducer is a device that converts the measurand into an _____.

(*ii*) The energy conversion process that takes place in a transducer is referred to as _____.

(*iii*) A spring is a mechanical transducer converting force to _____.

(*iv*) Passive transducers rely on an _____ for their operation.

(*v*)   The capacitance of a parallel plate capacitor can be varied by changing either the _____ or the _____ or the _____.

(*vi*) Piezoelectric transducers are made from natural crystals such as _____ or synthetic crystals such as _____.

(*vii*) Photoelectric transducers produce electrical signals in response to changes in the _____.

Answers:
(*i*) optical, mechanical, or electrical signal; (*ii*) transduction; (*iii*) displacement; (*iv*) external excitation voltage; (*v*) relative permittivity, overlapping (effective area), distance between the plates; (*vi*) quartz and Rochelle salt, lithium sulfate; (*vii*) intensity of incident light.

C. Indicate true or false in respect of the following statement. If false, into the correct statement:

(*i*)   High value of pot resistance leads to high sensitivity.

(*ii*) Capacitive transducers used for the measurement of liquid level operate or the principle of capacitance changes with the change of distance between plates.

(*iii*) Linear variable differential transformer (LVDT) is an active transducer working on the principle of variable resistance.

(*iv*) When a static force is applied to a piezoelectric transducer, there occur oscillations in the generated electric change.

(*v*)   Piezoelectric transducers produce an emf when an external magnetic field is applied across them.

(*vi*) The abbreviation LVDT stands for linear voltage differential transformer.

(vii) Piezoelectric crystals are used for the measurement of static as well as dynamic changes.

(viii) Hall effect transducers are highly sensitive to temperature variations.

(ix) Photoconductive transducer is a light-controlled variable resistor.

(x) The photovoltaic cell converts the light information to resistance change of the electric circuit.

Answers:

(i) True

(ii) False; change of dielectric strength

(iii) False; the principle of mutual inductances

(iv) True; the generated electric charge decreases over a period of time.

(v) False; when external mechanical force is applied

(vi) False; linear variable differential transformer

(vii) False; the piezoelectric transducers have a poor steady-state response and as such are used mainly for measuring dynamic quantities

(viii) True

(ix) True

(x) False; in a photovoltaic cell, the light strikes the junction of certain dissimilar metals and a potential difference is built up in the electric circuit.

D. Multiple choice questions:

1. Printout the device/devices that refer to self-generating transducers
   (a) resistive
   (b) capacitive
   (c) piezoelectric
   (d) thermoelectric

2. The active transducer which can be used for linear or angular velocity measurements depends on
   (a) generation of force by allowing current to flow through the conductor
   (b) variation in mutual inductance of the coils
   (c) movement of the conductor through a magnetic field
   (d) variation in a capacitance of a capacitor

3. The LVDT is an inductive transducer that functions due to
   (**a**) change in the air gap
   (**b**) change in the amount of core material
   (**c**) mutual inductance
   (**d**) variation in the position of the core

4. Specify the photoelectric device which converts the light information to resistance information
   (**a**) photoemissive cell
   (**b**) photoconductive cell
   (**c**) photovoltaic cell

5. When certain natural or artificial crystals are deformed, an electric charge is generated. This characteristic is referred to as
   (**a**) thermoelectric effect
   (**b**) capacitive effect
   (**c**) electromagnetic effect
   (**d**) piezoelectric effect

6. Specify the variable in a capacitive transducer that does not necessitate a physical contact between the transducer and the measurand
   (**a**) effective or overlapping area of plates
   (**b**) distance between plates
   (**c**) dielectric constant of the insulator

7. Specify the transducer which is generally used for a dynamic rather than for static measurements.
   (**a**) capacitive
   (**b**) resistive
   (**c**) piezoelectric
   (**d**) inductive transducer

8. A potentiometer produces a large variation in resistance by
   (**a**) moving a conductor through a magnetic field
   (**b**) moving a slider across a resistor
   (**c**) stretching a metal wire
   (**d**) thermally expanding a conductor

9. A piezoelectric transducer has all the following advantages except
   (**a**) small size and high output
   (**b**) negligible phase shift
   (**c**) good frequency response
   (**d**) capability to measure both static and dynamic quantities.

10. All of the following statements with reference to LVDT, are correct except one. Identify that statement. LVDT
   (**a**)   works on the principle of mutual induction
   (**b**)   is a self-generating type of transducer
   (**c**)   cannot be used for the measurement of static variables
   (**d**)   stands for linear variable differential transformer

11. The abbreviation LVDT stands for
   (**a**)   least varying differential transducer
   (**b**)   low varying digital transformer
   (**c**)   linear variable differential transformer
   (**d**)   linear voltage differential transformer

Answers:
   1. (*a*) and (*d*)   **2.** (*c*)   **3.** (*d*)   **4.** (*b*)   **5.** (*d*)   **6.** (*c*)   **7.** (*c*)
   8. (*b*)   **9.** (*d*)   **10.** (*b*)   **11.** (*c*)

CHAPTER 10

ACTUATION AND ACTUATING SYSTEMS

10.1. ACTUATOR AND ACTUATION

An **actuator** is an energy conversion device that makes some things move or operate. Essentially it uses a form of power/energy to convert a control signal into mechanical motion, and the process of energy conversion to mechanical form is called **actuation**.

The actuators are encountered at house and at workplace. For example:

- automatic opening of the door when a person enters a grocery shop, and

- forward or backward movement of a car seat.

Actuators are located in from electric door locks in automobiles to ailerons in air-crafts. Industrial plants use actuators to operate valves, dampers, fluid couplings, etc.

Actuators are classified by the type of motion and the power source. Linear actuators produce pull or push action and the linear motion may be along with one or more of the three axes X–X axis, Y–Y axis, and Z–Z axis.

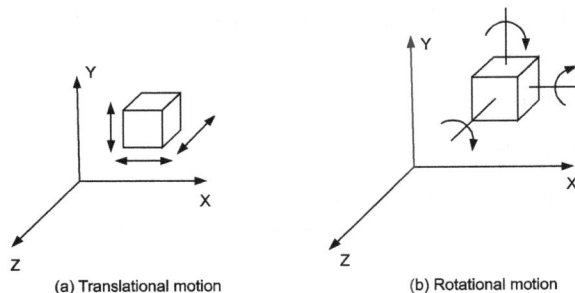

(a) Translational motion   (b) Rotational motion

FIGURE 10.1 Types of motion

The rotational motion is the motion of a rigid body along with one or more of the three-axis along the *X-X* axis, *Y-Y* axis, and *Z-Z* axis.

Depending upon the power source, actuators are categorized as:

- *Electromechanical actuators:* These actuators use AC, DC, or stepper electric motors to convert electrical energy into mechanical work.

- *Fluid power actuators:* These actuators use fluid energy that is transmitted through a fluid under pressure. The fluid used is a liquid (water or oil) in a hydraulic actuator and compressed air or some inert gas in a pneumatic actuator.

- *Active material-based actuators:* The operation of these transducers is based on the fact that some materials like piezoelectric and magnetostrictive materials undergo some change in their character when subjected to some physical interaction. The piezoelectric material undergoes a dimensional change when voltage is applied and that dimensional change is utilized to produce actuators. The magnetostrictive material changes its shape when it is subjected to a magnetic field.

Though a new field, the active material-based actuators have been successfully used in biomedical equipment, fluid control devices, aerospace, and automotive monitoring systems, precision manufacturing, and process monitoring equipment.

10.2. MECHANICAL ACTUATION SYSTEMS

Mechanical actuation systems are essentially the mechanisms formed by assembling a number of rigid bodies in such a way that the motion of one member causes the constrained and predictable motion of the other member. A modification in motion and its transfer from one location to another is achieved through the application of specially designed rigid interfacing components and units. Such rigid bodies are called *mechanical components* and these are categorized as:

- passive components, and

- active components.

The passive components do not transfer the mechanical power and include nut-bolt, screws, springs, and washers. The active components serve to transmit power in terms of force and torque, and motion regarding its speed and direction. Examples of active components are kinematic chains, belt and chain drive, gears and gear trains, cams, followers, bearings, ratchet-pawl mechanism, etc.

The different applications of these active mechanical components are:

- conversion of the reciprocating/linear motion of the piston into rotational motion of the crankshaft in an internal combustion engine,

- transformation of the rotational motion of the cam into the translatory motion of the follower in the cam-follower arrangement,

- change in magnitude and direction of speed by using gear train,

- transformation of motion in one direction into motion in another direction at a right angle to it by using bevel gear, and

- change of linear motion to circular motion and vice versa by the slider-crank mechanism.

The employment of these mechanisms within a mechanical system may also provide mechanical advantage, that is, amplifying the force. The term *loading* associated with these active components refers to several factors such as:

- force, torque, and speed of rotation, and

- accuracy, precision, and power consumption.

These units have been discussed in the following sections regarding their purpose, construction, and operation.

10.3. KINEMATIC LINK, KINEMATIC PAIR, MECHANISM, AND MACHINE

Each part of a machine that connects other parts having motion relative to it is called a kinematic link or simply a link. A link may also consist of a number of parts that are so connected that they form one unit and have no motion relative to each other.

With reference to the piston-cylinder arrangement of a reciprocating steam engine shown in Figure 10.2.

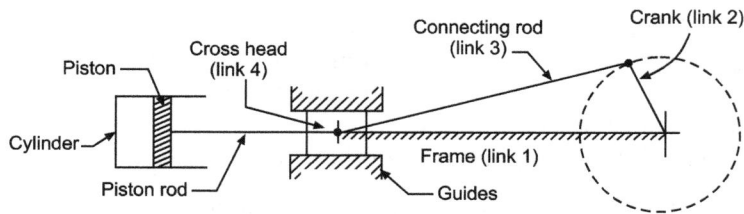

FIGURE 10.2 Schematics of a steam engine

i. Crank, connecting rod, cylinder, and piston of steam engine constitute four links. Each element is a separate link moving relative to one another.

    Link 1 is the fixed link and includes the engine frame and all other stationary parts, like cylinder, and main bearings.

    Link 2 includes crankshaft and flywheel, both having motion of rotation about the fixed axis.

    Link 3 is the connecting rod having oscillatory motion.

    Link 4 corresponds to the piston, piston rod and cross-head having reciprocating rectilinear translatory motion.

ii. Piston, piston rod and cross-head of a steam engine are rigidly fastened together and do not move relative to one another. They constitute one unit and are taken as one link.

iii. Connecting rod, big and small end bearings, caps, and bolts constitute one unit and are taken as one link.

The links are generally classified as:

i. **Rigid link:** a link that exhibits no relative deformation between two parts when it is acted upon by a force system. The distance between any two particles remains constant, that is, the size and the shape of the link do not change.

    A link need not be a rigid body but it must be a resistant body. A body is said to be resistant if it is capable of transmitting the required motion with negligible deformation.

ii. **Flexible link:** a link that gets only partly deformed while transmitting motion. However, a flexible link does not affect the transmission of motion which is either a pull or a push in one direction only. The typical examples of flexible links are the belts and springs.

iii. Fluid link: a link formed by having fluid in a container. The motion to other components is through the fluid by pressure or compression. Hydraulic brakes, jacks, and presses represent fluid links.

10.3.1. Mechanism, Machine, and Structure

Mechanism is an assemblage of the number of bodies (usually rigid) assembled in such a way that the motion of one causes constrained and predictable motion to the other. The function of a mechanism is to transmit and modify motion. Some examples of the mechanism are type writers, watches, clocks, and spring toys.

Machine is a mechanism or a combination of mechanisms that (apart from imparting definite motion to the parts) transmits and modifies the available energy into useful work. Steam engines, reciprocating compressors, and pumps are the machines derived from slider-crank mechanism.

Structure is a combination of a number of resistant links which are meant for carrying loads or are under the influence of forces having straining action. There is neither any relative motion between the links of a structure nor any useful energy is transmitted by it. Typical examples of a structure are the machine frame, bridge, and the structure of a roof.

10.3.2. Kinematic Pair and Types of Motion

A *kinematic pair* is a joint of two links or elements of a machine that are in contact and have relative motion between them.

With reference to the reciprocating engine, the kinematic pairs formed by different links are:

1. crankshaft with bearings which are fixed,

2. crank with connecting rod,

3. connecting rod with piston, and

4. piston with cylinder.

Further, the relative motion between the two links of a kinematic pair has to be completely or successfully constrained to make the required pair.

a. **Completely constrained motion:** Motion between the pair is limited to a definite direction, irrespective of the direction of the force applied. The constrained motion is completed by its own links.

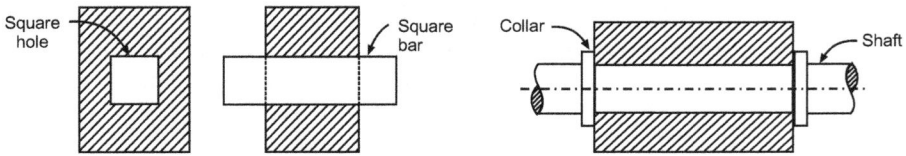

FIGURE 10.3 Completely constrained motion

Examples of a completely constrained motion are

— rectangular bar in a rectangular hole,

— shaft with collars at each end rotating in a round hole, and

— piston and a cylinder in a steam engine; here the piston has only a reciprocating motion relative to the cylinder irrespective of the direction of rotation of crank.

b. Incompletely constrained motion: Motion between the pair is in more than one direction. There occurs a change in the direction of relative motion with a change in the direction of an impressed force.

A circular bar moving in a round hole provides incompletely constrained motion. Here the bar can reciprocate or rotate, and both these motions are independent of one another.

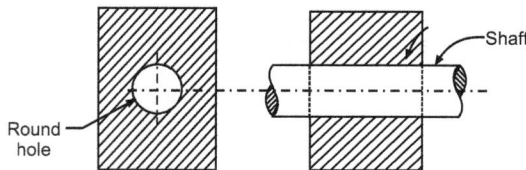

FIGURE 10.4 Incompletely constrained

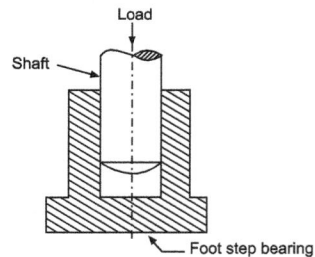

FIGURE 10.5 Partially constrained motion (foot step bearing)

c. Partially or successfully constrained motion: Motion between the elements forming the pair is not completed by itself but is done so by some other means.

The motion of a circular shaft in a foot step bearing is made completely constrained by applying adequate force as shown in Figure 10.5.

Other examples of partially constrained motion are:

i. rotor of a vertical turbine

ii. IC engine valves that are kept on their seats by a spring.

10.3.3. Classification of a Kinematic Pair

A *kinematic pair* may be classified according to type of relative motion, type of contact, and type of closure.

1. Type of relative motion

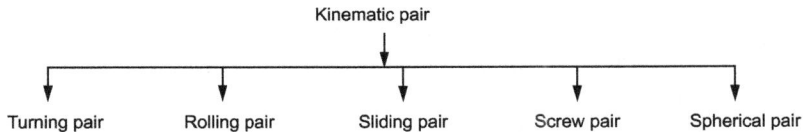

i.   *Turning pair:* The connection between the two links is such that one link can rotate or oscillate about the fixed axis of another link. The turning permits only one degree of freedom. This pair is also called a hinge, a pin joint, or a revolute pair. Examples are:

- lathe spindle supported in the head stock,

- cycle wheels turning on its axles, and

- shaft, with collars at both ends, fitted into a circular.

ii. *Rolling pair:* The connection between the two links in such a way that one link can roll over another link, which is fixed.

Ball and roller bearings, wheels of locomotive and castor wheel of trolleys, etc., constitute the rolling pair. In a ball bearing, the ball and the shaft constitute one rolling pair whereas ball and bearing is another rolling pair.

iii. *Slitting or prismatic pair:* The two elements of the pair are connected in such a way that one link can purely slide over another link. There is a relative motion of sliding only in one direction (along a line) and as such only one degree of freedom.

Examples of sliding pairs are:

- piston and cylinder,

- ram and its guides in a shaper,

- tailstock on the lathe bed, and

- cross-head and its guides in a reciprocating steam engine.

iv. *Screw pair:* The connection between the links is such that one link can turn about another link by means of screw threads. Both axial sliding and rotational motions are involved. However, the sliding and rotational motions are related through helix angle, and the pair is said to have one degree of freedom. The rotating lead screw operates in nuts for accurate transmission of motion as in lathes, machine tools, and measuring instruments from the screw pair.

v. *Spherical or global pair:* One of the elements is in the form of a sphere and it turns or swivels about the other element which is fixed. For a given position of spherical pair, the joint permits relative motion about three mutually perpendicular axes, and so has three degrees of freedom. The ball and socket joint, pen stand, and the attachment of a car mirror represent the spherical pairs.

vi. *Cylindrical pair:* The relative motion in a cylindrical pair is a combination of rotation and translation parallel to the axis of rotation between the contacting elements. Apparently, the cylindrical pair has two degrees of freedom.

    A shaft free to rotate in a bearing and also free to slide axially inside the bearing provides a cylindrical pair.

2. Type of contact

The links of a lower pair have surface contact while in motion and the relative motion is purely turning or sliding. Examples are:

- lathe spindle supported in headstock,

- shaft revolving in a bearing,

- straight-line motion mechanisms, and

- universal joint and automobile steering gear.

    The links of a *higher pair* have point or line contact, and the relative motion is a combination of sliding and turning. All sliding, screw, spherical, cylindrical, revolute pairs fall in the category of lower pairs. Examples are:

- ball and roller bearings,

- cam—follower,

- belt, rope, and chain drives, and

- meshing gear teeth.

    With reference to Figure 10.6 where link 1 is fixed:

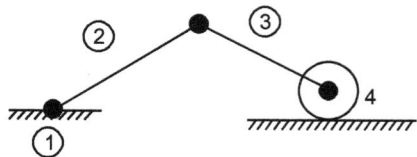

FIGURE 10.6 Lower and higher pair

- there is surface contact between links 1 and 2, and they constitute a lower pair,

- links 2 and 3 comprise a turning pair, and

- the pair of links 3 and 4 represent a higher pair due to a line contact between them.

3. Types of closure

A self-closed pair is formed when the elements constituting the pair are held together mechanically and only the required kind of motion occurs. The lower pairs represent self-closed pairs.

There is no mechanical connection between the two links of the unclosed or force-closed pair. The contact is maintained by an external force which may be either due to gravity or by spring action. Cam and follower are connected by forces exerted by spring and gravity and form an unclosed pair.

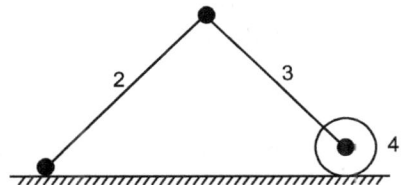

FIGURE 10.7 Self-closed and force-closed pair

With reference to Figure 10.7, we have:

- The links 1–2, 2–3, and 3–4 form a closed pair.

- The links 4 – 1 is a force-closed pair.

10.4. KINEMATIC CHAINS AND THEIR INVERSIONS

A kinematic chain is such a combination or coupling between kinematic pairs such that each link forms a part of two pairs and the relative motion between the links is completely or successfully constrained. The pairs are coupled in such a way that the last link joins the first link and a definite motion is transmitted.

Consider the following kinematic pairs formed by the links in the mechanism of a reciprocating engine:

- cylinder and piston,

- piston and the connecting rod,

- connecting rod and the crank, and

- crankshaft and the bearings.

The total combination of all these pairs can be referred to as a kinematic chain.

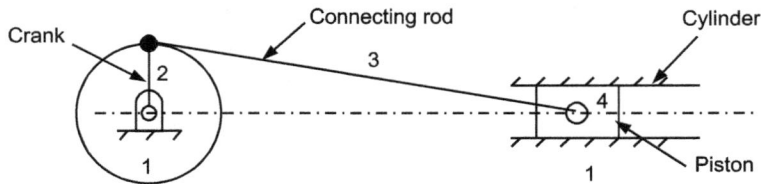

FIGURE 10.8 Splider crank chain

The following correlations exist between the number of pairs (p), number of links $(n$ or $l)$ and the number of joints (j) forming a kinematic chain:

$$n = 2p = 4 \text{ and } j = \frac{3}{2}n - 2 \qquad (10.1)$$

This is on the assumption that each link forms two pairs with adjacent links. Moreover, these relations are essentially valid when the chain comprises only lower pair. Further, the following relation has been suggested by AW Klien to determine the nature of chain.

$$j = \frac{h}{2} = \frac{3}{2}n - 2... \qquad (10.2)$$

where j is the number of joints, h is the number of higher pairs and n is the number of links.

When applying the above relations to a chain having ternary and quaternary joints

1 ternary joint = 2 binary joints
2 quaternary joint = 3 binary joints

- The chain is called a locked chain when the LHS is greater than RHS. The locked chain represents a frame or structure which is used in bridges and structures.

- The chain is called kinematic chain when LHS = RHS.

- The chain is called an unconstrained chain when LHS < RHS.

Consider an arrangement of three links

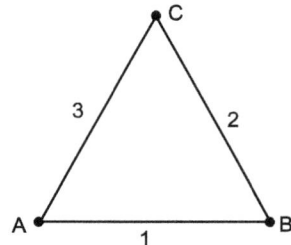

FIGURE 10.9 Three-link mechanism

AB, BC, and CA with pin joints at A, B, and C (Figure 10.9). Here

$$n = 3; p = 3 \text{ and } j = 3$$

Applying the condition: $n = 2p - 4$

$$3 = 2 \times 3 - 4; 3 = 2, \text{ that is, LHS} > \text{RHS}$$

Applying the condition: $j = \frac{3}{2}n - 2$

$$3 = \frac{3}{2} \times 3; 3 - 2 = 2.5, \text{ that is, LHS} > \text{RHS}$$

The arrangement of three links does not satisfy the necessary conditions and obviously does not form a kinematic chain. Such type of chain where LHS > RHS is called **locked chain**. This chain forms a rigid structure (no possibility of relative motion) and is used in trusses and bridges.

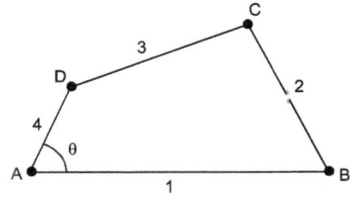

FIGURE 10.10 Four-link mechanism

Consider an arrangement of four links AB, BC, CD, and DA with joints at A, B, C, and D (Figure 10.10). Here

$$n = 4; p = 4, \text{ and } j = 4$$

Applying the conditions prescribed by Eq. (10.1), we get

$$n = 2p - 4; 4 = 2 \times 4 - 4 = 4, \text{ LHS} = \text{RHS}$$

and

$$j = \frac{3}{2}n - 2; 4 = \frac{3}{2} \times 4 - 2 = 4 \text{ LHS} = \text{RHS}$$

The arrangement of four links satisfies the necessary conditions and obviously forms a kinematic chain. Further, the position of a single link such as AD is sufficient to define the position of other links. The four-bar arrangement is then called a kinematic chain of one degree freedom.

When the conditions of Eqs. (10.1) and (10.2) are applied to an arrangement of five links (Figure 10.11), it will be found that LHS < RHS. The equality condition is thus not satisfied and accordingly, the five-bar arrangement does not form a kinematic chain.

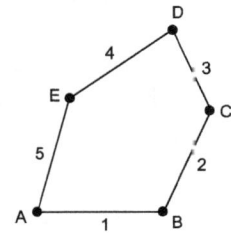

FIGURE 10.11 Five-link mechanism

Such an arrangement where LHS < RHS is called unconstrained chain, that is, the relative motion is not completely constrained.

An arrangement of six links would satisfy the necessary conditions (LHS = RHS) and therefore forms a kinematic chain.

A kinematic chain having more than four links is called the compound kinematic chain.

The different types of joints as found in a chain are classified as:

i. *Binary joint:* Two links are joined at the same connection. The chain shown in Figure 10.10 has four links and four binary joints at *A*, *B*, *C*, and *D*.

ii. *Ternary joint:* Three links are joined at the same connection. The chain shown in Figure 10.12(*a*) has six links, three binary joints (*A*, *B*, and *D*), and two ternary joints (*C* and *E*).

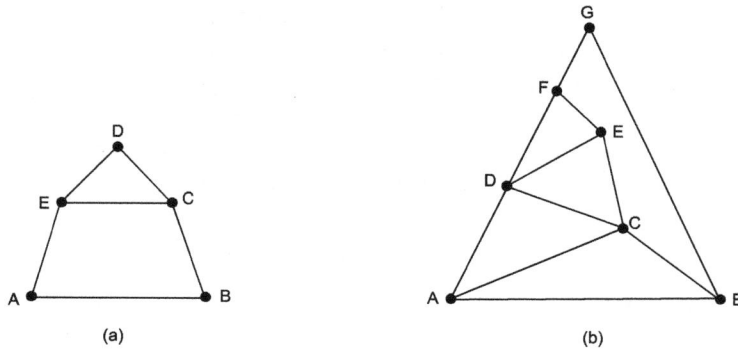

FIGURE 10.12 Ternary and quaternary joints

iii. *Quaternary joint:* Four links are joint at the same connection. The chain is shown in Figure 10.12(*b*) has 11 links, one binary joint (*G*), four ternary joints (*A*, *B*, *E*, and *F*), and two quaternary joints at (*C* and *D*).

EXAMPLE 10.1.
Check the nature of chain and identify the links as binary, ternary, or so with reference to the link arrangement as shown in Figure 10.13.

Solution: The given assemblage of links has:

Ten links, one binary joint (G), and six ternary joints (A, B, C, D, E, and F). Since ternary joint equals 2 binary joints, the total number of binary joints in the system is

$$1 + 2 \times 6 = 13$$

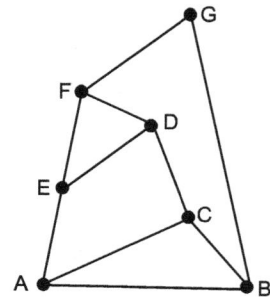

FIGURE 10.13

Applying the equation, $j = \dfrac{3}{2}n - 2$, we have

$$13 = \frac{3}{2} \times 10 - 2 = 13$$

Since left-hand side is equal to right-hand side, the given arrangement satisfies the necessary condition and obviously represents a kinematic or constrained chain.

Degree of Freedom

The **degree of freedom** of a mechanism is defined as the number of inputs that need to be independently controlled to have a constrained motion of the other links, that is, the mechanism can be brought to serve a useful technological purpose.

With reference to Figure 10.14(a), the angle θ is sufficient to define the relative motion of all links, and accordingly, this mechanism has a single degree of freedom. The five-link mechanism depicted in 10.14 (b), needs both θ_1 and θ_2 to define the motion of all links and so the degree of freedom of this arrangement is two.

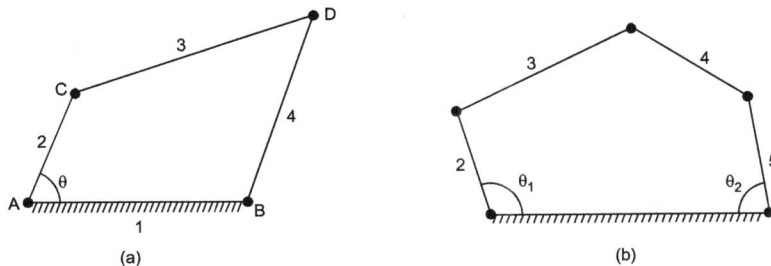

FIGURE 10.14 Concept of the degree of freedom

The number of freedom of some of the systems most commonly involved in mechanisms and machines are given below:

i.   A rigid body has no restraints and has six degrees of freedom.

ii. A circular shaft rotating in a hole and also having the motion of translation parallel to the axis of rotation has two degrees of freedom; angle turned and displacement.

iii. A rectangular bar sliding in a rectangular hole has one degree of freedom; linear displacement only.

iv. The resolute or turning pair has a single degree of freedom; this connection allows only a relative rotation between the elements 1 and 2.

v. A ball and a socket joint has three degrees of freedom.

vi. A screw pair has one degree of freedom; the relative motion between elements 1 and 2 is only rotational.

vii. The position of the crank of a slider-crank mechanism has one degree of freedom; position of crank is expressed by the angle through which it has turned.

Refer to Figure 10.15 for the different mechanisms that would be obtained when one link of a four-bar kinematic chain is fixed at a time. Though all the four inversions look identical, their mobility changes when the proportions of various links are suitably altered.

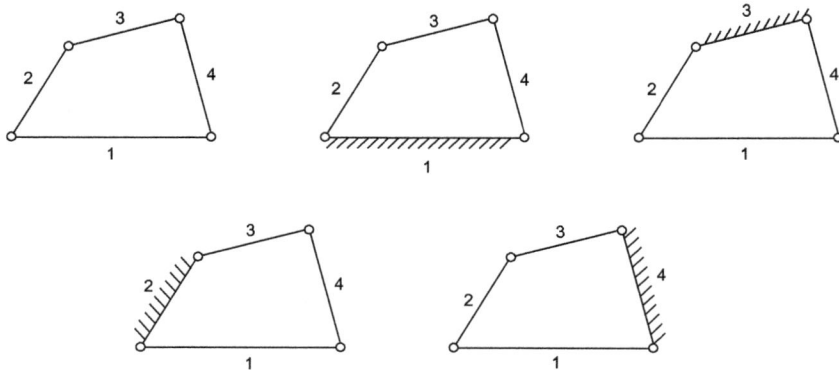

FIGURE 10.15 Inversions of a four-bar chain

The main properties of inversion are:

1. The number of inversions that are possible is equal to the number of links in the parent kinematic chain.

2. The relative motion (displacement velocity and acceleration) between different links is a property of the parent kinematic chain and as such, the relative motion between any two links would not change with inversion.

3. There may occur drastic change in the absolute motion of points on various links (measured with respect to the fixed link) from one inversion to the other.

A mechanism is formed by fixing one of the links of a chain and apparently different mechanisms result when different links of the same chain are chosen to become the fixed link. The process of choosing different links to become the fixed link is called kinematic inversion and the mechanisms obtained by fixing different links of a kinematic chain are called its *inversions*. Even though a four-bar chain with a single degree of freedom constitutes a kinematic chain, there are other chains also which provide inversions of practical utility.

A *quadratic cycle chain* is the simplest and basic kinematic chain consisting of four links which are connected in the form of quadrilateral by four-pin joints (turning pairs). The links form turning pairs and are capable of transmitting definite or completely constrained motion. Further, for continuous relative motion between the links, the sum of the lengths of the shortest and longest links is not to be greater than the sum of the lengths of the remaining two links.

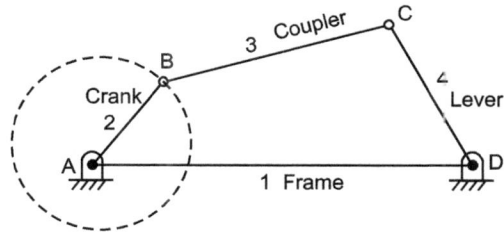

FIGURE 10.16 Quadratic cycle chain

Any of the links, in particular the shortest link, which may be able to make a complete rotation is known as crank or driver. With reference to Figure 10.16.

i. The link *AD* (link 1) is fixed and is referred to as *frame* of the mechanism.

ii. The link *AB* (link 4) makes a complete rotation and is the *crank*.

iii. The link *BC* (link 3) opposite to the fixed link is called the *coupler*.

iv. The link *BC* (link 2) makes a partial rotation, that is, oscillates and is known as *lever* or *rocker* or *follower*.

The coupler connects the crank and lever.

An important inversion of the quadratic cycle chain is the coupled wheel of a locomotive (double crank) meant for transmitting the rotary motion of one wheel to another.

The system has two cranks *AB* and *DC* which are of equal radius and connect with their center of wheels. The link *AD* is fixed and it maintains the center distance between the wheel centers. The couples link *BC* transmits the motion from one wheel to the other wheel.

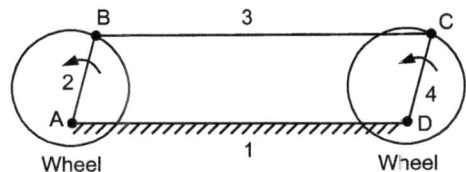

FIGURE 10.17 Coupled wheels of a locomotive

The mechanism couples the two driving wheels of a locomotive and servers to transmit the rotary motion of one wheel to the other wheel.

The *slider-crank chain* is another four-bar linkage consisting of one sliding pair and three turning pairs, and it is meant for converting the reciprocating motion of the piston of an IC engine into the rotary motion of the crank.

With reference to Figure 10.18, the four links are: link 1 (frame of the engine including cylinder and pivot) which is fixed, link 2 (crank), link 3 (connecting rod), and link 4 (piston).

The sliding pair is formed by cylinder and piston (links 1 and 4), and the three turning pairs are formed by frame and crank (links 1 and 2), crank and connecting rod (links 2 and 3), and connecting rod with piston (links 3 and 4).

FIGURE 10.18 Slider-crank chain

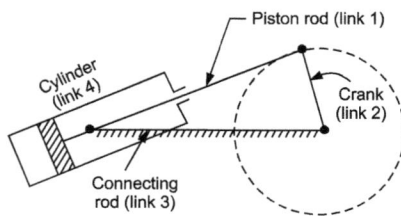

When the piston reciprocates inside the cylinder, the connecting rod oscillates and the crank rotates.

The two common inversions of slider-crank change are:

i. *Oscillating cylinder engine* that converts reciprocating motion to rotary one.

    The connecting rod (link 3) is fixed and is connected to the piston rod (link 1) at point *A*. When the piston attached to the piston rod link reciprocates, the cylinder oscillates about a pin pivoted to the fixed linked at *A* and the crank rotates.

ii. *Pendulum pump* used for supplying feed water to boilers. The mechanism is obtained by fixing the cylinder of the sliding pair of the basic slider-crank chain. When the crank (link 2) is made to rotate, the connecting rod (link 3) oscillates about a pin fitted to the stationary cylinder (link 4). The piston attached to the piston rod then reciprocates inside the cylinder.

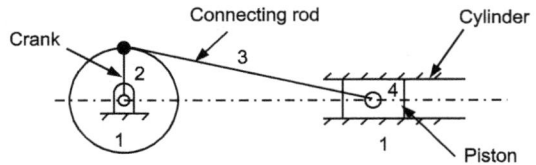

FIGURE 10.19 Oscillating cylinder engine

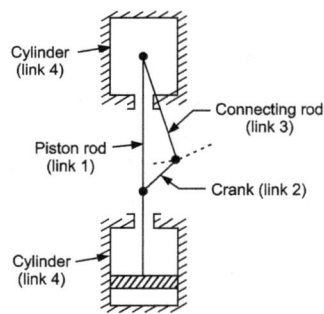

FIGURE 10.20 Pendulum pump

10.5. BELT AND CHAIN DRIVE

The transmission of power in factories from one rotating shaft to another that lies at a considerable distance is achieved through belts and ropes. The shafts are fitted with pulleys, the belt is wrapped around the pulleys and its ends are connected to form an endless connector. The belts and the pulley remain in contact by the frictional grip.

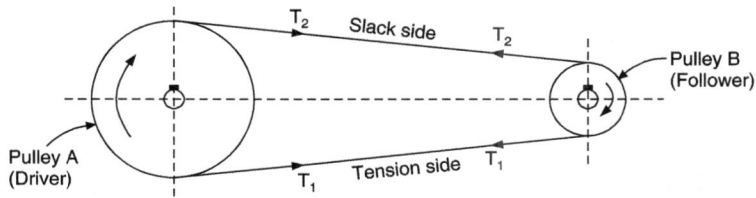

FIGURE 10.21 Belt drive-open system

With reference to Figure 10.21, pulley A which is connected to the rotating shaft is called the *driver*. The pulley B that needs to be driven is termed as *follower*. When the driver rotates, it carries the belt because of friction that exists between the pulley and the belt. The frictional resistance develops all along the contact surfaces that make the belt carry the follower which too starts rotating. The driving pulley pulls the belt from one side (called tension side) and delivers it to the other side (called slack side). The tension T_1 in the belt on the tension side is more than tension T_2 on the slack side.

Two parallel shafts may be connected by *open belt* (Figure 10.21) or by *cross belt* (Figure 10.22). In the open belt system, the rotation of both the pulleys is in the same direction. If a crossed belt system is used, the rotation of pulleys will be in the opposite direction.

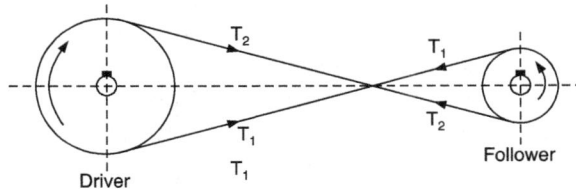

FIGURE 10.22 Belt drive-cross system

The angle of contact in this system of drive is more and accordingly it can transmit more power than an open belt drive system. However, the belt wears out fast at the places where crossing takes place in the crossed belt system.

Further, for a small center distance, the belt is not fully utilized because of its larger slanted runoff.

When a number of pulleys are used to transmit power from one shaft to another, a *compound* drive is used.

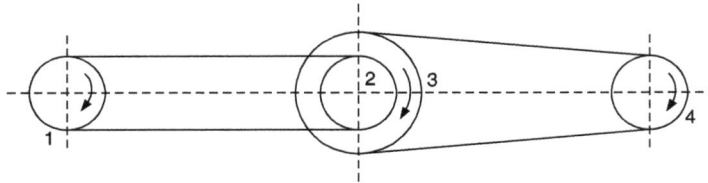

FIGURE 10.23 Belt drive-compound system

With reference to Figure 10.23, pulley 1 drives pulley 2. Since pulleys 2 and 3 are keyed to the same shaft, pulley 3 is also driven by pulley 1. Subsequently, pulley 3 drives pulley 4. The arrangement conforms to an extended open system and all the pulleys rotate in the same direction.

10.5.1. Belt Material and Sections

A belt is a continuous band of flexible material having a rectangular, trapezoidal, and round cross-section. A large variety of belts are available including those made of leather, fabric, rubber impregnated fabric, and synthetics. Belts and ropes obtain their flexibility from the distortion of the material of which they are made.

Flat Belt

A flat belt is a belt with a narrow rectangular cross-section. The flat belts are easier to use and are subject to minimum bending stress. The load-carrying capacity of a flat belt depends on its width.

Material used for belt is generally leather of various types having ultimate tensile strength between 4.5 and 7 N per cm width. For heavy-duty, two or three plies of leather are cemented and pressed one above the other. Such belts are called double or triple ply belts. Leather belts have the best pulling capacity, can be used both in dry and wet places at ordinary temperatures but are costly.

Belting is also made from plies of stitched canvas impregnated with rubber or balata gum and is correspondingly known as rubber belting or balata belting. **Rubber belts** give the best results in damp conditions as they do not absorb moisture as readily as leather. However, they are quite expensive and get ruined in the presence of oil and grease. **Balata belts** are acid and waterproof and are used in the heavily saturated steam laden atmosphere of dye house or where chemical fumes are likely to affect.

Steel belts are immune from stretching and slipping, remain unaffected by dampness or heat, and transmit more power per cm width.

The mid-section of flat pulleys is provided a slight dwell to prevent the belt from running off the pulley. This is referred to as **crowning** of pulleys and may be rounded or tapered.

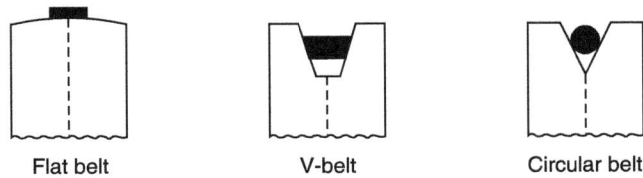

| Flat belt | V-belt | Circular belt |

FIGURE 10.24 Belt cross-section

V-belt

A V-belt is a belt of trapezoid section running on pulleys with grooves cut to match the belt. V-belts are usually made of cotton fabric, cords, and rubber which is vulcanized in molds of desired cross-section.

The salient features of V-belt are:

i. The groove angle in the pulley for running the belt is between 40° and 60°. The belt rests on both sides of the pulley but does not touch the bottom of the groove. The grooved pulleys are generally made of cast iron.

ii. The wedging action between the belt and sides of the groove increases the frictional grip and that reduces the chances of slipping; there is no possibility of belt coming out of the groove. Due to reduced slipping, V-belts offer a more positive drive.

iii. V-belts are used when the distance between the shafts is short, transmission members are large and large power is to be transmitted.

The ideal center distance for a V-belt drive is 1.25–1.5 times the diameter of the larger pulley.

iv. Multiple V-belts can be employed for greater power outputs. Further in the multiple drive system, the machine does not come to stop even if one belt fails.

v. V-belt drives run quietly at high speeds and are capable of absorbing high shock.

vi. Larger reductions in speed are possible in a single drive by using V-belts over small pulleys.

vii. There is better initial installation and replacement due to the standardization of V-belts.

V-belts are available in five sections designated as *A*, *B*, *C*, *D*, and *E*, and these are used in order of increasing loads. That is, section *A* is used for light loads only, and section *E* is used for heavy-duty machines. The angle of V-belt for all sections is about 40°.

Round Belt

The round cross-section belts are employed when low power is to be transmitted such as in instruments, household appliances, tabletop machine tools, and machinery of the clothing industry.

These belts are made of leather, canvas, and rubber. Their diameter is usually within the range of 4–8 mm, and the allowable ratio of the diameter of a smaller pulley to the belt diameter is about 20.

10.5.2. Velocity Ratio, Slip, and Creep

The velocity ratio of belt drives is the ratio of the speed of the driven pulley to that of the driving pulley.

Let d_1, d_2 = diameters of driver and driven pulleys,

ω_1, ω_2 = angular velocities of driver and driven pulleys, and

N_1, N_2 = rotational speeds of driver and driven pulleys; expressed in revolutions per second (rps).

$$\text{Linear speed of driving pulley} = \omega_1 \times \frac{d_1}{2}$$

$$\text{Linear speed of driven pulley} = \omega_2 \times \frac{d_2}{2}$$

Presuming that the belt is inelastic and there is sufficient friction to prevent any slip between the belt and pulleys, both the pulleys will have the same linear speed. That is

$$\omega_1 \times \frac{d_1}{2} = \omega_2 \times \frac{d_2}{2}$$

$$\therefore \qquad \frac{\omega_2}{\omega_1} = \frac{d_1}{d_2} \qquad \text{or} \qquad \frac{2\pi N_2}{2\pi N_1} = \frac{d_1}{d_2}$$

That gives: $\dfrac{N_2}{N_1} = \dfrac{d_1}{d_2}$ (10.3)

The ratio N_2/N_1 is a measure of the velocity ratio of the rotating pulleys. Further, it is apparent from Eq. (10.3) that the *speed of a pulley is inversely proportional to its diameter.*

If thickness t of the belt is taken into account, then

$$v = \omega_1 \times \frac{d_1 + t}{2} = \omega_2 \times \frac{d_2 + t}{2}$$

$$\frac{\omega_2}{\omega_1} = \frac{d_1 + t}{d_2 + t} \quad \text{or} \quad \frac{2\pi N_2}{2\pi N_1} = \frac{d_1 + t}{d_2 + t}$$

$\therefore$ Velocity ratio $\qquad \dfrac{N_2}{N_1} = \dfrac{d_1 + t}{d_2 + t}$ \hfill (10.4)

Slip and its effect on velocity ratio: When the frictional grip between the belt and pulley becomes insufficient, there occurs some forward motion of the driver without carrying the belt with it. The relative motion between the pulley and belt is called slip. The difference between the linear speeds of the pulley rim and belt is the measure of slip.

Let $\quad S_1$ = percentage slip between driver and the belt

$\qquad S_2$ = percentage slip between belt and the follower (driven pulley)

Linear velocity of driving pulley

$$v_1 = \omega_1 \times \frac{d_1}{2}$$

Due to a slip between the driving pulley and the belt, the velocity of belt will decrease

$$\text{Velocity of belt} = v_1 - v_1 \frac{S_1}{100} = v_1 \left(1 - \frac{S_1}{100} \right)$$

This will also be the velocity of the belt as it passes over the driven pulley. As there is slip at the driven pulley also, the velocity of the follower pulley will become less.

Linear speed of the driven pulley

$$= v_1 \left(1 - \frac{S_1}{100} \right) - v_1 \left(1 - \frac{S_1}{100} \right) \times \frac{S_2}{100} = v_1 \left(1 - \frac{S_1}{100} \right) \left(1 - \frac{S_2}{100} \right)$$

$$= v_1 \left(1 - \frac{S_1 + S_2 + 0.01 S_1 S_2}{100} \right) = v_1 \left(1 - \frac{S}{100} \right) = \omega_1 \times \frac{d_1}{2} \left(1 - \frac{S}{100} \right)$$

where $S = S_1 + S_2 + 0.01\, S_1 S_2$ is the percentage of total effective slip.

The linear speed of the driven pulley is also given by:

$$v_2 = \omega_2 \times \frac{d_2}{2}$$

$$\therefore \qquad \omega_2 \times \frac{d_2}{2} = \omega_1 \times \frac{d_1}{2} \times \left(1 - \frac{S}{100}\right)$$

or

$$\frac{\omega_2}{\omega_1} = \frac{d_1}{d_2}\left(1 - \frac{S}{100}\right)$$

or

$$\frac{2\pi N_2}{2\pi N_1} = \frac{d_1}{d_2}\left(1 - \frac{S}{100}\right)$$

$$\therefore \qquad \text{Velocity ratio } \frac{N_2}{N_1} = \frac{d_1}{d_2}\left(1 - \frac{S}{100}\right) \qquad\qquad (10.5)$$

It is apparent from equation 10.5 that the velocity ratio decreases due to slipping of belt.

If thickness t of the belt is also taken into account, then

$$\frac{N_2}{N_1} = \frac{d_1 + t}{d_2 + t} \times \left(1 - \frac{S}{100}\right) \qquad\qquad (10.6)$$

Creep: When the belt passes from the slack side to the tight side a certain portion of the belt extends, and when the belt passes from the tight to slack side the belt contracts. Due to these changes in length, there is relative motion between the belt and pulley surfaces.

The relative motion is termed as **creep** of the belt. Like slip, creep also reduces the velocity of the belt drive system.

EXAMPLE 10.2.

An engine shaft running at 240 rpm is required to drive a machine shaft by means of a belt. The pulley on the engine shaft is 600 mm in diameter. Determine the diameter of the pulley on the machine shaft if it is to run at 360 rpm under the following conditions:

a.   **the belt thickness is negligible and there is no slip,**

b.   **the belt thickness is 5 mm and slip is neglected,**

c.   **the belt is 5 mm tick, and a slip of 2% is allowed between the belt and each pulley.**

Solution: When the belt thickness is neglected and there is no slip, we have

a. $\qquad\qquad \dfrac{N_2}{N_1} = \dfrac{d_1}{d_2} \qquad \therefore d_2 = d_1 \times \dfrac{N_1}{N_2} = 600 \times \dfrac{240}{360} = \mathbf{400\ mm}$

b. When belt thickness $t = 5$ mm is considered and slip is neglected, we have

$$\frac{N_2}{N_1} = \frac{d_1 + t}{d_2 + t}$$

$$d_2 + t = (d_1 + t) \times \frac{N_1}{N_2} = (600 + 5) \times \frac{240}{360} = 403.3 \text{ mm}$$

$\therefore \qquad\qquad d_2 = 403.3 - 5 = \textbf{398.3 mm}$

c. When both belt thickness and slip are considered

$$\frac{N_2}{N_1} = \frac{d_1 + t}{d_2 + t}\left(1 - \frac{S}{100}\right)$$

Effective slip, $S = S_1 + S_2 - 0.01 S_1 S_2 = 2 + 2 - 0.01 \times 2 \times 2 = 3.96\%$

$$d_2 + t = (d + t) \times \frac{N_1}{N_2} \times \left(1 - \frac{S}{100}\right) = (600 + 5) \times \frac{240}{360} \times \left(1 - \frac{3.96}{100}\right) = 387.23 \text{ mm}$$

$\therefore d_2 = 387.23 - 5 = \textbf{382.2 mm}$

A shaft of 48 cm diameter is driven with the help of belt by an engine turning at 200 rev/min. The diameter of the engine pulley is 30 cm, the belt is 5 mm thick and a slip of 3% between the belt and each pulley is allowed. Determine the percentage of total effective slip and the speed of driven shaft.

Solution: Effective slip $S = S_1 + S_2 - 0.01 S_1 S_2 = 3 + 3 - 0.01 \times 3 \times 3 = \textbf{5.91\%}$

When both belt thickness and slip are considered,

$$\frac{N_2}{N_1} = \frac{d_1 + t}{d_2 + t}\left(1 - \frac{S}{100}\right)$$

$\therefore$ Speed of the driven shaft,

$$N_2 = N_1 \times \frac{d_1 + t}{d_2 + t} \times \left(1 - \frac{S}{100}\right) = 200 \times \frac{300 + 5}{480 + 5}\left(1 - \frac{5.91}{100}\right) = \textbf{1183.4 rev/min}$$

EXAMPLE 10.4.

Two pulleys A and B, with the sum of their diameters 100 mm, are connected by a belt. The driving pulley A rotates at 120 rpm while the driven pulley B turns 2400 rev/min. Presuming that the belt is 10 mm thick and the slip in the whole drive system is 5%, make calculations for the diameter of each pulley.

Solution: When the belt thickness and slip are considered,

$$\frac{N_a}{N_b} = \frac{D_b + t}{D_a + t} \times \left(1 - \frac{S}{100}\right)$$

where t is the belt thickness and S is the percentage slip in the entire drive system.

$$\therefore \quad \frac{1200}{2400} = \frac{D_b + 10}{D_a + 10} \times \left(1 - \frac{5}{100}\right)$$

$$0.5 = \frac{D_b + 10}{D_a + 10} \times 0.95; \quad \frac{D_b + 10}{D_a + 10} = 0.5263 \qquad (i)$$

Further $\qquad D_a + D_b = 100$ or $D_b = 100 - D_a \qquad (ii)$

From identities (i) and (ii),

$$\frac{100 - D_a + 10}{D_a + 10} = 0.5263$$

or $\qquad 110 - D_a = 0.5263\, D_a + 5.263$

$$D_a = \frac{110 - 5.263}{1.5263} = \textbf{68.62 mm}$$

and $\qquad D_b = 100 - 68.62 = \textbf{31.83 mm}$

10.5.3. Ratio of Tensions and Power Transmitted

The ratio of belt tensions T_1 on the tight side and tension T_2 on the slack side is

$$\frac{T_1}{T_2} = e^{\mu\theta} \text{ for flat belt drive, and}$$

$$\frac{T_1}{T_2} = e^{\frac{\mu\theta}{\sin\alpha}} \text{ for V-belt drive.}$$

where θ is the angle subtended at the center of pulley by the position of belt in contact with it and called the *angle of contact* or the *angle of lap*; μ is the coefficient of friction between the belt and pulley material; and α is half the angle of groove for the rope or V-belt drive,

Effective turning force $= (T_1 - T_2)$

$$= (T_1 - T_2)V$$

where V is the velocity of belt $= \omega r = \omega_1 r_1 = \omega_2 r_2$

Power $= (T_1 - T_2)V$ in watt

where values of T_1 and N_2 are in newton and the velocity V of the belt is in m/s.

EXAMPLE 10.5.
Calculate the force required to hold a weight of 10 kN suspended on a rope wrapped twice around a post. Take coefficient of friction $\mu = 0.2$.

Solution: $\mu = 0.2$ and $\theta = 2 \times 2\pi = 4\pi$ radian

Invoking the relation $\dfrac{T_1}{T_2} = e^{\mu\theta}$, we have

$$\frac{10}{T_2} = e^{0.2 \times 4\pi} = (2.718)^{2.512} = 12.326$$

$$\therefore \qquad T_2 = \frac{10}{12.36} = 0.811 \text{ kN} = \textbf{811 N}$$

EXAMPLE 10.6.
A belt is stretched over two identical pulleys of diameter D meter. The initial tension in the belt throughout is 2.4 kN when the pulleys are at rest. When this belt-pulley system is used to transmit power, it is found that an increase in tension on one side is equal to a decrease in tension on the other side. Find the maximum torque that can be transmitted by the belt drive. Take coefficient of friction between belt and pulley as 0.3

Solution: Since both the pulleys are of the same diameter, angle of the embrace (contact) $\theta = 180° = \pi$ radian.

Let increase/decrease in tension be Δt. Then
$$T_1 = 2.4 + \Delta t \text{ and } T_2 = 2.4 - \Delta t$$

Invoking the relation $\dfrac{T_1}{T_2} = e^{\mu\theta}$, we have

$$\frac{2.4 + \Delta T}{2.4 - \Delta T} = e^{0.3 \times \pi} = 2.57$$

or $\qquad 2.4 + \Delta t = 2.57 \times 2.4 - 2.57\Delta T; \ \Delta T = 1.06 \text{ kN}$

$\therefore \qquad\qquad T_1 = 2.4 + 1.06 = 3.46 \text{ kN}$

and $\qquad\qquad T_2 = 2.4 - 1.06 = 1.34 \text{ kN}$

Maximum torque that can be transmitted by the belt

$$= (T_1 - T_2) \times \frac{D}{2} = (3.46 - 1.34)\frac{D}{2} = \textbf{1.06 } \boldsymbol{D} \textbf{ kNm}$$

EXAMPLE 10.7.
Find out the number of turns a hauling rope must be wound round a rotating capstan in order to haul a load of 3 MN up to a gradient of 1 in 30. Presume the following data:

Resistance due to rolling = 0.00375 per newton load
Friction coefficient between the rope and drum = 0.35
Pull-on the free end of the rope = 250 N

Solution: The hauling is required up an inclined plane and for that T_1 has to overcome the frictional resistance as well as effort component of the incline. That is

$$T_1 = (3 \times 10^6 \times 0.00375) + \left(3 \times 10^6 \times \frac{1}{30} \right) = 111250 \text{ N}$$

$$T_2 = 250 \text{ N}$$

Invoking the relation $\dfrac{T_1}{T_2} = e^{\mu\theta}$, we have

$$e^{\mu\theta} = \frac{111250}{250} = 445$$

$$\mu\theta = \log_e 445; \ \theta = \frac{\log_e 445}{0.35} = 17.42 \text{ radians}$$

Number of turns $= \dfrac{17.42}{\pi} = 2.77$

Hence, three turns need to be given.

EXAMPLE 10.8.

A horizontal drum of belt drive carries the belt over a semicircle around it. The drum has a diameter of 1 m and when it rotates, 400 Nm of torque is transmitted. The drum has a mass of 25 kg and the coefficient of friction between the belt and drum is 0.25. Neglecting mass of belt, determine tensions induced in the limbs of the belt, and the vertical reaction on the bearings.

Solution: Since the drum carries the belt over a semicircle around it,
angle of contact = 180° = π radians

If T_1 and T_2 are the tensions on the tight and slack side of the belt, then

$$\frac{T_1}{T_2} = e^{\mu\theta} = e^{0.25 \times \pi} = 2.19 \qquad (i)$$

From the relation for torque transmitted: $T = (T_1 - T_2)\, r$

$$T_1 - T_2 = \frac{T}{r} = \frac{400}{0.5} = 800 \qquad (ii)$$

From expressions (i) and (ii)

$$2.19\,T_2 - T_2 = 800; \ T_2 = \frac{800}{1.19} = \textbf{672.27 N}$$

$$T_1 = 2.19 \times 672.27 = \textbf{1472.27 N}$$

Vertical reaction on the drum bearings,

$$= (T_1 + T_2) + W \text{ where } W \text{ is the weight of drum}$$
$$= (1472.27 + 672.27) + (25 \times 9.81)$$
$$= \textbf{2389.79 N}$$

EXAMPLE 10.9.
Determine the minimum value of weight W required to cause motion of a block that rests on a horizontal plane. The block weighs 300 N and the coefficient of friction between the block and plane is 0.6. The angle of wrap over the pulley is 90° and the coefficient of friction between the pulley and rope is 0.3.

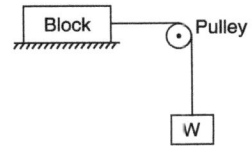

FIGURE 10.25

Solution: Since the weight W impends vertical motion in the downward direction, the tension on the two sides of the pulley will be as shown in Figure 10.26.

$$T_1 = W; \ \mu = 0.3; \ \theta = 90° = \frac{\pi}{2} \text{ radians}$$

Invoking the relation

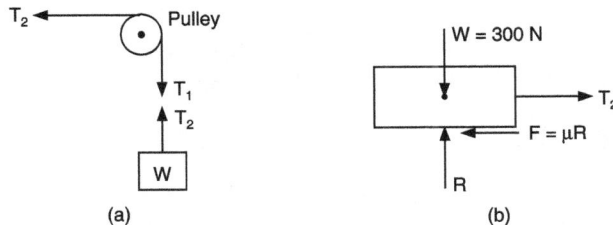

(a)   (b)

FIGURE 10.26

$$\frac{T_1}{T_2} = e^{\mu\theta}; \ \frac{W}{T_2} = e^{0.3 \times \pi/2} = (2.718)^{0.471} = 1.60$$

Considering equilibrium of block,

$$\sum F_x = 0; \ T_2 = F = \mu R$$
$$\sum F_y = 0; \ R = 300 \text{ N}$$

$\therefore$ $\qquad$ $T_2 = 0.6 \times 300 = 180$ N

Hence $\qquad$ $W = 1.6 \times 180 = \textbf{288 N}$

EXAMPLE 10.10.

A flat belt running on a pulley 1m in diameter is to transmit 7.5 kW power at a speed of 200 rev/min. The angle of embrace of belt and pulley is 170° and the coefficient of friction between the belt and pulley is 0.25. If the pull-in belt is not to exceed 200 N/cm, determine the width of belt.

Solution: Angle of embrace (lap)

$$= 170° = \frac{170}{180} \times \pi = 2.965 \text{ radians}$$

If T_1 and T_2 are the tensions on the tight and slack side of the belt, then

$$\frac{T_1}{T_2} = e^{\mu\theta} = e^{0.25 \times 2.965} = e^{0.741} = 2.1$$

$\therefore$ $\qquad$ $T_1 = 2.1 \, T_2$

Speed of belt, $V = \dfrac{\pi d_1 N}{60} = \dfrac{\pi \times 1 \times 200}{60} = 10.46$ m/s

Power transmitted by the belt

$$= (T_1 - T_2) \times V = (2.1 \, T_2 - T_2) \times 10.46 = 11.506 \, T_2 \text{ Nm/s}$$

$\therefore$ $\qquad$ $7.5 \times 10^3 = 11.506 \, T_2$

$\qquad\qquad T_2 = 651.83$ N and

$\qquad\qquad T_1 = 651.83 \times 2.1 = 1368.85$ N

Hence necessary width of belt $= \dfrac{1368.85}{200} = \textbf{6.84 cm}$

10.5.4. Chain Drive

The velocity ratio in belt and rope drives may vary due to slip, momentary overloads, or because of contact surface becoming slightly greasy. For constant velocity ratio positive drive with a short distance between the drive and driven shafts, one would use the chain drive.

The chains are made of rigid links which are hinged together. This hinging provides flexibility needed for wrapping them around the driving and follower (driven) wheels. The wheels (called sprockets) have projecting teeth that fit into the corresponding recesses in the links of the chain. The wheel and the chain move together without slip and a perfect velocity ratio is ensured.

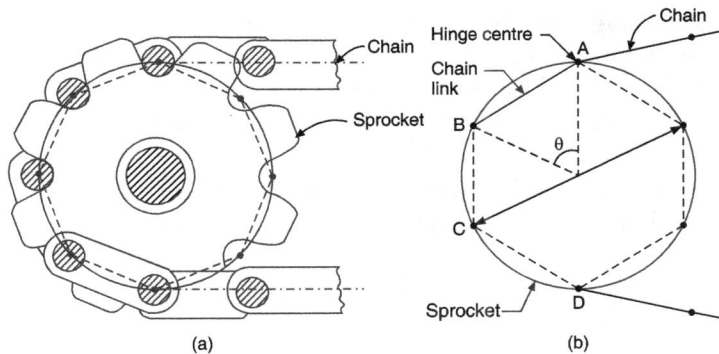

FIGURE 10.27 Sprocket and chain

Chain drive is used where the distance between shaft centers is short such as in cycles, motor vehicles, agricultural machinery, road rollers, heavy earth moving machinery, etc.

The advantages of the chain drive are:

- chain drive takes less space than a belt or rope drive,

- no slip takes place and that ensures a perfect velocity ratio,

- more suitable for transmission of power when the distance between the shafts is less,

- less load on the shaft,

- high transmission efficiency,

- capable of transmitting a good amount of power, and

- a single chain can transmit motion to several shafts.

However, the chain drive requires accurate mounting and careful maintenance, is relatively high in cost and is quite prone to velocity fluctuations particularly when they are overstretched. Further, gradual stretching of chains necessitates the removal of its links from time to time.

Classification of Chains

The chains are primarily classified into hoisting and hauling chains, conveyor chains, and power transmission chains.

(a) Chain with oval links   (b) Chain with square links

FIGURE 10.28 Hoisting and hauling chains

- **Hoisting and hauling chains:** The links in such chains are of oval or square shape. The joint of oval links are welded, and the sprockets (wheels) used for such chains have receptacles to receive the links. These chains operate at low speeds such as in chain hoists or in anchors for marine works.

  The chains with square links are used for hoists, cranes, and dredgers. These chains have low manufacturing costs but are easily prone to kinks on overloading.

- **Conveyor chains:** The links in such chains are either of hook joint type or of closed joint type.

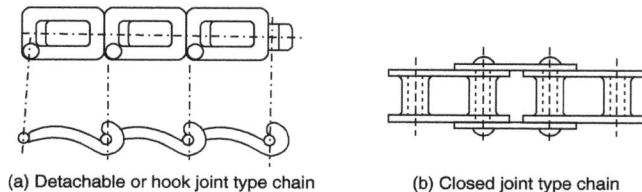

(a) Detachable or hook joint type chain   (b) Closed joint type chain

FIGURE 10.29 Conveyor chains

The conveyor chains are meant for continuous conveying and elevating the material. These chains run at average slow speeds of 1.75 m/s and lack smooth running qualities.

- **Power transmission chains:** These chains are available in the block, roller, and silent configuration. The link bearing surface for all these types is machined, hardened, and ground. These chains are essentially built for high-speed performance and have provision for efficient lubrication.

The **block chains** belong to the earliest stages of development in power transmission and are being put to some use as conveyor chains operating at comparatively low speeds. There occurs rubbing action between the teeth and links when approaching or leaving the teeth of the sprocket. This leads to noisy operations.

The **roller chain** assembly essentially consists of:

i.   roller link plate and the pin link plate, and

ii. pins, bushes, and rollers.

The roller is free to rotate on the bush which is secured in its hole. The pin passes through the hole and the roller is held by the roller link plates. The central pins are joined and held in position by the pin link plates provided on both sides. The outward lateral sliding of the pin link plates is prevented either by hammering the pin ends to rivet the head shape or using the split pins.

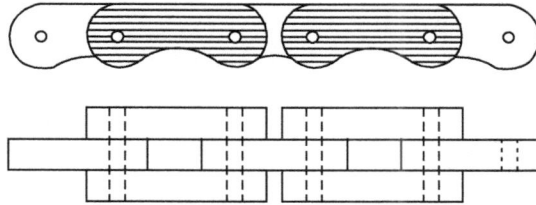

FIGURE 10.30 Block chain

The salient features of roller chains are:

- strong and simple construction,

- quieter operation; there is only little noise due to the impact of roller on the sprocket wheel teeth,

- provides good service even under severe working conditions, and

- requires little lubrication.

However, the wear and stretching of the parts leads to elongation of chain. That results in unequal fitting of rollers into the cavities of the wheel and the load has to be carried by one or only a few teeth.

10.6. GEARS AND GEAR DRIVE

A toothed wheel or gear is essentially a wheel with teeth cut on its periphery.

Power or motion is transmitted from one shaft to another with gear drive when

- center distances are relatively short,

- speed of the shaft is low and the use of a belt drive is not recommended,

- positive drive is necessary, that is, velocity ratio is fixed and known with certainty,

- there is a need to step up or step down the speed,

- high torque is to be transmitted, and

- precise timing is required.

The gear drive has a compact layout and it provides a highly efficient and reliable service. However, the operation tends to become noisy, the whole set is affected when one tooth gets damaged, and the manufacture of gears requires special tools and equipment.

10.6.1. Types of Gears

The commonly used forms of toothed gearing are:

- **Spur gear:** It is a cylindrical gear whose tooth traces are straight lines parallel to the gear axis. Further, the tooth profile is identical from one side of the face to the other.

  The spur gears are used for transmitting motion between two shafts whose axes are parallel and coplanar.

  These gears have a high (96–98%) efficiency of power transmission and are free

FIGURE 10.31 Spur gear

from any axial thrust during tooth engagement. Accordingly, they have been successfully employed as sliding gears for speed change mechanisms in gear boxes of the lathe. However, compared to gears of other types, spur gears are noisier in operation, wear out readily and develop backlash.

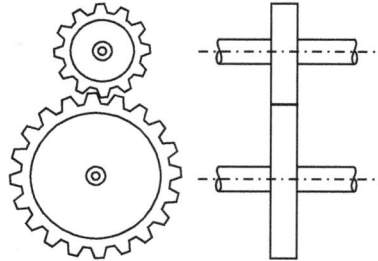

- **Helical gear:** It is a cylindrical gear whose tooth traces are straight helices, teeth are inclined at an angle to the gear axis. The teeth are thus of helical or screw form. This ensures smooth action and accurate maintenance of velocity ratio. However, lateral thrust is set up due to teeth being inclined. This lateral thrust gets neutralized by using double-helical gears called **herringbone gears**. The double-helical gears are the equivalent of two helical gears secured together, but they are manufactured as one piece. The teeth may be continuous or separated by

FIGURE 10.32 Helical gear

a small gap. Further, the two helices are of the opposite hand and meet at a common axis.

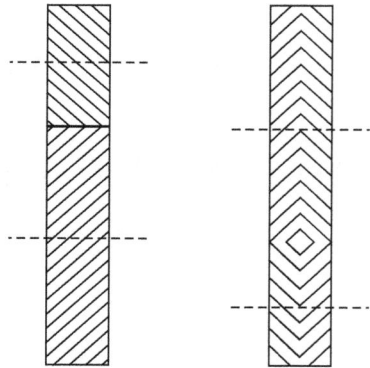

  The helical gears are used in automobile gear boxes, and in steam and gas turbines for speed reduction. The herringbone gears are used in machinery where large power is transmitted at low speeds.

- **Bevel gear:** The bevel gear wheels conform to the frusta of cones having a common vertex; tooth traces are straight-line generators of the cone.

The bevel gears are used to connect two shafts whose axes are coplanar but intersecting. When the shafts are at right angles and the wheels equal in size, the bevel gears are called *miter gears*. When the bevel gears have their teeth inclined to the face of the bevel, they are known as *helical bevel gears*.

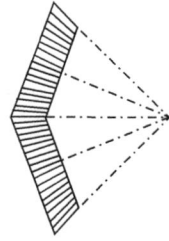

FIGURE 10.33 Bevel gear

- **Spiral gear:** These are identical to helical gears with the difference that these gears have a point contact rather than a line contact. These gears are used when the connection is to be made between intersecting and co-coplanar shafts.

- **Worm gear:** The system consists of a worm which is basically part of a screw. The warm meshes with the teeth on a gear wheel called the worm wheel.

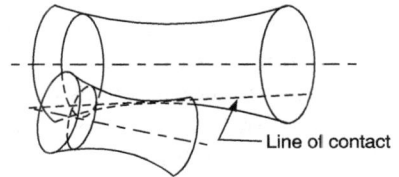

FIGURE 10.34 Spiral gear

The worm gear is used for connecting two nonparallel, non-intersecting shafts which are usually at right angles. This gearing system is smooth and quiet in operation and provides a high gear ratio; rotational speed of the worm is quite high compared to that of wheel. Their use is recommended when high-speed reduction (more than 10:1) is required.

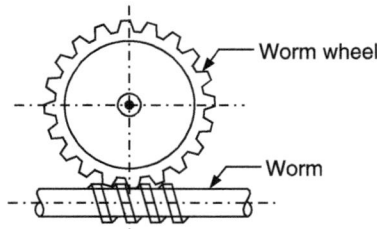

FIGURE 10.35 Worm gear

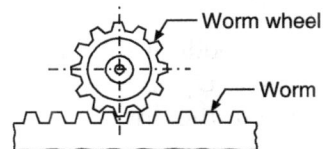

FIGURE 10.36 Rack and pinion

- **Rack and pinion:** Rack is a straight-line spur gear of infinite diameter. It meshes, both internally and externally, with a circular wheel called pinion. The arrangement finds application where linear motion is to be converted into rotary motion and vice versa.

- **Internal and external gearing:** Two toothed wheels on parallel shafts may gear either externally or internally.

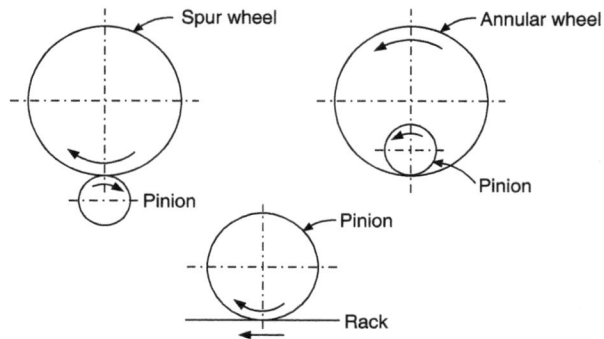

FIGURE 10.37 Internal and external gearing

In external gearing, the motion of the two wheels is unlike, that is, in the opposite direction. The larger wheel of the gear system is known as **spur** and the smaller wheel as **pinion**.

In internal gearing, the motion of the two wheels is alike; both rotate in the same direction. The external large wheel is called **annular wheel** and the small internal wheel as **pinion**.

The gears of shaft in the rack and pinion arrangement mesh externally and internally with the gears in a straight-line. With the help of rack and pinion, it is possible to convert linear motion into rotary motion and vice versa.

10.6.2. Gear Terminology

Figure 10.38 shows two gears 1 and 2 meshed against one another.

A tooth of the driving gear (center O_1) meshes against a tooth of the driven gear (center O_2) at point P. The point P is called the *pitch point*. The circles whose centers are O_1 and O_2, and radii O_1P and O_2P are called *pitch circles* of wheels 1

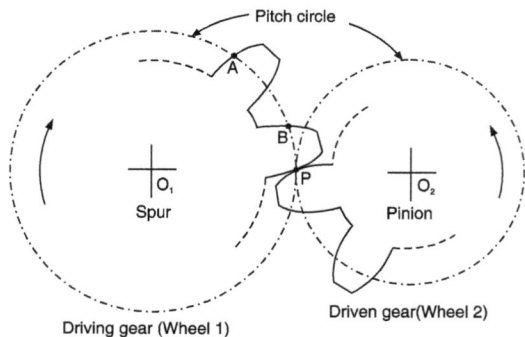

FIGURE 10.38 External gearing

and 2, respectively. A pitch circle is essentially an imaginary circle that by pure rolling action gives the same motion as the actual gear.

Further, on the pitch circle of wheel 1, let A be a point on one tooth and B be the corresponding point on the adjacent tooth. The distance AB (measured along the circular arc) is called *circular pitch* of wheel.

1. Circular pitch may then be defined as

"distance measured on the circumference of the pitch circle from a point of one tooth to the corresponding point on the next tooth"

$$\text{Circular pitch } p_c = \frac{\text{circumference of pitch circle}}{\text{number of teeth}} = \frac{\pi D}{T} \qquad (10.7)$$

where D is the pitch diameter and T is the number of teeth.

Two gears will mesh correctly if they have the same circular pitch. Therefore

$$p_c = \frac{\pi D_1}{T_1} = \frac{\pi D_2}{T_2} \text{ or } \frac{D_1}{T_1} = \frac{D_2}{T_2}$$

Thus the diameter of a wheel is proportional to the number of teeth on it.

Further, when the teeth of two or more gears mesh with one another, the linear speeds of the two gears will remain the same.

Linear speed of gear $1 = \pi D_1 N_1$

Linear speed of gear $2 = \pi D_2 N_2$

where D_1 and D_2 are the diameters, and N_1 and N_2 are the revolutions made by gears 1 and 2 per unit time.

Equating the linear speeds of the two gears,

$$\pi D_1 N_1 = \pi D_2 N_2; \qquad \frac{N_2}{N_1} = \frac{D_1}{D_2}$$

Since the diameter of a wheel is proportional to the number of teeth on it, we get

$$\text{velocity ratio} = \frac{N_2}{N_1} = \frac{T_1}{T_2}$$

Diametral pitch: It represents the number of teeth on a wheel per unit of its diameter.

$$\text{Diametral pitch } p_d = \frac{\text{number of teeth}}{\text{pitch circle diameter}} = \frac{T}{D} \qquad (10.8)$$

From Eqs. (10.7) and (10.8), we have

$$p_c \times p_d = \pi$$

Evidently, if the circular pitches of two wheels are equal, their diametral pitches are also equal.

Module: Module represents the ratio of pitch circle diameter (in mm) to the number of teeth. Obviously, the module is the reciprocal of the diametral pitch.

$$\text{Module } m = \frac{D}{T} \tag{10.9}$$

Refer to Figure 10.39 for some of the terms associated with profile of a gear tooth.

Addendum circle is the circle bounding the outer ends of the teeth and concentric with the pitch circle.

The radial distance between the pitch circle and addendum circle is called *addendum*.

Dedendum circle is the circle bounding the bottom of the tooth and concentric with the pitch circle.

The radial distance between the pitch circle and the dedendum circle is called *dedendum*.

The surface of tooth above the pitch surface is called the *face* of the tooth. *Flank* is the surface of tooth below the pitch surface. *Tooth space* is the width of tooth measured along the pitch circle. The length of the arc between the sides of a gear tooth, measured on the pitch circle is called *thickness* of *tooth* or *circular thickness*.

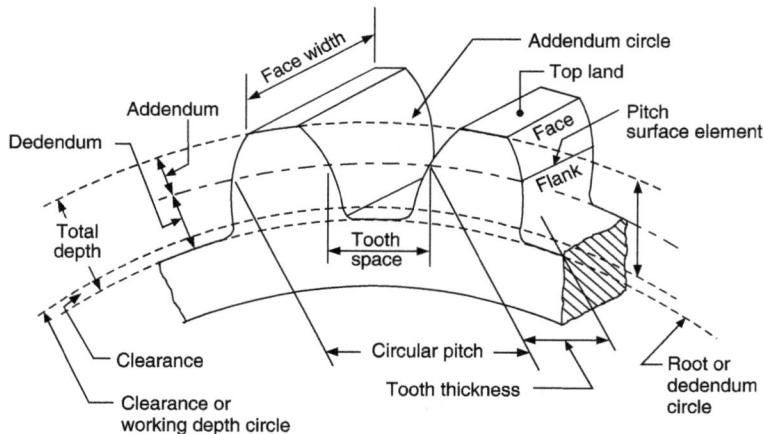

FIGURE 10.39 Gear terminology

The radial distance between the addendum circle and the dedendum circle is called the *total depth*. It equals the sum of the addendum and dedendum. *Working depth* equals the sum of the addenda of the two mating gears. The radial distance from the top of tooth to the bottom of tooth is called *clearance*. The standard value of clearance equals 0.157 *m* where *m* is the module. Then

dedendum = addendum + clearance = $m + 0.157\,m = 1.157\,m$
Addendum equals module and is always less than dedendum.

EXAMPLE 10.11.
A toothed gear is stated to have 48 teeth and circular pitch of 18. Determine the pitch diameter, diametral pitch, and module of the gear.

Solution:　Circular pitch $p_c = \dfrac{\pi D}{T}$

$\therefore$ pitch diameter $D = \dfrac{p_c \times T}{\pi} = \dfrac{18 \times 48}{\pi} = \textbf{275.16 mm}$

b. Invoking the relation: $p_c \times p_d = \pi$

$\therefore$ diametrical pitch $p_d = \dfrac{\pi}{p_c} = \dfrac{\pi}{18} = \textbf{0.174}$

c. Module, $m = \dfrac{D}{T} = \dfrac{275.16}{48} = \textbf{5.73 mm per tooth}$

EXAMPLE 10.12.
Two matching spur gears have 70 and 30 teeth, respectively. Corresponding to a module of 5 mm, determine the center to center distance between the gears.

Solution: Center to center distance between two matching spur gears is

$$l = \frac{D_1 + D_2}{2}$$

From the relation $m = D/T$; $D = mT$ and therefore

$$l = \frac{m(T_1 + T_2)}{2} = \frac{5(70 + 30)}{2} = \textbf{250 mm}$$

EXAMPLE 10.13.
Two parallel shafts are to be connected by spur gearing. One shaft runs at 100 rpm and the other shaft turns 300 rev/min. The appropriate center to center distance between the shafts is 500 mm. If the circular pitch for the toothed gears is 20 mm, determine the number of teeth on each wheel and the exact distance between the shaft centers.

Solution: Center to center distance between two mating spur gears is

$$l = \frac{D_1 + D_2}{2}; D_1 + D_2 = 2 \times 500 = 1000 \qquad (i)$$

For two meshing spur gears,

$$\frac{N_1}{N_2} = \frac{D_2}{D_1}; \quad \frac{100}{300} = \frac{D_2}{D_1}; \quad D_2 = 3D_1 \qquad (ii)$$

From the identities (i) and (ii)

$$D_1 + 3D_1 = 1000; \quad D_1 = 250 \text{ mm}$$

and

$$D_2 = 3D_1 = 3 \times 250 = 750 \text{ mm}$$

From the relation for circular pitch, $p_c = \dfrac{\pi D}{T}$, we get

Number of teeth on the first gear $T_1 = \dfrac{\pi \times 250}{20} = 39.25$

Number of teeth on the second gear $T_2 = \dfrac{\pi \times 750}{20} = 117.75$

However, the number of teeth on both the gears has to be an exact number. Accordingly, let it be presumed that the number of teeth on the first gear is 40. The second gear will then have 120 teeth. Then

Revised pitch circle diameter of first wheel $D'_1 = \dfrac{T_1 \times p_c}{\pi} = \dfrac{40 \times 20}{\pi} = 254.78$ mm

Revised pitch circle diameter of second wheel $D'_2 = \dfrac{T_2 \times p_c}{T_2} = \dfrac{120 \times 20}{\pi} = 764.34$ mm

$\therefore$ Correct distance between the shaft centers,

$$l_{correct} = \frac{D'_1 + D'_2}{2} = \frac{254.78 + 764.34}{2} = \mathbf{509.65 \text{ mm}}$$

10.6.3. Gear Trains

Gear train is any combination of gear wheels by means of which power and motion is transmitted from one shaft to another.

Various types of gear trains are:

1. simple gear train,   **2.** compound gear train,

3. reverted gear train, and   **4.** epicyclic gear train.

The nature of the train used depends upon the velocity ratio required and the relative position of the axes of the shafts.

- **Simple gear train:** Refer to Figure 10.40 for a simple gear train in which each shaft carries one wheel only. Between the driving gear 1 and the driven or follower gear 4, two intermediate gears 2 and 3 have been provided.

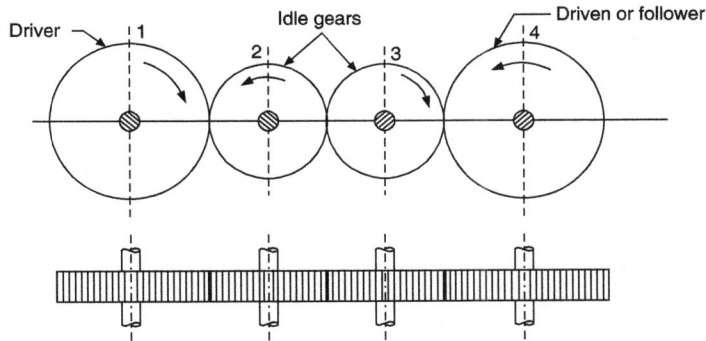

FIGURE 10.40 Simple gear train

Let,

N_1, N_2, N_3, and N_4 be the rotational speeds of gears 1, 2, 3, and 4
T_1, T_2, T_3, and T_4 be the number of teeth on gears 1, 2, 3, and 4.
The driving gear 1 is in mesh with the intermediate gear 2, and the speed ratio for these gears is

$$\frac{N_1}{N_2} = \frac{T_2}{T_1} \tag{10.10}$$

Likewise, the intermediate gears 2 and 3 are in mesh and accordingly

$$\frac{N_2}{N_3} = \frac{T_3}{T_2} \tag{10.11}$$

Further, the intermediate gear 3 is in engagement with the driven (follower) gear, and therefore

$$\frac{N_3}{N_4} = \frac{T_4}{T_3} \tag{10.12}$$

The speed ratio N_1/N_4 for the gear assembly can be obtained by multiplying identities (10.10), (10.11), and (10.12).

$$\frac{N_1}{N_2} \times \frac{N_2}{N_3} \times \frac{N_3}{N_4} = \frac{T_2}{T_1} \times \frac{T_3}{T_2} \times \frac{T_4}{T_3} \ \text{ or } \ \frac{N_1}{N_4} = \frac{T_4}{T_1}$$

$$\therefore \text{ Speed or velocity ratio} = \frac{\text{speed of the driving wheel}}{\text{speed of the driven wheel}}$$

$$= \frac{\text{number of teeth on the driven wheel}}{\text{number of teeth on the driving wheel}}$$

The train value of a gear train is defined as the reciprocal of the speed ratio of the gear train.

$$\text{Train value} = \frac{\text{speed of the driven wheel}}{\text{speed of the driving wheel}}$$

$$= \frac{\text{number of teeth on the driving wheel}}{\text{number of teeth on the driven wheel}}$$

Comments:

i. The speed ratio and the train value are independent of the size and number of intermediate gears.

ii. The intermediate gears help to obtain the desired direction of motion of the driven gear.

With odd number of idle gears, the direction of rotation of the driven wheel will be the same as that of the driving wheel. Both the driving and driven wheels will have opposite directions of motion when the number of idle gears happens to be even.

The idle gears are also used when shafts with large center distance are to be connected.

- **Compound gear train:** Refer to Figure 10.41 for a compound gear train in which all the gears do not mesh with one another. Between the driving gear 1 and the driven (follower) gears 6, two intermediate compound gears have been provided. A compound gear is a gear that carries two wheels mounted on the same shaft. Gears 2 and 3 constitute a compound gear mounted on shaft M, and the compound gear comprising gears 4 and 5 is mounted on shaft P.

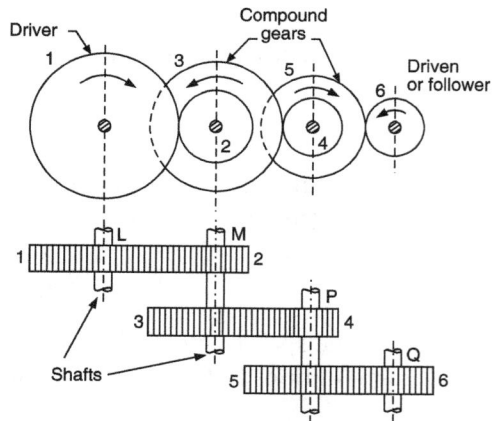

FIGURE 10.41 Compound gear train

In the compound gear train, gear 2 is a follower to the driving gear 1, and gear 3 is the driving wheel for gear 4. Obviously gears 1, 3, and 5 are the drivers, and gears 2, 4, and 6 are the drivens.

Let, N_1, N_2, N_3, N_4, N_5, and N_6 be the speeds of the respective gears

T_1, T_2, T_3, T_4, T_5, and T_6 are the number of teeth on respective gears. Gear 1 is in mesh with gear 2 and therefore its speed ratio is

$$\frac{N_1}{N_2} = \frac{T_2}{T_1} \qquad (a)$$

Gear 3 is in mesh with gear 4 and therefore the speed ratio is

$$\frac{N_3}{N_4} = \frac{T_4}{T_3} \qquad (b)$$

Similarly, for gears 5 and 6 which engage each other, the speed ratio is

$$\frac{N_5}{N_6} = \frac{T_6}{T_5} \qquad (c)$$

The speed ratio N_1/N_6 for the gear assembly can be worked out by multiplying identities (a) (b) and (c)

$$\frac{N_1}{N_2} \times \frac{N_3}{N_4} \times \frac{N_5}{N_6} = \frac{T_2 \times T_4 \times T_6}{T_1 \times T_3 \times T_5} \qquad (10.13)$$

Gears 2 and 3 are mounted on the same shaft M and therefore $N_2 = N_3$. Again gears 4 and 5 are mounted on the same shaft P and therefore $N_4 = N_5$. With these substitutions, equation 10.13 transforms to

$$\text{Speed or velocity ratio} = \frac{N_1}{N_6} = \frac{T_2 \times T_4 \times T_6}{T_1 \times T_3 \times T_5}$$

That is $\dfrac{N_1}{N_6} = \dfrac{\text{speed of the first driver}}{\text{speed of the last driven}}$

$$= \frac{\text{product of the number of teeth on drivens}}{\text{product of the number of teeth on drivers}} \qquad (10.14)$$

A compound gear train helps to obtain a large speed reduction from the first to the last shaft with an assembly of small gears. Usually for a speed reduction of the order of 8:1, compound gear train or worm gearing is employed.

- **Reverted gear train:** A reverted gear train manifests when the first driving gear and the last driven gear are on the same axis; axes are coincident and coaxial.

Refer to Figure 10.42 for the arrangement of a reverted gear train.

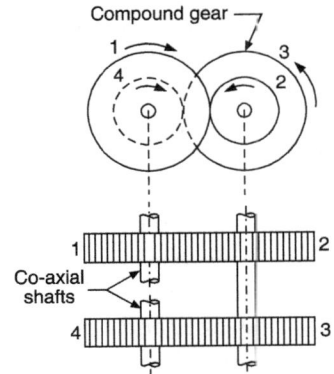

FIGURE 10.42 Reverted gear train

Gear 1 is the first driving gear (clockwise motion); it meshes with gear 2 and makes it turn is the opposite direction (counterclockwise motion). Gears 2 and 3 are mounted on the same shaft and constitute a compound gear. The motion of gear 3 is like that of gear 2 (counterclockwise). Gear 3 acts as a driver for gear 4. The direction of rotation of gear 4 will be clockwise, that is, opposite to that of gear 3. Apparently, in the reverted gear train, the motion of the first gear and the last gear is a like.

Let D_1, D_2, D_3, and D_4 be the pitch circle diameters of the respective wheels, and their corresponding speeds be N_1, N_2, N_3, and N_4.

The distance between shaft centers of gears 1 and 2, and that between gears 3 and 4 is the same. That gives

$$\frac{D_1 + D_2}{2} = \frac{D_3 + D_4}{2} \qquad \text{or} \qquad D_1 + D_2 = D_3 + D_4$$

Presuming that the circular pitch or module of all the gear wheels is the same, the number of teeth on each gear is proportional to its circumference or diameter.

That gives

$$T_1 + T_2 = T_3 + T_4$$

Now, Speed ratio $\dfrac{N_1}{N_4} = \dfrac{\text{number of teeth on drivens}}{\text{number of teeth on drivers}} = \dfrac{T_2 \times T_5}{T_1 \times T_3}$ \hfill (10.15)

The reverted gear train is used to connect hour hand to minute hand in a clock mechanism. Other applications lie in automotive transmissions, lathe back gears, and industrial speed reducers.

- **Epicyclic gear train:** In the simple and compound gear trains, the axis of the wheels remain fixed relative to one another. The epicyclic gear train is a special type of gear train in which the axis of rotation of one or more of the wheels is carried on an arm and this arm is free to rotate about the axis of rotation of one or the other wheels in the train.

  Refer to Figure 10.43 for the arrangement of epicyclic gear train. The gear A and the arm C can rotate about the axis at P, the gear B meshes with gear A and has its axis of rotation on the arm at Q. When gear A is fixed and arm C is made to rotate about P, gear B would be forced to roll around the outside of gear A.

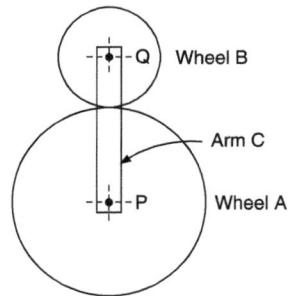

FIGURE 10.43 Epicyclic gear train

With N_C revolutions of the arm, the gear wheel B rotates about its axis by speed N_B given by

$$\frac{N_B}{N_C} = 1 + \frac{T_A}{T_B} \qquad (10.16)$$

The epicyclic gearing can bring about extensive speed reduction and is capable of transmitting high-velocity ratios with gears of moderate size in a comparatively short space. These aspects render it ideally suitable for use in differential gears of automobiles, wrist watches, hoists, pulley blocks, etc.

EXAMPLE 10.14.
A simple gear train consists of two gears only, each gear mounted on separate parallel shafts. The number of teeth on the driving and driven gears are 28 and 70, respectively. If the driving gear turns 1200 revolutions per minute, determine:

i. **speed ratio and train value of the gear train,**
ii. **speed of the driven gear, and**
iii. **direction of rotation of the driven gear**

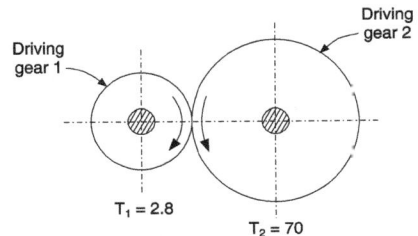

FIGURE 10.44

if the driving gear rotates clockwise

Solution: Refer to Figure 10.44 for the simple gear train consisting of two gears only. Gear 1 is driving gear 2.

$$\text{Speed ratio } \frac{N_1}{N_2} = \frac{T_2}{T_1} = \frac{70}{28} = \mathbf{2.5}$$

Train value of a gear train is defined as the reciprocal of the speed ratio of gear train

$$\text{Train value} = \frac{1}{\text{speed ratio}} = \frac{1}{2.5} = \mathbf{0.4}$$

ii. Speed ratio $\dfrac{N_1}{N_2} = 2.5$ as calculated above

∴ Speed of driven gear, $N_2 = \dfrac{N_1}{2.5} = \dfrac{1200}{2.5} = \mathbf{480\ rev/min}$

iii. In a simple gear train, the two meshing gears always move in opposite directions. As the driving gear is rotating clockwise, the direction of rotation of the driven gear would be counterclockwise.

EXAMPLE 10.15.

A simple gear train consists of three gears; each mounted on a separate parallel shaft as shown in Figure 10.45.

The number of teeth on gears 1, 2, and 3 are 28, 42, and 70, respectively, and the driving gear 1 rotates clockwise at speed of 1000 rpm. Calculate (*i*) the speed ratio of the gear train; (*ii*) speed and the direction of rotation of the follower.

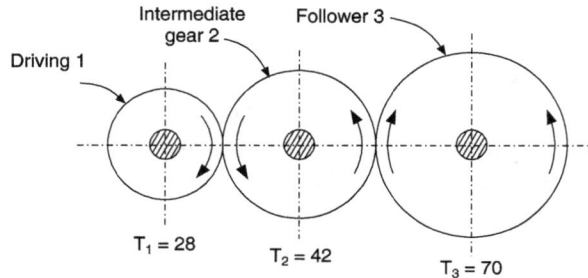

FIGURE 10.45

Solution: Speed ratio $= \dfrac{\text{speed of driver}}{\text{speed of follower}} = \dfrac{\text{number of teeth on follower}}{\text{number of teeth on driver}}$

$$= \frac{70}{28} = \textbf{2.5}$$

ii. Speed ratio $= \dfrac{N_1}{N_3} = 2.5$ as calculated above

iii. ∴ Speed of follower $N_3 = \dfrac{N_1}{2.5} = \textbf{400 rpm}$

With odd number of idle (intermediate) gears, the direction of rotation of the follower is the same as that of driving wheel. Here the driving gear is rotating clockwise and as such the direction of rotation of the follower would also be clockwise.

EXAMPLE 10.16.

A compound gear train consists of six gears and the number of teeth on the gears are as follows:

Gear	1	2	3	4	5	6
No. of teeth	35	80	40	125	45	115

Gears 2 and 3 are on the same shaft, and so are gears 4 and 5 on another shaft. Gear 1 drives gear 2, gear 3 meshes with gear 4, and gear 5 is in engagement with gear 6. Sketch the arrangement.

If the input shaft mounted on gear 1 turns 800 rev/min, determine the rotational speed of the output shaft mounted on gear wheel 6.

Solution: Refer to Figure 10.46 for the arrangement of the given compound gear train

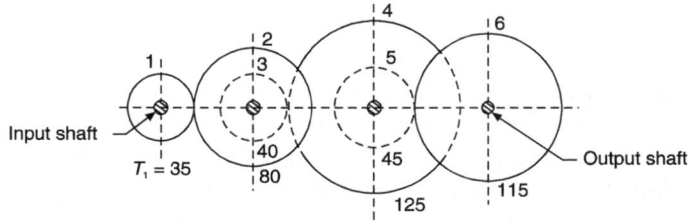

FIGURE 10.46

Driving gears are: 1, 3, and 5
Driven gears are: 2, 4, and 6.

For compound gear train,

$$\text{speed or velocity ratio} = \frac{\text{speed of the first driver}}{\text{speed of the last driven}}$$

$$= \frac{\text{product of the number of teeth on drivens}}{\text{product of the number of teeth on drivers}}$$

Thus,

$$\frac{N_1}{N_6} = \frac{T_2 \times T_4 \times T_6}{T_1 \times T_3 \times T_5} = \frac{80 \times 125 \times 115}{35 \times 40 \times 45} = 18.25$$

$\therefore$ speed of the last gear (gear wheel 6) $N_6 = \dfrac{800}{18.25} = \mathbf{43.83 \ rev/min}$

EXAMPLE 10.17.

A fixed gear having 200 teeth is in mesh with another gear having 50 teeth. The two gears are connected by an arm which makes one revolution about the center of the bigger gear.

Determine the number of turns made by the smaller gear.

Solution: Refer to Figure 10.43. The arrangement corresponds to an epicyclic gear train.

The ratio of the speed of wheel B to that of arm C is given by

$$\frac{N_B}{N_C} = 1 + \frac{T_A}{T_B} = 1 + \frac{200}{50} = 4$$

$\therefore$ $$N_B = 4N_C = 4 \times 1 = \mathbf{4}$$

10.7. CAMS AND FOLLOWERS

A cam-follower is a higher pair mechanism; its reciprocating or rotating element imparts the desired motion to another element. The desired motion may be reciprocating, rotating, or oscillatory in nature. The driving element is called *cam* and the driven member is referred to as *follower*. The direct point contact between the cam and the follower is ensured by a spring.

Generally, the cam is connected to a frame forming a turning pair, and the connection between the follower and the frame constitutes a sliding pair. As such the cam-follower mechanism is a three-link mechanism of the higher type. The three links are:

i.   cam which is the driving link and has a straight or curved surface,

ii. follower which is the driven link; it gets its motion due to its contact with the surface of cam, and

iii. frame which supports the cam and guides the follower.

Cam-follower mechanisms are quite impactful, easy to design and are used for generating complex coordinated movements. The common applications of the cam-follower mechanism are found in

• clocks and watches,

• IC engines for operating the valves,

• automatic screw-cutting machines, and

• printing machines and shoe-making machines.

Attention has been directed in this chapter to the study of various types of cams and followers, motion of follower with a given profile of cam, and to draw the profile of cam with the given motion of the follower.

10.7.1. Classification of Cams and Followers

Cams are classified according to shape, follower movement, and the manner of the constraint of the follower. However, the following two types are considered important:

a. Radial cams in which the follower reciprocates or oscillates in a plane that is perpendicular to the axis of cam.

b. Cylindrical cams in which the reciprocating or oscillatory movement of the cam is in a plane that lies parallel to the axis of cam.

The important shapes of the radial cams are:

- *Wedge cam* that has a specified contour. The reciprocating motion of the cam may impart reciprocating or an oscillatory motion to a knife-edged follower.

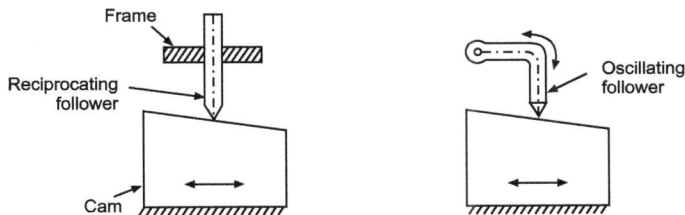

FIGURE 10.47 Wedge cams

- *Tangent cam* that has straight flanks and circular nose. The roller follower used in conjunction with such a cam may get reciprocating or oscillating motion.

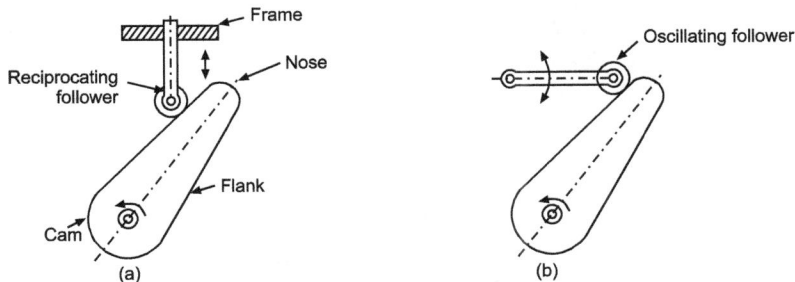

FIGURE 10.48 Tangent cams

- *Circular cam* that has a circular flank and circular nose. Figure 10.49 shows a circular cam operating with a flat-faced follower and an offset flat-faced follower. In the offset arrangement, the vertical center lines of the cam and the follower do not coincide.

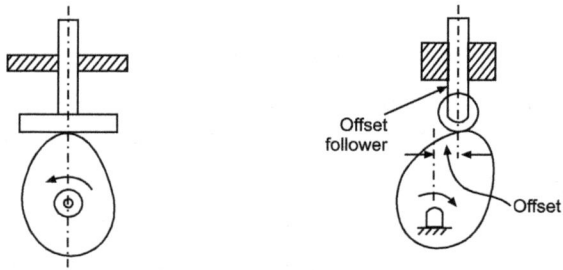

FIGURE 10.49 Circular cams

The cylindrical cam, also known as drum or barrel cam has a circumferential contour cut on the surface of a cylinder that rotates about its axis. The reciprocating or oscillating motion of the follower is in a plane parallel to that of the axis of the cam.

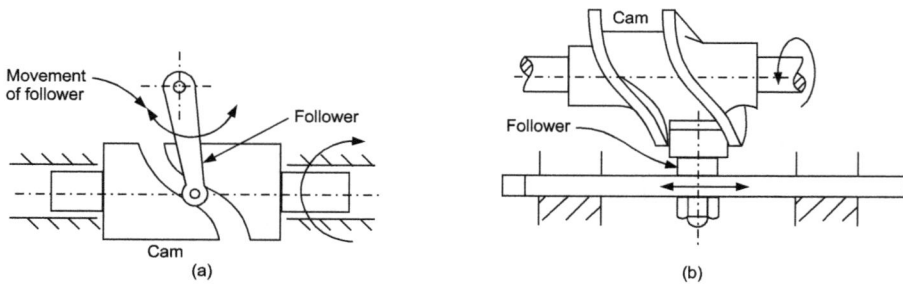

FIGURE 10.50 Cylindrical cams: (a) oscillating and (b) reciprocating

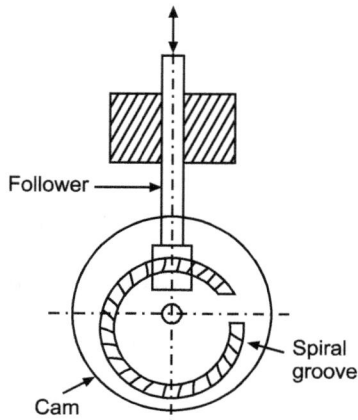

FIGURE 10.51 Spiral cam

The *spiral* or *face* cam consists of a circular plate having a spiral groove cut into it. There is engagement between the teeth cut on the spiral groove and the pin gear follower.

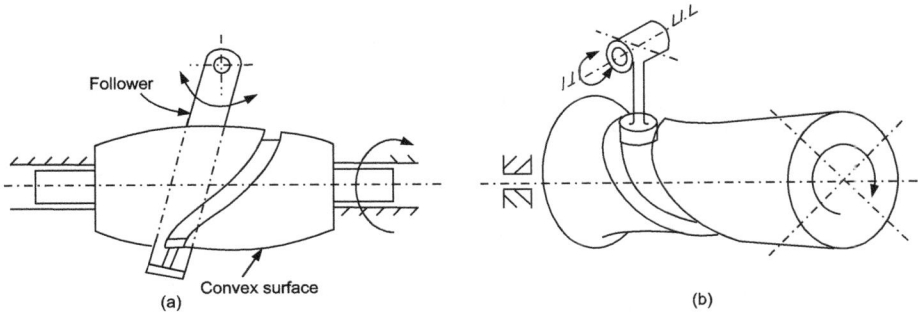

FIGURE 10.52 Convex and concave globoidal cam

A globoidal cam is constituted by a circumferential contour that is cut on a concave or convex surface. The motion of the follower in the globoidal cam is oscillatory in nature.

The follower in a cam-follower mechanism is classified either according to the type of motion of the follower or the nature of the surface in contact:

Based on the nature of the surface in contact, the followers belong to the following four categories:

a. **Knife-edge follower:** a follower that has a sharp pointed edge. There is a sliding motion between the contacting surfaces, that is, the knife-edge and the contacting surface. Though simple in construction, its utility is restricted due to considerable side thrust between the follower and guide and the high rate of wear.

b. **Roller follower:** is a follower that consists of a cylindrical roller that rolls over the cam surface. Due to the rolling motion between the contacting surfaces, there is much less wear as compared to that in a knife-edge follower. A side thrust, however, does exist between the follower and guide. Roller followers find favor for use in situations where more space is available such as in stationary gas or oil engine.

c. **Flat-faced or mushroom follower:** a follower that has a flat contacting surface. Quite often, the flat end of the follower is machined to a spherical shaft and the resulting follower is called a spherical-faced follower.

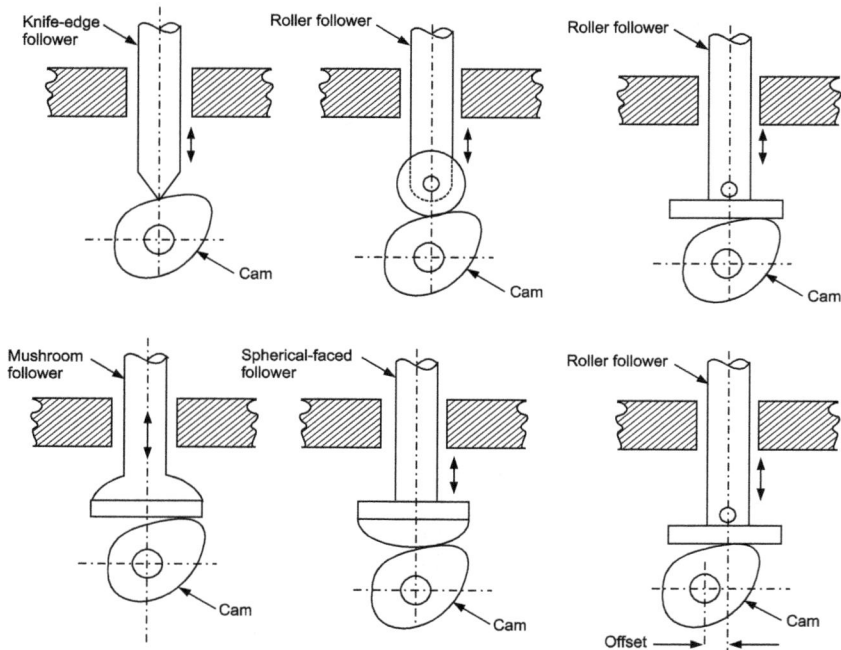

FIGURE 10.53 Types of cam followers

With these followers, there is less side thrust at the bearings and that implies reduced friction force and less changes of jamming in the bearings. These followers are generally used for operating the values of an automobile engine where space is limited.

Based on the path of the morion axis of the location of motion, followers are categorized as:

i. **Radial followers:** the motion of the followers is along an axis that passes through the center of the cam. The follower translates along a line passing through the axis of rotation of the cam (Figure 10.54).

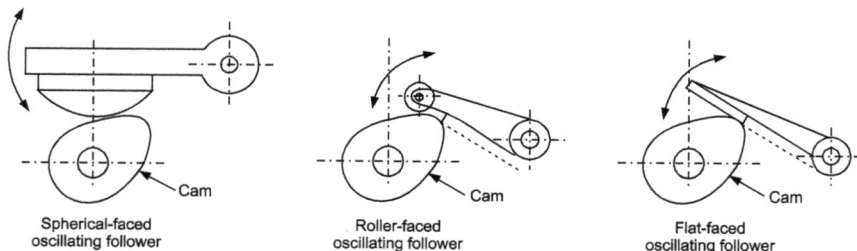

FIGURE 10.54 Types of oscillating followers

ii. Offset followers: the motion of the follower is away from the center of rotation of the cam.

Based on the nature of motion, there are the following two types of followers.

- *Reciprocating or translating followers:* the follower reciprocates in guides as the cam rotates.

- *Oscillating or rotating followers:* the follower oscillates about a hinge point as the cam rotates.

It would be appropriate to mention that the follower is always constrained to follow the cam, and objective is achieved by springs, gravity, or hydraulic means.

10.8. BEARINGS

A ***bearing*** is a machine element that supports another moving machine element (e.g., a rotating shaft) called a *journal*. While carrying the load, the bearing allows relative motion between the contact surfaces of the members.

In ***radial*** or ***journal*** bearings, the main load is perpendicular to the axis of rotation of the moving element.

The portion of the shaft laying within the bearing is known as a *journal*. The plain journal or sleeve bearings are classified as

i. *Full bearing*: The bearing completely envelops the journal.

ii. *Partial bearing*: The enveloping angle is not 360° but is 120°. The friction in a partial bearing is less than that in full bearing but its applications are limited to only those situations where the load is always in one direction, for example, rail road car axles. The full bearing and partial bearings are also known as clearance bearings since the journal size is less than the bearing bore.

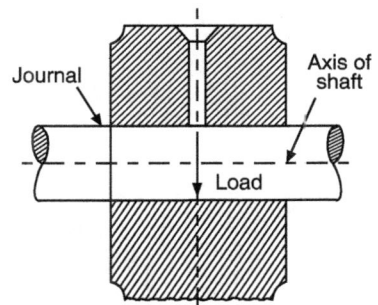

FIGURE 10.55 Journal bearing

iii. *Fitted bearing*: A special case of partial bearing in which the sizes of the journal and bearing are equal and hence there is no clearance.

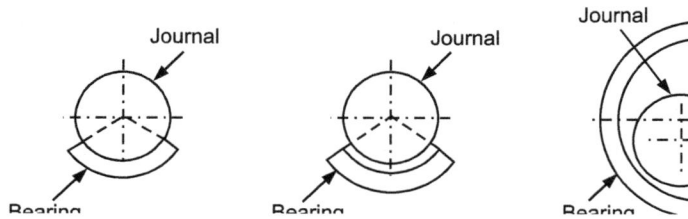

FIGURE 10.56 Types of journal bearing: (a) fitted bearing; (b) partial bearing; (c) full journal bearing

In the **collar thrust** bearings, the bearing pressure is parallel to the shaft axis and has end thrust.

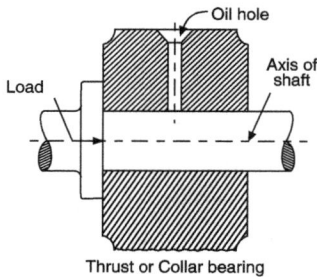

FIGURE 10.57

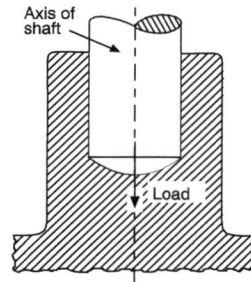

FIGURE 10.58 Footstep or pivot bearing

These bearings provide horizontal load reactions in machine tools and marine drive shaft.

In the **pivot** or **foot step** bearing, the bearing pressure is exerted parallel to the shaft whose axis is vertical. The end of the shaft rests within the bearing body.

Such bearings carry the load in vertical steam turbines, water turbines, motors, and pumps.

In **ball bearings**, the rolling element is a spherical ball. The self-alignment ball bearing permits inclination of the inner race or shaft axis with relation to the axis of the outer race with 2°–3°.

Refer to Figure 10.59 which shows a typical rolling bearing.

The unit consists of an inner ring, an outer ring, and the rolling elements (balls) placed at equal intervals in the open space between the two rings. The balls are usually made of steel or ceramic material. The rings are of uniform stiffness throughout and their surfaces, called the raceways, are quite hard. Depending upon the geometrical configuration of the rolling elements, the rolling bearings are classified as:

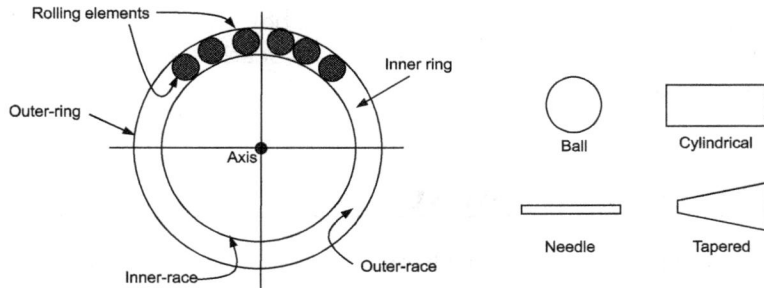

FIGURE 10.59 Typical rolling bearing and rolling elements

i. ball bearings,

ii. cylindrical roller bearings,

iii. needle-type roller bearings, and

iv. tapered roller bearings.

Ball bearings can handle both radial and thrust loads but these loads have to be relatively small or moderate.

The needle and the taper roller bearings belong to the category of roller bearings in which the rolling element is a roller that may be cylindrical, conical, spherical, or concave.

The sintered bearings are self-lubricated bearings suitable for applications where regular maintenance is difficult.

In the roller contact bearings, the contact between the bearing elements is rolling instead of sliding as in plain bearings. Since the rolling friction is very less as compared to the sliding friction, such bearings are also known as antifriction bearings. The roller contact bearings have the disadvantage of low resistance to shock loading.

The ball bearings have been standardized into four classes and designated by a number consisting of at least three digits

i. 100: extra light,

ii. 200: light,

iii. 300: medium, and

iv. 400: heavy.

Additional digits are added to specify special features. The last three digits denote the series and the bore of the bearings. The last two digits from 04 onwards, when multiplied by 5, give the bore diameter in mm. Thus if a bearing is designated by number 308, it is a bearing of medium series with bore equal to 5×08, that is, 40 mm.

10.9. RATCHET AND PAWL

The ratchet and pawl mechanism is a mechanical gearing system used to transmit intermittent linear or rotary motion in one particular direction and prevents motion in the opposite direction. The unit essentially comprises two components namely a ratchet and a pawl mounted on the base as shown in Figure 10.60.

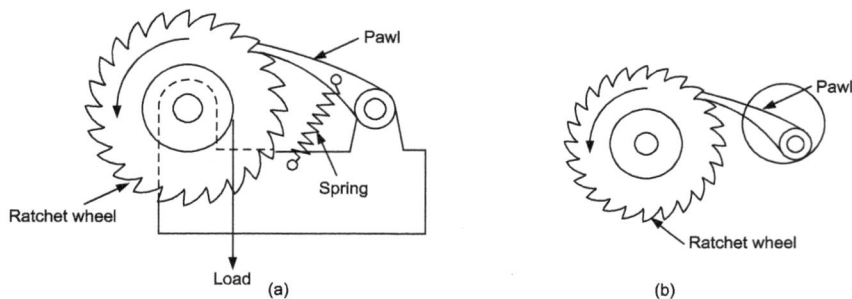

FIGURE 10.60 (a) Weighing mechanism; (b) Ratchet and Pawl

The *ratchet* is a gearing component having a star-shaped angled tooth around its outer periphery. The teeth are uniform but asymmetrical with each tooth having a moderate slope on one edge and a much steeper slope on the other edge. The ratchet receives an intermittent circular motion from the crank and serves to slow down the motion that happens in a jerky way. The pawl is a type of latch in the form of a teeth-shaped solid part that is pivoted at one end and rests against the ratchet at the other end. The pawl is usually spring-loaded and that ensures its automatic engagement with the ratchet.

When the ratchet rotates in one particular direction, the pawl rises and moves smoothly between the angled teeth of the ratchet. Subsequently, when the rotation of the ratchet stops, the pawl rests and gets jammed against the depression between the gear teeth, and that prevents any rotation of the socket in the backward direction.

As the ratchet can stop backward motion only at discrete points, that is, only at tooth boundaries, the ratchet does allow a limited amount of backward motion. This backward motion is limited is a maximum distance equal to the spacing between the teeth and is called backlash. In situations where the backlash is required to be minimum, use is sometimes made of a toothless ratchet with a high friction surface. The pawl then bears against the surface at an angle; any backward motion causes the pawl to jam against the surface and that prevents any further backward motion.

The various applications of ratchet and pawl assembly are:

- spanners, wrenches, and jacks,

- freewheel mechanism of bicycles,

- clocks, winches, turnstiles, and typewriters, and

- hoists and weight lifting machines.

In the weight lifting mechanism shown in Figure 10.49(b), a ratchet wheel is fixed to a shaft and a drum around which a rope is wound. The winding of the rope is done by rotating the ratchet in the counterclockwise direction. When the rotation stops, the load on the rope tries to unwind the rope. This tendency is, however, prevented by the action of ratchet and pawl.

Ratchet and pawls are usually made of steel, stainless steel, cast iron, brass, and other metallic materials. The product specifications for ratchet and pawl include the number of teeth, outside and bore diameter, face width, and pitch.

10.10. HYDRAULIC AND PNEUMATIC ACTUATING SYSTEMS

Pascal law states that "*intensity of pressure is transmitted equally in all directions that a mass of fluid in a confined place.*" This characteristic property of the fluid forms the working principle of fluid power systems. With reference to Figure 10.61, two cylinders of different cross-sectional arcs are interconnected at the bottom through a pipeline and are filled with some liquid

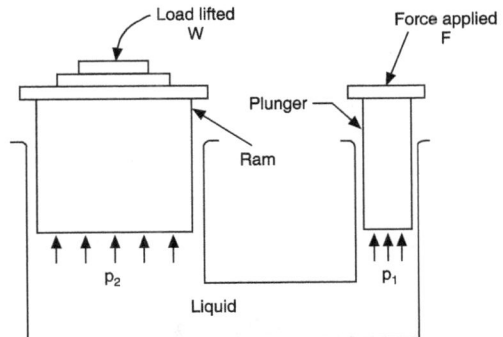

FIGURE 10.61 Working principle of hydraulic press

(oil or water). The larger cylinder contains a raw of area A and a plunger of the area a reciprocates inside the smaller cylinder. A force F_1 applied to the plunger produces an intensity of pressure p_1 which is transmitted in all directions through the liquid. If the plunger and ram and at the same level and if their weights are neglected then pressure intensity $p_2 = \dfrac{F_1}{a}$ and $p_2 = \dfrac{F_2}{A}$ where F_2 is the upward force acting on the ram must equal p_1.

Now, $p_1 = \dfrac{F_1}{a}$ and $p_2 = \dfrac{F_2}{A}$ where F_2 is the upward force acting on the ram. Since $p_1 = p_2$, we have

$$\frac{F_1}{a} = \frac{F_2}{A}; \ F_2 = F_1\left(\frac{A}{a}\right)$$

The above expression indicates that by applying a small downward force F_1 on the plunger, a large upward force F_2 acts on the ram by suitably selecting the ratio of the diameters of the two cylinders.

The fluid systems find applications in:

- hydraulic jack, hydraulic lift, hydraulic crane, etc.,

- measurement of process parameters and then using these parameters to act on necessary output, and

- carrying out mechanical work and using the resulting motion (linear or rotating in cutting operations (sawing, turning, milling, and drilling); feeding, sorting, and packaging; damping, shifting, and positioning.

- controlling of plant, process, and equipment where a hydraulic or pneumatic system. The condition of the process is sensed and the information is fed to the controller for taking the appropriate corrective action.

The performance indices of a fluid system include:

– flow rate of fluid and chamber capacity,

– pressure fatigue and bursting pressure, and

– cleaning, compressibility, and viscosity of the fluid.

The process of the pneumatic system is similar to that of water or oil, compressed air supplies the inlet power.

10.11. CONTROL VALVES: FUNCTIONS AND TYPE

A valve is a device that regulates, directs, or controls the flow of fluid (liquids or gases) by opening, closing, or partially obstructing the passageways. The valves way he operated manually either by handle, lever, pedal, or wheel. Modern control valves are automatically driven by changes in pressure, temperature, or flow rate. These changes act on a diaphragm or piston which then activates the valve. The automatic operation of the valve is based on an external input (flow regulation to a changing set-point) that requires an actuator. The actuator operates depending on its input and the set-point and that allows the valve to be positioned accurately and allows control over a variety of requirements.

Valves are found in virtually every industrial process and are used for:

- starting or stopping flow,

- preventing back flow, and

- relieving and regulating pressure in a fluid.

Valves are quite diverse and are generally classified by how they are actuated: hydraulic, pneumatic, manual, solenoid, and motor.

Further, depending upon how they work/operate, the valves are categorized as:

i.   normally open or normally closed valves. These valves are acted upon by some forces that keep them either open or closed.

ii.   throttling valves that are intermittently opened or closed.

iii. directional control valves are used to redirect the flow. These diverter-type valves are 2-way three-way or four-way valves.

- The two-way valves are the shut-off valves that have only the open and closed positions and have two ports called the inlet and outlet port.

- The three-way valves have three ports—an inlet port, the outlet port, and the exhaust port. The operation is in two conditions: (*i*) exhaust closed and the inlet open to outlet and (*ii*) inlet blocked and the outlet connected to exhaust.

- The 4-way valve arrangement consists of two 3-way valves which are operated by an actuator. The inlet port and the two exhaust ports are joined internally and that provides four ports comprising the inlet port, two outlet ports, and an exhaust port.

Further based upon construction, the control valves are categorized as: poppet valves (ball seat and disk seat type), spool valves (longitudinal slide and plate slide type), and rotary valves.

10.11.1. Pressure Control, Direction Control, and Sequence Valves

Pressure control valves are used virtually in every hydraulic system and they function to keep system pressure safely below a desired upper limit to maintain a set pressure in any part of the circuit. These valves are normally closed and have a restriction to produce the desired pressure.

The different types of pressure control valves are the relief, reducing, sequence, counter-balance, and unloading type. The *relief valve* shown in Figure 10.62 consists of a ball or poppet exposed to the system pressure on one side and opposed by a spring of

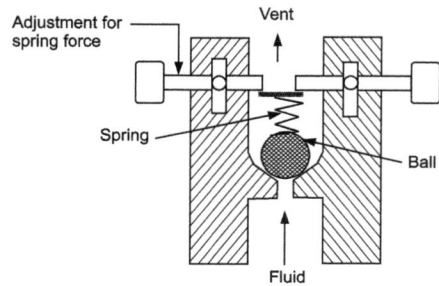

FIGURE 10.62 Relief valve

pre-set force on the other side. The spring holds the ball tightly seated and blocks the flow through the valve. When the force of system pressure rises and exceeds the spring force, the valve gets lifted from its seat and that lets the fluid out through the vent.

A *sequence valve* functions to provide a path of flow alternate and sequential to the primary circuit. The operation of these valves is controlled mechanically or by pressure. The pressure-operated sequence valve is normally a closed poppet or spool valve that opens at an adjustable set pressure.

Fluid at the inlet port of the valve will not pass to the secondary circuit or outlet port until the fluid pressure reaches the set pressure. When the set pressure is reached, the sequence valve directs the fluid to a second actuator or motor to do work in another part of the circuit.

Typically a sequential valve serves to operate multiple actuators and their sequence of operation.

The *direction control valves* (DCV's) are one of the most fundamental parts that are employed to start, stop and change the direction of fluid flow in a fluid control system. They allow the flow of working fluid (oil, gas) into different paths from one or more sources.

The DCV's are classified according to certain factors such as:

- number of ports or ways

- method of actuation: manual, mechanical, solenoid-operated, and pilot-operated.

- shape of valving element is mostly a ball, a sliding spool, or a rotary spool.

The controlling of the passage of a fluid signal is done by generating, canceling, or redirecting signals and categorization of DCV's is then done by signaling elements (input), processing elements, and power elements or final control elements.

The simplest direction control valve is a two-way valve that either stops flow or allows flow. The three-way valve allows fluid flow to an actuator in one position and exhausts the fluid flow from it in another position.

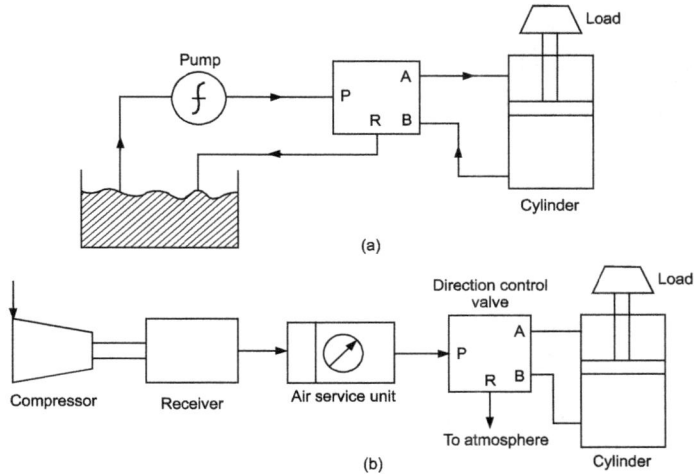

(a)

(b)

FIGURE 10.63 Four-way directional valve

Refer to the four-way directional valve which has four ports:

- Pressure port P in communication with a pump or compressor.

- Return or exhaust port R connected to the fluid reservoir or vented to the atmosphere.

- Two output ports A and B connected to the actuating system (load).

The internal operation of the valve has been shown in Figure 10.64 both for the extend (forward) and retract (backward) movement of the piston.

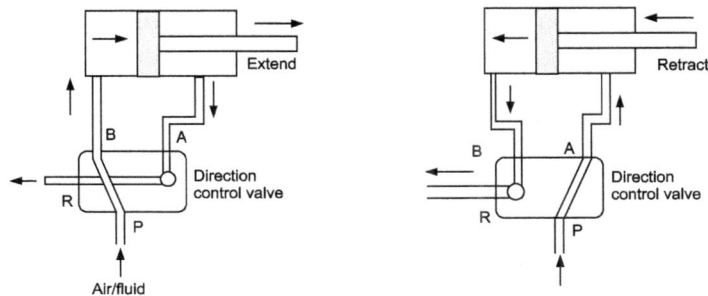

FIGURE 10.64 Internal operation of the valve.

During extended movement, the pressure port P communicates with the outlet port B and that allows the liquid/air pressure to be directed to the cylinder from the left side. The exhaust of liquid/air from the right side of the cylinder then occurs to the liquid tank/atmosphere via exhaust port R through port A. During retract movement of piston, ports P and A deliver the fluid to right of piston and the ports B and R provide a path for the return to liquid sump/atmosphere.

Since the valve has four ports and two control positions, it is known as a 4/2 valve.

10.12. ACCUMULATOR

An accumulator is a unit installed in a hydraulic system to store high-pressure fluid during idle periods and later make it available to supplements pump flow and serve as a backup during power failure.

Three common types of accumulators used in a hydraulic system are bladder, piston, and diaphragm type. The bladder types find favor and Figure 10.65 shows the constructional features of one such accumulator. There is a rubber member, called the bladder which separates gas from the oil. The gas is filled inside the bladder and the oil surrounds the bladder. Initially, the gas is pre-charged to around 80–90% of the system working pressure.

This expands the bladder to fill most of the accumulator space with only a small amount of oil remaining inside. During operation, the hydraulic pump raises system

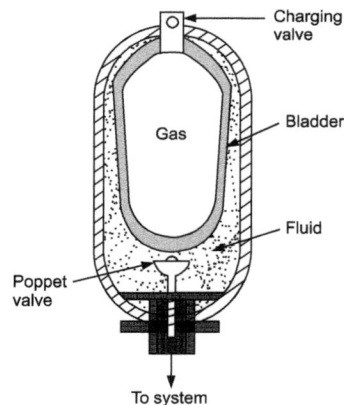

FIGURE 10.65 Bladder type accumulator

pressure and forces fluid to enter the accumulator. The bladder moves and compresses the gas valve because the oil pressure exceeds the pre-charge pressure. The bladder movement stops when the system and pressures become equal. When there is creation in the system demand, the hydraulic pressure falls and the stored pressurized oil is released to the circuit.

An accumulator is essentially an energy storage device that enables the hydraulic system to cope with extremes of demand and yet maintain specified pressure in the system. Other advantages are:

• smooth out pulsations and provide shock cushioning,

• supplement pump flow and allow the use of less powerful pump,

- act as an emergency standby power source,

- provide quick response to sudden enhanced demand, and

- compensate for any oil leakage, thermal expansion/contraction, and hold the required pressure in the circuit.

10.13. AMPLIFICATION

The signal from the detector transducer is normally very weak, and it needs amplification to a certain level where it can be detected for display or record. Amplification is also needed to transmit the signal over some distance. The device used to increase or augment the weak signal is referred to as an amplifier; it may operate on mechanical (levers, gears, etc.), optical, pneumatic, and hydraulic, or electrical and electronic principles.

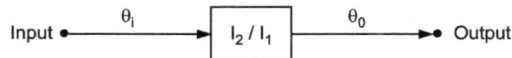

FIGURE 10.66 Amplifier unit

The ratio of the output signal (θ_0) to the input signal (θ_1) for an amplifier is generally referred to as **gain, amplification, or magnification**; the input to and output from the amplifier are related by the expression

$$\theta_0 = G\theta_1, \ G = \frac{\theta_0}{\theta_1}$$

where G is the gain or amplification. Since θ_0 and θ_1 are in the same unit, the gain is a dimensionless quantity.

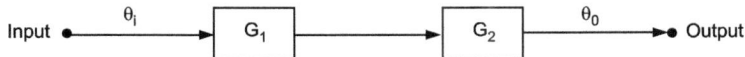

FIGURE 10.67 Series arrangement of amplifiers

Quite often, two or more amplifiers are arranged in series/cascades to get greater amplification. Presuming that no loading occurs, the overall gain of the arrangement is given by the product of individual gains of the amplifying units, that is,

$$\frac{\theta_0}{\theta_1} = G_1 \, G_2 \, G_3 \ldots$$

EXAMPLE 10.18.

**A measuring system has two amplifiers arranged in series correspond-
ing to an input of 10 units, the output from the system is 15,000 units.
If the first amplifier has a gain of 75, determine the gain requirements
of the second amplifier.**

Solution: For a measuring system with two amplifiers arrangement in series,

$$\frac{\theta_0}{\theta_1} = G_1 G_2$$

$$\frac{15000}{10} = 75 \times G_1; \ G_2 = \frac{15000}{10 \times 75} = 20$$

that is, the second amplifier should has a gain of 20.

10.13.1. Mechanical Amplifiers

Simple and Compound Levers

Figure 10.68 shows the arrangement and the associated block diagram for a
simple lever supported on a pivot.

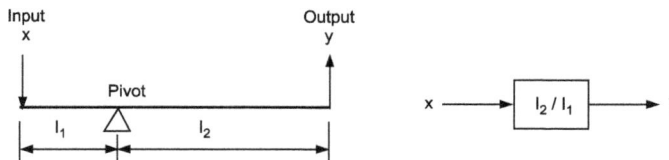

FIGURE 10.68 Simple lever signal amplifier

For an input displacement x, the simple lever causes the output end to
displace by an amount y; x and y are related to each other by the expression

$$\frac{y}{x} = \frac{l_2}{l_1}$$

where l_1 is a distance of input end from the pivot, and l_2 is a distance of output
end from the pivot. The ratio l_2/l_1 determines the displacement gain; amplification of the lever can be varied depending on the relative distances of input
and output ends from the pivot. With a simple lever, the input and out displacements are of opposite phase, that is, if the input x goes down the output
y moves up.

A greater increase in amplification can be achieved by having a compound-
lever system. The compound lever will have two or more levers linked together
so that the output from one lever provides the input to the other.

For the lever arrangement shown in Figure 10.69

$$\frac{y}{x} = \frac{l_2}{l_1} \text{ and } \frac{z}{y} = \frac{l_4}{l_3}$$

Therefore, the overall gain or amplification is

$$\text{overall gain } G = \frac{z}{y} = \frac{l_2}{l_1} \times \frac{l_4}{l_3}$$

and

$$\text{output } z = \left(\frac{l_2}{l_1} \times \frac{l_4}{l_3} \right) \times x$$

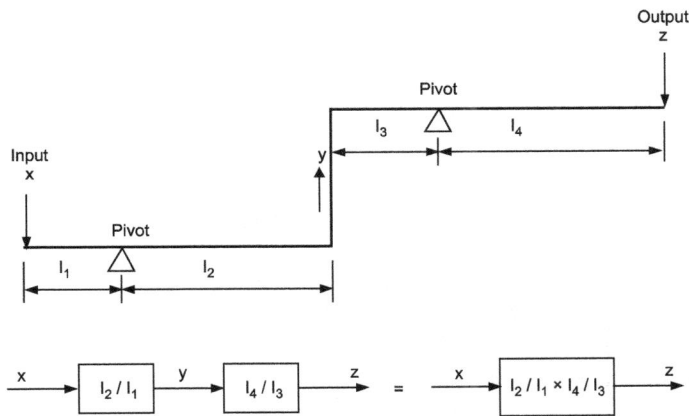

FIGURE 10.69 Compound-lever signal amplifier

Apparently, there is no problem of phase reversal with the compound lever; both the "d output y act along the same directions.

EXAMPLE 10.19.
Calculate the magnitude of the compound-lever arrangement illustrated in Figure 10.69. The length magnifying ratio of the levers is 2.5:1 and 4.5:1.

Solution: For the compound-lever arrangement,

$$\frac{\theta_0}{\theta_1} = \frac{l_2}{l_1} \times \frac{l_4}{l_3}$$

$$= \frac{2.5}{1} \times \frac{4.5}{1}$$

Thus the compound lever has a displacement gain of 11.25.

10.13.2. Simple and Compound Gears

The simple and compound gear trains are used quite frequently to provide mechanical amplification of either angular displacement or rotary speed.

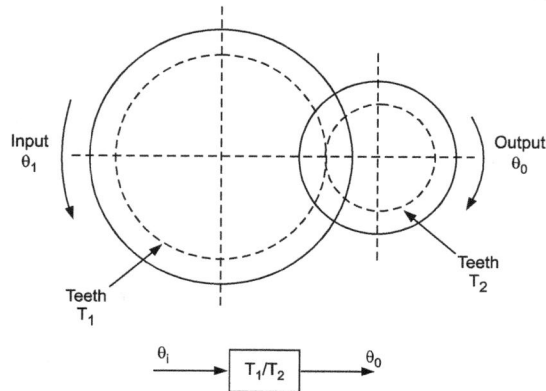

FIGURE 10.70 Simple gear signal amplifier

When the larger wheel (gear with teeth T_1) rotates with the input angular speed θ_i then each of its teeth fits into a corresponding space on the small wheel (gear with teeth T_2). That results in an angular speed θ_0 of the output gear; θ_i and θ_0 are related to each other by the expression.

$$\frac{\theta_0}{\theta_1} = \frac{T_1}{T_2} = \frac{\text{number of the teeth on input gear wheel}}{\text{number of teeth on output gear wheel}}$$

The ratio T_1/T_2 determines the gain; the amplification of a simple gear can be varied depending on the number of teeth on the input and output gear wheels. With a simple gear, the input and output displacements are of opposite phase, that is, if the input wheel rotates counterclockwise the output wheel turns clockwise.

A greater increase in amplification can be achieved by involving more gear wheels. In the compound gear train illustrated in Figure 10.71, the input rotational signal θ_i is applied to wheel A which has T_1 signal teeth. The teeth on this wheel mesh with teeth on wheel B which has T_2 teeth. One rotation of wheel A would result in (T_1/T_2) rotations of wheel B as well as wheel C. This is due to the fact that gear wheels B and C are mounted on the same shaft. The gear wheel C has T_3 teeth and it drives the gear wheel D which has T_4 teeth.

Thus one rotation of wheel C would give (T_3/T_4) rotations to wheel D, and (T_1/T_2) rotations of wheel C would amount to $(T_1/T_2 \times T_3/T_4)$ rotations of wheel D. Thus overall magnification of gear would be teeth on the wheel

$$\frac{\theta_0}{\theta_1} = \frac{T_1}{T_2} \times \frac{T_3}{T_4}$$

$$= \frac{\text{teeth on wheel } A}{\text{teeth on wheel } B} \times \frac{\text{teeth on wheel } C}{\text{teeth on wheel } D}$$

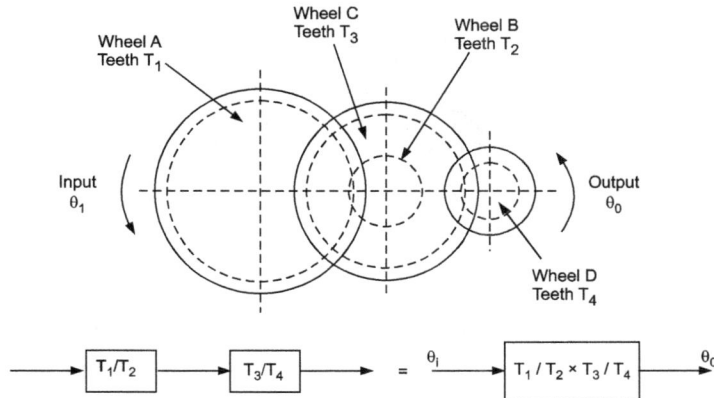

FIGURE 10.71 Compound gear trains

A compound gear train gives greater magnification with the additional advantage of no change in the direction of input signal. The gear trains are used for the magnification of displacement in the bourdon tube pressure gauge and in the dial-test indicator where linear movement is translated into rotation by means of rack and pinion.

The mechanical amplification usually suffers from errors caused by:

- internal loading,

- friction at the mating parts,

- elastic deformation, and

- backlash.

EXAMPLE 10.20.
Figure 10.72 illustrates the gear arrangement for a particular dial-test indicator. The plunger is a part of a rack and pinion mechanism so that the movement of the plunger vertically upwards or downwards rotates the pinion. The rack has a tooth pitch of 5 teeth per cm and engages with a pinion having 15 teeth. The gears A and C have

30 teeth, and the gears B and D have 15 teeth respectively. Calculate the magnification and the number of revolutions of the pointer as the plunger moves 30 mm.

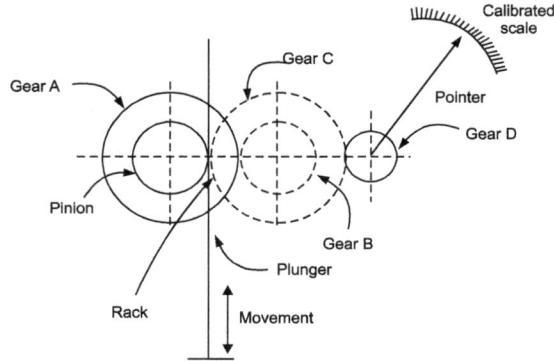

FIGURE 10.72 Basic features of dial-test indicator

Solution: Corresponding to a plunger movement of 30 mm, the rack (5 teeth/cm) turns 15 teeth. Since there are 15 teeth on the pinion which engage with the rack, the pinion will rotate through one revolution.

Now, one revolution of pinion corresponds to

$$= 1 \text{ revolution of gear } A = 1 \times \frac{30}{15} \text{ revolutions of gear } B$$

$$= 1 \times \frac{30}{15} \text{ revolutions of gear } C$$

$$= 1 \times \frac{30}{15} \times \frac{30}{15} \text{ revolutions of gear } D$$

$$= 4 \text{ revolutions of gear } D$$

Since the pointer is attached to gear D, it will turn four revolutions as the plunger moves 30 mm.

10.13.3. Fluid Amplifiers

Hydraulic amplifier: Refer 10.73.

When a small displacement x is applied to a piston operating inside a cylinder containing some liquid, there occurs a large displacement y of the liquid in the output tube which has a small diameter d. From volume balance of the liquid, displacement of liquid in the cylinder = displacement of liquid in the output tube

$$\frac{\pi}{4}D^2x = \frac{\pi}{4}d^2y$$

Therefore, the gain or amplification is

$$\text{gain } G = \frac{y}{x} = \left(\frac{D}{d}\right)^2$$

and output $y = \left(\dfrac{D}{d}\right)^2 x$

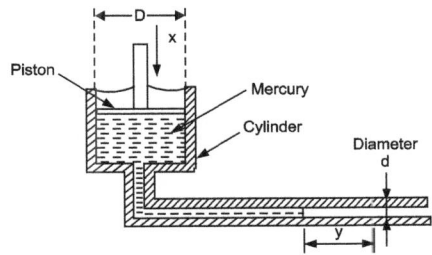

FIGURE 10.73 Schematics of hydraulic amplification

This principle is employed in the mercury-in-glass thermometers and in the single-column manometers.

EXAMPLE 10.21.

In a fluid amplification system (Figure 10.73), the amplification is to be 400. If the capillary tube is 1 mm diameter, calculate the required cylinder diameter.

Solution: Amplification $G = \dfrac{\text{movement of fluid meniscus in tube, } y}{\text{movement of pluger in cylinder, } x}$

From volume balance of liquid,

$$\frac{\pi}{4}D^2x = \frac{\pi}{4}d^2y; \quad \frac{y}{x} = \left(\frac{D}{d}\right)^2$$

$$\therefore \qquad G = \left(\frac{D}{d}\right)^2; \ 400 = \left(\frac{D}{d}\right)^2$$

$$\frac{D}{d} = 20; \ D = 20 \times 1 = 20 \text{ mm}$$

Thus the cylinder should have a diameter of 20 mm.

10.13.4. Pneumatic Amplifier

When pressure P_1 is applied to the bellows, the valve spindle moves toward right and that increases the gap between the valve and its seat. More of the high-pressure air (P_x) rushes inwards and results in an increase in the output pressure P_2. Normally a linear relationship exists over a part of the range of movement of the valve: $P_2 = KP_1$, where K is a constant.

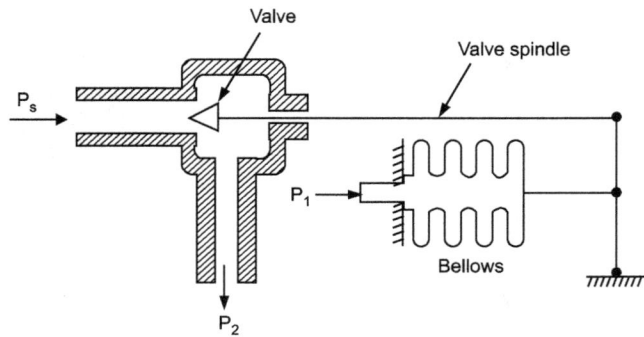

FIGURE 10.74 Schematics of pneumatic amplification

10.14. HYDRAULIC SYSTEMS

When the force required to affect a change in the flow variable is too large for self-actuation by the fluid system, a separate force augmenting system is utilized. The force augmenting system may be hydraulic (using a liquid such as oil) or pneumatic (using a gas, particularly air).

Hydraulic actuators use a liquid control medium to provide an output signal which is a function of an input error signal. The schematics of a hydraulic actuating system are illustrated in Figure 10.75; the major components are an error detector, an amplifier, a hydraulic control valve, and an actuator.

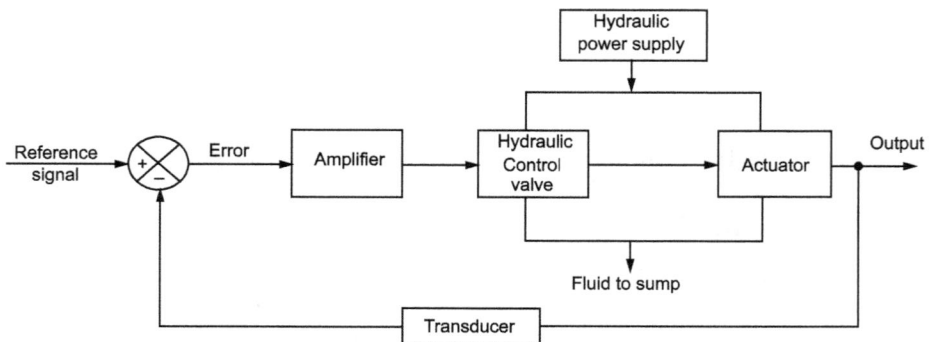

FIGURE 10.75 Schematics of a hydraulic actuating system

There are two main types of high-pressure hydraulic power supply systems:

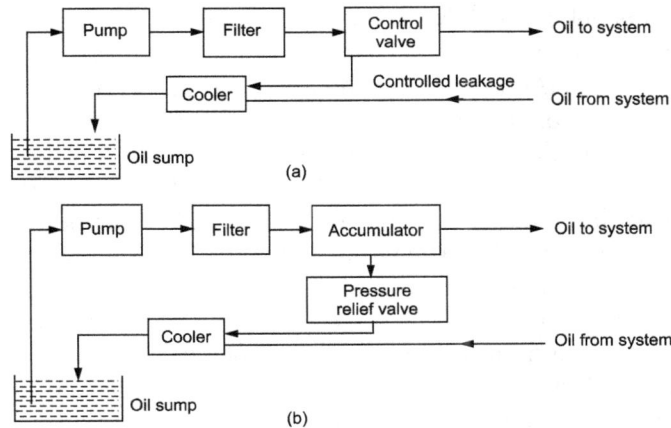

FIGURE 10.76 Constant-flow and constant-pressure hydraulic actuator systems

i. *Constant-flow arrangement* (Figure 10.76*a*): A constant displacement-type pump serves to draw oil from an oil sump and provides it at a constant-flow rate. The high-pressure oil is filtered and then led to a control valve with a controlled fluid leakage back to the sump. Under no-load conditions; the whole of the oil is returned to the sump. The arrangement works efficiently when the full flow of oil is to be used for longer periods.

ii. *Constant-pressure arrangement* (Figure 10.76*b*): An accumulator is used which maintains oil at constant-pressure, and also stores energy for peak power demands. At pressure values higher than the predetermined value, a pressure relief valve operates to bypass fluid between high pressure and return lines.

10.14.1. Hydraulic Pump

There are three main types of pumps used in hydraulic systems viz., the gear pump, the vane pump, and the piston pump.

Figure 10.77 shows the outline of a fixed displacement gear pump. The unit consists of two rotors that are mounted on separate parallel shafts. The oil trapped between the teeth of the suction port revolving gears and the metal casing is delivered from the suction port to the discharge port.

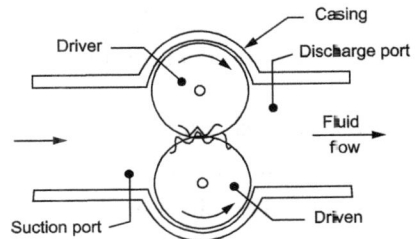

FIGURE 10.77 Gear pump

To check leakage between the suction and discharge side of the pump, the gear teeth are of special involute or cycloid construction. The meshing of gears serves both to transmit the drive and to maintain a seal between pressure and suction. Care is taken to ensure that the oil trapped between the successive lines of contact does not build up pressure. A change in flow direction can be affected by reversing the direction of the gear assembly.

The difficulties associated due to wear in gear pump are overcome in a vane pump which is shown in Figure 10.78. The unit consists of a slotted rotor with a number of spring-loaded radial vanes inserted in the slots. The rotor drum is mounted eccentrically in the casing and the vanes are allowed to slide into and out of the rotor as it rotates. The space between the vanes is filled with the liquid at the suctions side and is carried to the delivery side. The quantity of liquid pumped and the direction of flow can be controlled by affecting a change in the degree of eccentricity.

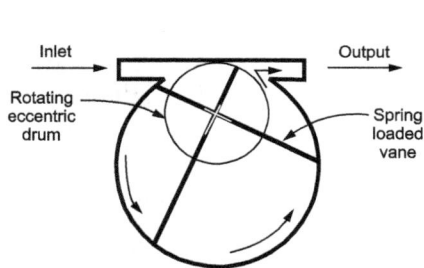

FIGURE 10.78 Vane pump

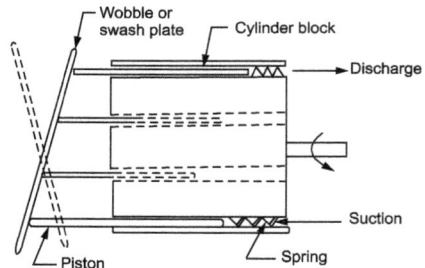

FIGURE 10.79 Axial flow piston pump

Figure 10.79 depicts a simplified version of an axial flow piston pump. The unit consists essentially of a cylindrical block that is rotated along with its piston. This rotation causes the piston to move back and forth parallel to the shaft, and butt against the non-metal wobble or swash plate. The angle of the wobble plate is either kept constant or is varied in accordance with whether the pump is to have a fixed displacement or a variable displacement. No flow takes place when the wobble plate is perpendicular.

10.14.2. Hydraulic Valves

These elements of a hydraulic actuating system function to regulate the flow of hydraulic fluid from the high-pressure side to the actuator, that is, the hydraulic motor. There are three main types of valves; the piston or spool-type, the flapper and nozzle type, and the jet-pipe valve. In all cases, the mechanical input motion can be controlled by manual operation, by a limited motion electric motor, or by the hydraulic pilot method. The output results in a change of hydraulic pressure.

Piston or *spool valve:* The commonly used spool valve is constructed in either a three-way or a four-way valve arrangement as shown in Figure 10.80.

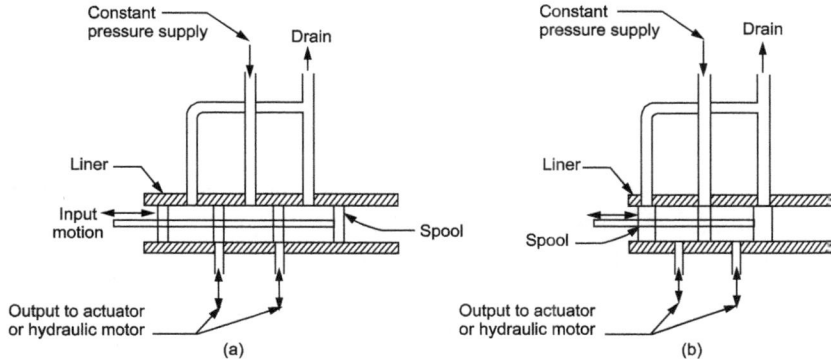

FIGURE 10.80 Cylindrical spool valves: (a) four-land and (b) thee-land

When the spool is in the neutral position, the oil flow to the actuator is completely blocked. Displacement of the spool to right and left causes alternately pressure in one port to be higher than that in the other port. This is because when one of the pipelines to the actuator gets connected to the constant-pressure supply, the other pipeline communicates with the drain. The differential pressure causes the hydraulic motor to rotate in a particular direction. The flow and, therefore, the motor speed is the function of the spool-valve opening, affected somewhat by the load pressure.

Flapper-nozzle valve: The operation of the unit is based on a variable leakage arrangement, which has the great virtues of simplicity and reliability. The unit incorporates two orifices in series, one of which is a fixed restriction and the other variable orifice consisting of a flapper and nozzle arrangement (Figure 10.81). The nozzle restriction is changed as the flapper is positioned closer to or farther from the nozzle. The fluid at constant-pressure passes through the restriction and a branch is led to the nozzle. When the flapper moves into a position that completely blocks the nozzle opening, there is very little leakage and the output pressure approaches that of the supply. With nozzle opening, there occurs an increase in pressure drop across the restriction, and consequently, the output pressure diminishes. Thus, the device produces a variable output pressure with a flapper position.

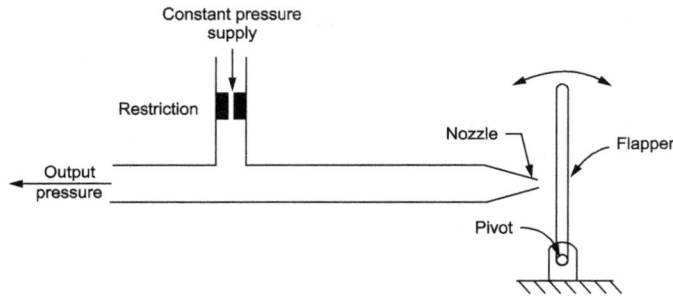

FIGURE 10.81 Flapper-nozzle valve

Jet-pipe valve: The device comprises a pivoted nozzle and two adjacent orifices. The nozzle converts the static pressure of the system into kinetic energy and then directs the high-velocity jet of hydraulic fluid toward the orifices. During flow through orifices, the kinetic energy is reconverted to pressures; the conversion being approximately 90% efficient at moderate supply pressure. This results from the act that friction can be reduced to a minimum by making the space between the nozzle and orifices relatively large.

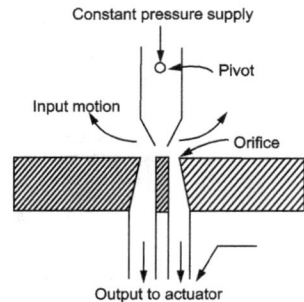

FIGURE 10.82 Jet-pipe valve

The flapper-nozzle and jet-pipe valve arrangements are frequently used as a preamplifier to a piston valve.

10.14.3. Desirable Characteristics of Hydraulic Fluids

- Low level of impurities and high value of the bulk modulus,

- sufficient film strength to prevent metal-to-metal contact between moving parts, that is, to avoid wear of control valve, motor, and pump elements,

- adequate viscosity to give good seal at piston, glands, and valves. At low viscosity, the moving parts would wear rapidly and there would be loss of fluid from the system. However, with highly viscous fluids, there would be excessive load to the moving parts and a considerable pressure drop along the feed lines,

- good chemical stability, that is, resistance to the formation of sludge and gum. The formation of the emulsion during the churning of fluid as it passes through pumps, valves, piping, and motors would reduce lubricity and often cause rusting, and

- freedom from acidity, a low pour point temperature, a high flash point, resistance to foaming, and anti-rust properties.

Petroleum-based oils, with or without additives, are the most common of hydraulic fluids because of their superior lubricating and corrosive protecting features. However, certain synthetic fluids like phosphate or silicate ester compounds, halogenated fluids, and silicon fluid are now being preferred because of their good fire resistance characteristics.

10.14.4. Advantages and Limitations of Hydraulic Systems

- High response due to effectively incompressible nature of the liquid control medium,

- high power gain due to readily conversion of liquids to high pressure or flows. The high energy liquid can be effectively piloted by the hydraulic controllers,

- simplicity of the actuator system,

- long life due to self-lubricating properties of the hydraulic liquids,

- requirements of proper seals and connection so as to prevent the leakage of hydraulic fluid,

- careful maintenance of the system to keep the fluid clean and pure, and

- stringent requirements for the hydraulic fluid to be fire-resistant, anti-corrosion, and self-lubricating.

Their high power-to-weight ratio results in their finding a wide range of use in machine tools, speed governing systems, and position control systems.

10.15. PNEUMATIC ACTUATORS SYSTEMS

Pneumatic controllers use an air control medium to provide an output signal which is a function of an input error signal. The schematics of a pneumatic control system are illustrated in Figure 10.83; the major components are an error detector, flapper-nozzle (controller mechanism), and an amplifier or pilot relay.

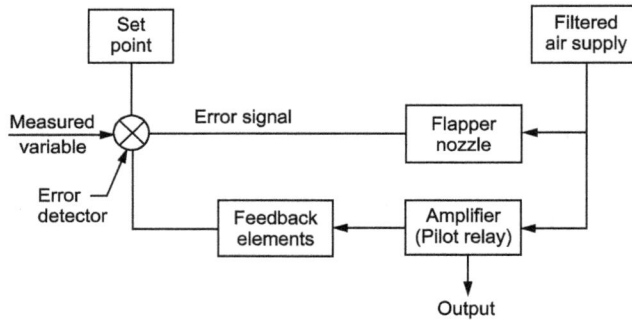

FIGURE 10.83 Schematics of a pneumatic actuating system

The pneumatic actuating mechanisms are of two types: force-balance and motion-balance. In a force-balance control (Figure 10.84a), the deviation of a diaphragm is proportional to the pressure difference $(P_s - P_0)$; P_s and P_0 being the set-point and output signal pressures, respectively, In a motion-balance controller (Figure 10.84b) the deviation signal is taken from a point on a mechanical linkage. A control on the deviation movement $(l_2 x - l_1 x)/(l_j + l_2)$ can be affected by altering the values of l_1 and l_2.

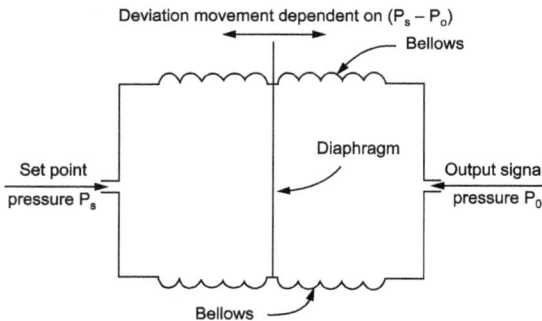

FIGURE 10.84 (a) Force-balance actuator

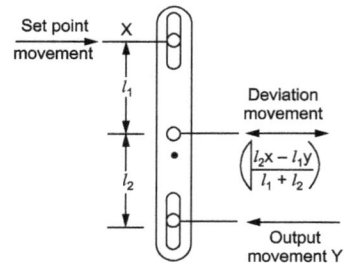

FIGURE 10.84 (b) Motion-balance actuator

10.15.1. Pneumatic Nozzle-flapper

A nozzle-flapper is a basic component of pneumatic and hydraulic measurements, control and transmission, and as a precision gauging equipment. Figure 10.85 shows a schematic diagram of a pneumatic nozzle-flapper amplifier.

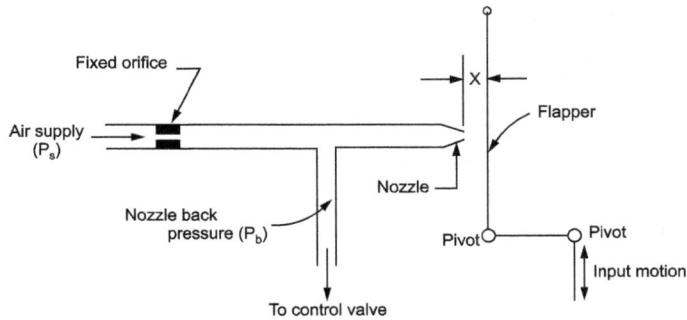

FIGURE 10.85 Pneumatic nozzle-flapper

The unit consists essentially of nozzle supplied with pressurized air (1.4 bar gauge) through an orifice restriction. The orifice diameter is approximately 0.25 mm and that of the nozzle is 0.675 mm. For proper functioning, the nozzle diameter must be larger than the orifice diameter. Just in front of the nozzle, there is a flapper which is positioned by the input motion. As the flapper approaches the nozzle, there is an increase in resistance to the flow of air through the nozzle; consequently, the nozzle back pressure increases. If the nozzle is completely closed by the flapper, air cannot escape through the nozzle.

Consequently, maximum air passes at the supply pressure P_s. When the nozzle-flapper distance x is made wider by moving the flapper away from the nozzle, there is practically no restriction and most of the air escapes to the atmosphere. The nozzle back pressure takes on a minimum value; the lowest possible value being the atmospheric pressure P_a. Thus a movement x of the flapper causes a proportional change in the nozzle back pressure P_b if the supply pressure P_s is kept constant. Figure 10.86 shows a typical curve relating the back pressure P_b to the nozzle-flapper distance x. The slope of the curve at any point is called the nozzle sensitivity or gain. For the commonly used 0.25 mm diameter orifice and 0.625 mm nozzle combination, 0.3625 mm motion of the baffle causes a change in the nozzle back pressure greater than 0.56 bar in the central portion of the curve. The steep and almost linear part of the curve is utilized in the actual operation of the nozzle-flapper unit.

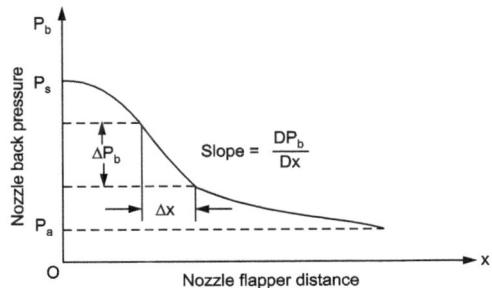

FIGURE 10.86 Flapper-valve characteristic

10.15.2. Pneumatic Relay

To increase gain and to have larger outputs required for the operation of large pneumatic actuators, a pneumatic relay is often used in conjunction with the nozzle-flapper unit. Figure 10.88 shows the outline of the most common form of a pneumatic relay widely used in process control. When the nozzle back pressure increases, the ball valve is forced toward the lower seat. The air supply is shut off and output pressure P_0 to the pneumatic valve drops. When the ball rests on its upper seat, the exhaust port is closed and the control pressure P_0 becomes equal to be supply pressure P_s. The control pressure can thus be varied from 0 bar to full supply pressure of 1.2 bar (gauge). The movement of the ball takes place due to a change in the value of nozzle back pressure which results from the displacement of a flapper in front of the nozzle. The relay system is of bleed type since, at all positions of the ball except at the top-seat, the air continues to bleed into the atmosphere.

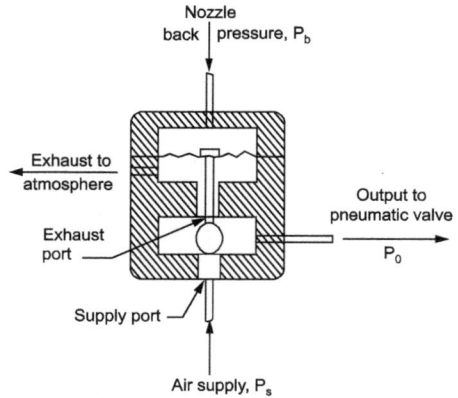

FIGURE 10.87 Pneumatic relay unit

10.15.3. Single-Acting and Double-Acting Pneumatic Actuators

In a pneumatic actuator, compressed air is the basic energy source. A reciprocating compressor sucks air from the atmosphere, compresses it to the requisite pressure, and the compressed air is stored into a pressure vessel called the air receiver. This air is required to be dry and clean free from dirt and dust. For that, a cooler is used to reduce the air temperature after compression and a separator is employed to remove any water vapor that may be present in the air.

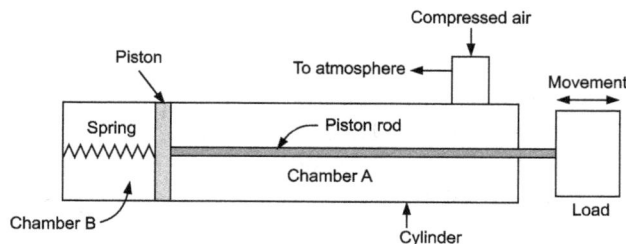

FIGURE 10.88 Single-acting pneumatic actuator

In the single-acting pneumatic actuator, spring force allows the piston to move linearly in one direction only into a cylinder which is a hollow chamber. The pressurized air from the receiver enters the cylinder into chamber A and overcomes the spring force. When the supply of compressed air is stopped and the air inside the chamber is allowed to escape to the atmosphere, the piston travels toward the right.

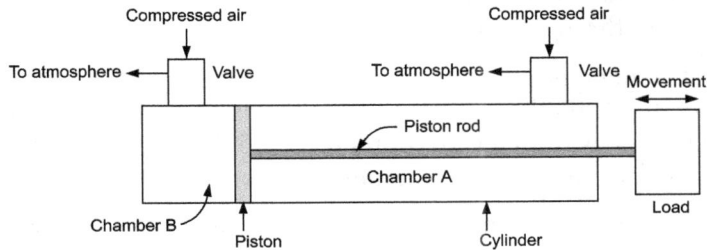

FIGURE 10.89 Double-acting pneumatic actuator

The double-acting pneumatic arrangement has two ports (valves) for compressed air supply and its subsequent release to the atmosphere. When pressurized air enters chamber A, there is an escape of air from chamber B, and the piston is pushed extreme on the left side of the cylinder. During the next stroke, the entry of pressurized air is chamber B, the escape of air to the atmosphere is from chamber A, and the piston movement is toward the right.

The single and double-acting linear pneumatic actuators have key advantages of:

- compact design and reliable operation,
- rugged construction and suitable for harsh environments,
- easy maintenance, and
- positioning accuracy and versatility of use and application.

These actuators have wide applications in industries where the load needs to be lifted or lowered, pushed or pulled, and positioned. The main industries using linear actuators are:

- – material handling and packaging,
- – food processing and pharmaceutical, and
- – machine tool, automotive, and defence.

The *rotary actuators* have been designed and developed for the applications requiring a rotary motion or torque to control the speed and rotation of the attached equipment. In these actuators:

- there is transformation of pneumatic, hydraulic or electric energy to mechanical rotation, and

- the linear motion in one direction gives rise to a turning or angular movement through a pre-set (defined) angle.

The pneumatic rotary actuators utilize the pressure of compressed air to generate the oscillatory rotary motion. For operation, the force is applied at a distance away from the axis of rotation and that causes a turning moment. The two most common configurations of pneumatic rotary actuators are:

1. **Rotary vane actuator:** The vane actuator consists of a vane mounted on a central shaft/spindle enclosed in a cylindrical chamber. A stream of pressurized air is made to impinge upon the vane and that push makes the spindle turn. There is a port through which the air behind the vane is released into the atmosphere. When the vane has turned through a specified angle, the airflow is reversed and the spindle rotates back to its original position. The stroke gets completed and subsequently, the process is repeated.
   The vane actuators are used for light loads.

2. **Rack and pinion actuator:** The rack and pinion actuator consist of a piston and rack that moves linearly and causes a pinion gear and output shaft to rotate. The rack is machined as a part of the piston rod of a double-acting linear cylinder. The pinion gear meshes with the rack and turns the spindle as the piston moves when air pressure is applied to it. The spindle is at right angles to the piston and rotates clockwise, then counterclockwise as the linear cylinder completes its double action.
   The rack and pinion actuators are used when more speed is required.

The key merits of pneumatic rotary actuators are: zero backlash, high force output, high repeatability and positioning accuracy, less wear and maintenance, and greater durability; ability to work effectively in hazardous environments.

These actuators are being successfully employed in the following industries for work with greater precision: robotics and CNC machines, aerospace and flight simulation, radar and monitoring systems, and medical industry.

10.15.4. Advantages and Limitations of Pneumatic Systems

- Simplicity of components and ease of maintenance,
- explosion-proof characteristics; freedom from the hazards,
- relatively high power amplification for operating the final control elements,
- relatively inexpensive power system; the abundance and free supply of air,
- slow response of final control elements, and transmission lag, and
- operation difficult under freezing conditions.

REVIEW QUESTIONS

A. Conceptual and conventional questions:

1. Define actuation and actuator. State the role of an actuator in a mechatronic system.

2. How actuators are classified?

3. State the characteristic properties which form the basis for the operation of a piezoelectric actuator and a magnetostrictive actuator.

4. List the passive and active mechanical components.

5. How does a mechanism differ from a machine? Give at least two examples for each.

6. What is a four-bar mechanism and a slider-crank mechanism?

7. What is meant by the term degree of freedom or mobility in relation to toothed gears.

8. **(a)** State the difference between an open belt and a crossed belt drive.
   (b) Give a brief description of the flat, V-shaped, and round belts.

9. Define the following terms in relation to bolt drive: (i) velocity ratio, (ii) crowning of pulley, (iii) creep.

10. What is slip and how does it affect the velocity ratio of the belt drive?

11. Present a brief account of the chain drive. Point out its advantages.

12. What purpose is served by a bearing? How bearings and classified?

13. State the conditions under which gear is used for power transmission.

14. Name the different types of toothed gears and point out their usual field of application.

15. Define circular pitch, diametral pitch, and module in relation to toothed gears.

16. What is a gear train? Name the different types of gear trains used in power transmission and explain their working.

17. Sketch the different types of cams and followers. Point out their applications.

18. What is a ratchet and pawl mechanism and what purpose does it serve?

19. State the purpose of pressure regulation in a fluid power system. How it is achieved?

20. Explain the working principle of fluid actuators.

21. (a) State the purpose of valve in a fluid actuation system.
    (b) What are two-way, three-way, and four-way valves?

22. State the purpose of an accumulator in a hydraulic control circuit. Sketch a typical accumulator and explain its working.

23. State the purpose of a pressure control valve in a hydraulic control circuit. Sketch a typical pressure control valve and explain its working.

24. State the objectives of direction control valves and pneumatic sequencing.

25. Define amplification and point out its necessity in a measurement/control system.

26. Explain the mechanical amplifier systems of:
    (a) single lever,
    (b) compound levers.

27. Explain the mechanical amplifiers systems of:
    (a) simple gear train,
    (b) compound gear train.

28. Sketch and explain the operating principle of:
    (a) hydraulic amplifiers,
    (b) pneumatic amplifiers.

29. Give the schematics of a hydraulic control system and describe the functions of its various components. In which way the constant-flow arrangement differs from the constant-pressure arrangement?

30. Sketch and explain the working of different types of pumps and control valves used in hydraulic control systems.

31. Give the schematics of a general pneumatic control system and describe the functions of various elements.
    How does the force-balance controller differ from a motion-balance controller?

32. State the essential difference between the hydraulic and pneumatic controllers. Enumerate the advantages and limitations of each of these controllers

33. (a) Sketch and explain the working of pneumatic nozzle-flapper and pneumatic relay as used in pneumatic control systems.
    (b) List the desirable characteristics of hydraulic fluids for use in hydraulics control systems.

34. Explain with neat sketches, the working of single-acting and double-acting pneumatic actuators.

B. Fill in the blanks with appropriate word(s)

(*i*) A structure has _____ degree of freedom.

(*ii*) A _____ is a joint of two links having relative motion between them.

(*iii*) Nut-bolt is a _____ mechanical component.

(*iv*) Materials that undergo some sort of transformations through physical interactions are referred to as _____.

(*v*) The relative motion between belt and pulley is called _____.

(*vi*) For constant velocity ratio positive drive with a short distance between the drive and driven shaft, one would use _____ drive.

(*vii*) Rack and pinion can be used to convert rotational motion into _____ motion.

(*viii*) Two nonparallel or intersecting shafts can be connected by means of _____ gear.

(ix) In _____ bearings, the pressure acts along or parallel to the shaft.

(x) In a ratchet and pawl mechanism, the ratchet wheel engages with a tooth-shaped short lever called _____.

(xi) A 4/2 direction control valve has _____ ports.

(xii) The return fluid is vented to _____ in a pneumatic system.

(xiii) The _____ is a storage device for high-pressure fluid and can store and release the hydraulic oil at the required system pressure.

(xiv) _____ valves are used to protect hydraulic systems from over pressures.

(xv) An _____ is a signal conditioner giving mechanical or electrical output larger than the output.

Answers:
(i) zero; **(ii)** kinematic pair; **(iii)** active; **(iv)** active materials; **(v)** slip; **(vi)** gear; **(vii)** linear; **(viii)** bevel; **(ix)** thrust; **(x)** pawl; **(xi)** four; **(xii)** atmosphere; **(xiii)** accumulator; **(xiv)** Relief; **(xv)** amplifier__

APPLICATIONS OF ROBOTS*

11.1. INTRODUCTION

The present-day applications for robots are much broader than most people realize. While the emphasis in robot development is on industrial robots, factories are not the only place where robots are used. Small, medium, and large companies in just about every industry are taking a fresh look at robots to see how this powerful technology can help them solve manufacturing challenges. In business offices and elsewhere, robots serve as mail delivery carts, promotional or show robots, laboratory assistants, hospital orderlies, and window washers. In general, industrial robots are best used for jobs that are dirty, dull, dangerous, or difficult, the types of jobs that humans do most poorly.

Robots are used in a wide range of industrial applications. The first commercial application of an industrial robot took place in 1961, when a robot was installed to load and unload a die-casting machine. This was an unpleasant task for human operators. Many robot applications took place in areas where a high degree of hazard or discomfort to humans existed, such as in welding, painting, and foundry operations. The earliest applications were in materials handling, spot welding, and spray painting. Robots were initially applied to jobs that were hot, heavy, and hazardous such as die-casting, forging, and spot welding. The reasons for using robots are:

1. *Reduced costs*—robots can perform tasks more economically than humans.

2. *Improved productivity*—robots are not only less expensive than human labor, but also have higher rates of output. This increase in productivity is due to robot's slightly faster work pace but much is the result of the robot's ability to work almost continually, without lunch breaks and rest periods.

---

* This chapter originally appeared in S. Siewert and J. Pratt. *Real-Time Embedded Components and Systems with Linux and RTOS*. ISBN: 978-1-942270-04-1. ©MERCURY LEARNING AND INFORMATION.

3. *Better quality*—robots have a distinct advantage of being able to perform repetitive tasks with a higher degree of consistency, which in turn leads to improved product quality.

4. Elimination of hazardous tasks.

11.2. ROBOT CAPABILITIES

The three important capabilities of robots that make them useful for applications are:

- Transport
- Manipulation
- Sensing

11.2.1. Transport

Material handling is one of the basic operations, which is performed on an object as it passes through the manufacturing process. The object is moved from one location to another to be stored, machined, assembled, or packaged. The robot's capability to acquire an object, move it through space, and release it makes it ideal for transport operations. Simple tasks such as part transfer from one conveyor to another may only require one- or two-dimensional movements, which are often performed by non-servo robots. Other parts handling operations such as machine loading and unloading and packaging may be more complicated and require varying degrees of manipulative capability. Servo-controlled robots perform these operations.

11.2.2. Manipulation

Another basic operation performed on an object as it is transformed from raw material to a finished product is processing, which generally requires some types of manipulation, i.e., workpieces are inserted, oriented, or twisted to be in proper position for machining, assembly, or some other operation. A robot's capability to manipulate both parts and tooling make it very suitable for processing applications. Examples in this regard include spot and arc welding, and spray painting.

11.2.3. Sensing

A robot's ability to react to its environment by means of sensory feedback is also important, particularly in applications like assembly and inspection.

These sensory inputs may come from a variety of sensor types, including proximity switches, force sensors, and machine vision systems.

In each application, one or more of the robot's capabilities of transport, manipulation, or sensing is employed. These applications make a robot ideal for many applications now performed manually.

11.3. APPLICATIONS OF ROBOTS

Manufacturing Applications

- Arc and spot-welding
- Spray painting
- Machine loading and unloading
- Machining
- Die casting
- Forging
- Investment casting
- Parts transferring
- Plastics molding
- Finishing
- Assembly
- Inspection

Materials Handling

- Transport goods
- Pick and place
- Palletizing

Space Industry

- Robot arms used as manipulator to handle bulky telescopes
- Mounted on space shuttle or repair craft

Military

- Remote bomb detonation
- Smart bombs

Medical Applications

- Intelligent wheelchairs
- Robot arms used to manipulate and handle patients

11.4. MANUFACTURING APPLICATIONS

11.4.1. Welding

Perhaps the most popular applications of robots are in industrial welding. The repeatability, uniformity quality, and speed of robotic welding are unmatched. A robot performing a welding operation is shown in Figure 11.1. The two basic types of welding are spot welding and arc welding, although laser welding is done. Some environmental requirements should be considered for a successful operation. The automotive industry is a major user of robotic spot welders. The other major welding task performed by robots is arc or seam welding. In this application, two adjacent parts are joined together by fusing them, thereby creating a seam.

Why Should Robots be Used for Welding?

A welding process that contains repetitive tasks on similar pieces might be suitable for automation. The number of items of any type to be welded determines how difficult automating a process will be or not. If parts normally need adjustment to fit together correctly, or if joints to be welded are too wide or in different positions from piece to piece, automating the procedure will be difficult or impossible. Robots work well for repetitive tasks or similar pieces that involve welds in more than one axis. The most prominent advantages of automated welding are precision and productivity. Robot welding improves weld repeatability. Once programmed correctly, robots will give precisely the same welds every time on workpieces of the same dimensions and specifications.

FIGURE 11.1 Robot Performing Welding Operation

Arc Welding

Arc or fusion welding is considered a major growth area for the application of robotics (Refer to Figure 11.2). The process is very hostile to the operator, generating noise, fumes, and intense light. Automation produces high quality welds with greater consistency and at a faster rate. In general, equipment for automatic arc welding is designed differently from that used for manual arc welding. Automatic arc welding normally involves high-duty cycles, and the welding equipment must be able to operate under

FIGURE 11.2 Robots for Arc Welding

those conditions. In addition, the equipment components must have the necessary features and controls to interface with the main control system.

A special kind of electrical power is required to make an arc weld. A welding machine, also known as a power source, provides the special power. All arc-welding processes use an *arc welding gun* or *torch* to transmit the welding current from a welding cable to the electrode. They also provide for shielding the weld area from the atmosphere. The nozzle of the torch is close to the arc and will gradually pick up spatter. A *torch cleaner* (normally automatic) is often used in robot arc welding systems to remove the spatter. All of the continuous electrode wire arc processes require an *electrode feeder* to feed the consumable electrode wire into the arc. The process is applied to automobile subassemblies mainly for reasons of strength, low distortion, high-speed applications where one-sided access is required, and sealing.

Spot Welding

Automatic welding imposes specific demands on resistance welding equipment. Often, equipment must be specially designed and welding procedures developed to meet robot-welding requirements. The spot welding robot (Refer to Figure 11.3) is the most important component of a robotized spot welding installation. Welding robots are available in various sizes, rated by payload capacity and reach. Robots are also classified by the number of axes. Spot welding involves applying a welding tool to some object, such as a car body, at specified discrete locations. A spot welding gun applies appropriate pressure and current to the sheets to be welded. This requires the robot to move its hand (end effector) to a sequence of positions with sufficient accuracy to perform the task properly. It is desirable to move at a high speed to reduce cycle time, while avoiding collisions and excessive wear or damage to the robot.

There are different types of welding guns, used for different applications, available. An automatic weld-timer initiates and times the duration of

FIGURE 11.3 Spot-welding Gun

current. A robot can repeatedly move the welding gun to each weld location and position it perpendicular to the weld seam. It can also replay programmed welding schedules. A manual-welding operator is less likely to perform as well because of the weight of the gun and monotony of the task. Spot welding robots should have six or more axes of motion and be capable of approaching points in the work envelope from any angle. This permits the robot to be flexible in positioning a welding gun to weld an assembly. A robot easily performs some movements that are awkward for an operator, such as positioning the welding gun upside down.

Typical components of an integrated robotic spot welding cell are:

- Spot welding robot
- Spot welding gun
- Weld timer
- Electrode tip dresser
- Spot welding swivel

Electron Beam Welding

Electron beam welding (EBW) is a fusion joining process that produces a weld by impinging a beam of high-energy electrons to heat the weld joint. Electrons are elementary atomic particles characterized by a negative charge and an extremely small mass. An electron beam-welding gun uses a high intensity electron beam to target a weld joint. The weld joint converts the electron beam to the heat input required to make a fusion weld. The electron beam is always generated in a high vacuum. The use of specially designed orifices separated a series of chambers at various levels of vacuum permits welding in medium and non-vacuum conditions. Although, high vacuum welding will provide maximum purity and high depth to width ratio welds. An electron beam robot welding system benefits the customer with a low contamination vacuum, narrow weld zone, uses low filler metal, and has low distortion.

MIG Welding

Gas metal arc welding (GMAW) is frequently referred to as MIG welding. MIG welding is a commonly used high deposition rate welding process. Wire is continuously fed from a spool. MIG welding is therefore referred to as a semiautomatic welding process. Robotic systems are integrated towards MIG welding applications on a consistent basis. With advances in technology, and the benefits of a GMAW robotic cell, factories and job shops large and small are investing in a robot. Return on investment of a robotic system is possible after just a few years. MIG welding robots, or GMAW cells have many benefits for customers. Robots are all position capable, have higher deposit rates than SMAW, need less operator skill, can perform longer welds without stopping, and have minimal post weld cleaning.

TIG Welding

Gas tungsten arc welding (GTAW) is frequently referred to as TIG welding. TIG welding is a commonly used high-quality welding process. TIG welding has become a popular choice of welding processes when high quality, precision welding is required. In TIG welding, an arc is formed between a non-consumable tungsten electrode and the metal being welded. Gas is fed through the torch to shield the electrode and molten weld pool. If filler wire is used, it is added to the weld pool separately. A TIG welding robot system has many benefits to customers. A robot produces high-quality welds, welds can be made with or without filler metal, variable precise controls, low distortion, and free of spatter.

11.4.2. Spray Painting

Another popular and efficient use for robots is in the field of spray painting. The consistency and repeatability of a robot's motion have enabled near perfect quality while at the same time wasting no paint. The spray painting applications seems to epitomize the proper applications of robotics, relieving the human operator from a hazardous, albeit skillful job, while at the same time increasing work quality, uniformity, and cutting costs. In spraying applications, the robot manipulates a spray gun, which is used to apply some material such as paint, stain, or plastic powder, to either a stationary or moving part. These coatings are applied to a wide variety of parts, including automotive body panels, appliances, and furniture. The spray-painting environment constitutes a fire hazard and is also dangerous to human worker's respiration. An early paint-spraying machine was built by Pollard in the 1930s. Today, this machine would be called an industrial robot. Robots perform painting, coating, and dispensing

FIGURE 11.4 Robots for Spray Painting

jobs in many industries today. Companies making products such as motorcycles, bicycles, boats, jet skis, and cars are using painting automation to their advantage. Painting robots are generally equipped with six axes, three for the base motions and three for applicator orientation. Some units incorporate machine vision for guidance or to check application quality. Typically, these painting robots are electrically driven, rather than hydraulically or pneumatically powered. Advantages of using spray-painting robots (Figure 11.4) include:

1. Humans are removed from a hostile environment.

2. Less energy is needed for fresh air requirements and the need for protective clothing is reduced.

3. The quality of the painting is improved, reducing works and warranty costs.

4. Less paint and other materials are used.

5. Direct labor costs are reduced.

11.4.3. Machine Tending

Machine loading and unloading is a more complex application than basic material handling; for this application, the robot provides both manipulative and transport capabilities (Refer to Figure 11.5). Robots can be used to grasp a workpiece from a supply point (*e.g.*, a conveyor belt), transport it to a machine, orient it, and then insert it into the machine work holder. This may require that the robot signal the machine tool when the workpiece is in the correct position, so that the part can be secured in the work holder. The robot then releases the part and withdraws the arm so that machining can begin. Upon completion of the machining, the robot

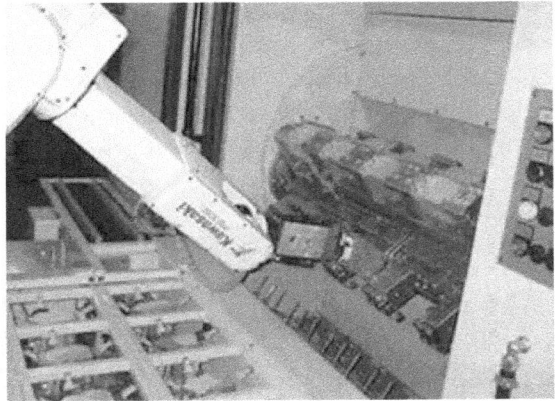

FIGURE 11.5 Machine Tending Operation

unloads the workpiece and transfers it to another machine or conveyor. In a robotic cell, a single robot can service several machines. The single robot may be used to perform other operations while the machines are performing their primary functions. This may require that the robot be able to exchange end-effectors. Examples of machine tending functions include the following:

- Exchanging machine tools, such as lathe and machining centers.
- Stamping press loading and unloading.
- Tending plastic injection molding machines.
- Holding a part for a spot-welding operation.
- Loading hot billets into forging presses.
- Loading auto parts for grinding.
- Loading gears onto CNC milling machines.

11.4.4. Forging

Robots have been applied in many different types of forging applications such as hammer forge operations, upsetter operations, roll forges, hot forming presses, and draw bench applications. In some cases, a robot acts as a forging machine operator or as a role of forge helper.

Forge hammers: Forging hammers are either hydraulic, steam hydraulic, or air driven. One-half of the forming die is on the anvil, and the other half of the die is on a ram that moves up and down, either under force by air, steam, or gravity. Under the control of an operator, the hammer is allowed to strike the part that is lying between the two dies a certain number of times. Depending on the observation of the forging operator, the operator determines when to take a part out of one die and moves it into the next. The function of the robot in this application can be to act as a forge helper. When working heavier parts, the robot can be used to load and unload furnaces and process the billets to the forging bed, where the operator can take over and process the parts through the various forming cycles. The robot then can maneuver the finished product to a trim operation.

11.4.5. Die Casting

Die casting involves forcing nonferrous metals into dies under high pressure to form parts of a desired shape. A typical die-casting task involves unloading a part from the die-casting machine, quenching the part, and then disposing of the part on a conveyor belt or into a bin. A possible robot task cycles in die-casting could include any of the following:

1. Alternately loading two or more die casting machines.

2. Unloading, quenching, trimming, and disposing of the part.

3. Unloading the die casting machine and preparing the dye for the next casting cycle. (This would require another gripper or attachment to spray the die.)

4. Loading an insert into the die casting machine and unloading the finished part.

11.4.6. Plastic Molding

The plastic molding process is typically used for thermoplastic materials. The material to be molded is supplied in a granule form and moved from a hopper to a cylinder, from which a plunger forces the granules through a heat chamber into the mold. Then, the mold half opens, and the products are withdrawn. Many automotive parts are injection molded today, as well as many parts utilized in household appliances. The robot is typically employed to remove the part from the mold either by grasping a sprue and runner assembly. A robot is typically used at an injection molding machine workcell where parts must not be dropped because of fragility, or where runs are so short that it is not economic to build a totally automatic mold to drop the part through the bottom of the machine.

11.4.7. Assembly

This is the process of robot manipulation of components, resulting in a finished assembled product. Assembly (Refer to Figure 11.6) is the process of fitting and holding together parts and assemblies; generally performed by means of nuts, bolts, screws, fasteners, or snap-fit joints. Examples of assembly operations include:

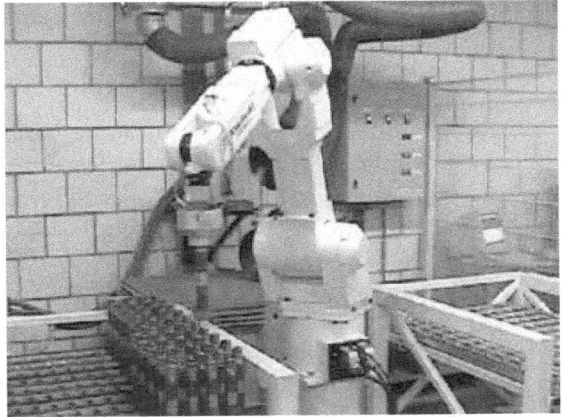

FIGURE 11.6 Robot Doing Assembly Operation

- Assembly of computer hard drives.
- Insertion of lamps into instrument panels.
- Insertion and placement of components onto printed circuit boards.
- Automated assembly of small electric motors.
- Furniture assembly.

11.4.8. Sealing/Dispensing

For dispensing applications, the robot manipulates a dispenser or gun to apply a material such as paint, adhesive, sealant, or washing solution to a stationary or moving part. Additional equipment used to complete a dispensing system may include material containers, pumps, and regulators. It is very important that, for operations that involve the application of material that produces flammable or explosive fumes, the robot be sealed and have a system that purges the robot's internal cavities. If the robot is not sealed and does not have a purging system, the possibility of the ignition of flammable or explosive fumes by the arcing of the robot's internal electrical components (i.e., motors, electronic components, and electrical connections) exists.

In applications where the part is on a conveyor line, the robot's motion is coordinated with the motion of the conveyor. The manipulative capability of the robot is the primary function that makes a robot especially well suited for dispensing applications. The major benefit of using a robot (Figure 17.7) for dispensing applications is that a robot provides uniform application of material (repeatability). Other benefits include

FIGURE 11.7 Robot Performing Sealing Operation

a reduction of labor costs, coating material waste, and exposure of workers to hazardous materials.

Examples of dispensing applications include:

- Painting parts on an automated line.
- Application of adhesive and sealant to car bodies.
- Application of thermal material to rockets.
- Washing.

11.4.9. Inspection and Measurement

With a growing interest in product quality, the focus has been on zero defects. However, the human inspection system has somehow failed to achieve its objectives. Robot applications of vision systems have provided services in part location, completeness and correctness of assembly products, and collision detection during navigation. Machine vision applications require the ability to control both position and appearance in order to become a productive component of an automated system.

11.4.10. Material Removal

Robotic material removal is a new application that has many uses in industrial automation. The robot can grind, roll, and file metal parts to precision. The robotic system can remove material with quality every single time it is being used. The robot can work long hours and produce more throughputs. Robotics has become more affordable and many factories are looking to buy a robotic deburring cell for their operation. With expert engineering, any material removal application can bring your company a good return on investment (ROI). A material removal robot can be very beneficial to an operation because of the robot's precision and quality. With a material removal robot system technology can be as precise as removing spots on jewellery. Material removal can be described as a deburring process, and can be integrated to the most precise applications. Companies can automate a material removal process and gain many benefits.

11.4.11. Deburring

Robotic deburring is the use of a robot to remove burrs, sharp edges, or fins from metal parts. The robot can grind, roll, and file metal parts to precision (Refer to Figure 11.8). The robotic system can produce a deburring application, *i.e.*, quality every single time it is being used. The robot can work long hours and produce more throughputs. Robotics has become more affordable and many factories are looking to buy a robotic deburring cell for their operation. A robotic deburring cell can work long hours without fatigue removing rough edges. The quality of the process, as well as the long hours of a robotic cell is unmatched.

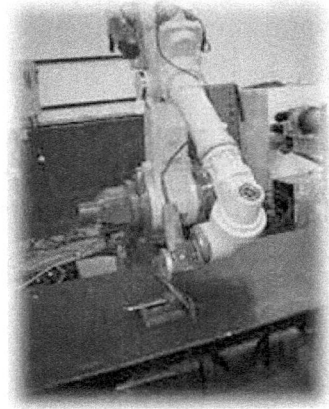

FIGURE 11.8 Robot Performing Deburring Operation

11.4.12. Grinding

Manual grinding is tough, dirty, and noisy work. The metal dust produced by grinding is harmful to employee's eyes and lungs. Grinding robots remove excess material from the surface of machined parts/products quickly and efficiently. A robot can perform work with more consistency and higher quality results.

11.4.13. Drilling

With the automation of a drilling robot system, companies can improve accuracy and repeatability. Drilling robots can work 24 hours a day without worries or fatigue thus enhancing operational output. (Refer to Figure 11.9.)

FIGURE 11.9 Robot Performing Drilling Operation

11.5. MATERIAL HANDLING APPLICATIONS

Using robots for handling materials are an essential component of today's automated manufacturing systems. Material handling and logistics is the movement, protection, storage, and control of materials and products throughout the process of their manufacture and distribution, consumption, and disposal. It is a repetitive operation; carried out often under unpleasant and hard working conditions, which requires little skill. This application makes use of the robot's capability to transport objects. Figure 11.10 shows a robot doing material handling operation.

The term "material handling" covers a lot of ground in the world of robotics: tiny work pieces that people can't handle very well, if at all; large, heavy parts like engine blocks and wheels; bulky items like bags and boxes; delicate and expensive electronic components; medical equipment. The list is extensive. Robotic material-handling applications range from tending injection-molding machines and machine tools, to reorienting parts between processes, to packaging and palletizing.

FIGURE 11.10 Material Handling Robot

By fitting the robot with an appropriate end-effector (*e.g.*, gripper), the robot can grasp the object that needs to be moved. The robot may be mounted either stationary on the floor or on a traversing unit, enabling it to move from one workstation to another. The robot can also be ceiling mounted. The primary benefits of using robots for material handling are reduction of direct labor costs and removal of humans from tasks that may be hazardous, tedious, or fatiguing. Also, the use of robots for moving fragile objects results in less damage to parts during handling. Robots that are used for material handling can interface with other material handling equipment such as containers, conveyors, guided vehicles, monorails, and automated storage/retrieval systems. A handling process consists of eight sequences:

1. Transfer of the robot arm up the workpiece

2. Fine motion approaching the workpiece

3. Grasping

4. Fine motion uprising the workpiece

5. Transfer the workpiece to the desired position

6. Fine motion down to the destination position

7. Release the workpiece

8. Fine motion upward

The gripping sequence is the more delicate part of handling material. The gripper's and the robot's positions have to be checked for collision with other objects in the environment. The limitation of the robot's workspace can also be a problem for all other planning sequences of the handling.

The following are examples of material handling applications:

- Transferring parts from one conveyor to another.
- Transferring parts from a processing line to a conveyor.
- Loading bins and fixtures for subsequent processing.
- Moving parts from a warehouse to a machine.
- Transporting explosive devices.
- Transfer of parts from a machine to an overhead conveyor.

• Parts Transferring

Parts transferring refer to removing parts from pallets and placing them in bins or on conveyor belts, or removing parts from bins and conveyor belts and placing them on pallets. Part transfer applications are generally referred to as material handling robot applications. With the advancements in end-of-arm tooling, and technology companies are taking a closer look into part transfer robotics. As robots become more afford-able, and competition becomes fierce, many companies will look into a part transfer robot to automate their press operation. An integrated part transfer robot (Refer to Figure 11.11) system is easy to install and brings many benefits to our customers. A robot moves a part in and out of a press, with ease and no fatigue. The robot can transfer a part 24 hours a day, giving more flexibility to the company.

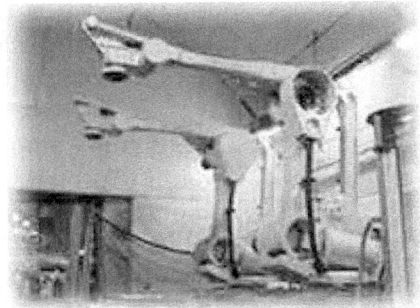

FIGURE 11.11 Robot used for Parts Transferring

• Palletizing

Palletizing is the act of loading or unloading material onto pallets. A robotic palletizing system (Refer to Figure 11.12) allows more flexibility to run more products for longer periods of time. With the advancements in end-of-arm tooling, robot-palletizing workcells have been integrated in many factories. The use of robots in palletizing is a popular material-handling operation, particularly where more than one type of packaging is being handled. The newspaper industry has been particularly hard hit by increased labor costs. Part of the solution to this problem was to use robots like Cincinnati Milacron Robot being used to palletize advertising inserts for a newspaper. Robotic palletizing technology can help with productivity and profitability. The robotic workcells can be integrated towards any project. The savings of using a robot for longer hours without fatigue is always a consideration for plant managers. Robots have become more affordable in recent years and can be paid off in just a few years worth of work. Many factories,

FIGURE 11.12 Robots Performing Palletizing Operation

food processing plants, and palletizing plants have automated their application with a palletizing robot.

- **Pick and Place Operations**

  Industrial robots also perform what are referred to as pick and place operations. Pick and place is the name commonly given to the operation of picking up a part and placing it appropriately for subsequent operations. Pick-and-place operations have some requirements. The part must not be dropped. It must be held securely enough to prevent it from slipping in the gripper but gently enough to avoid damage. In addition, care must be taken to avoid disturbing the part during approach and departure. A pick and place robot is a material handling robot that can work 24 hours a day without worries or fatigue. The consistent output of a robotic system along with quality and repeatability are unmatched. Among the most common of these operations is loading and unloading pallets, used across a broad range of industries. This requires relatively complex programming, as the robot must sense how full a pallet is and adjust its placements or removals accordingly. Robots have been vital in pick and place operations in the casting of metals and plastics. In the die casting of metals, for instance, productivity using the same die-casting machinery has increased up to three times, the result of robots' greater speed, strength, and ability to withstand heat in parts removal operations.

  Pick and place robotic cells are being integrated into factory floors all over the world. They are a more favorable solution to production lines because they move at faster speeds and have more flexibility than ever before. Managers are realizing the long-term savings with a pick and place robotic workcell rather than the operation they are currently doing. An increase in output with a material handling robotic system has saved factories money. With the advancements in technology, and robotics becoming more affordable pick and place robotic cells are being installed for many different pick and place automation applications.

- **Machine Tool Loading and Unloading**

   Machine loading and unloading applications are generally referred to as material handling. With the advancements in end-of-arm tooling, and technology, companies are taking a closer look into the benefits of machine loading robotics. As robots become more affordable, and competition becomes fierce, many companies will look into machine loading robotics to automate their operation. A robot can be given a longer reach than a human worker and might be able to load and unload more then one machine. An integrated machine-loading robot is easy to install and brings many benefits to our customers. A robot loads a part into the press, with ease and no fatigue. The robot can load parts 24 hours a day, giving more flexibility to the company.

- **Order Picking**

   A robot order picking system can be programmed to do multiple tasks. With the advancements in end-of-arm tooling, order-picking application can be used in most companies. A robotic system benefits a firm with flexibility, repetitive quality and no fatigue. A robot order picking process would most likely be associated with material handling.

11.6. CLEANROOM ROBOTS

Cleanroom robots are specifically sealed and isolated from dust or any kind of air particles to perform tasks in an isolated atmosphere. Mostly used in medical or lab application to handle parts, tend machinery, and dispense drugs. A robot is a valuable tool in a clean room setting because it can provide the same quality results without error for a long period of time. A robotic system allows laboratories and other clean room robot applications much flexibility. (Refer to Figure 11.13.)

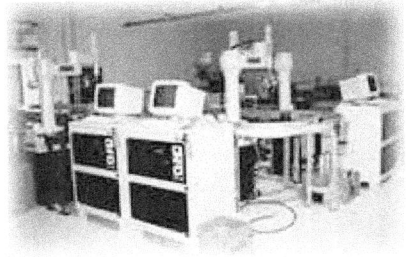

FIGURE 11.13 Clean Room Robot Cell

REVIEW QUESTIONS

1. Name some industrial applications of robots.

2. Name some industrial applications of robots.

3. What are the reasons for successful application of robots in manufacturing industries?

4. Identify some typical applications of robots in the industry.

5. Write an explanatory note on the industrial applications of robots.

6. Explain the use of a robot for industrial material handling with the help of some examples.

7. Explain the use of robot for machine loading/unloading.

8. Prepare a list of any four industrial operations that can be performed by robot.

9. List various types of assembly tasks, which can be performed by robots.

10. What are the reasons for using robots in industries?

11. List the capabilities of a robot.

12. Discuss the material handling application of robots.

13. What is machine tending in industrial robots?

14. Write a short note on sealing/dispensing application of robots.

15. Write an essay on the robots used in spray painting applications.

16. Write a short note on the following:

   Material handling.
   Welding operation.

INDEX

www.ingramcontent.com/pod-product-compliance
Lightning Source LLC
Chambersburg PA
CBHW061922190326

41458CB00009B/2624